Plant Parasitic Nematodes

Reddy P. Parvatha

Plant Parasitic Nematodes

Diagnosis and Management

Springer

Reddy P. Parvatha
ICAR - Indian Institute of Horticultural Research
Bengaluru, Karnataka, India

ISBN 978-981-95-9305-7 ISBN 978-981-95-9306-4 (eBook)
https://doi.org/10.1007/978-981-95-9306-4

This Springer imprint is published by the registered company Springer Nature Singapore Pte Ltd.
The registered company address is: 152 Beach Road, #21-01/04 Gateway East, Singapore 189721, Singapore

Preface

Plant-parasitic nematodes are costly burdens on crop production. Ubiquitous in nature, phytoparasitic nematodes are associated with nearly every important agricultural crop and represent a significant constraint on global food security. Root-knot nematodes (*Meloidogyne* spp.), cyst nematodes (*Heterodera* and *Globodera* spp.), lesion nematodes (*Pratylenchus* spp.), and reniform nematodes (*Rotylenchulus* spp.) rank at the top of the list of the most economically and scientifically important species due to their intricate relationship with the host plants, wide host range, and the level of damage caused by infection. Estimated overall average annual yield loss of the world's major crops due to damage by plant-parasitic nematodes is 12.3%. In India, plant-parasitic nematodes cause 21.3% crop losses amounting to 102039.79 million (US$ 1577 million) annually. In view of limitations on the use of chemical pesticides, there has been a growing awareness and increasing demand for novel and improved nematode diagnosis and management approaches to guarantee effective and sustainable food production, which offers new opportunities.

There is no book at present which comprehensively covers the most advanced technologies for nematode diagnosis and management. Hence, the present book, entitled *Plant Parasitic Nematodes: Diagnosis and Management*, comprehensively deals with the most recent frontier technologies, such as molecular diagnostics, remote sensing, agro-ecological approaches, less toxic nematicides, microbial bio-protectants, precision nematicides application, nanotechnology, transgenic nematode resistance, genome editing, bio-intensive integrated nematode management using decision support systems, climate-resilient agriculture, and the use of artificial intelligence and machine learning for proper nematode identification and management. The book also outlines the vision for future nematode management strategies.

The book is divided into 21 chapters under 8 parts. Part I deals comprehensively with an overview of frontier technologies for nematode diagnosis and management and emerging nematode problems under Indian and international scenarios. Eco-friendly agro-ecological approaches for nematode management such as microbial bio-protectants and bio-intensive integrated nematode management are discussed in Part II. Part III deals with emerging novel low-toxic nematicides. Innovative approaches such as precision agriculture, nanotechnology, decision support systems, and remote sensing/drone technology are dealt with in Part IV. Part V discusses molecular diagnostics, genomics and proteomics, genome editing, and transgenic nematode resistance. Effects of climate change variables, impacts of climate change-induced consequences, development of prediction models, and climate change adaptation and mitigation strategies are discussed in Part VI. Part VII deals with the utilization of artificial intelligence for nematode counting, identification, and management. The vision for future nematode diagnosis and management is outlined in Part VIII.

This book will be of immense value to the scientific community involved in teaching, research and extension activities related to the use of frontier technologies for diagnostics and management of phytonematodes. The material can be used for teaching postgraduate courses in the field of Plant Nematology, Plant Pathology, Agricultural Entomology, and crop protection. The book can also serve as a very useful reference to policymakers and practicing farmers. Suggestions to improve the contents of the book are most welcome (E-mail: reddypp42@gmail.com). The publisher deserves commendation for their professional contribution.

Bengaluru, Karnataka, India Reddy P. Parvatha

Competing Interests The author has no competing interests to declare that are relevant to the content of this manuscript.

Contents

About the Author

P. Parvatha Reddy obtained his PhD degree jointly from the University of Florida, USA, and the University of Agricultural Sciences, Bangalore.

Dr. Reddy served as the Director of the prestigious ICAR-Indian Institute of Horticultural Research (IIHR) at Bangalore from 1999 to 2002, during which period the Institute was honored with the "ICAR Best Institution Award." He also served as the Head, Division of Entomology and Nematology at IIHR and gave tremendous impetus and direction to research, extension, and education in developing bio-intensive integrated pest management strategies in horticultural crops. These technologies are being practiced widely by the farmers across the country since they are effective, economical, eco-friendly, and residue-free. He has about 34 years of experience working with horticultural crops and was involved in developing an F1 tomato hybrid "Arka Varadan" resistant to root-knot nematodes.

Dr. Reddy has over 250 scientific publications to his credit, which also include 60 books. He has guided two PhD students at the University of Agricultural Sciences, Bangalore.

Dr. Reddy is among the Top 2% Scientists of the World in the Stanford University Rankings based on standardized citation metrics that reflect the most influential researchers globally during 2023–24. He is serving as Senior Scientific Advisor, Dr. Prem Nath Agricultural Science Foundation, Bangalore. He also served as Chairman, Research Advisory Committee (RAC), Indian Institute of Vegetable Research, Varanasi; Member, RAC of National Centre for Integrated Pest Management, New Delhi; Member of the Expert Panel for monitoring the research program of National Initiative on Climate Resilient Agriculture (NICRA) in the theme of Horticulture including Pest Dynamics and Pollinators; Member of the RAC of the National Research Centre for Citrus, Nagpur; and the Project Directorate of Biological Control, Bangalore. He served as a Member, QRT to review the progress of the Central Tuber Crops Research Institute, Trivandrum; AICRP on Tuber Crops; AICRP on Nematodes; and AINRP on Betel vine. He is the Honorary Fellow of the Nematological Society of India, New Delhi, and the Society for Plant Protection Sciences, New Delhi; Fellow of the Indian Phytopathological Society, New Delhi; and Founder President of the Association for Advancement of Pest Management in Horticultural Ecosystems (AAPMHE), Bangalore.

Dr. Reddy has been awarded the prestigious "Association for Advancement of Pest Management in Horticultural Ecosystems Award," "Dr. G.I. D'Souza Memorial Award," "Prof. H.M. Shah Memorial Award," "Hexamar Agricultural Research and Development Foundation Award," and "PNASF Award for Independent Author of Large Number of Books (more than 60) on Horticulture and Nutrition" for his unstinted efforts in developing sustainable, bio-intensive, and eco-friendy integrated pest management strategies in horticultural crops.

Dr. Reddy has organized "Fourth International Workshop on Biological Control and Management of Chromolaena odorata," "National Seminar on Hi-Tech Horticulture," "First National Symposium on Pest Management in Horticultural Crops: Environmental Implications and Thrusts," and "Second National Symposium on Pest Management in Horticultural Crops: New Molecules and Biopesticides."

Part I

Introduction

Plant Parasitic Nematodes—Diagnosis and Management: An Overview

1

Abstract

Frontier technologies like molecular diagnostics, remote sensing, artificial intelligence (AI), precision agriculture, nanotechnology, biotechnological approaches, and biological control, are advancing plant nematode diagnostics and management. Molecular methods like isothermal amplification, high-throughput sequencing, PCR, and CRISPR/Cas12a, improve detection accuracy. The evidence leans toward remote sensing and AI enhancing large-scale monitoring and automation. Genetic engineering and new nematicides offer sustainable nematode management. There is ongoing debate about the environmental impact of new nematicides, balancing efficacy and safety, cost-effectiveness, and scalability of some advanced techniques.

Keywords

Diagnosis · Management · Artificial Intelligence · Precision agriculture · Biotechnological approaches

1.1 Learning Objectives

- Recognize the challenges in identifying nematode species using conventional approaches.
- Describe modern molecular diagnostic approaches for nematode detection.
- Compare traditional morphological/biometric methods with innovative AI- and imaging-based diagnostics.
- Summarize recent eco-friendly and sustainable nematode management methods (biological control, botanicals, and biofumigants).
- Understand how rapid diagnostics inform timely and precise management decisions.
- Identify the potential of AI, machine learning, and big data in linking diagnosis with predictive management.

R. P. Parvatha, *Plant Parasitic Nematodes*,
https://doi.org/10.1007/978-981-95-9306-4_1

- Anticipate future trends, challenges, and opportunities for sustainable nematode control.
- Appreciate the importance of interdisciplinary collaboration for advancing nematode diagnostics and management.

1.2 Introduction

Plant-parasitic nematodes (PPNs), such as root-knot nematodes (*Meloidogyne* spp.), cyst nematodes (*Heterodera* spp.), lesion nematodes (*Pratylenchus* spp.), and reniform nematodes (*Rotylenchulus reniformis*) are microscopic soil-dwelling organisms that feed on plant roots, leading to reduced yield and quality of the produce. Over 4100 PPN species have been described, with their impact varying by crop type, nematode density, and environmental conditions. Accurate and rapid diagnosis is essential for effective management, as traditional morphological identification can be labor-intensive and prone to errors due to interspecific similarities and intraspecific variations. Management strategies have historically relied on chemical nematicides, but regulatory restrictions and environmental concerns have spurred the development of alternative approaches, including precision farming, nanotechnology, remote sensing, artificial intelligence, biological control, and advanced diagnostics.

Plant-parasitic nematodes (PPNs), microscopic roundworms that attack plant roots and other tissues, pose a significant threat to global agriculture, causing substantial crop losses estimated to vary considerably from US$ 80 billion to US$ 157 billion per year (Abad et al. 2008), and affecting a wide range of crops, including cereals, pulses, oil seeds, and horticultural species. Traditional approaches for diagnosing and managing these pests such as chemical nematicides, crop rotation, and visual inspections, have limitations in precision, sustainability, and effectiveness. With the growing need for sustainable agriculture and the phasing out of traditional chemical nematicides due to environmental concerns, frontier technologies have emerged as critical tools for diagnosing and managing nematode infestations. Frontier technologies, representing the latest innovations in this field, are transforming how we diagnose and manage nematode infestations. This chapter provides a comprehensive overview of cutting-edge solutions for both diagnosis and management strategies, highlighting their potential for applications, advantages, limitations, and future direction in order to revolutionize agricultural practices.

1.3 Molecular Diagnosis

Early and accurate detection of nematodes is essential for effective management. Frontier technologies in diagnosis leverage advancements in science and technology to provide faster, more precise identification of these pests, making it easier to spot them early.

Molecular diagnostics, like polymerase chain reaction (PCR) and CRISPR, help identify nematodes by their DNA, even in small numbers. AI tools can analyze images to count nematodes quickly, while remote sensing with drones can spot stressed plants from above, helping farmers to act fast. New apps are also emerging to make identification easier for everyone.

Molecular diagnostics have revolutionized nematode identification by targeting their DNA. Key techniques include:

1.3.1 Polymerase Chain Reaction (PCR) and Next-Generation Sequencing (NGS)

These methods enable the identification of nematode species in low populations or complex soil samples, far surpassing traditional microscopy. They are highly sensitive and can detect species like *Meloidogyne* spp., *Heterodera* spp., and *Radopholus similis*.

1.3.2 CRISPR/Cas12a-Based Detection

This technology combines CRISPR with isothermal amplification methods like Loop-mediated Isothermal Amplification (LAMP) and Recombinase Polymerase Amplification (RPA). Loop-Mediated Isothermal Amplification (LAMP) has been used to detect species like *Meloidogyne javanica*, *M. enterolobii*, and *Bursaphelenchus xylophilus*, with applications in soil and root samples. Recombinase Polymerase Amplification (RPA), often combined with CRISPR-Cas12a for visualization, has shown higher sensitivity than PCR or qPCR for detecting *Heterodera schachtii* and *H. avenae*. It offers detection sensitivity at the attomolar level, suitable for field use without thermal cycling. For example, it has been successfully used to test *H. schachtii*, *H. avenae*, and *H. filipjevi* (Yao et al. 2021; Shao et al. 2023). These methods, often combined with Lateral Flow Dipstick (LFD), allow for naked-eye visualization of results, enhancing field applicability. These techniques are fast (10–100 times faster than PCR) and cost-effective, though they require careful primer design to avoid cross-contamination.

1.3.3 Matrix-Assisted Laser Desorption/Ionization Time-of-Flight Mass Spectrometry (MALDI-TOF MS)

This technique identifies nematodes by their protein profiles, offering another molecular approach for species identification (Perera et al. 2005; Rivero et al. 2022).

1.3.4 High-Throughput Sequencing and Meta-Barcoding

High-throughput sequencing analyzes large volumes of genetic data, enabling the study of nematode communities and their interactions with plants. It has been applied to investigate soil ecology in crops like banana and coffee, and to identify microRNAs in root-knot nematodes (*M. incognita*). Meta-barcoding enhances taxonomic resolution at the genus and species levels, resolving discrepancies with morphological methods. For example, it has been used to identify *Heterodera* species in soil samples, though challenges include limited reference databases and PCR primer binding issues that may fail to detect all taxa.

1.3.5 Droplet Digital PCR (ddPCR)

This technology quantifies low-abundance targets without manual extraction, offering resistance to PCR inhibitors and direct copy number determination. It has been used to detect *Paratrichodorus allius* in soil, though its dynamic range is limited compared to other methods.

The advantages include high specificity, sensitivity, and suitability for field diagnostics, with integrative approaches combining molecular methods recommended for accuracy (Shao et al. 2023).

1.3.6 Case Studies

The following Table 1.1 summarizes the case studies in molecular diagnostic technologies:

1.4 Monitoring

1.4.1 Remote Sensing and Hyper-Spectral Imaging

Remote sensing technologies provide noninvasive, large-scale monitoring capabilities:

- Drones equipped with multispectral or hyperspectral cameras detect nematode-induced stress in plants by analyzing changes in leaf color, moisture levels, or temperature. This is particularly useful for species like *Radopholus similis*, *Globodera rostochiensis*, *G. pallida*, *H. glycines*, and *H. schachtii* (Norman and Fritz 1965; Hillnhütter et al. 2011a, b).
- Spectroscopy (visible, infrared, hyperspectral) and imaging help observe above-ground symptoms, though accuracy can be challenged by similar symptoms from other stressors. Advances in data analysis are improving differentiation (Shao et al. 2023). It has achieved 78% accuracy in detecting root-knot nema-

Table 1.1 Technologies for molecular diagnostics

Technology	Description	Applications	Advantages	Limitations	Examples/ studies
High-throughput sequencing	Sequences DNA/RNA, detects variants/mutations, analyzes large gene volumes	Investigates nematode communities, soil ecology (banana, coffee), miRNA in RKN, SCN	Analyzes hundreds to millions of targets, explores parasitism	Limited primers/extraction kits, challenges in DNA analysis	miRNA in RKN (Zhang et al. 2016), SCN in soybean (Li et al. 2012), *M. incognita*/ *M. javanica* in tomato (Kulshrestha et al. 2020)
Meta-barcoding	Enhances taxonomic assignments at genus/species level, combines with morphology	Identifies soil nematode communities, e.g., *Heterodera* species	Improved precision, resolves taxonomic discrepancies with high-throughput sequencing	Fails to detect all taxa due to PCR primer binding issues, limited reference DB	*Heterodera* species (Huston et al. 2022), Atlantic soil meiofaunal diversity (Müller et al. 2019)
Isothermal amplification	Amplifies DNA/RNA without thermal cycling, e.g., LAMP, RPA	Detects *Ditylenchus destructor*, *Meloidogyne* spp., *P. allius*, *H. schachtii*	Fast (10–100x faster than PCR), easy operation, high sensitivity (e.g., RPA > PCR/qPCR in 1 h)	Sensitive to cross-contamination, requires checking for inhibitors, RPA dependent on primers	LAMP for *M. enterolobii* (Niu et al. 2012), RPA for *P. allius* (Goraya et al. 2024), *H. schachtii* (higher sensitivity than PCR/qPCR) (Yao et al. 2021)
Droplet digital PCR (ddPCR)	Quantifies low-abundance targets, eliminates manual extraction	Detects/quantifies *Paratrichodorus allius* in soil	Resistant to PCR inhibitors, direct copy numbers, no Ct values needed	Limited dynamic range	*P. allius* detection (Lawaju et al. 2023)

todes (*Meloidogyne* spp.) in coffee and 72% accuracy for *H. schachtii* in sugar beet, with kappa values of 0.71 and 0.65, respectively. Hyperspectral imaging, which analyzes molecular-level responses to light, has detected *Meloidogyne* in cotton with 95% accuracy (350–1000 nm) and differentiated *Meloidogyne* from water stress in tomatoes with 90–100% accuracy (400–2500 nm). These methods are labor- and time-efficient, but face challenges like resolution limitations and weather dependency. This technology enables early detection and large-scale field monitoring, crucial for timely interventions.

1.4.2 Convolutional Neural Networks (CNNs)

Artificial intelligence, particularly Convolutional Neural Networks (CNNs), automates nematode detection from microscopic images. For instance, CNNs have identified *Globodera* species with 0.83 accuracy in training and detected root-knot nematode galls in cucumber with 94.8% mean average precision (mAP) using YOLOv5-CMS. The NemaRec app has achieved 60–97% accuracy in identifying nematode genera. These technologies handle complex cases like overlapped nematodes but require large datasets and high computational resources, with poor performance in some scenarios like burrowing nematode (*R. similis*) detection.

AI and ML are transforming nematode counting and identification:

- Convolutional Neural Networks (CNNs) and deep learning algorithms, such as YOLOv8x, are used for automated image analysis. For instance, the SEGNEMA GUI, developed in 2024, achieves 94% accuracy for egg detection and 93% for *Meloidogyne* juveniles, processing images in 4–5 s compared to 2.5 h manually (Shao et al. 2023).
- These tools are trained on datasets like 650 images for eggs and 1339 for juveniles, annotating 8655 eggs and 4742 juveniles, with plans to expand to other genera through global collaboration.
- Data generation remains a bottleneck, but advancements in hyperspectral imaging and CNNs are enhancing detection capabilities, particularly for species like *M. enterorlobii* and soybean nematodes (Lu et al. 2021; Abade et al. 2022).

AI reduces reliance on manual inspections, accelerating diagnosis, especially in areas with limited access to expert nematologists.

1.4.3 Mobile Applications

Recent developments include mobile applications for nematode identification, enhancing accessibility. For example, an X post by @monalisacdc on June 2, 2025, promoted a new app for identifying plant-parasitic nematodes.

1.4.4 Artificial Intelligence (AI)

Accurate diagnosis and identification of plant-parasitic nematodes (PPNs) are critical for effective management of these cryptic pests, which collectively cause billions of dollars in crop losses annually. Traditional detection methods such as morphological and molecular assays, while reliable, are labor-intensive, time-consuming, prone to observer bias, and require specialized expertise. Recent advances in artificial intelligence (AI), particularly machine learning (ML) and deep learning (DL), offer automated, high-throughput alternatives for nematode detection, classification, and quantification. This chapter synthesizes current AI-driven approaches in PPN diagnosis and identification, covering image-based methods, hyperspectral and remote sensing integration, convolutional neural networks (CNNs), object detection architectures (YOLO variants), and decision-support tools, highlighting their methodologies, performance, and practical deployment. Challenges such as domain shifts, annotation requirements, and hardware constraints, and proposed future directions toward universal, collaborative nematode-counting platforms are also discussed.

1.4.5 Case Studies

The following Table 1.2 summarizes the case studies on monitoring technologies:

1.5 Nematode Management

Sustainable management of nematode populations is critical for modern agriculture. Frontier technologies focus on innovative, eco-friendly strategies that minimize the use of harmful chemicals.

For managing nematodes, genetic engineering creates resistant crops, reducing the need for chemicals. Precision agriculture uses GPS to apply treatments only where needed and new less toxic nematicides are designed to be safer for the environment. Biocontrol agents, like certain fungi and bacteria, also offer a natural way to control nematodes.

1.5.1 Biotechnology and Genetic Engineering

Biotechnological applications, such as RNAi and miRNA, show promise for enhanced resistance, though adoption varies. Gene editing technologies, such as CRISPR-Cas9, are used to confer resistance to nematode infestations by modifying crop genomes. This approach targets specific genes to enhance plant defenses against PPNs, offering a long-term solution. Biotechnology breakthroughs, such as CRISPR-Cas9 gene editing, allow scientists to develop nematode-resistant crop

Table 1.2 Monitoring technologies for plant parasitic nematodes

Technology	Description	Applications	Advantages	Limitations	Examples/studies
Convolutional neural network (CNN)	Automates detection/counting from microscopic images, handles complex cases	Identifies *Globodera* species, nematodes in soybean, RKN root galls in cucumber	Rapid, accurate, handles overlapped nematodes, 0.83–0.88 accuracy in training	Requires large data, high computational resources, poor detection for BKW	*Globodera* species (0.83 accuracy) (Thevenoux et al. 2021), NemaRec app (60–97% genera) (Qing et al. 2022), YOLOv5-CMS (94.8% mAP) (Wang et al. 2022)
Remote sensing	Uses electromagnetic radiation, hyperspectral imaging for noninvasive detection	Detects RKN in coffee (78% accuracy, kappa 0.71), *P. thornei* in wheat ($R^2 = 0.92$), *H. Schachtii* in sugar beet (72% accuracy, kappa 0.65)	Noninvasive, monitors spatial/temporal crop conditions, high-throughput	Limited by resolution, illumination, cloudy weather, requires optimal UAV parameters	Coffee RKN (Martins et al. 2017), wheat *P. thornei* (Robinson et al. 2019;Thompson et al. 2020), sugar beet *H. schachtii* (Hillnhütter et al. 2011a, b), pine wilt disease (94.13% accuracy) (Syifa et al. 2020)
Hyperspectral imaging	Spectroscopy/radiometric methods detect molecular-level responses to light	Detects RKN in cotton (95% accuracy, 350–1000 nm), differentiates RKN/water stress in tomatoes (90–100% accuracy, 400–2500 nm)	Labor/time-efficient, detects chemical constituents, early diagnosis	Generates high-dimensional data, requires efficient algorithms, calibration needed	Cotton RKN (Samiappan et al. 2021), tomato RKN/water stress (Susič et al. 2018)

varieties. By introducing or enhancing resistance traits, these genetically modified plants offer a durable, long-term solution to nematode damage.

Biotechnology offers long-term solutions by enhancing crop resistance:

- **CRISPR-Cas9 Genome Editing**: This technique allows scientists to develop nematode-resistant crop varieties by introducing or enhancing resistance traits. It targets specific genes to combat nematode feeding sites, offering a durable solution (Westphal 2011).
- **Genomic Analysis**: Technologies like laser capture micro-dissection, microarray analysis, and genomic analysis identify genes specific to resistant reactions. These efforts aim to create publicly available, easy-to-mine gene annotation databases for functional genomics analyses, facilitating the development of resistant varieties (Lawrence and Lawrence 2020).

These advancements reduce the need for repeated chemical applications, promoting sustainability.

1.5.2 Agro-Ecological Approaches

Agro-ecological approaches such as crop rotation with resistant cultivars can reduce nematode populations effectively. The cover crops and biofumigation, like using *Brassica* species, help manage nematodes sustainably. The mycorrhizal fungal inoculations improve plant vigor and indirectly aid in nematode control. Reduced tillage appears to enhance soil suppressiveness, lowering nematode densities. Biological control using organisms like *Purpureocillium lilacinum, Pochonia chlamydosporia, Trichoderma* spp., *Pseudomonas fluorescens, Bacillus subtilis,* and *Pasteuria penetrans* may biologically manage nematodes, though effectiveness varies. Organic amendments, such as farmyard manure and compost, likely improve soil health and reduce nematode impact. Solarization, especially with cover crops, seems effective for killing nematodes in upper soil layers. Phytochemical-based strategies, like essential oils, might suppress nematodes, but effectiveness can be inconsistent. The use of marine organisms, like seaweeds, offers a novel approach, with potential for organic systems, but needs more research.

1.5.3 Precision Agriculture

Precision agriculture optimizes resource use for nematode control:

- **GPS-Guided Machinery and Soil Sensors**: These enable variable-rate nematicide applications based on nematode spatial distribution, using grid sampling (0.025–0.5-acre grid size) or zone sampling. Remote sensing creates contour and prescription maps without physical contact, enhancing efficiency (Lawrence and Lawrence 2020).

- **Soil Electrical Conductivity Measurement**: Using systems like Veris with GPS, soil conductivity (0–12 inches shallow, 0–36 inches deep) influences application rates, with less nematicide applied in high clay content zones. Smaller grid sizes increase map detail but raise lab costs, requiring a balance between precision and economics. This approach minimizes environmental impact and reduces waste, aligning with sustainable farming practices.

1.5.4 Novel Less Toxic Nematicides

Recent developments in less toxic nematicides offer improved efficacy and safety:

- **Fluensulfone (Nemitz®, registered 2014):** Fluroalkenyl sulfone with EC50 0.12 mg/L (48 hr.) for *M. incognita*, ovicidal at >50 mg/L, application rate 1.9–3.3 kg/ha, half-life in soil 36 days.
- **Fluopyram (Verango®, Velum®, Indemnify®, 2012):** A nematicide-fungicide inhibiting succinate dehydrogenase (SDH), with EC50 0.7 mg/L for *M. incognita* J2, nematostatic against *H. schachtii* at 20 mg/L.
- **Cyclobutrifluram (Vaniva®, Tymirium®, Victrato®, 2013):** Trifluoromethylpyridine, also inhibits SDH, effective against PPNs and fungi.
- **Fluazaindolizine (Salibro™, registered July 2021):** N-phenylsulfonylimidazopyridine-2-carboxamide for carrots, squash, tomatoes, etc., with LC50 for root-knot nematodes (RKNs) 177 mg/L (72 h.), ovicidal at >500 mg/L (82% reduction *M. incognita* eggs, 3 days).
- **Tioxazafen (NemaStrike™ ST, registered 2017):** A phenyloxadiazole seed nematicide for cotton, corn, and soybean, with EC50 for *M. incognita* and *R. reniformis* ~ 60 mg/L (24 h), ovicidal at >3.0 mg/L. Withdrawn by the manufacturer since 2020 but synthetically accessible (Páez 2023).

These nematicides have mechanisms like ribosomal disruption (tioxazafen), pantoate synthase inhibition (fluazaindolizine), and SDH inhibition (fluopyram, cyclobutrifluram), with varying ovicidal effects and environmental impacts. For instance, fluensulfone is less toxic to humans (LD50 in rats 671 mg/kg) than older nematicides like aldicarb (1 mg/kg), but toxic to aquatic organisms (Páez 2023).

1.5.5 Nanotechnology

Nanotechnology provides innovative approaches to address the challenges posed by plant-parasitic nematodes, including targeted nematicide delivery, early detection through nanosensors, enhanced plant resistance, and improved soil health. These advancements have the potential to revolutionize nematode management, making it more effective and sustainable than traditional methods. However, realizing this potential requires ongoing research to fully understand environmental impacts, reduce costs, and meet regulatory standards. As the field progresses, nanotechnology could become a cornerstone of integrated pest management strategies in plant nematology, benefiting agriculture worldwide.

Nanotechnology aids in controlling plant nematodes primarily through two methods: enhancing RNA interference (RNAi) for targeted gene silencing and using nanoparticles with direct nematicidal effects. For RNAi, nanomaterials deliver double-stranded RNA (dsRNA) to nematodes, protecting it from degradation and ensuring it reaches the target, which can disrupt nematode reproduction and survival. Nanoparticles, such as silver, titanium, and silicon, have shown potential to kill nematodes directly, offering an alternative to toxic chemical nematicides.

Nanoparticles can be engineered to deliver nematicides directly to nematodes or their infection sites, minimizing the impact on non-target organisms and the environment. This precision reduces the amount of chemicals needed and enhances sustainability. Early detection of nematode infestations is critical to prevent widespread crop damage. Nanosensors, capable of detecting specific biomolecules or physiological changes in plants, enable farmers to identify nematode presence before visible symptoms emerge. Nanotechnology can contribute to developing plants that are more resilient to nematode attacks, reducing reliance on external control measures. Certain nanomaterials can improve soil conditions by indirectly suppressing nematode populations and supporting plant health.

1.5.6 Artificial Intelligence (AI)

Management strategies are also enhanced by AI, which recommends tailored control methods based on data analysis. For example, AI can suggest biological controls like biopesticides (beneficial fungi or bacteria) or crop rotation, informed by local and historical data. Precision agriculture is further supported by autonomous robots and drones equipped with AI, which apply nematicides precisely to infested areas, minimizing chemical use. Smart irrigation systems, integrated with AI, adjust watering to create unfavorable conditions for nematodes, conserving water while protecting crops.

AI contributes to developing sustainable and precise control measures for nematodes:

- **Precision Agriculture**: AI optimizes the application of nematicides, biocontrol agents, or resistant crop varieties by identifying high-risk zones through spatial analysis. This reduces chemical use by up to 30% while maintaining efficacy.
- **Biocontrol Optimization**: AI models predict the efficacy of biocontrol agents, such as nematophagous fungi or bacteria or predatory nematodes, by analyzing environmental factors and nematode life cycles. For example, reinforcement learning has been used to optimize the release timing of nematodes.
- **Genomic and Breeding Support**: AI accelerates the development of nematode-resistant crop varieties by analyzing genomic data. Machine learning identifies genetic markers associated with resistance, reducing breeding timelines by 20–30%.

1.5.7 Biological Control

Beyond diagnostics and monitoring, biological control represents a frontier in nematode management. The use of fungi, bacteria, and predatory nematodes as biocontrol agents has shown promise, with in vitro, greenhouse, and field experiments demonstrating their ability to regulate nematode development. For example, certain fungi and bacteria have been effective against ecto- and endoparasitic nematodes, offering a sustainable alternative to chemical nematicides.

Natural enemies like fungi (e.g., *Purpureocillium lilacinum, Pochonia chlamydosporia,* and *Trichoderma harzianum*) and bacteria (e.g., *Pseudomonas fluorescens, Bacillus subtilis,* and *Pasteuria penetrans*) target nematodes specifically, preserving beneficial soil ecosystems. These agents are increasingly used as sustainable alternatives to chemical nematicides, reducing environmental harm (Saikai et al. 2023).

- **Fungal Biocontrol Agents:** Nemaphagous fungi encompass a diverse group, including trapping fungi (e.g., *Arthrobotrys* spp.), egg-parasitic fungi (e.g., *Purpureocillium lilacinum*), and endoparasitic fungi (e.g., *Hirsutella rhossiliensis*). These organisms employ mechanisms such as adhesive nets, specialized spores, and toxin secretion to capture, penetrate, and kill nematodes. Under elevated temperatures and CO_2, studies report variable impacts: while some *Trichoderma* strains maintain antagonism under heat stress, others show diminished spore germination and VOC production, affecting volatile-mediated nematode suppression (Ayaz et al. 2024). Similarly, moisture deficits can reduce the formation of trapping structures by *Arthrobotrys* spp., highlighting the need for robust isolates adapted to drier soils.
- **Bacterial Biocontrol Agents:** Rhizobacteria and endophytes such as *Bacillus* spp., *Pseudomonas fluorescens*, and *Pasteuria penetrans* are among the most studied bacterial BCAs. They employ multiple modes of action: direct parasitism (e.g., *P. penetrans* attaching to J2 nematodes), antibiosis via secondary metabolites (e.g., hydrogen cyanide, lytic enzymes), competition for niche and nutrients, and induction of systemic resistance (ISR) in host plants. Multi-omics approaches have identified novel nematicidal metabolites—such as dimethyl disulfide and 2-undecanone—from *Bacillus atrophaeus*, which retain activity at varied temperature ranges and soil moisture levels (Ayaz et al. 2024). Yet, the interplay between elevated CO_2-driven root exudation changes and bacterial colonization patterns warrants further investigation to ensure consistent performance under future climate scenarios.

1.5.8 Decision Support Systems

Decision support systems (DSS) for plant-parasitic nematodes are an active area of development. DSS integrate data acquisition, modeling, and recommendation modules to aid farmers in timely and precise interventions. Current tools combine agronomic models (rotations, damage thresholds) with growing computational support (machine learning, simulation). Real-world deployments are most advanced in high-value crops like potato and sugar beet in Europe, but new sensing and AI technologies promise broader applicability. Continued integration of diverse data (soil, remote sensing, genomics, and climate) and user-friendly platforms (web/mobile apps) is driving DSS toward making nematode management both more precise and more practical for growers (Molendijk and Sikora 2022). This chapter surveys the development of DSS for PPN management, categorizes major system architectures, highlights recent AI- and simulation-based tools, and discusses case studies such as NemDST, NemaDecide, and SBN-Watch. Key challenges and future research directions are identified to guide the deployment of efficient and scalable DSS solutions.

1.5.9 Genomics, Proteomics, and Metabolomics

Genomics has revolutionized our understanding of nematode biology, offering novel tools for precise diagnosis, population monitoring, and targeted control strategies. This chapter synthesizes recent advances in nematode genomics, focusing on diagnostic marker development, resistance gene discovery, and genomic-based management approaches. Key case studies, discuss current challenges, and propose future directions for integrating genomics into sustainable nematode management are highlighted.

While genomics and transcriptomics have advanced our understanding of PPN biology, proteomics offers direct insight into functional proteins that mediate host–pathogen interactions. The latest proteomic tools and approaches for PPN detection, characterization of nematode effectors, and identification of host resistance factors, as well as how proteomics informs diagnostic assays and targeted management strategies, are highlighted.

Metabolomics bridges the gap between genotype and phenotype, offering powerful tools for early diagnosis of plant parasitic nematode infections and for the development of sustainable, targeted management strategies.

1.5.10 Remote Sensing/Drone Technology

Remote sensing offers a noninvasive approach to diagnose and manage plant-parasitic nematodes (PPNs) in agriculture, addressing the significant economic losses they cause, estimated at $80–157 billion annually. PPNs, such as root-knot (*Meloidogyne* spp.) and cyst nematodes (*Heterodera* and *Globodera* spp.), are challenging to detect due to their microscopic size and often nonspecific symptoms like stunted growth or chlorosis. Traditional methods rely on labor-intensive soil sampling and microscopic analysis, but remote sensing provides a scalable, efficient alternative by detecting plant stress and nematode infestations through spectral reflectance and imaging technologies.

Drone technology has emerged as a transformative tool in agriculture, particularly for addressing plant nematode diagnostics and management. Drones can help diagnose plant nematodes by detecting stress symptoms early. They aid nematode management by monitoring treatment effectiveness and soil health. The evidence leans toward drones being useful for large-scale agriculture, but direct nematode targeting is challenging. This note provides a comprehensive overview, drawing on recent research to explore how drones are utilized, their effectiveness, and future potential, with a focus on plant parasitic nematodes.

1.5.11 Climate Change

Climate change encompasses a suite of environmental alterations—rising temperatures, elevated atmospheric CO_2 concentrations, shifting precipitation regimes, and increased frequency of extreme weather events—that collectively reshape soil ecosystems. Plant-parasitic nematodes (PPNs), as soil-dwelling obligate parasites dependent on root systems, are highly sensitive to these abiotic shifts. Changes in nematode abundance in soil have profound implications for crop health and yield, as higher PPN populations amplify root damage and nutrient imbalances, whereas reductions in their numbers may temporarily relieve plant stress but disrupt soil food-webs.

Climate-induced shifts in temperature, CO_2, and precipitation exert multifaceted impacts on soil-borne PPN abundance. While warming tends to accelerate nematode life cycles within optimal ranges, excessive heat can curtail populations. Elevated CO_2 often fosters PPN proliferation through greater root biomass, yet plant defense changes and biocontrol dynamics may moderate this effect. Altered precipitation—especially drought—has the most pronounced negative impact on soil nematode abundance, whereas sustained increases in rainfall may modestly favor root feeders. Predictive models and region-specific studies are essential to anticipate PPN population dynamics and to tailor integrated management strategies in a changing climate.

1.6 Benefits, Challenges, and Integration

These frontier technologies bring significant benefits, including greater diagnostic accuracy, reduced environmental harm, and improved crop productivity. However, challenges include high initial costs (e.g., reagents for qPCR, instrumentation for AI), the need for technical expertise, and regulatory barriers, particularly for genetically modified crops. For example, the environmental impact of new nematicides like fluensulfone, while improved, remains a concern due to toxicity to aquatic organisms. Addressing these hurdles will be the key to widespread adoption.

While these technologies show promise, their cost and complexity can limit accessibility for small-scale farmers. Research continues to focus on making them more affordable and integrating them for comprehensive solutions, balancing innovation with practical application.

An integrative approach is recommended, combining molecular diagnostics (e.g., PCR, CRISPR) for accurate species identification with remote sensing for field-scale monitoring, and management strategies like biotechnology (resistant varieties), precision agriculture, and biocontrol agents. This holistic strategy leverages the strengths of each technology to address the complex challenge of plant nematodes effectively. Combining these technologies can lead to better, more sustainable control, considering how nematodes, plants, and control agents interact, and how soil conditions affect them.

Despite these advancements, challenges remain. The cost of reagents, instrumentation, and computational resources for technologies like high-throughput sequencing, ddPCR, and AI can be prohibitive, particularly for small-scale farmers. Remote sensing and hyperspectral imaging are limited by factors like resolution, weather conditions, and the need for optimal UAV parameters. Moreover, the integration of these technologies into practical, field-applicable systems requires further research, especially to reduce contamination risks in nucleic acid detection and improve scalability.

Future trends include the development of integrated nucleic acid detection systems for field environments, combining molecular diagnostics with remote sensing for comprehensive diagnostics. There is also a need to upgrade knowledge about host-specific susceptibility/resistance mediated by PPN effectors, resistance genes, or quantitative trait loci, leveraging advances in genome sequencing, RNA interference (RNAi), and gene editing for sustainable nematode control.

1.7 Case Studies

The following Table 1.3 summarizes frontier technologies for nematode diagnosis and management.

This table summarizes the key technologies, their features, and relevant examples, providing a structured overview for researchers and practitioners.

1.8 Conclusion

Frontier technologies for plant nematode diagnostics and management, including molecular diagnostics, precision agriculture, nanotechnology, remote sensing, AI, and biological control, are transforming agricultural practices by offering innovative tools to combat a persistent agricultural challenge. They offer rapid, accurate, and noninvasive solutions to identify and manage PPNs, addressing the global challenge of food security and sustainable agriculture. However, their adoption requires addressing cost, accessibility, and integration challenges to ensure widespread benefits for farmers worldwide. By integrating molecular diagnostics, AI, genetic engineering, and precision techniques, these advancements promise more resilient and

Table 1.3 Frontier technologies for nematode diagnosis and management

Category	Technology	Key features	Examples of PPNs	References
Diagnosis	Molecular diagnostics (PCR, NGS)	High sensitivity, DNA-based identification	*Meloidogyne* spp., *Heterodera* spp.	Shao et al. 2023
Diagnosis	CRISPR/Cas12a	Attomolar sensitivity, field-friendly, visual results with LFD	*Heterodera schachtii*, *H. avenae*, *H. filipjevi*	Yao et al. 2021; Shao et al. 2023
Diagnosis	Remote sensing	Noninvasive, large-scale monitoring, hyperspectral imaging	*Radopholus similis*, *Globodera rostochiensis*	Shao et al. 2023
Diagnosis	AI and machine learning	Automated counting, 94% accuracy for eggs, 93% for juveniles, SEGNEMA GUI	*Meloidogyne* spp., soybean nematodes	Saikai et al. 2023
Diagnosis	Mobile applications	Enhances accessibility for identification	Plant-parasitic nematodes	X post
Management	Genetic engineering (CRISPR-Cas9)	Develops resistant varieties, targets feeding site genes	Various PPNs	Saikai et al. 2023
Management	Precision agriculture	Variable-rate nematicide, GPS-guided, soil conductivity measurement	Various PPNs	Lawrence and Lawrence 2020
Management	New nematicides	Eco-friendly, e.g., tioxazafen, fluazaindolizine, fluensulfone, SDH inhibitors	*Meloidogyne incognita*, *R. reniformis*	Páez 2023
Management	Biocontrol agents	Natural enemies like fungi, bacteria, sustainable alternative	Various PPNs	Saikai et al. 2023

sustainable farming systems. As they continue to develop, they hold immense potential to enhance food security and reduce agriculture's ecological footprint, shaping a brighter future for global crop production.

1.9 Suggested Readings

- Abd-Elgawad, M.M.M. 2025. Integrated nematode management strategies: Optimization of combined nematicidal and multi-functional inputs. *Plants* 14(7): 1004. https://doi.org/10.3390/plants14071004.
- Afzal, A. and Mukhtar, T. 2024. Revolutionizing nematode management to achieve global food security goals - An overview. *Heliyon* 10(3): e25325. https://doi.org/10.1016/j.heliyon.2024.e25325.
- Ayaz, M., Zhao, W., Chi, Y.K. et al. 2024. Biocontrol of plant parasitic nematodes by bacteria and fungi: a multi-omics approach for the exploration of novel nematicides in sustainable agriculture. *Frontiers in Microbiology* 15: 1433716.
- Khan, R.M. 2023. *Novel Biological and Biotechnological Applications in Plant Nematode Management*. Springer.
- Páez, A.R. 2023. Plant-parasitic nematodes and their management: A focus on new nematicides. In: S. Mukherjee and S. Ray (eds.) *Nematodes - Ecology, Adaptation and Parasitism*. IntechOpen. DOI: 10.5772/intechopen.1002237.
- Pun, T.B., Thapa M.R., Koech, R. et al. 2024. Emerging trends and technologies used for the identification, detection, and characterisation of plant-parasitic nematode infestation in crops. *Plants* 13(21): 3041. https://doi.org/10.3390/plants13213041.
- Shao, H., Zhang, P., Peng, D. et al. 2023. Current advances in identification of plant nematode diseases: From lab assays to in-field diagnostics. *Front. Plant Sci.* 14: 1106784. https://doi.org/10.3389/fpls.2023.1106784.

1.10 Review Questions

- What are the major challenges in diagnosing plant-parasitic nematodes?
- What role do CRISPR/Cas technologies hold in the future of nematode management?
- How are genomics and proteomics being applied in understanding nematode–host interactions?
- Discuss the role of artificial intelligence (AI) and machine learning in nematode detection and counting.
- Evaluate the potential of nanotechnology in nematode diagnostics and control.
- How can beneficial microbes (e.g., *Pochonia chlamydosporia*, *Pasteuria penetrans*) contribute to sustainable nematode management?

References

Abad P, Gouzy J, Aury JM et al (2008) Genome sequence of the metazoan plant parasitic nematode *Meloidogyne incognita*. Nat Biotechnol 26(8):909–915

Abade A, Porto L, Ferreira P, Vidal F (2022) NemaNet: a convolutional neural network model for identification of soybean nematodes. Biosyst Eng 213:39–62

Ayaz M, Zhao W, Chi YK et al (2024) Biocontrol of plant parasitic nematodes by bacteria and fungi: a multi-omics approach for the exploration of novel nematicides in sustainable agriculture. Front Microbiol 15:1433716

Goraya M, Yan G, Whitworth J, Grimm KS (2024) Advancing nematode identification on potato: an isothermal recombinase polymerase amplification assay for stubby root nematode, Paratrichodorus allius. Am J Potato Res 101:52–64

Hillnhütter C, Mahlein AK, Sikora R, Oerke EC (2011a) Remote sensing to detect plant stress induced by Heterodera schachtii and Rhizoctonia solani in sugar beet fields. Field Crop Res 122:70–77

Hillnhütter C, Mahlein AK, Sikora R, Oerke EC (2011b) Use of imaging spectroscopy to discriminate symptoms caused by *Heterodera schachtii* and *Rhizoctonia solani* on sugar beet. Precis Agric 13:17–32

Huston DC, Khudhir M, Hodda M (2022) Reliability and utility of standard gene sequence barcodes for the identification and differentiation of cyst nematodes of the genus Heterodera. J Nematol 54:20220024. https://doi.org/10.2478/jofnem-2022-0024

Kulshrestha K, Parihar A, Parihar P (2020) Next generation sequencing based transcriptome analysis for nematode resistance in different species of tomato. Plant Gene 24:100255. https://doi.org/10.1016/j.plgene.2020.100255

Lawaju BR, Yan G, Whitworth J (2023) Development of a droplet digital PCR assay for detection and quantification of stubby root nematode, Paratrichodorus allius in soil. Plant Dis 107:3344–3353

Lawrence KS, Lawrence GW (2020) Plant-parasitic nematode management. In: Bergtold J, Sailus M (eds) Conservation tillage Systems in the Southeast Production, profitability and stewardship, SARE Handbook Series, vol 15. SARE, pp 164–180

Li X, Wang X, Zhang S et al (2012) Identification of soybean microRNAs involved in soybean cyst nematode infection by deep sequencing. PLoS One 7:e39650. https://doi.org/10.1371/journal.pone.0039650.71

Lu X, Wang Y, Fung S, Xue Q (2021) A biological image dataset for nematode recognition. I-Nema 4:1–13

Martins GD, Galo MDLBT, Vieira BS (2017) Detecting and mapping root-knot nematode infection in coffee crop using remote sensing measurements. IEEE J Sel Top Appl Earth Obs Remote Sens 10:5395–5403

Molendijk LPG, Sikora RA (2022) Decision support systems in integrated nematode management: the need for a holistic approach. In: Sikora RA, Desaeger J, Molendijk LPG (eds) Integrated nematode management: state-of-the-art and visions for the future. CAB Int., Wallingford, UK, pp 428–438

Müller CA, de Mattos Pereira L, Lopes C et al (2019) Meiofaunal diversity in the Atlantic Forest soil: a quest for nematodes in a native reserve using eukaryotic metabarcoding analysis. For Ecol Manag 453:117591. https://doi.org/10.1016/j.foreco.2019.117591

Niu J, Jian H, Guo Q et al (2012) Evaluation of loop-mediated isothermal amplification (LAMP) assays based on 5S rDNA-IGS2 regions for detecting Meloidogyne enterolobii. Plant Pathol 61:809–819

Norman GG, Fritz NL (1965) Infrared photography as an indicator of disease and decline in citrus trees. Proc Fla State Hort Soc 12:31–41

Páez AR (2023) Plant-parasitic nematodes and their management: a focus on new nematicides. In: Mukherjee S, Ray S (eds) Nematodes - Ecology, Adaptation and Parasitism. IntechOpen. https://doi.org/10.5772/intechopen.1002237

Perera MR, Vanstone VA, Jones MGK (2005) An ovel approach to identify plant parasitic nematodes using matrix-assisted laser desorption/ionization time-of-flight mass spectrometry. Rapid Commun Mass Spectrometry 19:1454–1460
Qing X, Wang Y, Lu X et al (2022) NemaRec: a deep learning-based web application for nematode image identification and ecological indices calculation. Eur J Soil Biol 110:103408. https://doi.org/10.1016/j.ejsobi.2022.103408
Rivero J, Zurita A, Cutillas C, Callejon R (2022) The use of MALDI-TOF MS as a diagnostic tool for adult *Trichuris* species. Front Veterinary Sci 9:867919. https://doi.org/10.3389/fvets.2022
Robinson NA, Sheedy JG, MacDonald BJ et al (2019) Tolerance of wheat cultivars to root-lesion nematode (Pratylenchus thornei) assessed by normalized difference vegetation index is predictive of grain yield. Ann Appl Biol 174:388–401
Samiappan S, Bheemanahalli R, Zhou M et al (2021) Early detection of root-knot nematode (Meloidogyne incognita) infestation in cotton using hyperspectral data. In: Proceedings of the 2021 IEEE International Geoscience and Remote Sensing Symposium IGARSS, Brussels, Belgium. 11–16 July 2021, pp 5849–5852
Saikai KK, Oduori C, Situma E. et al. (2023) Biocontrol-based strategies for improving soil health and managing plant-parasitic nematodes in coffee production. Front Plant Sci 141196171. https://doi.org/10.3389/fpls.2023.1196171
Shao H, Zhang P, Peng D et al (2023) Current advances in identification of plant nematode diseases: from lab assays to in-field diagnostics. Front Plant Sci 14:1106784. https://doi.org/10.3389/fpls.2023.1106784
Susič N, Žibrat U, Širca S et al (2018) Discrimination between abiotic and biotic drought stress in tomatoes using hyperspectral imaging. Sens Actuators B Chem 273:842–852
Syifa M, Park SJ, Lee CW (2020) Detection of the pine wilt disease tree candidates for drone remote sensing using artificial intelligence techniques. Engineering 6:919–926
Thevenoux R, Van Linh L, Villessèche H et al (2021) Image based species identification of Globodera quarantine nematodes using computer vision and deep learning. Comput Electron Agric 186:106058. https://doi.org/10.1016/j.compag.2021.106058
Thompson JP, Sheedy JG, Robinson NA (2020) Resistance of wheat genotypes to root-lesion nematode (Pratylenchus thornei) can be used to predict final nematode population densities, crop greenness, and grain yield in the field. Phytopathology 110:505–516
Wang C, Sun S, Zhao C et al (2022) A detection model for cucumber root-knot nematodes based on modified YOLOv5-CMS. Agronomy 12:2555. https://doi.org/10.3390/agronomy12102555
Westphal A (2011) Sustainable approaches to the management of plant-parasitic nematodes and disease complexes. J Nematol 43(2):122–125
Yao K, Peng D, Jiang C et al (2021) Rapid and visual detection of Heterodera schachtii using recombinase polymerase amplification combined with Cas12a-mediated technology. Int J Mol Sci 22:12577. https://doi.org/10.3390/ijms222212577
Zhang Y, Wang Y, Xie F et al (2016) Identification and characterization of microRNAs in the plant parasitic root-knot nematode Meloidogyne incognita using deep sequencing. Funct Integr Genom 16:127–142

Emerging Nematode Problems and Their Management

2

Abstract

Advances in global trade, coupled with warming temperatures, have facilitated the invasion of several nematode species into regions where they were previously absent. Climate change is expanding nematode ranges and increasing crop damage. Emerging nematode problems are driven by climate change, intensified farming practices, global trade, and evolving nematode populations. Emerging issues include climate-driven range expansion and evolving nematode resistance, threatening food security. The resistance-breaking nematode strains are challenging existing crop defenses. These emerging challenges highlight the urgent need to diversify nematode management approaches. Management focuses on integrated approaches, blending resistant crops, natural enemies, and sustainable farming practices. Integrated nematode management (INM) is effective, combining biological, cultural, and chemical strategies. Sustainable practices, like crop rotation and biological controls, are gaining traction due to environmental concerns.

Keywords

Climate change · Changing cropping patterns · Introduction of new crops · Crop diversification · Agronomic practices

2.1 Learning Objectives

- Identify key emerging plant-parasitic nematodes, their drivers (climate change, globalization, intensive farming), and their impacts on crops.
- Explain the factors contributing to the emergence and spread of new emerging plant-parasitic nematode problems worldwide.
- Identify the impact of globalization, climate change, and intensive agriculture on nematode dynamics.

R. P. Parvatha, *Plant Parasitic Nematodes*,
https://doi.org/10.1007/978-981-95-9306-4_2

- List and describe key emerging nematode species (e.g., *Meloidogyne enterolobii*, *Globodera pallida*, *Aphelenchoides besseyi*, *Rotylenchulus reniformis*, invasive *Bursaphelenchus* spp.).
- Evaluate cultural, biological, and chemical methods for sustainable nematode management.
- Discuss quarantine regulations, phytosanitary measures, and international cooperation in managing invasive nematodes.
- Explore integrated and innovative management strategies, including biocontrol, resistance breeding, and quarantine measures.

2.2 Introduction

Plant-parasitic nematodes are microscopic worms that pose significant threats to global agriculture, causing estimated annual crop losses of $157 billion. The changing cropping patterns, introduction of new crops, crop diversification, and agronomic practices change the spectrum of pests and pathogens, including plant-parasitic nematodes. For example, the adoption of water saving techniques like System of Rice Intensification (SRI) in rice and drip irrigation in horticultural crops; diversification toward horticultural crops, particularly protected cultivation systems; and widespread and unchecked movement of planting materials from horticultural nurseries have led to the emergence of new nematode problems in newer areas and intensification of existing nematode problems. A number of new nematode problems have emerged in the intensive cropping systems.

Climate change is likely accelerating nematode life cycles, with warmer soils boosting reproduction and damage. New resistant strains in *Meloidogyne incognita* are breaking through crop defenses, especially in regions like Africa. Global trade and organic farming also face challenges, with nematodes adapting to new hosts and limited chemical options.

2.3 Causes for Emergence of New Nematode Problems

2.3.1 Climate Change Impacts

Warmer temperatures and altered precipitation patterns expand the geographic range of nematode species like *Meloidogyne* (root-knot nematodes) and *Ditylenchus* (stem and bulb nematodes), enabling them to infect new regions. Prolonged growing seasons increase nematode reproduction cycles, amplifying population densities and damage. Shifts in soil moisture affect nematode survival, with some species thriving in wetter or drier conditions.

Climate change is altering nematode ecology, with warmer soil temperatures accelerating life cycles and increasing reproduction rates. For instance, studies on

Rotylenchulus reniformis and *Meloidogyne floridensis* show enhanced survival and virulence at elevated temperatures (Khanal and Land 2023). This facilitates infection by altering host plant physiology and expanding nematode ranges into new regions.

Altered precipitation patterns and soil moisture levels further exacerbate the problem, creating favorable conditions for root-knot nematodes like *Meloidogyne* spp. to thrive in previously unaffected areas.

2.3.2 Resistance-Breaking Nematode Strains

Nematodes are evolving resistance-breaking pathotypes/isolates, challenging existing host plant resistance. For example, *Meloidogyne incognita* and *M. javanica* are affecting resistant genotypes of crops like tomato, pepper, cowpea, and potato (Abd-Elgawad 2024). This evolution necessitates continuous breeding efforts to identify new resistance genes.

The emergence of species like *Meloidogyne enterolobii* (guava root-knot nematode) is particularly concerning due to its high virulence and resistance to traditional controls, impacting crops globally.

2.3.3 Challenges in Organic and Sustainable Agriculture

Organic farming systems face unique challenges, as chemical nematicides are often restricted. Nematodes like *Meloidogyne* spp. (root-knot nematodes) and *Pratylenchus* spp. (root-lesion nematodes) are particularly problematic, with limited alternative control options (Hotte et al. 2022). This is compounded by the wide host range of nematodes, making crop rotation less effective in some cases.

The reliance on sustainable practices in organic systems highlights the need for robust biological and cultural controls to manage nematode populations.

2.3.4 Global Spread and New Host Ranges

Global trade and climate shifts are facilitating the spread of nematodes to new regions and crops. For instance, *Meloidogyne enterolobii* is emerging as a significant threat in sub-Saharan Africa, where technological gaps hinder control efforts (Abd-Elgawad 2024). Nematodes are also adapting to infect new host plants, expanding their impact across diverse agricultural systems.

The microscopic size and rapid life cycle of nematodes, combined with nondescript foliar symptoms, often allow their movement to go unnoticed until populations are well-established, complicating early intervention.

2.3.5 Economic and Food Security Implications

Nematode infestations increase production costs, particularly in developing nations, leading to food shortages and reduced export earnings. For example, potato cyst nematodes (*Globodera pallida* and *G. rostochiensis*) cause around 9% loss of global potato crops, with eggs surviving in soil for years. The technological gap between developed (e.g., USA, Europe) and developing countries (e.g., sub-Saharan Africa) exacerbates these challenges, necessitating cooperative programs for global integrated nematode management (INM) (Hotte et al. 2022).

2.3.6 Intensified Agricultural Practices

Monoculture and continuous cropping deplete soil health, creating favorable conditions for nematode species like *Heterodera* (cyst nematodes) and *Pratylenchus* (root-lesion nematodes). Reduced crop rotation diversity limits natural suppression of nematode populations. Overuse of chemical inputs disrupts beneficial soil microbes, indirectly boosting nematode virulence.

2.3.7 Global Trade and Invasive Species

Increased international trade introduces invasive nematodes, such as *Bursaphelenchus xylophilus* (pine wood nematode), to new ecosystems. Contaminated planting material, soil, or equipment spreads nematodes across borders. Emerging species like *Meloidogyne enterolobii* (guava root-knot nematode) show high virulence and resistance to traditional controls.

2.3.8 Nematode Adaptation and Resistance

Nematodes are developing resistance to nematicides due to repeated applications, reducing chemical efficacy. Host plant resistance breakdown occurs as nematodes evolve to overcome resistant crop varieties, particularly in *Meloidogyne* and *Globodera* species. Coinfections with other pathogens (e.g., fungi or bacteria) enhance nematode damage, complicating management.

2.3.9 Urbanization and Land Use Changes

Conversion of natural landscapes to agriculture increases nematode pressure on crops unaccustomed to such pests. Soil disturbance from construction or irrigation projects spreads nematodes to new areas.

2.4 Case Studies on Emerging Nematode Problems: Indian Scenario

Numerous challenges in the discipline of Nematology have opened up new opportunities to prove nematodes as important limiting factors in agricultural production: root-knot nematode problem on rice, wheat, pomegranate, guava, acid lime, and vegetables and flowers grown under protected cultivation; cyst nematodes problem on potato; floral malady on tuberose and many others; impact of climate change; threat to the introduction of new nematode pest under globalization of agricultural produce; and lack of management strategy in standing crops are some of the challenges in near future.

2.4.1 Rice Root-Knot Nematode (*Meloidogyne graminicola*)

Among the emerging biotic constraints affecting rice productivity, the rice root-knot nematode (*Meloidogyne graminicola*) has gained considerable importance over the last two decades. Once confined primarily to upland rice ecosystems, the nematode is now increasingly reported from irrigated and direct-seeded rice systems across diverse agroecological regions of India. The nematode is widely distributed across rice-growing regions of Eastern India (West Bengal, Odisha, Bihar, and Assam), Southern India (Karnataka, Andhra Pradesh, Tamil Nadu, and Telangana), Northern plains (Uttar Pradesh and adjoining areas), and Hill and upland rice ecosystems. Yield losses ranging from 10% to 80% have been documented under severe infestations.

Nematode's adaptation to fluctuating flooded and aerobic soil environments, rapid life cycle, and wide host range have facilitated its emergence as a major yield-limiting pathogen. Several agricultural transitions responsible for its increasing prevalence include expansion of direct-seeded rice systems, reduced continuous flooding, climate warming trends, intensive rice–rice and rice–vegetable cropping sequences, and movement of infected seedlings and contaminated soil.

Aboveground symptoms include stunted plant growth, chlorosis and poor tillering, patchy crop appearance, early wilting under moisture stress, and yield reduction resembling nutrient deficiency. Belowground symptoms include hook-shaped or bead-like galls on roots, reduced root system, excess lateral root formation, and root tip swelling (Fig. 2.1).

Management Cultural practices include continuous flooding (where feasible) to suppress infection; crop rotation with non-hosts; timely transplanting to avoid nursery damage; soil solarization or soil application of organic amendments.

Resistant varieties include RP-5219 series, NPT-10, MTU 1390, Kushiari, Sugandha-3, and HR-12 (Gautam et al. 2025). Breeding programs focus on these for sustainable resistance.

Biological control agents like *Pochonia chlamydosporia, Purpureocillium lilacinum, Bacillus firmus* and *Trichoderma* spp. were found effective.

Fig. 2.1 *Left*—Healthy and root-knot infected rice seedlings. *Right*—Closeup of infected seedling roots (Note hook-shaped or bead-like galls)

2.4.2 Groundnut Root-Knot Nematodes, *Meloidogyne arenaria*

Among soil-borne pathogens, root-knot nematodes (RKNs) have emerged as a serious constraint to sustainable groundnut production. *Meloidogyne arenaria*, commonly referred to as the peanut root-knot nematode, is increasingly reported as a damaging species in sandy and light-textured soils where groundnut is intensively cultivated. The nematode is particularly problematic because it causes chronic yield decline and often remains unnoticed until significant economic loss occurs. Major groundnut-growing areas of India infested with *M. arenaria* include Gujarat, Andhra Pradesh, Karnataka, Tamil Nadu, and Rajasthan.

Several agronomic and environmental factors such as continuous monocropping of groundnut, expansion into light sandy soils favorable for nematode multiplication, climate warming enhancing nematode reproduction cycles, reduced crop rotation diversity, and increased cultivation of susceptible varieties are contributing to its increasing importance.

Aboveground symptoms include stunted growth, yellowing (chlorosis), and wilting, often resembling nutrient deficiencies. Belowground, galls form on roots, pegs, and pods, reducing plant efficiency and yields by up to 10% or more in infested fields (Fig. 2.2). Damage is exacerbated in sandy soils and under drought stress. *M. arenaria* causes substantial economic losses by impairing root function, reducing pod quality, and facilitating secondary infections from fungi and other pathogens.

Management Cultural methods like crop rotation with non-hosts like cotton (for *M. arenaria*), bahia grass, millet, or sorghum-Sudan grass can help to starve populations. Fallow periods reduce densities but require weed control; cover crops like sun hemp or marigolds offer nematicidal effects. Exclusion prevents spread via contaminated equipment. Application of organic amendments like neem cake, castor cake, and farmyard manure enriched with biocontrol fungi improve soil microbial antagonism and reduce nematode populations.

Fig. 2.2 *Left*—Galling of peanut roots and pods due to infection by *Meloidogyne arenaria*. *Right*—irregularly shaped growths (galls) on peanut pods caused by root-knot nematodes

Non-fumigants such as fluopyram, fluensulfone, and biologicals like *Purpureocillium lilacinum* (Melocon) suppress root-knot nematode populations.

Breeding has incorporated resistance from wild *Arachis* species into cultivars like COAN, NemaTAM, Tifguard, Georgia-14 N, Webb, TifNV-High O/L, TifNV-HG, Georgia-22MPR, TifJumbo, and Georgia-19HP. These reduce reproduction and galling, often yielding higher than susceptible standards like Georgia-06G. For example, TifNV-High O/L showed superior vigor and yields.

Groundnut Stunt Nematode, *Tylenchorhynchus brevilineatus*

During 1975–1976, a severe disease of groundnut characterized by reduction in pod size and brownish discoloration of pod surface was noticed near Kalahasti village of Andhra Pradesh, and since then it is popularly known as *Kalahasti* malady. However, a "pod scab" is the most characteristic symptom of infection. The "Kalahasti Malady" is a serious disease of groundnut in parts of Chittoor and Nellore districts of Andhra Pradesh, India. This disease was reportedly caused by the nematode *Tylenchorhynchus brevilineatus* (Reddy et al. 1984). The nematode has also been reported on CO-2 variety of groundnut in Bhavanisagar, Tamil Nadu (Sharma et al. 1996). Yield losses ranging from 20% to 60% have been reported under severe infestation, making it a significant constraint in endemic areas.

Field symptoms include patchy distribution of affected plants in fields, stunted growth with comparatively dark or pale green foliage, and reduced plant vigor and uneven crop stand. Peg and pod symptoms (diagnostic) include small brownish-yellow lesions on pegs and developing pods, slightly raised lesion margins due to host cell proliferation, reduced peg length, pod surface becomes blackened or scabbed at advanced stages, and pods become small, shriveled, and malformed (Fig. 2.3). Root symptoms include mild discoloration and limited damage compared to pods and reduced functional efficiency of roots.

Fig. 2.3 Groundnut plant infected with *Tylenchorhynchus brevilineatus* showing small severely discolored pods

Management

(i) **Cultural Methods:**

- Crop rotation with rice, marigold, and *Tagetes erecta* (poor host) significantly reduces nematode populations (Naidu et al. 2000).
- Avoid continuous groundnut cultivation.
- Application of gypsum during earthing-up improves pod health.
- Adequate irrigation reduces stress and symptom severity.

(ii) **Chemical Methods**

Soil application of carbofuran at 1.0 kg a.i. /ha 25–30 days after sowing along with irrigation water reduced nematode populations.

(iii) **Host Resistance**

High yielding groundnut line TCG 1518 was found resistant (Mehan et al. 1993).

(iv) **Integrated Methods**

- Resistant cultivars + crop rotation.
- Organic amendments (FYM, neem cake) + biological control agents.
- Monitoring nematode population before sowing.
- Avoid susceptible intercrops that maintain inoculum.

Guava Root-Knot Nematode, *Meloidogyne enterolobii*

Unlike traditional root-knot nematodes such as *M. incognita* or *M. javanica*, *M. enterolobii* is characterized by extremely high virulence, wide host range, ability to overcome host resistance, and rapid reproduction under tropical conditions. Because of these traits, it is now regarded as an emerging biosecurity threat to horticulture in India. *M. enterolobii* has progressively spread within the country, largely through infected nursery planting material. It was first detected in Tamil Nadu around 2016 (Poornima et al. 2016), and has since spread to states like Karnataka, Gujarat (Ghule et al. 2020), Madhya Pradesh (Khan et al. 2022), Uttarakhand (Kumar and Rawat 2018), and Uttar Pradesh (Vishwakarma et al. 2023), likely through infested planting materials. It contributes to rapid plant decline and yield losses of up to 80% in severely affected areas. The nematode often interacts synergistically with fungal pathogens such as *Fusarium* spp., exacerbating guava wilt complex.

Typical symptoms included leaf bronzing, root galling, and plant wilting (Fig. 2.4). Severe infestations were reported in Kerala guava plantations intercropped with coconut, where molecular diagnostics confirmed >99% genetic

Fig. 2.4 *Left*—Guava plants showing symptoms of root-knot nematode infection (stunted growth and leaf yellowing). *Right*—Root galling on guava roots

similarity with reference populations. The nematode has expanded beyond guava to other crops such as watermelon, demonstrating increasing host adaptation and epidemiological risk. The spread is strongly associated with movement of infected nursery saplings, emphasizing the need for quarantine surveillance.

Management Cultural practices like removing infested plants and using non-host rotations (e.g., with garlic or maize) help reduce populations. Biological agents, such as the fungus *Purpureocillium lilacinum* or bacteria like *Pseudomonas fluorescens*, have shown promise in suppressing egg hatching and juvenile mortality, with efficacy up to 100% in lab tests. Chemical options include nematicides like fluopyram, but environmental concerns limit their use. Screening for resistant guava genotypes offers long-term potential, though *M. enterolobii*'s virulence often bypasses common resistance.

2.4.3 Pomegranate Root-Knot Nematode, *Meloidogyne incognita*

Recent surveys indicate that the root-knot nematode, *M. incognita*, is frequently reported as the dominant species and is widespread in commercial orchards, where it is associated with declining productivity and orchard longevity. The nematode pest has transitioned from a localized problem to an emerging national constraint due to expansion of pomegranate cultivation into semiarid regions, use of infected planting material, continuous cropping without rotation, drip irrigation maintaining favorable soil moisture, and increased nursery-based propagation. The infestation can lead to significant yield reductions, often ranging from 17% to 40%, alongside reduced fruit quality, though impacts vary based on local factors like soil type and management practices.

The root-knot nematode has become an emerging threat to pomegranate production, particularly in arid and semiarid regions where pomegranate is a key commercial crop. The crop's export value and domestic demand make nematode infestations a critical concern, as they can lead to substantial yield declines and quality degradation.

The emergence of *M. incognita* as a problem in Indian pomegranate orchards is linked to several factors. Intensive monocropping, use of infested planting material, and changing climatic conditions—such as warmer temperatures and altered irrigation practices—have facilitated its spread. In arid zones with light sandy soils, which are ideal for pomegranate but also highly conducive to nematode activity, infestations are particularly severe. For instance, surveys in Rajasthan's Barmer and Jodhpur districts revealed that 100% of pomegranate orchards were infested, with 13% showing severe incidence (>40% plants affected), 78% moderate (10–40%), and 9% mild (0–10%). Similarly, in Himachal Pradesh (Kullu, Mandi, Solan districts), average infestation rates varied from 0.36% to 9.25%, with the highest at Nauni in Solan. These variations highlight the influence of local microclimates and orchard management on nematode dynamics (Singh et al. 2022; Singh et al. 2019; Mukesh and Sharma 2024).

Fig. 2.5 *Left*—Root-knot nematode infected pomegranate trees. *Right*—Galling symptoms on pomegranate roots

Affected pomegranate plants typically exhibit yellowing of leaves, reduced leaf and fruit size, stunting, partial wilting, and gradual decline. Below ground, characteristic root galls or knots disrupt nutrient and water uptake (Fig. 2.5). Economic repercussions include yield losses of 17–40%, smaller fruits, and increased production costs, potentially threatening orchard sustainability if unmanaged. The nematode's wide host range, including weeds, further complicates control efforts.

Management Cultural methods include using nematode-free seedlings, crop rotation with non-hosts, sanitation (destroying infested roots/weeds), and improving drainage via trenches to reduce soil moisture.

Biological control involves bioagents like *Purpureocillium lilacinum* (formerly *Paecilomyces lilacinus*), *Pochonia chlamydosporia*, *Trichoderma* spp., *Pseudomonas fluorescens*, and AMF like *Glomus fasciculatum*. Organic amendments such as neem cake (2 tons/ha) enhance efficacy when combined with bioagents (Yadav and Patil 2021).

Chemical options include nematicides like Fluopyram (400 SC at 1250 ml/ha or 2 ml/L drench), Fluensulfone (2% GR at 2 g/ml), and Carbofuran. Field trials in Maharashtra showed that Fluopyram reducing nematode populations by 49% at intermediate stages, while bioagent-neem combinations (e.g., *P. fluorescens* + neem) achieved 35–40% reductions at termination. In Himachal Pradesh, Fluopyram drench reduced J2 by 98–99% and galling by 59–66%, outperforming Fluensulfone (89–95% J2 reduction, 44–55% galling). In Rajasthan, combined applications (e.g., Carbofuran 20 g + Fluensulfone 20 g per plant) reduced galls by 50%, with neem cake + *P. lilacinum* + nematicides yielding similar results (Binage et al. 2025; Mukesh and Sharma., 2021; Singh et al. 2022).

2.4.4 Citrus Root-Knot Nematode, *Meloidogyne indica*

Meloidogyne indica is an emerging threat to citrus crops in India. It has emerged as a regionally important and under-recognized nematode pathogen associated with citrus decline in India. Unlike widely studied species such as *M. incognita* and *M. javanica*, *M. indica* shows strong adaptation to perennial hosts and orchard ecosystems, making it a growing concern in long-duration citrus plantations. Reports indicate occurrence of *M. indica* in several citrus-growing belts, particularly in North Gujarat (e.g., on acid lime in districts like Banaskantha, Mahesana, and Anand, where numerous small to large galls lead to poor root systems and plant decline), Meghalaya (West Khasi Hills on Khasi mandarin, with monolocular galls but incomplete maturation, resulting in a gall index of 3.5 on a 0–5 scale), Tamil Nadu (incidences on citrus and linked to increasing wilt in mango-citrus areas) and Andhra Pradesh (in acid lime orchards) suggesting spread through infected planting material or soil (Patel et al. 1999; Phani et al. 2018; Walia et al. 2018; Krishna Kumar and Balasubramanian 2019).

Aboveground symptoms include progressive yellowing of foliage, reduced vigor and stunted growth, twig drying and canopy thinning, poor flowering and fruit set, and decline resembling nutrient deficiency or dieback (Fig. 2.6). Affected orchards often appear as patches of weak or nonproductive trees. Belowground symptoms include characteristic root galls, reduction in feeder roots (Fig. 2.6), impaired nutrient and water uptake, and secondary infection by soil pathogens. Infected citrus plants frequently show a "sick orchard" appearance with negligible yield.

Fig. 2.6 *Left*—Citrus (Mandarin) tree infected with root-knot nematode showing leaf yellowing symptoms. *Right*—Mandarin roots infected with root-knot nematode, *Meloidogyne indica*

Management Cultural practices are foundational: strict vigilance during seedling procurement, culling of affected plants, clean cultivation to remove root debris post-harvest, and deep summer plowing to expose nematodes to desiccation. Crop rotation is limited for perennial citrus but can incorporate non-hosts in nurseries. Sanitation, including avoiding contaminated soil or water, is crucial to prevent spread via saplings.

Biological control has advanced significantly as an eco-friendly alternative to chemicals, which face resistance and environmental backlash. Key biological control agents include fungi like *Purpureocillium lilacinum* (egg-parasitic, reducing populations by 50–80% in field trials), *Pochonia chlamydosporia* (targeting eggs and females), and *Trichoderma* species (promoting plant growth while antagonizing nematodes). Bacteria such as *Bacillus firmus, Pseudomonas fluorescens*, and *Pasteuria penetrans* offer suppressive effects through toxin production or parasitism. Mycorrhizal fungi like *Glomus intraradices* enhance root resilience. At least 16 microbial formulations are registered for citrus in various countries, with field efficacy improved by soil amendments like organic matter to boost indigenous antagonists. For *M. indica* specifically, these biocontrols show promise, though inconsistent performance due to soil conditions remains a challenge.

Chemical options, used judiciously in IPM, include nematicides like fluopyram (34.48% SC) at 2 ml/L, which achieved 98–99% larval reduction, and fluensulfone (2% GR). These induce nematode paralysis and death but should be rotated to avoid resistance.

2.4.5 Potato Cyst Nematodes, *Globodera rostochiensis, G. pallida*

Potato cyst nematodes (PCN), encompassing the species *G. rostochiensis* (commonly known as the golden cyst nematode) and *G. pallida* (the pale cyst nematode), represent a significant biotic constraint to potato production worldwide, including in India. PCN are mainly confined to cooler, high-elevation zones suitable for their survival. In southern India, infestations span about 4000 hectares in the Nilgiri hills of Tamil Nadu (Krishna Prasad 1986), extending to nearby areas in Kodai Hills, Kerala (Ramana and Mohandas 1988). Northern states like Himachal Pradesh, Uttarakhand, and Jammu and Kashmir show patchy distribution, often higher in government seed farms than in private fields. Altitudes above 1950–2000 meters above mean sea level are most affected (Ganguly et al. 2010; Walia 2024; Chandel et al. 2020). In India, PCNs pose an emerging threat due to their quarantine status, persistent soil viability (cysts can survive up to 30 years without a host), and adaptation to local agro-climatic conditions, exacerbated by factors like seed trade, climate variability, and incomplete regulatory enforcement (He et al. 2022; Walia 2024). PCN can reduce potato yields by 27–87% depending on infestation levels.

Aboveground symptoms include patchy crop growth, chlorosis and stunting, premature senescence, and poor canopy development (Fig. 2.7). Belowground symptoms include reduced root system and presence of lemon-shaped white/yellow/brown cysts attached to roots (Fig. 2.8).

Fig. 2.7 *Left one row*—potato plants infected with cyst nematodes (Note stunting of plants and yellowing of leaves). *Right*—Healthy plants

Fig. 2.8 Potato roots infected with cyst nematodes (*Left—Globodera rostochiensis*, *Right—G. pallida*)

Management Cultural methods that include 3–4-year rotation with non-host crops (cabbage, cauliflower, beetroot, radish, beans, carrots) reduces population density. Trap cropping with *Solanum sisymbriifolium* generates high mortality in PCN but requires a full growing cycle to be effective. Incorporation of Brassica residues (radish, cabbage, cauliflower) suppresses PCN populations through glucosinolate breakdown products (Umamaheswari et al. 2015). Phytosanitary measures include certified nematode-free seed, equipment sanitation, and quarantine enforcement.

Effective biological control components include *Pseudomonas fluorescens, Pochonia chlamydosporia,* and *Purpureocillium lilacinum*. Addition of neem cake and organic amendments enhances the biocontrol activity.

In the USA, both the USDA and Cornell University scientists have evolved two commercial potato varieties, viz., "Peconic" and "Wauseon" which were resistant to the cyst nematode (Fig. 2.9). "Peconic" derives its resistance from the subspecies andigena. "Wauseon" is a selection from a cross between B 4159–8 and Katahdin.

Potato cv. Kufri Swarna developed by Ooty Regional Research Station of Central Potato Research Institute, Shimla, is found resistant to PCN.

An integrated system utilizing resistant cultivars, non-host crops, and susceptible cultivars with chemical treatment (Carbofuran) will successfully manage densities of *G. rostochiensis* at desired levels.

Fig. 2.9 Host resistance in potato. *Left 2 rows*—susceptible potato variety. *Right*—Resistant variety

2.4.6 Root-Knot Nematodes Polyhouse Grown Vegetables and Flowers

Root-knot nematodes (*Meloidogyne* spp.) have emerged as one of the most serious biological constraints in protected cultivation systems in India, particularly in polyhouse vegetables and floricultural crops. While polyhouse technology enhances productivity, it simultaneously creates an ideal ecological niche for rapid nematode multiplication, transforming localized infestations into chronic production problems. Among them, root-knot nematodes (RKN) are the dominant and most destructive group affecting tomato, bell pepper, cucumber, brinjal, gherkins, gerbera, carnation, rose, and other ornamentals. Yield losses under polyhouse conditions may reach 25–100%, depending on infestation level and crop susceptibility (Kamra et al. 2024). Intensive monocropping, continuous irrigation, and stable microclimatic conditions promote severe buildup of soil-borne pathogens, especially plant-parasitic nematodes.

The dominant species in Indian polyhouses are *Meloidogyne incognita* (most prevalent), *M. javanica, M. arenaria*, and occasionally *M. hapla* in cooler regions. Other nematodes like reniform (*Rotylenchulus reniformis*) and lesion (*Pratylenchus* spp.) may co-occur but are less dominant. Secondary infections with fungi (*Fusarium oxysporum*, *Rhizoctonia solani*, *Macrophomina* spp., *Thielaviopsis* spp.) or bacteria (*Ralstonia solanacearum*) often lead to wilt complexes and plant mortality.

Polyhouses offer advantages like higher yields per unit area and protection from external weather, but they also create ideal conditions for soil-borne pests. Root-knot nematodes, once introduced via infested soil, seedlings, or tools, multiply rapidly due to consistent moisture from drip irrigation, temperatures ranging from 15–30 °C, and uninterrupted host availability from long-duration varieties. Population densities can surge from 1 to 3 second-stage juveniles (J2) per cubic centimeter of soil at planting to up to 80 J2/cc by crop maturity, completing multiple life cycles (25–30 days each) within a single season (Kamra et al. 2024; Patil and Yadav 2021).

Symptoms often start subtly with wilting or chlorosis but progress to visible root galls (Fig. 2.10), which may invite fungal or bacterial diseases. For visual reference, typical damage appears as swollen, knotted roots that impair plant health. Yield reductions vary, but studies show losses up to 80% in some polyhouse settings, with economic implications for farmers relying on protected cultivation for higher productivity. The issue is more pronounced in regions like Haryana and Punjab, where over 60% of polyhouses may be affected.

Why Polyhouses Favor Nematode Outbreaks Polyhouse ecosystems unintentionally provide near-optimal biological conditions for nematode multiplication.

(i) **Favorable Microclimate:** Favorable conditions include soil temperature: 25–30 °C (ideal for root-knot development), continuous moisture availability, and absence of natural climatic stress.

Fig. 2.10 Severe root galling due to the root-knot nematode infection on tomato (*Left*) and carnation (*Right*)

(ii) **Continuous Host Availability:** Long-duration hybrids and year-round cultivation allow uninterrupted feeding and reproduction cycles.
(iii) **Rapid Population Explosion:** Studies show nematode populations increasing from 1 to 3 J2/cc soil at planting to ~80 J2/cc soil at crop maturity under polyhouse conditions.
(iv) **Soil Reuse:** Unlike open fields, polyhouse soils are reused repeatedly without sufficient fallow periods.
(v) **Restricted Natural Suppression:** Restricted nematode suppression measures include reduced antagonistic microflora and limited ecological dilution effects.

Management Preventive measures like using nematode-free soil and seedlings, integrated strategies, such as grafting onto resistant rootstocks or applying biocontrol agents, can help manage populations without heavy reliance on chemicals. Regular monitoring is key to early intervention (Kamra et al. 2024).

(i) **Cultural and Physical Methods**: Deep tillage disturbs populations; soil solarization (25–30 μm polyethylene sheets for 4–8 weeks, achieving 40–50 °C) reduces nematodes by up to 92% (Kumar et al. 2019). Crop rotation with non-hosts (e.g., marigold, garlic) or trap crops (e.g., lettuce, radish) lowers densities, though limited in polyhouses. Organic amendments (e.g., neem cake at 20 tons/ha, vermicompost) suppress nematode population by 36–89% (Gowda et al. 2019; Patil and Yadav 2021).

(ii) **Biological Control**: Agents like fungi (*Purpureocillium lilacinum*, *Pochonia chlamydosporia*, and *Trichoderma* spp.) and bacteria (*Pseudomonas fluorescens*, *Bacillus* spp., and *Pasteuria penetrans*) reduce galling and reproduction; when applied every 45 days in high-infestation areas. VAM fungi (*Glomus* spp.) enhance plant tolerance.

(iii) **Genetic and Biotechnological Approaches**: Grafting onto resistant rootstocks (e.g., *Solanum torvum* for tomato, *Cucurbita* spp. for cucumber) yields 30–50% higher; rotate to prevent resistance breakdown. Advanced tools include RNAi to silence nematode genes and CRISPR-Cas9 for editing crop genomes to incorporate resistance.

(iv) **Chemical Options**: Metam sodium, fluopyram (Velum Prime), and fluensulfone provide initial control but populations rebound after 3–4 months. Use sparingly due to microbiome disruption.

(v) **Integrated Methods**: Combining solarization with biocontrol and resistant varieties can reduce populations by 65% and boost yields by 56–117% in crops like tomato and chili. Ongoing research emphasizes monitoring, farmer education, and sustainable practices to curb this emerging threat.

2.4.7 Tuberose Foliar Nematode, *Aphelenchoides besseyi*

Among emerging nematode threats affecting tuberose, the foliar nematode *Aphelenchoides besseyi* has gained importance due to its destructive impact on flower quality and marketability. This nematode is well known as the causal organism of white tip disease of rice, but in tuberose it induces a distinct disorder called floral malady, which severely reduces spike production and commercial value. The nematode survives in a quiescent state in planting material and becomes active under moist conditions, migrating upward through water films on plant surfaces. Tuberose infestations of *Aphelenchoides besseyi* have been documented mainly in West Bengal and Odisha. The increasing movement of infected bulbs and intensified ornamental cultivation have contributed to its emergence as a significant nematode problem in tuberose production systems (Khan et al. 2012). It may contribute to yield losses of 10–60% in affected crops.

Foliar symptoms include browning and necrosis of leaves, yellowing followed by drying, distorted and twisted foliage, and retarded plant growth. Floral symptoms (diagnostic) include suppression or delay of spike emergence, brown streaks on bracts and petals, brittle and malformed flowers, flower buds fail to open under severe infection, and prickle-like structures appear on scapes (Fig. 2.11). Severely infected plants become stunted and produce unmarketable flowers. Symptom intensity increases directly with nematode population density (Kadam et al. 2024).

Fig. 2.11 Field infestation of foliar nematode on tuberose spike

Management Preventive strategies emphasize nematode-free planting material: bulbs should be sourced from certified nurseries and treated via hot water immersion at 52–54 °C for 10–15 min or solar exposure on concrete for 4–6 h daily during summer.

Crop rotation with non-hosts (e.g., brassicas), soil solarization (using LDPE sheets for 6–8 weeks in hot months), and sanitation (destroying infested debris) are perennial nature.

Seed soaking in 0.2% carbosulfan 25EC for 12 h has proven effective in rice, adaptable to tuberose bulbs. Curative options include foliar sprays, with field trials in West Bengal demonstrating superior results from monocrotophos 36 SL (0.05%, applied three times at 15-day intervals), reducing PDI to 12–33% and boosting yields with a cost-benefit ratio of 1:13. Challenges include chemical resistance and regulatory gaps—new nematicides like fluensulfone are promising but unregistered for many Indian crops.

Botanical alternatives, such as neem seed kernel extract (5% dip for 6 h) followed by azadirachtin 1% sprays, offer moderate suppression (infestation <38%), appealing for organic farming. Integrated treatments combining bulb dipping in benomyl 50 WP (3 g/L) with azadirachtin sprays, or soil application of carbofuran 3G (1 kg a.i. /ha), achieve 50–80% population reduction.

Biological controls, including fungi like *Purpureocillium lilacinum* and bacteria such as *Pseudomonas fluorescens*, enrich organic amendments (e.g., neem cake, vermicompost) to promote antagonism.

Farmer education on early detection and IPM adoption is crucial, as untreated fields show 32–62% infestation versus 13–38% in treated ones.

2.5 Emerging Nematode Problems: International Scenario

Studies on new nematode species associated with agricultural crops are usually limited to a description of the new species, rarely examining the extent of damage and yield loss inflicted. As a result, the pest status of most of these new nematode species is unknown, frequently because of the lack of human resources and financial support.

2.5.1 *Meloidogyne enterolobii* (Syn. *Meloidogyne mayaguensis*)

M. enterolobii was first described in 1988 (Rammah and Hirschmann 1988) from eggplant (*Solanum melongena*) in Puerto Rico. This nematode is now considered to be one of the most important root-knot nematode species, posing a threat to agriculture in the subtropics and tropics because of its wide geographical distribution, broad host range, and ability to break resistance genes to the major root-knot nematode species (*M. incognita*, *M. javanica*, and *M. arenaria*) in tomato, sweet potato, soybean, and pepper (Fargette 1987; Prot 1984).

M. enterolobii has been reported from West Africa (Senegal, Ivory Coast, Burkina Faso), South Africa, Malawi, the Caribbean (Puerto Rico, Cuba, Dominican Republic, Guadeloupe, Martinique, Trinidad), Brazil, Florida, and in glasshouses in France (Blok et al. 2002; Brito et al. 2004; Carneiro et al. 2000; Carneiro et al. 2001; Decker and Rodriguez-Fuentes 1989; Fargette et al. 1994; Trudgill et al. 2000; Willers 1997).

It infects numerous crops including vegetables, ornamentals, potato, sweet potato, soybean, tobacco, and coffee and tropical fruit trees (guava).

2.5.2 *Meloidogyne paranaensis*

M. paranaensis was first described in 1996 (Carneiro et al. 1996) from coffee in Brazil. In contrast to *M. enterolobii*, the known geographical distribution of *M. paranaensis* is restricted and its known host range is small (Campos and Villain 2005).

Until recently, this nematode species was thought to occur only in Brazil, but Carneiro and coworkers demonstrated that it is also the most widely distributed *Meloidogyne* species on coffee in Guatemala (Carneiro et al. 1996; Herve et al. 2005). Moreover, in Colombia, a root-knot nematode population was found to exhibit a perineal pattern similar to that of *M. incognita* but produced a response to the North Carolina differential host test similar to that reported for *M. paranaensis* (Campos and Villain 2005). *M. paranaensis* was also recently detected in soybeans in Brazil (Roese et al. 2004). The geographical distribution and host range of *M. paranaensis* may therefore be much larger than currently assumed.

The frequency of occurrence of this nematode species in most coffee-growing plantations in Brazil is so high (Campos and Villain 2005) that it must have been identified as *M. incognita*, with which it frequently occurs, and have been there unnoticed for a very long time. The fact that this nematode species does not produce typical root-knot nematode galls on coffee (nematode feeding causes the tissues around the giant cells to die, resulting in necrotic spots where females are located) (Carneiro et al. 1996) probably contributed to its hidden status.

2.5.3 *Heterodera ciceri*

H. ciceri was first described in 1985 (Vovlas et al. 1985) from chickpea and other leguminous crops in northern Syria. It has subsequently also been found on leguminous crops (lentil, pea, faba bean, etc.) in the Mediterranean Basin (Spain, Italy, Turkey, Lebanon, and Jordan). In some provinces in Syria, 30% of the surveyed fields were infested with *H. ciceri* (Greco et al. 1992). The damage caused by this nematode can be very high: The tolerance limit of chickpea to *H. ciceri* is only 1 egg/cm^3 of soil.

2.5.4 *Radopholus* spp.

Radopholus citri was described in 1996 (Machon and Bridge 1996) from roots of citrus seedlings in Indonesia. *R. duriophilus* was described in 2003 (Nguyen et al. 2003) from roots of durian in the Western Highlands of Vietnam, where there was tree decline and death of young trees. At only 1000 nematodes per plant caused severe root degradation and significantly reduced plant growth.

2.5.5 *Achlysiella williamsi*

Achlysiella williamsi, another potentially damaging nematode species (Luc et al. 2005), was originally reported as *Radopholus similis*, which attacks sugarcane in Mauritius. *A. williamsi* has also been found in many localities in Papua New Guinea, a wide area of the Pacific Region (Fiji, Samoa, Tonga, Tuvalu, Vanuatu), Queensland, and Australia.

2.5.6 *Hirschmanniella miticausa*

Hirschmanniella miticausa was described in 1983 (Bridge et al. 1983) from roots of taro in the Solomon Islands. This nematode causes a corm rot disease known as "miti-miti" (the name in pidgin English given to affected corms because of their similarity to uncooked fatty meat). "Miti-miti" disease makes taro corms inedible and, when severe, can destroy almost all consumable corm tissue of the crop. The risk potential of *H. miticausa* is high because this nematode can cause heavy damage and can easily spread with infected planting material. Early infection of a corm by "miti miti" nematode causes the decay mostly in the outer parts of the corm. Secondary rot-causing organisms are responsible for complete decay of the corm during later stages of infection (Fig. 2.12).

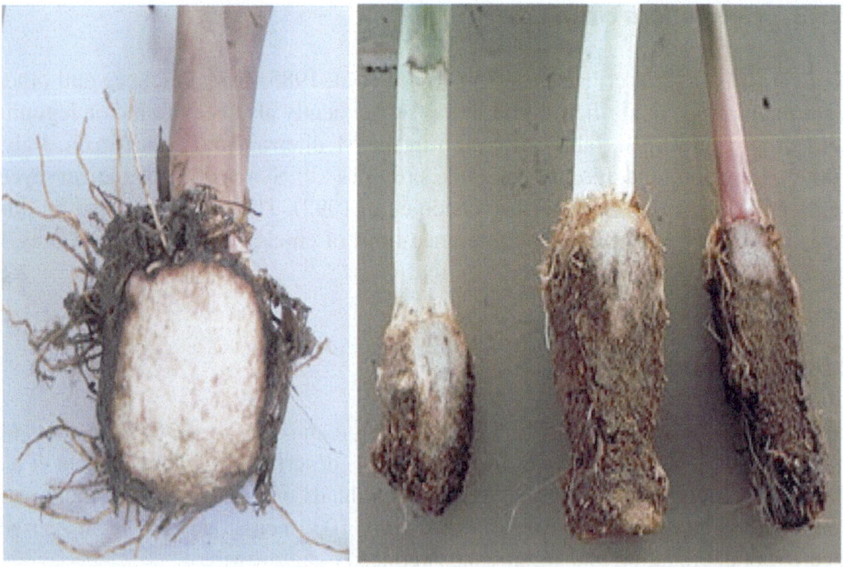

Fig. 2.12 *Left*—Early infection of a corm by "miti miti" nematode, *Hirschmanniella miticausa* (decay is mostly in the outer parts of the taro corm). *Right*—Late symptoms of "miti miti" infection by the nematode, and secondary rot-causing organisms, has caused complete decay of the corm

The disease and nematode have been reported from four islands in the Solomon Islands group (Mortimer et al. 1981) and the highlands of Papua New Guinea (Bridge and Page 1984).

2.5.7 Ditylenchus africanus

Ditylenchus africanus, the peanut pod nematode, was first found in hulls and seeds of peanut in South Africa in 1987. Nematode-infected pods show a black discoloration resembling black hull or black pod rot disease caused by the fungus *Chalara elegans* (Jones and De Waele 1988). Of 877 seed samples examined, 73% were infected with *D. africanus* (De Waele et al. 1989). *D. africanus* affects the yield of peanut seeds by increasing premature germination before harvest, reducing fresh seed weight, and decreasing seed quality by increasing the number of blemished and unsound seeds (Venter et al. 1991).

In 1995, peanut seeds infected with a *Ditylenchus* species, originating from the Pacific, were intercepted in India. The nematode remained unnoticed in South Africa for so long because the symptoms it causes are very similar to those caused by the fungus *C. elegans* and the nematodes themselves have a weak stylet and resemble harmless fungivorous nematodes.

It has been found in all major peanut-producing areas of South Africa. Its widespread distribution in South Africa suggests that it may also be present in other southern African countries, especially those neighboring South Africa. Peanut pods showing symptoms typical of *D. africanus* have been reported from Mozambique, Malawi, and Congo (De Waele et al. 1997).

2.6 Challenges, Recent Developments, and Future Directions

2.6.1 Diagnostic Limitations

Accurate identification of nematode species requires expertise and access to molecular tools, which may be unavailable to resource-poor regions.

2.6.2 Sustainability

Overreliance on chemical controls threatens long-term soil health and ecosystem balance.

2.6.3 Climate Adaptation

Management must adapt to shifting nematode distributions and behaviors driven by climate change.

2.6.4 Research Needs

Investment in breeding programs is helpful for developing durable resistance, scalable biocontrols, and climate-smart practices. For specific crops or regions, nematode management should be customized based on soil type, climate, and dominant nematode species. Regular soil testing and collaboration with local agricultural experts are critical for success.

2.6.5 Technological Advancements

Gene editing tools like CRISPR-Cas9 are revolutionizing resistance breeding, while nanoencapsulation reduces nematicide environmental impact. Decision support systems like Nematool enhance precision agriculture, optimizing management decisions (Abd-Elgawad 2024).

2.6.6 Global Challenges

The technological gap between developed and developing countries, exacerbated by climate changes, requires cooperative programs and standardized procedures for global INM. Sub-Saharan Africa, for instance, faces significant hurdles due to limited access to advanced tools (Abd-Elgawad 2024).

2.6.7 Research Focus

Investment in nematology research is crucial for identifying new resistance genes, developing biological control agents, and understanding nematode ecology. This includes studying behavior, distribution, and interactions to inform targeted strategies (Afzal and Mukhtar 2024).

2.6.8 Policy and Education

Strengthen quarantine measures to limit invasive nematode spread via trade. Educate farmers on nematode identification, monitoring, and sustainable practices through extension services. Support research on climate-resilient crops and novel control methods.

2.7 Conclusion

In recent years, with changes in the cropping systems and climate, and introduction of highly input intensive high yielding varieties/hybrids, a shift in nematode status has been observed. New and reemerging plant nematodes have raised fear of their potential impact on livelihoods, food security, and global markets. There were outbreaks of the root-knot nematode on rice, wheat, groundnut, citrus (acid lime), guava, pomegranate, vegetable, and flower crops grown under protected cultivation systems; cyst nematodes, *G. rostochiensis* and *G. pallida* on potato; and foliar nematode on tuberose. Key species like *Meloidogyne* spp., *Heterodera* spp., and *Pratylenchus* spp. continue to challenge crop production, particularly in developing regions. Management strategies are evolving toward integrated approaches, combining host resistance, biological control, cultural practices, and biotechnological tools. The emerging and future nematode problems have to be tackled with integrated nematode management approach as a part of a sustainable crop production technology. Recent advancements, such as CRISPR-Cas9 for resistance breeding and nano-encapsulated nematicides, offer promising solutions. However, addressing these challenges requires continued research, global cooperation, and tailored strategies for different regions and farming systems, ensuring sustainable agriculture in the face of evolving nematode threats. There is an urgent need to redefine strategies for the management of the above nematodes in different crops on war footing to meet future food requirements.

2.8 Suggested Readings

- Das T.K. and Khan, M.R. 2007. Occurrence and distribution of white tip nematode, *Aphelenchoides besseyi* in West Bengal, India. *Indian Journal of Nematology* 27: 94–97.
- Ganguly, S., Singh, M. and Ganguly, A.K. 2010. Record of potato cyst nematode, *G. rostochiensis* and *G. pallida* in Shimla, Himachal Pradesh, India. *Indian J Nematol* 40: 96–102.
- Kantor, M., Handoo, Z., Kantor, C. and Carta, L. 2022. Top ten most important U.S.-regulated and emerging plant-parasitic nematodes. *Horticulturae* 8: 208. https://doi.org/10.3390/horticulturae8030208.
- Parvatha Reddy, P. 2017. *Emerging Crop Pest Problems: Redefining Management Strategies*. Scientific Publishers, Jodhpur, India, 383 pp.
- Patel, B.A., Patel, D.J., Sharma, S.B. and Patel, H.V. 1996. Nematode problems of groundnut and their management in Gujarat, India. *International Arachis Newsletter* 16: 38–39.
- Pun, T.B., Thapa Magar, R., Koech, R. et al. 2024. Emerging trends and technologies used for the identification, detection, and characterization of plant-parasitic nematode infestation in crops. *Plants* 13(21): 3041. https://doi.org/10.3390/plants13213041.
- Somasekhar, N. and Prasad, J.S. 2009. Root-knot nematode *Meloidogyne graminicola* –An emerging threat to rice cultivation. *DRR Newsletter* 7: 3–4.
- Yadav, S. and Patil, J.A. 2021. The emerging nematode problems in horticultural crops and their management. In Cristiano and T.K. Kaspary (eds.) *Nematodes - Recent Advances, Management and New Perspectives*. IntechOpen. https://doi.org/10.5772/intechopen.99292.

2.9 Review Questions

- What are the major emerging plant-parasitic nematodes of economic importance worldwide?
- What are the factors influencing the emergence of new nematode problems in crops?
- Give examples of newly reported nematode species or populations that are gaining importance in horticultural crops.
- Critically evaluate the global distribution and recent outbreaks of *Meloidogyne enterolobii* as an emerging nematode problem.
- How should future research priorities be structured to address the problem of emerging plant nematodes in the context of global food security?

References

Abd-Elgawad MMM (2024) Upgrading strategies for managing nematode pests on profitable crops. Plants (Basel) 13(11):1558. https://doi.org/10.3390/plants13111558

Afzal A, Mukhtar T (2024) Revolutionizing nematode management to achieve global food security goals - an overview. Heliyon 10(3):e25325. https://doi.org/10.1016/j.heliyon.2024.e25325

Binage BP, Palande PR, Walunj AR et al (2025) Management of root-knot nematode (*Meloidogyne incognita*) infesting pomegranate. International Journal of Advanced Biochemistry Research 9(4):559–562

Blok VC, Wishart J, Fargette M et al (2002) Mitochondrial DNA differences distinguishing *Meloidogyne mayaguensis* from the major species of tropical root-knot nematodes. Nematology 4:773–781

Bridge J, Page SLM (1984) Plant nematode pests of crops in Papua New Guinea. J Plant Prot Tropics 1:99–109

Bridge J, Mortimer JJ, Jackson GVH (1983) *Hirschmanniella miticausa* n. sp. (Nematoda: Pratylenchidae) and its pathogenicity on taro (*Colocasia esculenta*). Rev N'ematol 6:285–290

Brito J, Powers TO, Mullin PG et al (2004) Morphological and molecular characterization of *Meloidogyne mayaguensis* isolates from Florida. J Nematol 36(3):232–240

Campos VP, Villain L (2005) Nematode parasites of coffee and cocoa. In: Luc M, Sikora RA, Bridge J (eds) Plant parasitic nematodes in subtropical and tropical agriculture, 2nd edn. CABI Publ., Wallingford, UK, pp 529–579

Carneiro RMDG, Carneiro RG, Abrantes IMO et al (1996) *Meloidogyne paranaensis* n. sp. (Nemata: Meloidogynidae), a root-knot nematode parasitizing coffee in Brazil. J Nematol 28:177–189

Carneiro RMDG, Almeida MRA, Qu'en'eherv'e P (2000) Enzyme phenotypes of *Meloidogyne* spp. populations. Nematology 2:645–654

Carneiro RMDG, Moreira WA, Almeida MRA, Gomes ACMM (2001) First record of *Meloidogyne mayaguensis* on guava in Brazil. Nematol Bras 25:223–228

Chandel YS, Surendra Singh B, Rajan S et al (2020) Prevalence and spread of potato cyst nematodes, *Globodera* spp. in northern hilly areas of India. Curr Sci (12):1946–1952

De Waele D, Jones BL, Bolton C, Vandenberg E (1989) *Ditylenchus destructor* in hulls and seeds of peanut. J Nematol 21:10–15

De Waele D, Venter C, Alexander A (1997) *The Peanut Pod Nematode*, Ditylenchus africanus. Nematology Circular No. 208, Fla. Dept. Agric. & Consumer Services, Divn. of Plant Industry, 6 pp

Decker H, Rodriguez-Fuentes ME (1989) The occurrence of root gall nematodes *Meloidogyne mayaguensis* on *Coffea arabica* in Cuba. Wiss Ztg Wilhelm Pieck Univ Rostock Naturwiss Reihe 38:32–34

Fargette M (1987) Use of esterase phenotype in taxonomy of the genus *Meloidogyne*. 2. Esterase phenotypes observed in west African populations and their characterization. Rev. N'ematol 10:45–55

Fargette M, Davies KG, Robinson MP, Trudgill DL (1994) Characterization of resistance breaking *Meloidogyne incognita*-like populations using lectins, monoclonal-antibodies and spores of *Pasteuria penetrans*. Fund Appl Nematol 17(6):537–342

Ganguly S, Singh M, Ganguly AK (2010) Record of potato cyst nematode, *Globodera rostochiensis* and *G. pallida* in Shimla, Himachal Pradesh, India. *Indian J Nematol* 40:96–102

Gautam V, Garg V, Dashrath HA et al (2025) Evaluation of resistance in Indian rice to root-knot nematode (*Meloidogyne graminicola*): insights from field and histopathological studies. Rice (NY) 18(1):93. https://doi.org/10.1186/s12284-025-00852-x

Ghule TM, Phani V, Somvanshi VS et al (2020) Further observations on *Meloidogyne enterolobii* (Nematoda: Meloidogynidae) infecting guava (*Psidium guajava*) in India. J Nematol 52:1–9

Gowda MT, Sellaperumal C, Rai AB, Singh B (2019) Root-knot nematodes menace in vegetable crops and their management in India: a review. Veg Sci 46(1&2):1–16

Greco N, Di Vito M, Saxena MC (1992) Plant parasitic nematodes of cool season food legumes in Syria. Nematol Mediterr 20:37–46

He Y, Wang R, Zhao H et al (2022) Predicting potential global distribution and risk regions for potato cyst nematodes (*Globodera rostochiensis* and *Globodera pallida*). Sci Rep 12:21843. https://doi.org/10.1038/s41598-022-26443-0

Herve G, Bertrand B, Villain L et al (2005) Distribution analyses of *Meloidogyne* spp. and *Pratylenchus coffeae sensu lato* in coffee plots in Costa Rica and Guatemala. Plant Pathol 54:471–475

Hotte H, Neveux MS, Ollivier F et al (2022) Can quarantine of plant-parasitic nematodes within wastes be managed by useful tools in a circular economy approach? J Environ Manag 323:116–184

Jones BL, De Waele D (1988) First report of *Ditylenchus destructor* in pods and seeds of peanut. Plant Dis 72:453

Kadam V, Jeevan H, Bablu H et al (2024) Floral infestation of tuberose with *Aphelenchoides* and its management. Journal of Scientific Research and Reports 30(11):316–325

Kamra A, Sidhu AJ, Mallikarjun G et al (2024) Nematode management in polyhouse cultivation. Indian J Nematol 53(Spl. Issue):43–48

Khan MR, Handoo ZA, Rao U et al (2012) Observations on the foliar nematode, *Aphelenchoides besseyi*, infecting tuberose and rice in India. Indian J Nematol 44(4):391–398

Khan RM, Poornima K, Somvanshi VS, Walia RK (2022) *Meloidogyne enterolobii*: a threat to guava (*Psidium guajava*) cultivation. Arch Phytopathol Plant Protect 55(17):1961–1997

Khanal C, Land J (2023) Study on two nematode species suggests climate change will inflict greater crop damage. Sci Rep 13:14185. https://doi.org/10.1038/s41598-023-41466-x

Krishna Kumar S, Balasubramanian P (2019) First report on occurrence of citrus root-knot nematode, *Meloidogyne indica* on Khasi mandarin in Meghalaya. J Entomol Zoology Stud 7(6):418–419

Krishna Prasad KS (1986) Potato nematodes. In: Swarup G, Dasgupta DR (eds) Plant parasitic nematodes in India. Indian Agri Res Inst, New Delhi

Kumar S, Rawat S (2018) First report on the root-knot nematode *Meloidogyne enterolobii* (Yang and Eisenback, 1988) infecting guava (*Psidium guajava*) in Udham Singh Nagar of Uttarakhand. Int J Curr Microbiol App Sci 7:1720–1724

Kumar A, Patil JA, Verma KK (2019) Management of root-knot nematode, *Meloidogyne* spp. in vegetable crops grown under protected cultivation through fumigants. Indian J Nematol 49(2):125–130

Luc M, Bridge J, Sikora RA (2005) Reflections on nematology in subtropical and tropical agriculture. In: Luc M, Sikora RA, Bridge J (eds) Plant parasitic nematodes in subtropical and tropical agriculture, 2nd edn. CABI, Wallingford, UK, pp 1–10

Machon J, Bridge J (1996) *Radopholus citri* n. sp. (Tylenchida: Pratylenchidae) and its pathogenicity on citrus. Fund Appl Nematol 19:127–133

Mehan VK, Reddy DDR, McDonald D (1993) Resistance in groundnut genotypes to Kalahasti malady caused by the stunt nematode, Tylenchorhynchus brevilineatus. Int J Pest Mangmt 39:201–203

Mortimer JJ, Bridge J, Jackson GVH (1981) *Hirschmanniella* sp. an endoparasitic nematode associated with miti-miti disease of taro corms in the Solomon Islands. FAO Plant Protection Bulletin 29(1/2):9–11

Mukesh KK, Sharma SK (2024) Understanding the dynamics of *Meloidogyne incognita* infestation in pomegranate orchards of Himachal Pradesh, India (year 2018, 2019 and 2021) and its management strategies. Heliyon 10(15):e34752. https://doi.org/10.1016/j.heliyon.2024.e34752

Naidu PH, Mosas GJ, Reddy DDR (2000) Influence of intercropping on Kalahasti malady (*Tylenchorhynchus brevilineatus*) in groundnut. J Mycol Plant Pathol 30:207–209

Nguyen CN, Subbotin SA, Madani M et al (2003) *Radopholus duriophilus* sp. n. (Nematoda: Pratylenchidae) from Western highlands of Vietnam. Nematology 5:549–558

Patel, D.J., Patel B.A., Patel, S. *et al.* 1999. Root-knot nematode, *Meloidogyne indica* on kagzi lime in North Gujarat. Indian J Nematol 29: 197–197

Patil, J.A. and Yadav, S. 2021. Root-knot nematodes a major peril to protected cultivation system in India: current status and its management. IntechOpen. Available at: https://doi.org/10.5772/intechopen.100541

Phani V, Bishnoi S, Sharma A et al (2018) Characterization of *Meloidogyne indica* (Nematoda: Meloidogynidae) parasitizing neem in India, with a molecular phylogeny of the species. J Nematol 50(3):387–398

Poornima K, Suresh P, Kalaiarasan P et al (2016) Root-knot nematode, *Meloidogyne enterolobii* in guava (*Psidium guajava* L.). A new record from India. Madras Agric J 103(10–12):359–365

Prot JC (1984) A naturally occurring resistance breaking biotype of *Meloidogyne arenaria* on tomato. Reproduction and pathogenicity on tomato cultivars Roma and Rossol. *Rev N'ematol* 7:23–28

Ramana KV, Mohandas C (1988) Occurrence of potato cyst nematode *Globodera pallida* in Kerala. Indian J Nematol 18:141

Rammah A, Hirschmann H (1988) *Meloidogyne mayaguensis* n. sp. (Meloidogynidae), a root-knot nematode from Puerto-Rico. J Nematol 20:58–69

Reddy DDR, Subrahmanyam P, Sankara Reddy GH et al (1984) A nematode disease of peanut caused by *Tylenchorhynchus brevilineatus*. Plant Dis 68:526–529

Roese AD, Oliveira RD, Lanes FF (2004) Reaction of soyabean (*Glycine max* L. Merril) cultivars to *Meloidogyne paranaensis*. Nematol Bras 28:131–135

Sharma SB, Varaprasad KS, Balasubramanian P, Siva Kumar CV (1996) 'Kalahasti malady' on groundnut outside Andhra Pradesh. India Int Arachis Newsletter 16:33

Singh T, Prajapati A, Maru AK et al (2019) Root-knot nematodes (*Meloidogyne* spp.) infecting pomegranate: a review. Agric Rev 40(4):309–313

Singh A, Kaul RK, Khapte PS et al (2022) Root-knot nematode presence and its integrated management in pomegranate orchards located in Indian arid areas. Horticulturae 8:160. https://doi.org/10.3390/horticulturae8020160

Trudgill DL, Bala G, Blok VC et al (2000) The importance of tropical root-knot nematodes (*Meloidogyne* spp.) and factors affecting the utility of *Pasteuria penetrans* as a biocontrol agent. Nematology 2:823–845

Umamaheswari R, Aarti B, Venkatasalam EP et al (2015) Effect of biofumigation on potato cyst nematodes. Potato J 4(2):124–129

Venter C, De Waele D, Meyer AJ (1991) Reproductive and damage potential of *Ditylenchus destructor* on peanut. J Nematol 23:12–19

Vishwakarma PK, Chinnaiyan V, Umamaheswari R, Shivashankara KS (2023) Screening of *Psidium* species and their interspecific hybrid progenies for resistance to *Fusarium oxysporum* f. sp. *psidii* and southern root-knot nematode. S Afr J Bot 155:249–260

Vovlas N, Greco N, Vito M (1985) *Heterodera ciceri* sp. n. (Nematoda: Heteroderidae) on *Cicer arietinum* from northern Syria. Nematol Bras 13:239–252

Walia RK (2024) Potato cyst nematodes—the policy paralysis. Indian J Nematol 53(Special Issue):21–29

Walia RK, Kranti KVVS, Kumar V (eds) (2018) Consolidated biennial report (2015–17), XX Biennial Group Meeting, All India Coordinated Research Project on Nematode in Cropping Systems, AAU, Anand, February 15–17, 2018

Willers P (1997) First record of *Meloidogyne mayaguensis* Rammah and Hirschmann, 1988: Heteroderidae on commercial cops in the Mpumalanga province, South Africa. Inligtingsbull Inst Trop Subtrop Gewasse 294:19–20

Yadav S, Patil JA (2021) The emerging nematode problems in horticultural crops and their management. In: Bello C, Kaspary TE (eds) Nematodes – recent advances, management and new perspectives. https://doi.org/10.5772/intechopen.99292

Part II

Agro-Ecological Approaches

Agro-Ecological Approaches: Promising Solution for Nematode Management

3

Abstract

Agro-ecological approaches such as crop rotation with resistant cultivars can reduce nematode populations effectively. This chapter provides a detailed examination of agro-ecological approaches for sustainable plant nematode management, focusing on methods that minimize environmental impact and promote soil health, particularly in organic and sustainable agriculture systems. The cover crops and biofumigation, like using *Brassica* species, help manage nematodes sustainably. The mycorrhizal fungal inoculations improve plant vigor, indirectly aiding nematode control. Reduced tillage appears to enhance soil suppressiveness, lowering nematode densities. Biological control using organisms like *Purpureocillium lilacinum, Pochonia chlamydosporia, Trichoderma* spp., *Pseudomonas fluorescens, Bacillus subtilis,* and *Pasteuria penetrans* may biologically manage nematodes, though effectiveness varies. Organic amendments, such as farmyard manure and compost, likely improve soil health and reduce nematode impact. Solarization, especially with cover crops, seems effective for killing nematodes in upper soil layers. Phytochemical-based strategies, like essential oils, might suppress nematodes, but effectiveness can be inconsistent. The use of marine organisms, like seaweeds, offers a novel approach, with potential for organic systems, but needs more research. These methods aim to reduce nematode populations while promoting soil health and minimizing environmental impact, making them ideal for organic and sustainable farming.

Keywords

Crop rotation · Biofumigation · Biological control · Organic amendments · Solarization

R. P. Parvatha, *Plant Parasitic Nematodes*,
https://doi.org/10.1007/978-981-95-9306-4_3

3.1 Learning Objectives

- Explain the principles of agro-ecology and their relevance to sustainable nematode management.
- Understand the ecological interactions between nematodes, soil microbes, plants, and the agro-ecosystem.
- Differentiate agro-ecological approaches from conventional chemical-based nematode control strategies.
- Analyze how crop diversity, rotations, and soil health influence nematode population dynamics.
- Identify and evaluate cultural practices such as crop rotation, intercropping, cover crops, green manuring, and organic amendments for nematode suppression.
- Discuss the role of soil biodiversity, beneficial microbes, and natural enemies (e.g., predatory fungi, bacteria, and nematophagous organisms) in nematode regulation.
- Assess how agro-ecological approaches reduce reliance on synthetic nematicides and minimize environmental risks.
- Explore the potential of agro-ecological nematode management to support climate-resilient and resource-efficient farming systems.
- Critically evaluate scientific literature and case studies demonstrating successful agro-ecological nematode management.
- Propose innovative, ecosystem-based strategies for sustainable crop protection against nematodes.

3.2 Introduction

Plant-parasitic nematodes (PPNs) threaten global food security by dramatically lowering both quantity and quality of harvested produce. PPNs, such as root-knot nematodes (*Meloidogyne* spp.) and cyst nematodes (*Heterodera* spp.), are significant agricultural pests, causing 10–25% global crop yield losses annually (Veronico and Melillo 2021). A recent report of such losses for the 40 life-sustaining crops with impressing figures for food and export recorded an average of 13.5% retardations, equaling USD 358.24 billion annually (Abd-Elgawad and Askary 2015). Traditional chemical nematicides face restrictions, necessitating sustainable alternatives that align with agro-ecological principles. These approaches integrate ecological processes, enhance biodiversity, and reduce reliance on synthetic inputs, addressing both economic and environmental concerns.

Agro-ecological approaches use natural processes to manage pests like nematodes, which are tiny worms that can damage plant roots and reduce crop yields. These methods focus on working with the environment rather than relying heavily on chemicals, making them sustainable for farmers and the planet. They include practices like planting specific crops to deter nematodes, adjusting farming techniques to improve soil health, and using natural pest controls.

The emerging agro-ecological tactics and strategies can use certified seed/planting material (Sikora et al. 2022), more timely and optimized soil solarization for

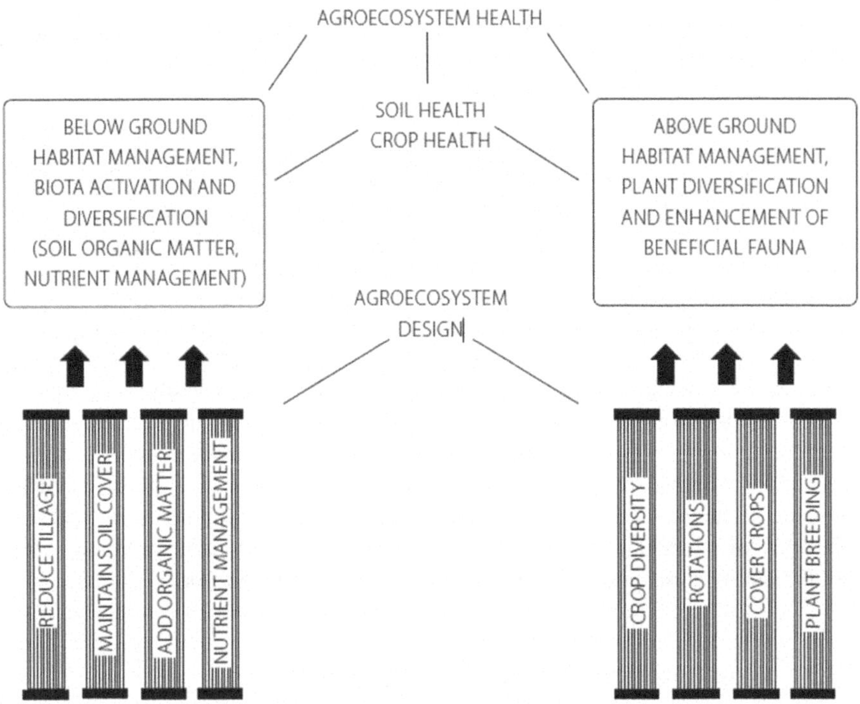

Fig. 3.1 Agro-ecological approaches for nematode management

PPN disinfestation (Hammam et al. 2022), and novel bionematicides (D'Addabbo et al. 2023) along with plowing, crop sequencing (crop rotation), fallowing, cover cropping, intercropping, trap cropping, planting time, mulching flooding, and/or soil amendments (Fig. 3.1). Developing these novel strategies is necessary to address major nematode problems (Sikora et al. 2022; Abd-Elgawad 2022; Burns et al. 2023).

3.3 Agro-Ecological Approaches

3.3.1 Reduced Tillage

Reducing tillage enhances soil suppressiveness and lowering nematode densities. In long-term trials, minimum tillage in corn-soybean rotations reduced soybean cyst nematode, *Heterodera glycines* densities, and foliar sudden death syndrome (SDS) symptoms compared to plow or chisel plow systems. A 1-year soybean-corn rotation did not reduce *H. glycines* or SDS, indicating the need for longer rotations (Krueger et al. 2022). This approach minimizes soil disturbance, preserving beneficial organisms and supporting sustainable soil management.

3.3.2 Solarization

Solarization involves covering of moist soil with a Linear Low-Density Polyethylene (LLDPE) plastic film (25–30 μm thick) during periods of intense sunshine and heat, for at least 4 weeks, heating soil to 35–60 °C to kill nematodes in the top 10–30 cm. It captures radiant heat energy from the sun, causing physical, chemical, and biological changes in the soil. Combining with cover crops like cowpea enhances results, comparable to methyl bromide fumigation, though it impacts beneficial organisms and generates plastic waste (Westphal 2011). This method is effective for upper soil layers but may require reuse of plastic sheets for sustainability.

Although solarization can provide excellent soil disinfestation under suitable conditions, it has significant limitations and should not be considered a cure-all or universal replacement for methyl bromide. For example, solarization is most effective close to the surface of the soil under climatic and weather conditions of high air temperature and long days for soil heating. It will not control all pest organisms, may require that land be taken out of production for 3- to 6-week treatments during the summer months, and requires disposal of used plastic film. Therefore, its practical value to the user must be assessed by several factors, including extent and predictability of pesticidal efficacy, effect on crop growth and yield, economic cost/benefit, and personal pest management philosophy (Stapleton 1997).

3.3.3 Crop Rotation and Host Management

- *Principle*: Rotating to non-host or poor-host crops breaks the nematode life cycle. Because most PPNs cannot survive long without a host, planting a different crop family starves the nematode and reduces its population. Studies show that alternating crop families "can reduce the size of nematode populations, thus mitigating establishment" (deGives 2022). For example, rotating corn, cotton, or small grains (which are non-hosts) into fields infested with soybean cyst nematode (SCN) will sharply lower SCN levels (Kirkpatrick and Thomas 2014). Likewise, alternate-season planting of non-solanaceous crops can keep root-knot nematodes in check under tomatoes or bell peppers.
- *Benefits*: Rotation is a time-tested, low-cost method that fits into Integrated Pest Management. It also diversifies the farm ecosystem. Long-term rotations can greatly suppress specialized PPNs (e.g., root-knot, cyst nematodes) once non-hosts dominate the field (Briar et al. 2016).
- *Challenges*: Effective rotation requires knowing the local nematode species and their host range (Briar et al. 2016). Some nematodes (e.g., lesion nematodes) have very wide host ranges, making rotation harder. Economically, rotation may force a change in the cash crop plan (farmers must "consider changing and alternating their crops" (deGives 2022) affecting short-term profitability. Proper sanitation (clean equipment, nematode-free planting stock) and weed control are also needed, since alternate hosts (even weeds) can undermine rotation benefits (Abd-Elgawad 2024).
- *Crop Rotation with Resistant Cultivars*: Crop rotation with resistant or non-host cultivars is a cornerstone of sustainable nematode management. Research suggests

rotating crops like tomatoes or eggplants with resistant varieties can suppress *Meloidogyne* spp. populations. For instance, eight soybean cultivars were tested for resistance to both *M. incognita* and *H. glycines*, with some showing suppression post-early or late planting, and two planted in fumigated control plots with 1, 3-dichloropropene (Krueger et al. 2022).
- Rotating crops with varieties resistant to nematodes, like certain tomatoes or cowpeas ("Mississippi Silver"), can significantly lower nematode numbers. For example, research shows soybean cultivars resistant to *Meloidogyne incognita* and *Heterodera glycines* can suppress populations, with costs for seedless watermelon transplants at around $0.28 each (Westphal 2011).

3.3.4 Cover/Green Manure Crops and Biofumigation

Cover crops, particularly *Brassica* species (e.g., "Boss" oilseed radish), can reduce nematode damage through resistance or biofumigation effects. Replacing cereal cover crops like rye and wheat, which host *M. incognita*, with resistant options like "Boss" oilseed radish or canola (*Brassica napus*) for overwintering can reduce galling. A micro-plot trial showed "Boss" had less galling than susceptible variety "Siletina" in February, with the lowest *M. incognita* juveniles in May after bare fallow, though bare fallow is unacceptable for erosion control (Mukherjee 2020). Canola is recommended for overwintering and enhancing sustainability (Mukherjee 2020). This approach enhances soil health and nematode suppression, aligning with sustainable practices.

Leguminous green manure crops like sun hemp (*Crotalaria juncea*) or hairy indigo (*Indigofera hirsuta*) are inherently poor hosts for most PPNs. Sun hemp in Florida exemplifies this—it resists a wide range of nematodes and fixes N, making it a high-value summer cover for sandy soils (Noling 2012). Another classic crop is marigold (*Tagetes* spp.), whose roots exude alpha-terthienyl. Marigold can suppress ~14 nematode genera (especially root-knot and lesion nematodes) (Krueger et al. 2019). For example, French marigold cultivars are often planted pre-cropping to reduce *Meloidogyne* levels.

Cover crops thus "trap" or kill nematodes before the cash crop is planted. In practice, mixed cover-crop strategies (e.g., brassicas plus legumes) can combine soil enrichment with nematode suppression.

- *Principle*: Grow plants between cash crops that either do not host the nematode or actively suppress it. Cover crops and green manures (grown and then tilled in) can improve soil fertility while starving or killing PPNs. A classic example is marigold (*Tagetes* spp.), whose roots exude alpha-terthienyl—a naturally nematicidal compound (Briar et al. 2016). Marigold rotations have been used in vegetable and orchard systems for generations to reduce root-knot and lesion nematode levels. Other cover crops with proven effects include sun hemp (*Crotalaria juncea*) and velvet bean (*Mucuna pruriens*), both of which significantly suppress root-knot nematodes when incorporated (Briar et al. 2016). Pearl millet and certain sorghum/Sudan grass cultivars can reduce *Pratylenchus* (lesion nematodes) or *Meloidogyne* population densities (Briar et al. 2016).

- *Benefits*: Cover crops prevent soil erosion and add biomass. When non-hosts (or poor hosts) are chosen, nematode food sources are cut off and the nematode population "declines below damaging levels" (Briar et al. 2016). Allelopathic species like marigold or brassica (mustard, radish) varieties also release biofumigant chemicals that directly kill or deter PPNs (Briar et al. 2016). Surface mulching of cover crops (instead of deep plowing) can suppress nematodes while preserving soil biota.
- *Challenges*: Timing and management are critical. Many effective cover crops require termination (mechanical or chemical) before they seed, and incorporation often means tillage that can disturb soil predators (Briar et al. 2016). Poor choice of cover can backfire: a crop that is a resistant trap for one PPN may actually host another. Also, cover cropping ties up land and may delay cash crops, which farmers must balance economically (Briar et al. 2016).

3.3.5 Intercropping and Push-Pull Strategies

- *Principle*: Grow two or more crops together so that one repels or traps nematodes from the other. Intercropping with nematode-suppressive plants can reduce damage to the main crop. For instance, marigolds interplanted with peppers or tomatoes are reported (in some regions) to lower root-knot nematode infestation, likely via their nematicidal exudates (Briar et al. 2016). Similarly, integrating legumes or cereals as intercrops can dilute nematode hosts in the field. A broad review notes that strip cropping or intercropping with "poor/non-host crops…may sustain nematode-resistant genes" and significantly reduce PPN density (Abd-Elgawad 2024). Good weed control is part of this strategy, since weeds often serve as nematode reservoirs (Abd-Elgawad 2024).
- *Benefits*: Diversified plantings can enhance soil cover and encourage natural enemy diversity above and below ground. By increasing plant diversity, intercropping often improves overall system resilience and can result in yield advantages (the "biotic resistance" effect).
- *Challenges*: Intercropping requires careful planning. Crops must be compatible (similar needs, not outcompeting each other). Spatial arrangement and timing become more complex. The nematode suppression effect can be variable and is still an active research area; it often works best when the intercrop is a proven trap or repellent species.

3.3.6 Organic Amendments

Organic amendments (compost, vermicompost, farmyard manure, biochar, cover-crop residues), improve soil health and indirectly reduce nematode impact by supporting beneficial organisms. Sun hemp decomposition, completed within 2 weeks, increased bacterivore and omnivore nematodes but did not directly affect plant-parasitic nematodes (Krueger et al. 2022).

To maximize benefits, organic amendments should be used with crop rotations and biocontrols (Briar et al. 2016). Some innovations bridge amendments and biofumigation: for example, incorporating marigold or mustard meals (seed meal amendments) can suppress nematodes while adding organic matter (Batistič et al. 2025). Cover-crop residue management (timing of incorporation) also matters, as shown in trials where late-incorporated cruciferous covers gave the most nematode reduction.

- *Principle*: Adding composts, manures, or other organic materials enhances beneficial soil microbes and often suppresses PPNs. Rich organic amendments boost soil health and can release natural nematicidal compounds as they decompose (Abd-Elgawad 2024). Composted manures (poultry, livestock) are widely used: numerous studies report that adding compost is positively correlated with reduced root-knot and root-lesion nematode populations (Briar et al. 2016). Some amendments (e.g., brassica green manures, neem cake, chitin) have specific bioactive properties that kill nematodes or stimulate chitinolytic microbes.
- *Benefits*: Organic matter improves soil structure, water retention, and nutrient cycling while fostering diverse microbial life. Increased microbial biomass can "release nematode-regulating compounds" (antibiotics, volatile fatty acids, etc.) that suppress PPNs (Abd-Elgawad 2024). Over time, this builds a more resilient soil food web that competes with or preys upon nematodes.
- *Challenges*: The suppressive effect of amendments can be inconsistent. Efficacy depends on amendment type, rate, timing, and existing soil biology (Briar et al. 2016). Some organic materials can transiently deplete nitrogen or introduce weed seeds if not well composted. Large applications (to kill PPNs) may be costly or logistically difficult on a commercial scale.

3.3.7 Biological Control

Biological control uses living organisms like *Pasteuria penetrans*, which parasitizes *Meloidogyne* spp., or nematode-trapping fungi (*P. lilacinum*), to manage nematodes. A 7-year tobacco experiment showed *P. penetrans* suppressed *M. incognita* and *M. javanica*, reducing galls, egg masses, and eggs (Mukherjee 2020). However, generalist predators may affect beneficials, and commercial predatory nematodes or mites are not widely available for augmentative releases, limiting scalability. These methods are effective but may impact beneficial organisms (Veronico and Melillo 2021).

These BCAs work best in healthy soils with adequate organic matter, and often complement other IPM measures. Field trials in Africa, Asia, and the Americas have demonstrated significant nematode suppression by combining BCAs (fungi or bacteria) with reduced chemical input (Ayaz et al. 2024). For instance, coating seed with *Trichoderma* or adding *Bacillus*-based bioinoculants is common in vegetable and cereal systems to manage nematodes and other soil-borne pathogens. While BCAs rarely eliminate nematodes alone, they offer safe, environment-friendly control that can be amplified through repeated or multi-agent applications.

- *Principle*: Employ beneficial organisms (microbes or predators) that attack PPNs. The soil hosts a variety of natural nematode enemies. Beneficial bacteria (e.g., *Bacillus*, *Pseudomonas*) and fungi (e.g., *Purpureocillium* formerly *Paecilomyces*, *Pochonia*, *Trichoderma*) can parasitize or inhibit nematodes. A classic bacterial parasite is *Pasteuria penetrans*, which infects *Meloidogyne* eggs and juveniles, dramatically reducing root-knot populations (Briar et al. 2016). Nematode-trapping fungi (e.g., *Arthrobotrys* spp.) form hyphal networks or adhesive knobs that ensnare nematodes (Briar et al. 2016). Many of these organisms are formulated as biocontrol products: for instance, *Bacillus firmus* (Nortica®) and *P. lilacinum* (MeloCon®) are registered nematicidal agents (Abd-Elgawad 2024).
- *Benefits*: Biocontrol is highly specific to nematodes and environmentally benign. Unlike chemicals, these agents pose little risk to non-targets. Augmenting soils with nematode-antagonistic microbes can provide on-going, self-sustaining suppression. In some cases, combinations of bacteria and fungi are used together for synergistic effects.
- *Challenges*: Biocontrol agents often work more slowly than chemical nematicides and can be sensitive to field conditions. Their efficacy can be inconsistent due to competition with native soil microbes, and they may require specific temperature/moisture conditions to establish themselves successfully (Briar et al. 2016). Production and formulation of biocontrol products is complex and costly—as one review notes, "the introduction of beneficial organisms via augmentative biological control has had limited success" (Briar et al. 2016). Currently, only a few PPN-specific BCAs are commercially available, and farmers must often use them as part of an integrated approach.

3.3.8 Soil Health Innovations

Recent innovations link nematode suppression with overall soil health. Biofumigation turns cover crops into nematicidal soil treatments, and is being refined through selection of high-glucosinolate Brassica lines. Conservation agriculture (no-till, permanent soil cover) tends to favor a more complex soil food web. Studies show reduced tillage and cover-crop mulching often increase populations of predatory/omnivorous nematodes and microbial antagonists, which can indirectly help suppress PPNs (though effects vary by system). Practices like compost tea application or vermicomposting are being explored for nematode-suppressive biota. Anaerobic soil disinfestation (ASD), where organic matter is solarized under plastic, has shown promise against nematodes by creating toxic fermentation byproducts. Across systems, building soil organic carbon and diversity (through composts, cover crops, reduced pesticides) is a key strategy for long-term nematode suppression and resilience. For example, fields with rich organic amendment histories often exhibit "suppressive" soils where nematode damage remains low, even without heavy chemical inputs (Batistič et al. 2025; McSorley 2011).

Fig. 3.2 Biological control of root-knot nematodes using new nematicide fertilizer. A—normal fertilizer, B—humic acid, C—neem oil granule type, D—simulated type of A + B + C mixed (Nematicide fertilizer)

Methods like mycorrhizal fungal inoculations, reduced tillage, and organic amendments (e.g., yard waste compost, sun hemp) improve soil health, indirectly managing nematodes. For instance, minimum tillage in corn-soybean rotations reduced *H. glycines* densities, and amendments can support beneficial organisms, though effects may vary (Krueger et al. 2022).

- *Nematicide fertilizers*: Nowadays, granule type of plant extracts can be mixed with fertilizers or humic acid. This new fertilizer type would play the role of a nematicide (Fig. 3.2). Nematicide fertilizer would be an effective method to control nematodes because it supplies humic acid and natural pesticides simultaneously. Humic acid is able to help biological activities and increase high organic materials in soil, which bring the biochemical defense mechanisms. The application of nematicide fertilizer including natural pesticides and humic acid would increase defense mechanisms in plant roots (Fig. 3.3).

3.3.9 Resistant and Tolerant Varieties

Genetic resistance is a powerful, specific tool in many cropping systems. Numerous crop varieties have been bred for nematode resistance. Such varieties prevent nematode reproduction in roots, effectively reducing nematode pressure on subsequent plantings (Noling 2012). Resistant cultivars are crop- and pest-specific, for example, a tomato variety may resist *M. incognita* but not a lesion nematode—so choice of variety must align with the known nematode species. Resistance can also be overcome: high soil temperatures or evolution of new virulent nematode races can reduce efficacy (Batistič et al. 2025). Thus, resistant cultivars are most effective when integrated with other tactics (rotation, biocontrol) to slow resistance breakdown (Briar et al. 2016).

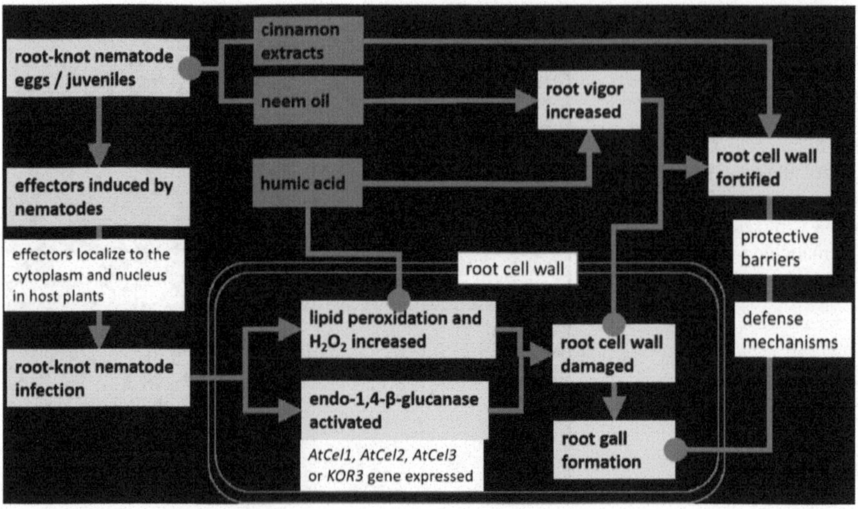

Fig. 3.3 In a nematicide fertilizer, each component plays the role of a nematicide and the combination effect led to highly effective nematicidal activities

- *Principle*: Use plant genotypes that naturally impede nematode reproduction or damage. Many crops have resistant cultivars carrying genes that arrest PPN development. For example, the tomato *Mi* gene (*Mi-1*) provides strong resistance to several *Meloidogyne* (root-knot) species (Briar et al. 2016). Other examples include the *Hs1^pro-1* gene in sugar beet (resisting *Heterodera schachtii* cyst nematode) and the *Gpa2* gene in potato (against *Globodera pallida* cyst nematode) (Briar et al. 2016). These resistance genes often trigger localized cell death at the nematode's feeding site, preventing its development (Briar et al. 2016).
- *Benefits*: Resistance is usually very effective, specific, and cost-efficient for farmers once the cultivar is in hand. It requires no additional inputs during the season. For some nematodes, even tolerant rootstocks (e.g., grafted tomatoes or melons on resistant rootstock) can greatly reduce damage.
- *Challenges*: Resistance is often species- or race-specific (Briar et al. 2016). A cultivar resistant to one nematode (or race) may be susceptible to another or to new virulent races. Resistance genes can "break down" if nematodes evolve (e.g., some SCN-resistant soybeans no longer control local nematode populations). Moreover, many resistant cultivars have been developed via biotechnology or conventional breeding that may not meet organic standards (e.g., GM herbicide-tolerance stacked with resistance) (Briar et al. 2016). Also, developing new resistant varieties is time-consuming and costly, and current resistance sources are limited for many crops.

3.3.10 Mycorrhizal Fungal Inoculations

Plant parasitic nematodes and arbuscular mycorrhizal fungi (AMF) are sharing plant roots as a resource for food and space. The interest in AMF-nematode interactions lies in the possibility of enhanced resistance or tolerance of AMF-infected plants to nematodes and the potential value of this for control of crop pests. As the market for beneficial microbial inocula is growing steadily (Glare et al. 2012) and the development of beneficial microbial inocula for large-scale field application is moving forward quickly (Ijdo et al. 2011; Salvioli and Bonfante 2013), field applications of AMF might be realistic. Thus far, few research about AMF-mediated biocontrol involved "omics" tools and systems biology approaches (Salvioli and Bonfante 2013) but this will increase over time and provide more detailed insights in the complex mechanisms underlying AMF-mediated biocontrol. These insights might in turn lead to the effective application of AMF in the field. The comparatively underappreciated AM function of nematode pathogen protection should become the target of much more intense ecological and functional mycorrhizal research.

Mycorrhizal fungal inoculations involve amending potting mixes during transplant production to improve seedling vigor and early fruit yield, indirectly aiding nematode management. Three commercial formulations were tested on watermelon, with one increasing early vine length and another enhancing early fruit yield compared to controls. The transplant cost is $0.28, with inoculum cost varying from a few cents to near the transplant cost, making it economically feasible for high-value crops (Westphal 2011). This method supports plant health, potentially reducing nematode impact through improved root systems.

3.3.11 Phytochemical-Based Strategies

Phytochemical-based strategies use plant extracts or essential oils with nematicidal properties, such as those from Compositae or citral and menthol, to suppress *M. incognita* infection. These methods are pest-specific and nontoxic, aligning with sustainable practices, though effectiveness can be slower than synthetic chemicals and may vary by application (Mukherjee 2020).

3.3.12 Marine Organisms

Marine organisms, particularly seaweeds like *Ascophyllum nodosum* and *Ecklonia maxima*, provide a novel approach, reducing RKN infestations with low environmental impact. For instance, *A. nodosum* at 3.6% Maxicrop reduced *M. javanica* eggs, and *E. maxima* at Kelpak 66 at 1:500 v/v reduced galling, enhancing plant growth (Table 3.1) (Veronico and Melillo 2021). Cyanobacteria and sponges show potential but require further research, with innovative formulations like nanoencapsulation suggested for commercialization.

Table 3.1 Seaweed, *Ascophyllum nodosum* tested for the control of plant-parasitic nematodes

Nematode	Type of study	Material	Dose	Inoculum	Host plant	Reference
M. javanica	In vitro, in vivo (P)	Maxicrop	3.6%	J2	Tomato	Whapham et al. (1994)
M. javanica	In vivo (P)	Maxicrop	3.6%	J2	Tomato	Wu et al. (1997)
M. javanica	In vivo (A)	Maxicrop	0.075%	J2	*Arabidopsis*	Wu et al. (1998)
M. incognita	In vivo (P)	Algaefol	10–25 mL/kg soil	Eggs	Tomato	Radwan et al. (2012)
M. javanica	In vivo (P)	Extract	1%	Eggs	Basil	Nour El-Deen et al. (2013)
M. chitwoodi, M. hapla	In vitro, in vivo (P)	OSMO	5 mL/L	J2	Tomato	Ngala et al. (2016)
Radopholus similis	In vivo (P)	Powder extract	2.24 kg/ha	NI	Citrus	Tarjan (1977)
Belonolaimus longicaudatus	In vivo (F)	Maxicrop, Seaborn	2.25–4.5–6.75 kg/ha, 2.47–4.94–7.42 L/ha	–	–	–

3.3.13 Integrated Nematode Management (INM)

These agro-ecological tactics are most effective when combined in an INM framework. INM for nematodes involves regular soil and root sampling to identify nematode species and population thresholds, then applying multiple complementary controls. For example, one might rotate to a non-host cover crop, apply compost and biofertilizer, use resistant varieties, and release BCAs in a single season. Research and extension emphasize that no single "silver bullet" exists: nematode management in organic or low-input systems relies on a holistic approach (Briar et al. 2016). Indeed, integrated strategies (rotation + green manure + soil amendment + weed control) have proven more reliable than single practices (Briar et al. 2016). Decision-support tools and farmer training are increasingly important to tailor INM plans by region and cropping system. For example, INM guidelines may recommend pasture rotations for subtropical vegetables in Latin America, or relay intercropping of eggplant with Solanaceous-resistant varieties in Asia, depending on local pest pressures.

3.4 Case Studies

3.4.1 Reduced Tillage

Zero till rice and zero till wheat with and without crop residues integrated with *Sesbania* sp. showed low population densities of rice root-knot nematode, *M. graminicola*, root gall index, and nematode to root biomass ratio and thereby serve as better option for the nematode management (Table 3.2) (Upadhyay et al. 2014).

Table 3.2 Effect of resource conservation practices on population of *Meloidogyne graminicola* on rice plants grown in soil samples collected from wheat fields (Upadhyay et al. 2014)

Resource conservation practices	J2 population at harvest	Nematode biomass (NB) (μg)	Root biomass (RB) (μg)	NB: RB ratio	Root gall index
Zero till rice + zero till wheat (without residue)	4317	2426	156,973	1:65	5.0
Zero till rice + zero till wheat (with residue)	4102	1639	60,110	1:37	5.0
Zero till rice + zero till wheat with *Sesbania*	800	787	79,043	1:100	2.5
CD (P = 0.05)	*15.28*	*17.21*	*15.52*	–	–

3.4.2 Crop Rotation

Rotating corn or sorghum after soybean dramatically cuts soybean cyst nematode populations (Kirkpatrick and Thomas 2014). In vegetable systems, a fallow or small grain year can reduce soil populations of root-knot nematodes before planting a susceptible crop. Rotating maize with sainfoin (*Onobrychis viciifolia*) in South Africa reduced root-knot nematode (*M. enterolobii*) populations by 60–90% relative to continuous maize (Rashidifard et al. 2021). Similarly, cereal cyst nematodes (*Heterodera* spp.) in wheat are managed by rotations including non-hosts like legumes or rapeseed (Noling 2012).

3.4.3 Cover/Green Manure Crops

In carrot production, a winter cover of oilseed radish has been used to reduce lesion nematode damage. Feld trials show that mustard cover-crop residues (as biofumigant green manure) significantly lower nematode counts in subsequent potato crops. In tropical systems, sun hemp planted during the off-season of tomatoes dramatically cuts *Meloidogyne* numbers.

3.4.4 Intercropping

In some tropical banana plantations, understory planting of chili pepper has been cited (by growers) as deterring nematodes and other pests. In smallholder gardens, mixed stands of legumes (e.g., cowpea) and cereals (e.g., millet) have been observed to suffer less nematode damage on the main crop.

3.4.5 Organic Amendments

Incorporating decomposed poultry manure into vegetable beds has been shown to reduce *Meloidogyne* (root-knot) reproduction. Mustard and brassica cover-crop residues (rich in glucosinolates) hydrolyze into isothiocyanates that kill nematodes and soil pathogens. Chitin amendments (from crustacean waste) promote *Pasteuria* and other biocontrol organisms that attack nematode eggs.

3.4.6 Biological Control

Soil drenches of *Bacillus subtilis* or *B. firmus* have been shown to reduce root-knot galls on tomato. Commercial products like *Trichoderma harzianum* or *P. lilacinum* spores are applied to lettuce and potato soils to cut down lesion nematodes. Organic growers sometimes inoculate seedlings with beneficial rhizobacteria (PGPR) that induce systemic resistance, making the plant less hospitable to nematodes (Briar et al. 2016). *Trichoderma* and *P. lilacinum* formulations have successfully reduced *M. hapla* on coffee, improving yields under nematode pressure (Abd-Elgawad 2024).

3.4.7 Resistant and Tolerant Varieties

Commercial peanut varieties with the nematode resistance gene *nemA* limit root-knot in warm climates. Many soybean varieties carry SCN-resistance QTLs (like *rhg1* and *Rhg4*) that keep cyst nematodes at bay. In vegetables, resistant cucumber and pepper cultivars resist common lesion nematodes. Tomato and sweet pepper cultivars carrying the *Mi-1* gene (resistant to most root-knot nematodes), barley and oat varieties with genes against cereal cyst nematodes, potato lines with *H1* resistance to golden nematode (*G. pallida*), and banana hybrids resistant to burrowing nematode are available for cultivation.

3.5 Comparative Analysis

The following Table 3.3 summarizes key agro-ecological approaches, their effectiveness, crop applicability, and practical considerations:

Each strategy's impact depends on local conditions (soil type, climate) and must be adapted to each crop system (Briar et al. 2016). Worldwide case studies (Africa, Asia, Europe, A mericas) confirm that combining crop rotations, cover crops, organic amendments, resistant varieties, and biocontrol agents in INM can markedly suppress nematodes while maintaining yield (Rashidifard et al. 2021; Krueger et al. 2015; Ayaz et al. 2024).

Policymakers should support farmer adoption of these agro-ecological methods through research funding, extension services, and incentives. For instance, subsidies for cover-crop seed or organic soil amendments, regulatory fast-tracks for biocontrol products, and strict limits on nematicide residues can accelerate sustainable practice. Agricultural regulations (e.g., IPM directives, organic standards) should prioritize integrated nematode management and monitor PPN quarantine pests (Pires et al. 2022). International cooperation (EPPO, FAO, APHIS) should continue updating pest lists and guidelines to reflect emerging nematode threats and promote eco-friendly control.

3.6 Recent Innovations and Research Trends

Research is rapidly expanding the agro-ecological toolbox for nematode management. Advances in plant and microbial genomics are yielding new resistant varieties and biological products. For example, researchers are applying RNA interference (RNAi) and CRISPR Cas9 genome editing to engineer crop resistance or silence nematode genes (Abd-Elgawad 2024). On the biocontrol front, "multi-omics"

Table 3.3 Key agro-ecological approaches, their effectiveness, and practical considerations

Strategy	Mechanism and effectiveness	Crop applicability	Regions/examples
Crop rotation	Non-host/poor-host crops starve nematodes; breaks pest life cycles (Krueger et al. 2015). In trials, rotation reduced RKN >60% (e.g., maize→sainfoin) (Rashidifard et al. 2021). Often combined with fallow periods or cover crops for added effect	Cereals (wheat, maize), legumes, vegetables (tomato, potato), orchards (perennial rotations)	Widely used: e.g., cereal–pulse rotations in Asia, maize rotations in Africa, US grain–legume rotations
Cover crops (green manures)	Allelopathic or toxic cover crops (Brassicas, Crotalaria, Tagetes) reduce nematodes. Brassica biofumigation releases isothiocyanates (Batistič et al. 2025); marigolds release α-terthienyl (Krueger et al. 2015). Legume covers (sun hemp) add biomass and fix N. Cover crops boost soil health, indirectly suppressing nematodes	Vegetable beds, orchards, fallow fields, any rotation requiring erosion control	E.g., mustard cover in potato fields (Europe, Asia); sun hemp in tropical vegetable production; marigolds in home gardens and field trials globally (Krueger et al. 2015; Batistič et al. 2025)
Organic amendments	Composts, manures, and mulch improve fertility and soil biota. Some release nematicidal compounds (NH_3, fatty acids) during decomposition (McSorley 2011). Effects on PPNs are variable—often inconsistent suppression but clear crop vigor gains	High-value vegetables, orchards, field crops	Common in organic/ sustainable farms worldwide (e.g. compost in fruit orchards, manure in vegetable plots); Florida trials with compost saw growth improvement but mixed nematode control (McSorley 2011)
Reduced tillage	Moderate, enhances soil suppressiveness	Long-term benefits, initial adjustment needed	–
Solarization	High for upper layers, impacts beneficials	Plastic waste, seasonal application	–
Phytochemical-based strategies	Moderate, pest-specific, slower	Application method, effectiveness varies	–

(continued)

Table 3.3 (continued)

Strategy	Mechanism and effectiveness	Crop applicability	Regions/examples
Resistant varieties	Crops bred with nematode-resistance genes prevent reproduction. Very effective against target nematodes (e.g., *Mi-1* tomato vs. RKN, *H1* potato vs. PCN). Must be matched to local nematode species; often used in high-value crops	Vegetables (tomato, pepper, potato), cereals (wheat, barley), legumes (soybean), fruits (banana)	Used globally where available: e.g., SCN-resistant soybeans (USA), CCN-resistant wheat (Europe), RKN-resistant tomatoes (worldwide) (Noling 2012)
Biological controls	BCAs like predaceous fungi (*Trichoderma, Pochonia*), bacteria (*Pasteuria, Bacillus*), nematode-trapping fungi, and entomopathogenic nematodes (against insects) suppress PPNs. Combined BCAs can markedly cut pest levels—e.g., *Trichoderma* + *P. lilacinum* cut coffee RKN by >70% (McSorley 2011). Suitable for organic/IPM use	All crop types, especially high-value (coffee, banana, vegetables) and in nurseries	Commercial biofungicides sold in Europe, Asia, Americas; farmers apply *Pochonia* in banana plantations (Latin America), *Purpureocillium* in tomato (Africa), *Pasteuria* in potato soils (Europe). See global review (McSorley 2011; Abd-Elgawad 2024)
AMF inoculations	Moderate, improves plant vigor	Application during transplant, cost varies	–
Marine organisms	Promising, low impact, needs research	Novel, formulation challenges	–
Habitat/IPM practices	Includes sanitation (clean planting material), weeding, intercropping, and monitoring. Intercropping with antagonistic species (e.g., maize–pumpkin, mulberry under coconut) can reduce nematode incidence. Conservation tillage preserves predators/microbes. IPM integration (e.g., rotating resistant cultivars with cover crops and BCAs) is critical	All systems (organic and conventional)	E.g., border rows of marigold or *Tagetes* in orchards (Europe); IPM programs in California and India require combining tactics; certified organic systems worldwide employ multifaceted nematode IPM

approaches (metagenomics, metabolomics) are discovering novel nematicidal compounds from soil microbes (Abd-Elgawad 2024). Biotechnology firms are developing more effective formulations: nanoencapsulation of bio-nematicides is being tested to improve stability and delivery (Abd-Elgawad 2024). Digital tools (e.g., decision-support apps) and precision sampling are also emerging to optimize when and where to use these methods. Even plant-derived nematicides are making a comeback—for instance, bioactive garlic extracts (e.g., Nemguard®) are now registered against PPNs.

Overall, combining these eco-friendly tactics in an integrated approach ("Push-Pull" strategies, INM) is the trend. Ongoing field trials continue to refine which crop rotations, cover-crop mixes, or biocontrol strains work best in specific regions. The goal is a sustainable nematode management system that maintains high yields with minimal chemical inputs.

3.7 Ecological Benefits

Agro-ecological nematode control enhances overall soil health and sustainability. Practices like crop rotation, cover cropping, and organic amendments increase soil organic matter, which improves structure, moisture retention, and nutrient availability for all crops (Abd-Elgawad 2024). A healthy, biologically active soil food web (with diverse bacteria, fungi, and predatory nematodes) better regulates nematodes, pests, and diseases. By relying on plants and microbes instead of synthetic fumigants, these methods avoid toxic residues in soil and water, protect non-target wildlife (like earthworms and beneficial insects), and lower greenhouse gas emissions from chemical manufacturing. For example, repeated cover crops have been shown to increase populations of beneficial nematodes (bacterivores, fungivores, predators) relative to harmful ones, indicating a more balanced ecosystem.

3.8 Policy and Regulatory Frameworks

On the policy front, many regions now restrict synthetic nematicides and encourage INM. International agreements (e.g., the Montreal Protocol) led to phase out of methyl bromide, and EU regulations (e.g., Regulation 2019/2072 and the Sustainable Use Directive) list major nematodes as quarantine pests and mandate INM adoption (Pires et al. 2022). For example, the EU's list of regulated pests includes *M. chitwoodi*, *G. pallida*, *Heterodera glycines*, etc. (Pires et al. 2022), compelling member states to monitor and restrict their spread. Many national plant protection agencies (USDA-APHIS, EPPO, African, and Asian phytosanitary bodies) similarly regulate nematodes and promote biological or cultural controls. Organic agriculture standards (USDA, EU) explicitly forbid most synthetic nematicides, effectively requiring farmers to use agro-ecological methods. Policy recommendations include subsidizing cover crops and resistant seed, funding nematode extension services, and streamlining registration of biopesticides. Strengthening research-to-farm pathways (Farmer Field Schools, Krishi Vigyan Kendras, IPM training) ensures these ecologically based methods reach growers. In sum, regulatory trends worldwide increasingly favor habitat management, breeding, and biological solutions over chemical fumigants, aligning with global sustainability goals.

Governments and organizations can support these methods by:

- Approving safer, eco-friendly nematicides to replace harmful chemicals.
- Promoting research into nematode-resistant crops and natural pest controls.
- Including these practices in official pest management guidelines to encourage farmers to adopt them. Such policies, seen in places like the European Union, aim to balance farm productivity with environmental health, though their adoption varies globally.

3.9 Challenges and Considerations

While ecologically sound, these methods can be more complex to implement than a single pesticide application. They often require good agronomic knowledge and monitoring. For instance, successful crop rotation demands accurate nematode diagnostics and careful planning of alternative crops (Kirkpatrick and Thomas 2014). Organic amendments and cover crops may tie up land or nutrients if not well managed. Some practices (like deep plowing in green manures) can temporarily disturb soil biota (Briar et al. 2016). In practice, farmers must balance immediate economic needs against long-term gains: a rotational scheme is "a long-term, ongoing strategy" that needs annual reevaluation and may involve added costs for new equipment or herbicides for cover crops. Finally, the effects of agro-ecological tactics can be slower and more variable than chemicals; multiple seasons of consistent management may be needed before nematode pressure is measurably reduced.

3.10 Future Directions

These approaches are most effective when integrated, considering the entire cropping system. For instance, combining crop rotation with cover crops and organic amendments can enhance soil health and nematode suppression. Future research should focus on scaling biotechnological applications and marine organism use, addressing formulation challenges for commercialization. The emphasis on soil biological and ecological factors, as highlighted in recent studies, underscores the need for holistic management strategies to achieve sustainable agricultural production with minimal environmental impacts.

An X post from @daanmoca linked to a study by Prof. Yuanmei Zuo's team at China Agricultural University that uses soil nematodes as indicators of soil and plant health in continuous dragon fruit cropping systems, highlighting ongoing research in agroecological contexts.

3.11 Conclusion

An array of agro-ecological strategies exists to manage nematodes across crops and regions. No single method suffices on its own; success comes from well-planned integration. By rotating crops, using suppressive covers and amendments, planting resistant varieties, and encouraging beneficial soil organisms, farmers can sustainably reduce nematode damage. This integrated approach not only controls pests but also builds soil health and resilience, aligning nematode management with broader goals of agro-ecosystem sustainability (Pires et al. 2022; Briar et al. 2016).

Future studies should focus on environmentally friendly approaches, which are based on multidisciplinary strategies that can fill the gaps of single-sided management methods. Such approaches will also reduce the chance of resistance as the complexity of different nematode management components would make resistance highly improbable. Whatever strategy is devised, future attempts should focus on important factors such as synergism between nematode antagonists, environmental conditions, sustainability, studying the effect of new treatments on non-target organisms, and association of individual plants with nematode antagonists of interest. In summing up, a sustainable management of plant-parasitic nematodes is feasible when two or more compatible tactics are applied concurrently while appraising environmental protection.

3.12 Suggested Readings

- Abd-Elgawad MMM (2024) Upgrading strategies for managing nematode pests on profitable crops. Plants 13(11):1558. https://doi.org/10.3390/plants13111558.
- Norton DC (1978) Ecology of plant-parasitic nematodes. Wiley-Interscience, New York. (Classic text explaining ecological interactions of nematodes in soil systems).
- Sikora RA, Bridge J, Starr JL (2005). Management practices: an overview of integrated nematode management technologies. In Luc M, Sikora RA, Bridge J (eds) Plant parasitic nematodes in subtropical and tropical agriculture. CABI Bioscience, Egham, UK, pp 793–826. (Chapter on cultural and ecological management practices).
- Stirling GR (2014) Biological control of plant-parasitic nematodes: soil ecosystem management in sustainable agriculture. CABI Publishing. (Comprehensive overview of agro-ecological and biological approaches).
- Mukherjee A (2020) Eco-friendly management of plant parasitic nematodes. Ind J Sci Technol 13(28):2883–2891.
- Westphal A (2011) Sustainable approaches to the management of plant-parasitic nematodes and disease complexes. J Nematol 43(2):122–125.

3.13 Review Questions

- Define agro-ecology. How does it differ from conventional pest management?
- List the major agro-ecological approaches used for nematode management.
- What role do organic amendments play in reducing nematode infestation?
- Mention at least three nematode-suppressive crops and their mode of action.
- How does crop rotation help in nematode management? Give suitable examples.
- What is the significance of cover crops and green manures in nematode control?
- Explain the mechanisms by which soil biodiversity contributes to nematode suppression.
- Discuss the role of antagonistic fungi and bacteria in agro-ecological nematode management.
- How do trap crops work against plant-parasitic nematodes? Provide examples.
- Describe how habitat diversification in agro-ecosystems affects nematode dynamics.
- Why agro-ecological approaches are considered more sustainable than chemical nematicides?

References

Abd-Elgawad MMM (2022) Exploiting plant-phytonematode interactions to upgrade safe and effective nematode control. Life 12:1916

Abd-Elgawad MMM (2024) Upgrading strategies for managing nematode pests on profitable crops. Plants 13(11):1558. https://doi.org/10.3390/plants13111558

Abd-Elgawad MMM, Askary TH (2015) Impact of phytonematodes on agriculture economy. In: Askary TH, Martinelli PRP (eds) Biocontrol agents of phytonematodes. CAB International, Wallingford, pp 3–49

Ayaz M, Zhao JT, Zhao W et al (2024) Biocontrol of plant parasitic nematodes by bacteria and fungi: a multi-omics approach for the exploration of novel nematicides in sustainable agriculture. Front Microbiol 15:1433716

Batistič L, Bohinc T, Trdan S (2025) Biofumigation with *brassica* species and their derivatives: a comprehensive review of an innovative pest control strategy targeting wireworms (Coleoptera: Elateridae). Agronomy 15(4):967. https://doi.org/10.3390/agronomy15040967

Briar SS, Wichman B, Reddy GVP (2016) Plant-parasitic nematode problems in organic agriculture. In: Nandwani D (ed) Organic farming for sustainable agriculture, sustainable development and biodiversity. Springer, Cham, pp 107–122

Burns AR, Baker RJ, Kitner M et al (2023) Selective control of parasitic nematodes using bioactivated nematicides. Nature 618:102–109

D'Addabbo T, Ladurner E, Troccoli A (2023) Nematicidal activity of a garlic extract formulation against the grapevine nematode *Xiphinema index*. Plants 12:739

deGives PM (2022) Soil-borne nematodes: impact in agriculture and livestock and sustainable strategies of prevention and control with special reference to the use of nematode natural enemies. Pathogens 11(6):640. https://doi.org/10.3390/pathogens11106064

Glare T, Caradus J, Gelernter W et al (2012) Have biopesticides come of age? Trends Biotechnol 30:250–258

Hammam MMA, Abd-El-Khair H, El-Nagdi WMA, Abd-Elgawad MMM (2022) Can agricultural practices in strawberry fields induce plant–nematode interaction towards *Meloidogyne*-suppressive soils? Life 12:1572

Ijdo M, Cranenbrouck S, Declerck S (2011) Methods for large-scale production of AM fungi: past, present, and future. Mycorrhiza 21:1–16

Kirkpatrick TL, Thomas AC (2014) Crop rotation for management of nematodes in cotton and soybean. University of Arkansas Cooperative Extension Service http://www.uaex.edu/Other_Areas/publications/PDF/FSA-7550.pdf.FSA7550

Krueger DH, Fourie JC, Malan AP (2015) The effect of cover crops and their management on plant-parasitic nematodes in vineyards. S Afr J Enol Vitic 36(2):195–209

Krueger R, Dover KE, McSorley R, Wang KH (2019) *Marigolds* (Tagetes *spp.*) *for Nematode Management*. Publication #ENY-056. https://doi.org/10.32473/edisng045-2007

Krueger R, McSorley R, Grabau JG (2022) Nematode management in organic agriculture. Department of Entomology and Nematology, UF/IFAS Extension

McSorley R (2011) Overview of organic amendments for management of plant-parasitic nematodes, with case studies from Florida. J Nematol 43(2):69–81

Mukherjee A (2020) Eco-friendly management of plant parasitic nematodes. Indian J Sci Technol 13(28):2883–2891

Ngala MB, Valdes Y, dos Santos G et al (2016) Seaweed-based products from *Ecklonia maxima* and *Ascophyllum nodosum* as control agents for the root-knot nematodes *Meloidogyne chitwoodi* and *Meloidogyne hapla* on tomato plants. J Appl Phycol 28:2073–2082

Noling JW (2012) Nematodes and their management. Pubn. ENY625, Department of Horticultural Sciences, UF/IFAS extension. http://edis.ifas.ufl.edu

Nour El-Deen AH, Abdel-Kafie OM, El-Ghareb NM (2013) Evaluation of seaweed extract and various plant products against *Meloidogyne incognita* on basil. J Agric For 16:29–34

Pires D, Vicente CSL, Menéndez E et al (2022) The fight against plant-parasitic nematodes: current status of bacterial and fungal biocontrol agents. Pathogens 11(10):1178

Radwan MA, Farrag SAA, Abu-Elamayem MM, Ahmed NS (2012) Biological control of the root-knot nematode, *Meloidogyne incognita* on tomato using bioproducts of microbial origin. Appl Soil Ecol 56:58–62

Rashidifard M, Ashrafi S, Claassens S et al (2021) A pilot approach investigating the potential of crop rotation with sainfoin to reduce *Meloidogyne enterolobii* infection of maize under greenhouse conditions. Front Plant Sci 12:659322

Salvioli A, Bonfante P (2013) Systems biology and "omics" tools: a co-operation for next-generation mycorrhizal studies. Plant Sci 203–204:107–114

Sikora RA, Molendijk LPG, Desaeger J (2022) Integrated nematode management and crop health: future challenges and opportunities. In: Sikora RA, Desaeger J, Molendijk LPG (eds) Integrated nematode management: state-of-the-art and visions for the future. CAB International, Wallingford, pp 3–10

Stapleton JJ (1997) Solarization: an implementable alternative for soil disinfestation, Biological and cultural tests. APS Press, St. Paul

Tarjan AC (1977) Kelp derivatives for nematode-infected citrus trees. J Nematol 9:287

Upadhyay V, Vaish SS, Bhardwaj NR (2014) Effect of resource conservation practices and conventional practices on population dynamics of *Meloidogyne graminicola* (Golden and Birchfield, 1965) under rice-wheat cropping system. Afr J Microbiol Res 8(32):3025–3028

Veronico P, Melillo MT (2021) Marine organisms for the sustainable management of plant parasitic nematodes. Plants 10(2):369. https://doi.org/10.3390/plants10020369

Westphal A (2011) Sustainable approaches to the management of plant-parasitic nematodes and disease complexes. J Nematol 43(2):122–125

Whapham CA, Jenkins T, Blunden G, Hankins SD (1994) The role of seaweed extracts *Ascophyllum nodosum*, in the reduction in fecundity of *Meloidogyne javanica*. Fund. Appl. Nematol. 17:181–183

Wu Y, Jenkins T, Blunden G et al (1997) The role of betains in alkaline extracts of *Ascophyllum nodosum* in reduction of *Meloidogyne javanica* and *M. Incognita* infestations of tomato plants. Fund Appl Nematol 20:99–102

Wu Y, Jenkins T, Blunden G et al (1998) Suppression of fecundity of the root-knot nematode, *Meloidogyne javanica*, in monoxenic cultures of *Arabidopsis thaliana* treated with an alkaline extract of *Ascophyllum nodosum*. J Appl Phycol 10:91–94

4 Microbial Bioprotectants: Nature's Nematode Defense Mechanism

Abstract

Biological control of plant-parasitic nematodes uses natural enemies like fungi, bacteria, and nematodes. These agents reduce nematode damage by parasitism, toxins, and inducing plant resistance. Fungal species like *Trichoderma, Pochonia, Purpureocillium*; mycorrhizal fungal species of *Glomus*; and bacterial species of *Bacillus, Pseudomonas, Pasteuria* are effective for the management of plant-parasitic nematodes through parasitism and induced resistance, but efficacy varies by soil and crop. There is some controversy over inconsistent results and challenges like cost and regulatory hurdles. The key biological control agents, their mechanisms, practical applications, and associated challenges are discussed, based on insights from recent studies.

Keywords

Nematophagous fungi · Antagonistic bacteria · Predatory nematodes · Arbuscular mycorrhizal fungi · Mechanisms of action

4.1 Learning Objectives

- Define biological control in the context of nematode management.
- Explain how it differs from chemical and cultural control strategies.
- Identify different groups of organisms used in nematode biocontrol, such as nematophagous fungi, bacteria, predatory nematodes, viruses, and endophytes.
- Describe their modes of action (parasitism, predation, antibiosis, competition, induced resistance).
- Discuss factors influencing the success of biocontrol agents (soil type, environmental conditions, host plant, nematode species).
- Compare the relative effectiveness of different biocontrol strategies.

R. P. Parvatha, *Plant Parasitic Nematodes*,
https://doi.org/10.1007/978-981-95-9306-4_4

- Understand the importance of combining biological control with cultural, chemical, and host resistance strategies (Integrated Nematode Management).
- Highlight examples of successful field applications.
- Learn about molecular and biotechnological tools in enhancing biocontrol efficiency (genetically improved strains, microbial consortia, formulations).
- Recognize constraints such as inconsistency in field performance, formulation difficulties, and environmental dependence.
- Discuss strategies to overcome these challenges.
- Identify crops and nematode species where biological control is most successfully applied.

4.2 Introduction

Plant-parasitic nematodes, such as root-knot (*Meloidogyne* spp.) and cyst (*Heterodera* spp.) nematodes, pose significant threats to global agricultural production by feeding on plant roots, leading to yield losses. Traditional chemical nematicides, while effective, are increasingly restricted due to environmental and health concerns, prompting a shift toward biological control. Biological control of plant-parasitic nematodes involves using natural enemies or antagonists (fungi, bacteria, and predatory nematodes) to suppress nematode populations, reducing their damage to crops (Fig. 4.1). This approach is sustainable and minimizes reliance on chemical nematicides, which can harm the environment. This approach leverages natural enemies, primarily filamentous fungi and bacteria, to regulate nematode populations and mitigate damage. This chapter provides a detailed examination of the methods, mechanisms, and practical applications, drawing from recent research and authoritative sources to offer a thorough understanding for agricultural practitioners and researchers.

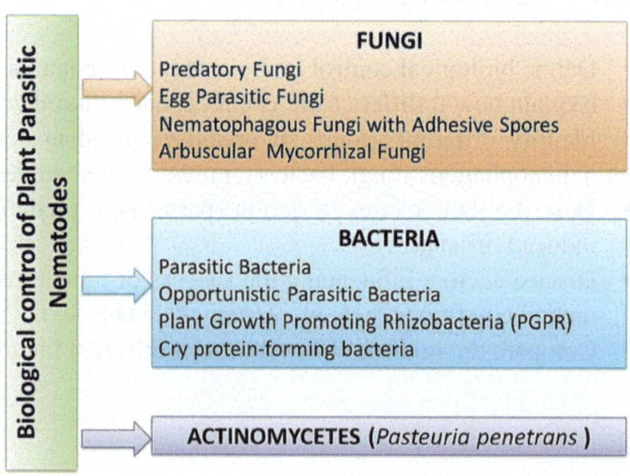

Fig. 4.1 The components of biological control of plant parasitic nematodes

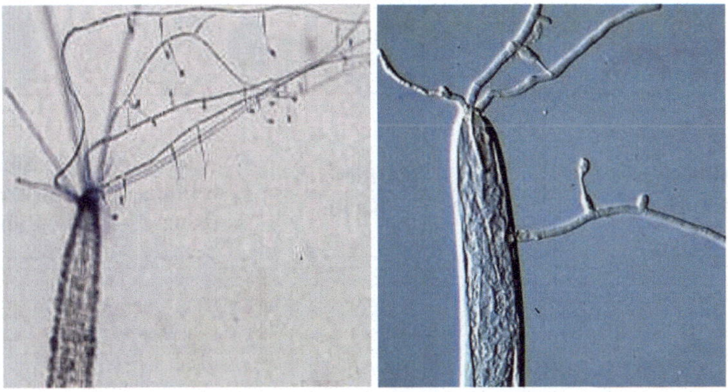

Fig. 4.2 Nematode infected with *Hirsutella rhossiliensis*

4.3 Nematophagous Fungi

Biological control of plant-parasitic nematodes using antagonistic fungi represents a promising and sustainable approach to address the significant economic losses caused by these pests in agricultural systems. This approach is seen as environmentally friendly, but its effectiveness can vary, and ongoing studies aim to refine its application. Antagonistic fungi, often termed nematophagous fungi, offer eco-friendly strategies by targeting nematode eggs, juveniles, and adults through diverse mechanisms such as trapping structures, enzymatic degradation, and toxin production (Al-Ani et al. 2022; Ur Rahman et al. 2023).

Nematophagous fungi encompass a diverse group, including trapping fungi (e.g., *Arthrobotrys* spp.), egg-parasitic fungi (e.g., *Purpureocillium lilacinum*), and endoparasitic fungi (e.g., *Hirsutella rhossiliensis*) (Fig. 4.2). These organisms employ mechanisms such as adhesive nets, specialized spores, and toxin secretion to capture, penetrate, and kill nematodes. Under elevated temperatures and CO_2, studies report variable impacts: while some *Trichoderma* strains maintain antagonism under heat stress, others show diminished spore germination and VOC production, affecting volatile-mediated nematode suppression (Ayaz et al. 2024). Similarly, moisture deficits can reduce the formation of trapping structures by *Arthrobotrys* spp., highlighting the need for robust isolates adapted to drier soils.

Antagonistic fungi can be broadly classified into five ecological groups based on their interaction with nematodes:

4.3.1 Nematode-Trapping (Predacious) Fungi

The nematode trapping fungi such as *Dactylella oviparasitica, Nematophthora gynophila* and *Arthrobotrys robustus* have been reported suitable for practical utilization. These fungi form specialized trapping devices—adhesive nets, constricting rings, or adhesive branches—to capture motile nematodes (Fig. 4.3). *A. oligospora*,

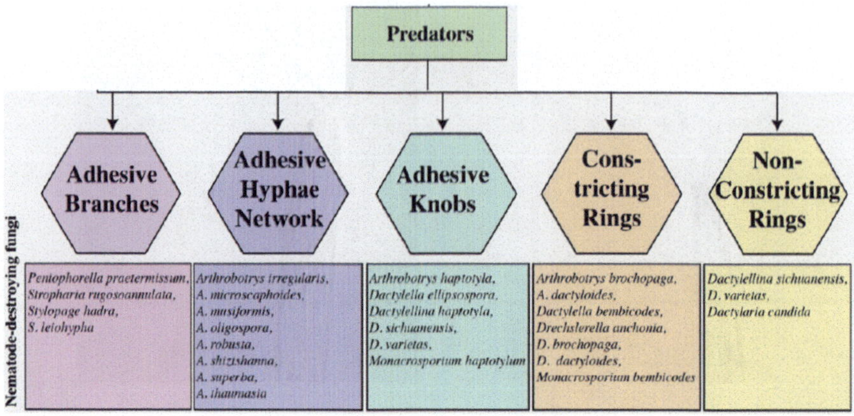

Fig. 4.3 Modes of action of nematode trapping fungi

Fig. 4.4 Sticky nets of *Arthrobotrys oligospora*

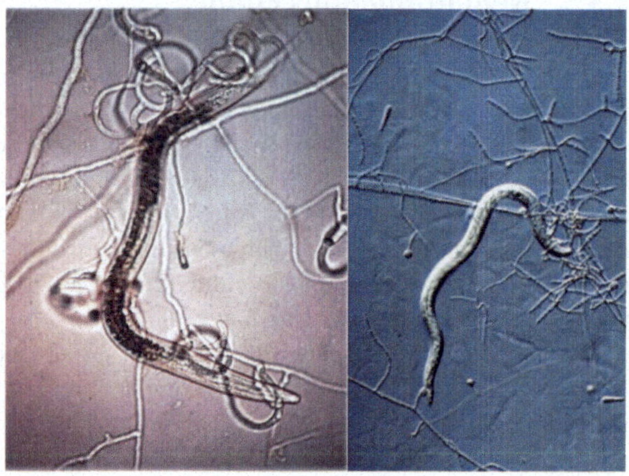

for instance, produces adhesive nets that immobilize nematodes before enzymatic penetration (Fig. 4.4) (Ur Rahman et al. 2023); *A. dactyloides* uses constricting rings (Fig. 4.5), *D. candida* make non-constricting rings and adhesive knobs (Fig. 4.6), *Monacrosporium cionopagum* forms adhesive branches and a two-dimensional adhesive network, while *M. ellipsosporium* traps nematodes with adhesive knobs or adhesive branches that may form loops. Their ability to produce endogenous nemin makes them potential biocontrol agents.

4.3.2 Egg- and Ovicidal Fungi

Ovicidal fungi like *Pochonia chlamydosporia* (Fig. 4.7), *Verticillium lecanii*, *Trichoderma harzianum*, and *Purpureocillium lilacinum* (Fig. 4.8) parasitize nematode eggs by forming appressoria, secreting cell-wall-degrading enzymes, and

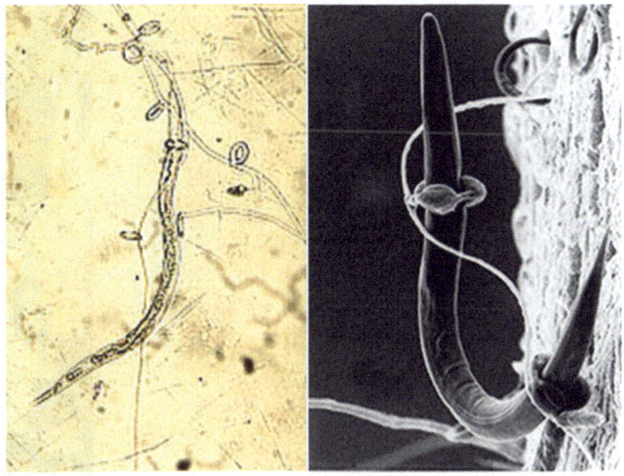

Fig. 4.5 Constricting ring traps—*Arthrobotrys dactyloides*

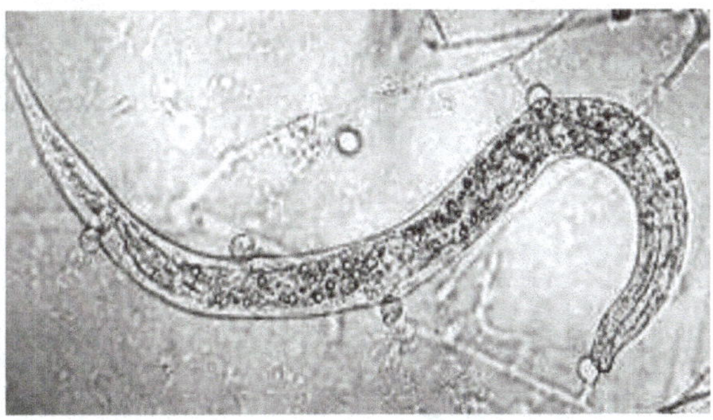

Fig. 4.6 Adhesive knobs—*Dactylella candida*

digesting the eggshell to access the juvenile within (Al-Ani et al. 2022). The feasibility of using selected opportunistic soil fungi such as *P. lilacinum*, *P. chlamydosporia*, *V. lecanii*, *T. harzianum*, *T. viride*, and *T. virens* have shown some effect in controlling *M. incognita*, *T. semipenetrans*, and *Rotylenchulus reniformis*. Species of *Verticillium*, *Trichoderma*, and *P. lilacinum* were recovered from females obtained in galls of plants.

4.3.3 Endoparasitic Fungi

Species such as *Hyalorbilia oviparasitica* infect nematodes internally, proliferating within the host body and eventually leading to nematode death.

4.3.4 Toxin-Producing Fungi

Certain basidiomycetes, notably *Pleurotus ostreatus*, secrete nematotoxic metabolites (e.g., trans-2-decanoic acid), which immobilize nematodes prior to fungal invasion (Luo et al. 2004).

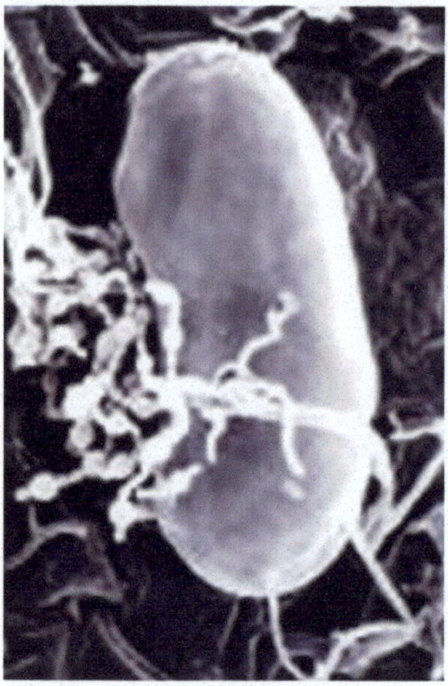

Fig. 4.7 *Pochonia chlamydosporia* infecting nematode egg

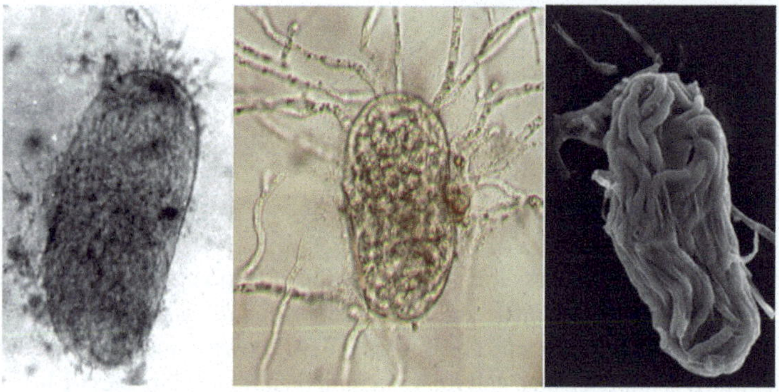

Fig. 4.8 *Purpureocillium lilacinum* infecting the eggs of *Meloidogyne enterolobii*

4.3.5 Opportunistic (Saprophytic) Fungi

Fungi like *Trichoderma* spp. are not obligate nematode antagonists but can inhibit nematode development via antibiosis, competition for nutrients, or induction of plant defenses (Luo et al. 2004; Al-Ani et al. 2022).

4.3.6 Key Antagonistic Fungi

- *Trichoderma* spp.: Combines mycoparasitism, antibiosis, and plant-growth promotion; widely commercialized as seed coatings and soil amendments (Luo et al. 2004; Al-Ani et al. 2022).

- *Pochonia chlamydosporia*: Specialist egg parasite; effective in diverse crops when applied as a soil drench or seed treatment (Al-Ani et al. 2022).
- *Purpureocillium lilacinum*: Effective against root-knot and cyst nematodes; isolates vary in field efficacy due to environmental sensitivity.
- *Arthrobotrys oligospora* and *Arthrobotrys* spp.: Model nematode-trapping fungi, forming nets or rings; promising for biopesticide development (Ur Rahman et al. 2023).
- *Metarhizium anisopliae*: An entomopathogen with opportunistic nematophagous activity; produces chitinases and proteases against nematodes (Luo et al. 2004).

4.3.7 Arbuscular Mycorrhizal Fungi (AMF)

Arbuscular mycorrhizal fungi are the most spectacular and widespread type of mycorrhizas present in nearly all major agronomic crops (vegetables, fruit crops, flowers, forest trees, and plantation crops). The invading endomycorrhizal hyphae grow on the root surface, forms an appressorium, and penetrates into the cortex. Hyphal growth in root cortical cells is both inter-and intra-cellular where they form the two diagnostic structures, vesicles and arbuscules (Fig. 4.9). Many species do not form vesicles, therefore, the new name arbuscular mycorrhizal fungi (AMF).

Arbuscular mycorrhizal fungi have shown antagonistic influence on the population of plant parasitic nematodes (Fig. 4.10), which may be due to increased amino acid contents (especially arginine) in the mycorrhizal roots (Nemec and Meridith 1981) or the presence of an additional amount of phosphorus in roots (Fig. 4.11). AMF can markedly alter the plant response to nematodes.

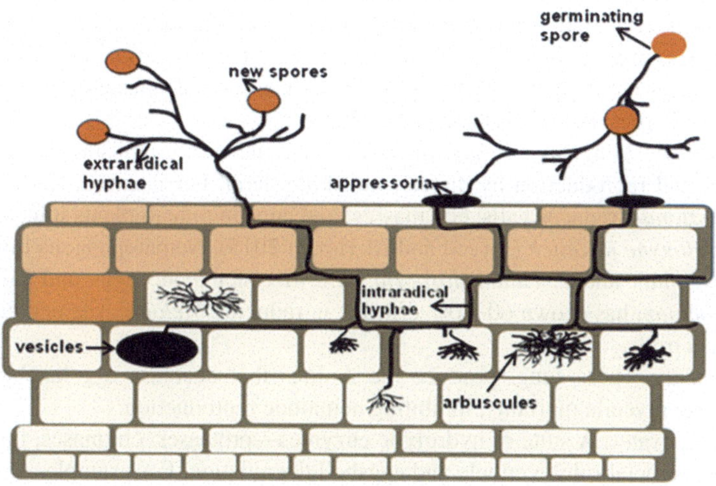

Fig. 4.9 Diagrammatic representation of mycorrhizal association within the plant root

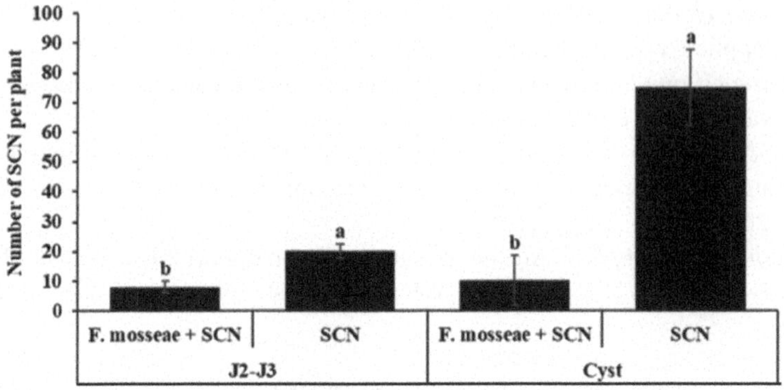

Fig. 4.10 Effect of *Funneliformis mosseae* on juveniles (J2–J3) and cysts of soybean cyst nematode. Bars with different letters indicate means differ significantly ($P = 0.05$)

4.3.8 Mechanisms of Action

Filamentous fungi, including *Trichoderma*, mycorrhizal, and endophytic fungi, are prominent biological control agents (BCAs) against plant-parasitic nematodes. A schematic representation of different mechanisms of biocontrol is shown in Fig. 4.12. Even though a large number of biocontrol agents have been revealed as potential pesticides, only a few have been commercialized (Roberts et al. 2005).

Their mechanisms can be categorized into direct and indirect actions:

Direct Mechanisms

- *Mechanical Trapping*: Nematode-trapping fungi deploy physical traps: adhesive networks, constricting rings, or sticky knobs, which immobilize J2 (second-stage juveniles) and facilitate enzymatic penetration. The formation and function of these traps are regulated by environmental cues such as nutrient limitation and nematode presence (Ur Rahman et al. 2023)
- *Parasitism*: Fungi like *T. harzianum* reduce nematode establishment, development, and reproduction by directly attacking them. For instance, studies show *T. harzianum* reduces galls, egg masses, and eggs in tomato plants infected with *Meloidogyne javanica* (Javeed and Al-Hazmi 2015). Nematophagous fungi like *P. lilacinum* and *P. chlamydosporia* parasitize nematode eggs and juveniles. *P. lilacinum* has shown 60–70% efficacy in reducing *Meloidogyne* egg hatching in field trials.
- *Antibiosis*: These fungi produce metabolites that decrease egg hatching and increase juvenile mortality, inhibiting nematode reproduction.
- *Lytic Enzymes*: A suite of hydrolytic enzymes—proteases, chitinases, lipases—plays a crucial role in cuticle and eggshell degradation. For example, *A. oligospora* produces serine proteases that breach the nematode cuticle within hours, while *M. thaumasium* secretes chitinases to degrade chitin-rich eggshells (Al-Ani et al. 2022). Enzymes such as chitinase and protease penetrate nematode eggs,

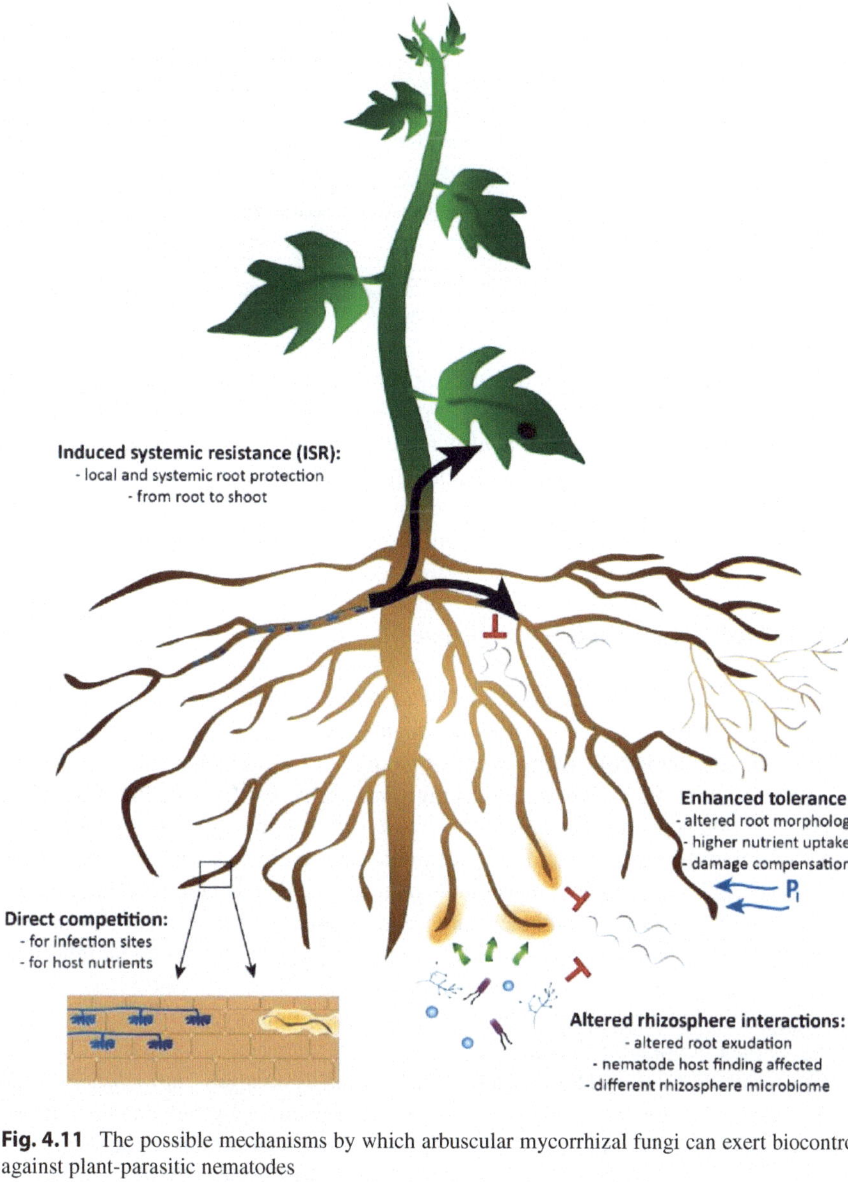

Fig. 4.11 The possible mechanisms by which arbuscular mycorrhizal fungi can exert biocontrol against plant-parasitic nematodes

reducing hatching rates. For example, *T. longibrachiatum* has been effective against *Heterodera avenae*. *Trichoderma* spp. produces enzymes and secondary metabolites that degrade nematode cuticles or inhibit their reproduction. For example, *T. harzianum* reduces root-knot nematode (*Meloidogyne* spp.) populations by up to 50% in some studies.

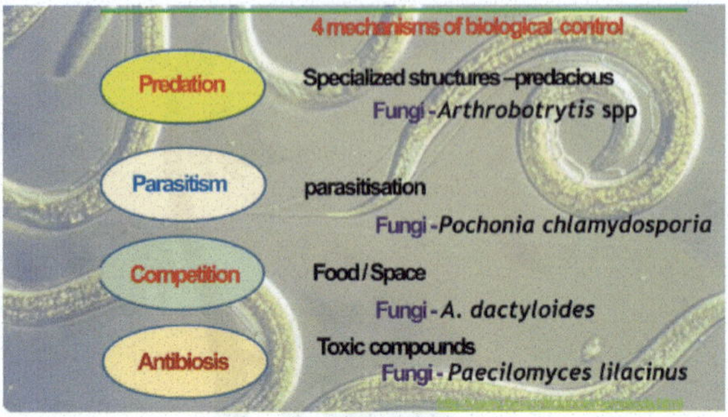

Fig. 4.12 An overview of different mechanisms of biological control

- *Toxin Production*: Toxin-producing fungi release secondary metabolites (e.g., cyclodepsipeptides, peptaibiotics) that paralyze nematodes. Basidiomycetes like *Pleurotus* spp. secrete linoleic acid-derived compounds toxic to nematodes, acting at a distance before fungal invasion (Luo et al. 2004)
- *Space/Resource Competition*: Fungi colonize the rhizosphere, competing with nematodes for nutrients and space, limiting their access to host plants.

Indirect Mechanisms (Induced Resistance)

- *Systemic Acquired Resistance (SAR) and Induced Systemic Resistance (ISR)*: Fungi activate plant defense pathways, such as jasmonic acid (JA)/ethylene (ET) and salicylic acid (SA) pathways. For example, *T. harzianum* induces ISR in tomato against *M. javanica*, enhancing plant defenses. Certain fungi, notably *Trichoderma* spp. and *Pochonia chlamydosporia*, colonize plant roots endophytically, eliciting salicylic-acid and jasmonic-acid pathways that enhance plant defenses against nematode invasion (Al-Ani et al. 2022). Endophytic fungi (e.g., *Fusarium oxysporum*) colonize plant roots, inducing systemic resistance and releasing nematicidal compounds. Resistance induced by fungi can be inherited by plant offspring, as observed with *T. atroviride*.
- *Epigenetic Changes*: Fungi induce small RNAs and methylome alterations, further enhancing plant resistance.
- *Mycorrhizal Fungi*: These fungi, such as *Rhizophagus intraradices* and *Funneliformis mosseae*, improve plant competitiveness by enhancing nutrient and water uptake, modifying root morphology, and altering rhizosphere interactions. They also induce Mycorrhiza-Induced Resistance (MIR), which is JA-mediated and enhances defense compounds like phytoalexins and pathogenesis-related (PR) proteins, reducing nematode penetration (Schouteden et al. 2015).
- *Endophytic Fungi*: These fungi, such as *Acremonium implicatum*, colonize roots and destroy nematode eggs, produce nematicidal metabolites like chaetoglobosin A/B from *Chaetomium globosum* against *M. incognita*, and secrete lytic enzymes. They also induce SAR and ISR through JA and SA pathways, enhancing chemical defenses (Poveda et al. 2020).

4.3.9 Case Studies

Ur Rahman et al. (2024) reported the control of root-knot nematodes with *Metarhizium anisopliae* in tomatoes and cucumbers, cyst nematodes with *P. lilacinum* reducing hatching rates, and root-lesion nematodes with *H. rhossiliensis* achieving 24–25% reduction in *Pratylenchus penetrans* in potatoes. Numerous studies illustrate the efficacy of these fungi:

- *Parasitic Fungi Effects*: *T. harzianum* reduces *M. javanica* in tomato by decreasing galls, egg masses, and egg hatching, while *T. longibrachiatum* inhibits *H. avenae* cyst hatching, and *T. atroviride* induces heritable resistance (Poveda et al. 2020). Soil drench with *P. chlamydosporia* at 10^6 cfu g^{-1} reduced galling by 60% and egg-mass production by 55% on tomato under greenhouse conditions (Prasad et al. 2020). Cucumber seed coating with *T. harzianum* improved plant growth by 25% and suppressed nematode reproduction by 45% in field trials (Li et al. 2018). Granular formulation of *P. lilacinum* applied in-furrow led to a 50% reduction in soybean cyst nematode counts and 15% yield gain (Jones and Akhtar 2019). Soil amendment with *A. oligospora*-colonized rice straw decreased nematode density by 70% and improved root health of banana.
- *Mycorrhizal Effects*: *R. intraradices* and *F. mosseae* reduce *Nacobbus aberrans* in tomato, and *Glomus* spp. affect *Meloidogyne javanica* in peach, though effects on *Pratylenchus brachyurus* in maize can vary, with increases noted in some cases.
- *Endophytic Effects*: *A. implicatum* parasitizes *M. incognita* eggs, *Fusarium oxysporum* paralyzes *Pratylenchus goodeyi*, and *C. globosum* produces nematicidal compounds against *M. incognita*.

A detailed Table 4.1 summarizing specific studies is provided below:

4.3.10 Advantages of Fungal Biocontrol

Fungal BCAs offer several benefits:

- *Selectivity and Environmental Safety*: They target nematodes specifically, reducing harm to non-target organisms and lowering the risk of resistance development (Ur Rahman et al. 2024).
- *Long-Term Effectiveness*: Fungi persist via spores and sclerotia, inducing systemic resistance and integrating well with other strategies like crop rotation and resistant varieties.
- *Sustainability*: Economically viable and less labor-intensive, they maintain soil health, contributing to sustainable agriculture.
- *Compatibility*: Can be combined with chemical nematicides and biostimulants, enhancing integrated pest management (IPM) approaches.

The modes of action for common fungal bioprotectants are summarized in Table 4.2.

Table 4.1 Effect of nematophagous fungi for the management of plant nematodes

Fungus type	Nematode targeted	Plant host	Key findings/ Mechanisms	Reference
Trichoderma harzianum	*Meloidogyne javanica.*	Tomato	Reduces galls, egg masses, egg hatching via parasitism	Poveda et al. (2020)
T. longibrachiatum	*Heterodera avenae*	Wheat	Inhibits cyst hatching, parasitic effects	Poveda et al. (2020)
Funneliformis mosseae	*M. incognita, P. penetrans*	Tomato	Reduces infection via enhanced defenses, JA pathway	Poveda et al. (2020)
Pochonia chlamydosporia	*Globodera pallida*	–	34–49% pathogenicity against eggs, uses appressoria	Ur Rahman et al. (2024)
Metarhizium anisopliae	Root-knot nematodes	Tomato, cucumber	Effective control, detailed in case studies	Ur Rahman et al. (2024)

4.4 Antagonistic Bacteria

With chemical nematicides facing restrictions due to environmental and health concerns, biological control using antagonistic bacteria has emerged as a promising solution. Antagonistic bacteria, naturally residing in the rhizosphere and capable of suppressing nematode populations through diverse mechanisms, are among the most promising biocontrol agents.

Rhizobacteria and endophytes such as *Bacillus* spp., *Pseudomonas fluorescens*, and *Pasteuria penetrans* are among the most studied bacterial BCAs. They employ multiple modes of action: direct parasitism (e.g., *P. penetrans* attaching to J2 nematodes), antibiosis via secondary metabolites (e.g., hydrogen cyanide, lytic enzymes), competition for niche and nutrients, and induction of systemic resistance (ISR) in host plants. Multi-omics approaches have identified novel nematicidal metabolites such as dimethyl disulfide and 2-undecanone from *Bacillus atrophaeus*, which retain activity at varied temperature ranges and soil moisture levels (Ayaz et al. 2024). Yet, the interplay between elevated CO_2-driven root exudation changes and bacterial colonization patterns warrants further investigation to ensure consistent performance under future climate scenarios.

4.4.1 Major Antagonistic Bacteria

- *Bacillus* spp.: *Bacillus* species are gram-positive, endospore-forming bacteria widely studied for PPN control. They exert nematicidal effects via lipopeptide production (surfactins, fengycins, and iturins disrupt nematode cuticle integrity, causing cell lysis); Hydrolytic enzymes (Chitinases and proteases degrade eggshells and cuticles, reducing juvenile viability); Induced systemic resistance (ISR) [Activation of jasmonic acid (JA) and salicylic acid (SA) pathways

Table 4.2 Modes of action of fungal bioprotectants against phytonematodes (Askary and Martinelli 2015)

Bioprotectants	Mode of action
Aspergillus niger	It is an egg parasite and induces systemic resistance against plant-parasitic nematodes. The fungus coming in contact with a cyst or an egg mass begins to grow rapidly. It colonizes the eggs where larval formation has not been completed, thus providing early protection to the growing plants against nematodes
Purpureocillium lilacinum	It is mainly an egg parasite. The fungus produces antibiotics, viz., leucinostatin and lilacin and enzymes such as protease and chitinase. Protease has nematicidal activity, causes degradation of the eggshell, and inhibits hatching. Chitinase breaks down the eggshell making the route for the fungus to pass through. The decomposition of chitin releases ammonia, which is toxic to second-stage juveniles of root-knot nematode (RKN). Its hypha enters the vulva and anus of RKN females. The fungus penetrates the egg and develops profusely inside and over the eggs, completely inhibiting juvenile development. The infected eggs swell and buckle. As penetration continues, the vitelline layer of the egg splits into three bands and a large number of vacuoles; lipid layer disappears at this stage. The developing juvenile inside the egg is destroyed by the rapidly growing hyphae. Many conidiophores are produced and the hypha moves to the adjacent eggs
Trichoderma harzianum	Secretes many lytic enzymes like chitinase, glucanases, and proteases, which help parasitism of *Meloidogyne* and *Globodera* eggs. The chitin layer is dissolved through enzymatic activity. The hyphae of *T. harzianum* penetrate the eggs and juvenile cuticle, proliferate within the organism, and produce toxic metabolites
T. viride	Produce antibiotics like trichodermin, dermadin, trichoviridin, and sesquiterpene heptalic acid, which are involved in the suppression of nematodes
Pochonia chlamydosporia	Parasitizes the eggs and adult females of plant-parasitic nematodes. The root-knot and cyst nematodes are the primary hosts of this fungus, but it is also known to parasitize citrus, burrowing, and reniform nematodes. The fungus enters the nematode cysts either through natural openings or it may directly penetrate the wall of the cyst. It forms a branched mycelia network when in close contact with the smooth eggshell. The fungus produces an appressorium that adheres to the eggshell by mucigens and from which an infection peg develops and penetrates the eggshell. Penetration also occurs from lateral branches of the mycelium. This results in disintegration of the eggshell's vitelline layer and also partial dissolution of the chitin and lipid layers, possibly due to the activity of exoenzymes. Egg hatching is inhibited due to toxins secreted by the fungus

enhances plant defenses against nematode penetration] and Volatile organic compounds (VOCs) (Some strains produce VOCs that are toxic or repellent to nematodes) (Vasantha-Srinivasan et al. 2025).

- Notable species include:
  - *B. subtilis*: Demonstrated reductions of over 60% in *M. incognita* gall formation in tomato by combined lipopeptide and ISR mechanisms (Vasantha-Srinivasan et al. 2025). Multiplication rate of *Rotylenchulus reniformis* was

reduced significantly (44.5%) in tomato when soil was drenched with *B. subtilis* (isolate Bs_t) due to induced systemic resistance (Niknam and Dhawan 2001).

- *B. cereus*: Produces metalloproteinases (e.g., bacillolysin) and surfactins; field trials show significant yield improvements in various crops (Vasantha-Srinivasan et al. 2025).
- *B. amyloliquefaciens*/*B. velezensis*: Polyketides and lipopeptides effectively suppress nematode populations and promote beneficial rhizosphere communities (Vasantha-Srinivasan et al. 2025).
- *Pseudomonas fluorescens*: *P. fluorescens* is a well-studied plant growth-promoting rhizobacterium (PGPR). Its biocontrol efficacy against PPNs is attributed to:
  - *Antibiotic production*: Phenazines, pyoluteorin, and 2, 4-diacetylphloroglucinol are toxic to nematodes.
  - *Siderophore-mediated competition*: Iron chelation limits nutrient availability for nematodes (Antil et al. 2023).
  - *ISR induction*: Stimulates plant defenses through ACC deaminase activity, reducing ethylene and enhancing stress tolerance (Antil et al. 2023).

Greenhouse trials using seed coatings and soil drenches have reported reductions of 50–70% in root-knot nematode reproduction on tomato and cucumber crops (Antil et al. 2023). Application of *P. fluorescens* at 4 g/plant significantly reduced root galling (40%), egg mass production (60%), increased root colonization by *P. fluorescens,* and fruit yield in grapevine (166% over control) (Shanthi et al. 1998). DAPG producing *P. fluorescens* F113 is proposed as a potential biocontrol inoculant for the protection of potato crop against the potato cyst nematode, *Globodera rostochiensis*.

Pasteuria penetrans: *Pasteuria penetrans* is an obligate bacterial parasite that specifically targets root-knot nematodes:

- *Endospore adhesion*: Spores attach to nematode cuticles, germinate, and invade the host body cavity (Fig. 4.13).
- *Proliferation inside nematodes*: Leads to sterilization and death of juveniles and adult females.
- *Long-term suppression*: Persistence of spores in soil can provide multiyear control (Antil et al. 2023).

Field applications via soil amendment with infested root material have shown up to 80% reduction in *Meloidogyne* populations, although commercial adoption is limited by challenges in mass-producing infective spores. Weibelzahl-Fulton et al. (1996) showed that *M. incognita* and *M. javanica* were suppressed exclusively by *P. penetrans* in tobacco. *P. penetrans* offers potential for the biocontrol of *Tylenchulus semipenetrans* infecting citrus (Mani 1988). *P. penetrans* has been demonstrated to exert effective biocontrol of root-knot nematode on tomato (Rao et al. 1997). Black

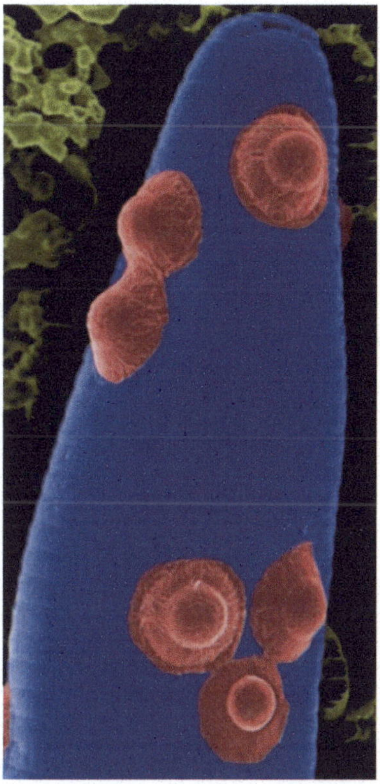

Fig. 4.13 Attachment of endospores of *Pasteuria penetrans* to cuticle of root-knot nematodes larva

pepper plants inoculated with *P. penetrans* showed reduction in *M. incognita* population and root gall indices and improved plant growth parameters (Sosamma and Koshy 1995).

4.4.2 Mechanisms of Action

Pasteuria penetrans attaches to nematode cuticles, penetrating and killing them. It is highly effective against *Meloidogyne* spp. but requires specific soil conditions. *Bacillus* spp. (e.g., *B. subtilis, B. firmus*) produce toxins and enzymes that disrupt nematode development. Field applications have reduced cyst nematode (*Heterodera* spp.) densities by 40–60%. Rhizobacteria like *P. fluorescens* promote plant growth and secrete metabolites toxic to nematodes. *Streptomyces* spp. produces nematicidal compounds like avermectins, effective against a broad range of nematodes.

Antagonistic bacteria suppress PPNs through two primary modes:

Direct Mechanisms

- *Extracellular Enzymes*: Bacteria produce enzymes like proteases, chitinases, collagenases, lipases, glucanases, cellulases, and pectinases that disrupt nematode exoskeletons and eggshells. For example, *Serratia marcescens* produces glucanases, cellulases, and pectinases to control *M. incognita.*
  - *Brevibacillus laterosporus* G4 penetrates nematode cuticles and digests them using proteases, as identified through reverse genetics studies.
- *Nematicidal Toxins and Secondary Metabolites*: *Bacillus thuringiensis* produces Cry proteins (e.g., Cry5, Cry6, Cry12, Cry13, Cry14, Cry21, Cry55) that damage nematode internal organs. Other examples include sphingosine from *B. cereus*, causing oxidative damage to *M. incognita*, and lipo-peptides like surfactin from *Bacillus* spp., inhibiting *M. incognita.*
  - *B. subtilis* OKB105 and *B. cereus* 09B18 show 95% nematicidal activity against *M. javanica* and *H. filipjevi* juveniles, while *S. plymuthica* M24T3 achieves 100% mortality against pinewood nematodes. These bacteria produce toxins, antibiotics, and volatile organic compounds (VOCs) that are lethal to nematodes.
- *Volatile Organic Compounds (VOCs)*: Bacteria like *B. atrophaeus* produce VOCs such as dimethyl disulfide (DMDS), 2-undecanone (2-UD), and methyl isovalerate (MIV), showing up to 90% nematicidal activity against *M. incognita. Virgibacillus dokdonensis* produces acetaldehyde, DMDS, ethylbenzene, and 2-butanone, affecting *M. incognita* J_2s via attraction, repellence, fumigation, and egg-hatching inhibition (Vasantha-Srinivasan et al. 2025).
- *Endoparasitism*: Bacteria like *Pasteuria penetrans* attach to nematode cuticles, penetrate, and destroy them, with ongoing genome sequencing to understand nematicidal activity. *P. penetrans*, *P. thornei*, and *P. nishizawae* parasitize 323 nematode species across 116 genera, including major economic PPNs. For instance, *P. penetrans* attaches to and infects nematodes, leading to their death.
- *Nutrient Competition and Niche Exclusion*: Rhizobacteria, such as *P. fluorescens* and *Bacillus* species (>29 genera, with *Bacillus* including >15 species and *Pseudomonas* >11 species), compete for nutrients and space, reducing nematode survival. This is evident in commercial bionematicides like Deny (*Burkholderia cepacia*), Bio Yield™ (*Paenobacillus macerans, B. amyloliquefaciens*), and BioNem (*B. firmus*, 3% lyophilized spores, 97% nontoxic additives).

Indirect Mechanisms

- *Induced Systemic Resistance (ISR)*: Plant Growth Promoting Rhizobacteria (PGPR) like *P. fluorescens, B. amyloliquefaciens*, and *B. cereus* induce ISR by regulating defense-related genes and producing compounds like indole acetic acid (IAA), siderophores, lipopolysaccharides, exopolysaccharides, N-acyl-homoserine lactones, and enzymes (ascorbate peroxidase, β-1,3-glucanase, chitinase, catalase, lipoxygenase, phenylalanine ammonia-lyase, polyphenol oxidase, superoxide dismutase). For example, *B. cereus* B1 and *P. fluorescens* P2 increased resistance to *H. trifolii* in white clover.

- *Competition for Nutrients and Space*: PGPR colonize roots, competing with nematodes for resources, reducing their populations.
- *Plant Growth Promotion*: Bacteria enhance root development and nutrient uptake, indirectly reducing nematode damage. This is supported by studies on rhizobacteria promoting plant health, such as reducing ethylene levels through ACC deaminase expression and altering nematode behavior (e.g., feeding, recognition, sex ratio).

Table 4.3 summarizes key bacterial groups, their target nematodes, and mechanisms.

The modes of action for common fungal and bacterial nematicides are summarized in Table 4.4.

4.4.3 Case Studies

- Tomato–*Meloidogyne incognita*: Combined application of *B. subtilis* and *P. fluorescens* reduced root galling by 75% and increased fruit yield by 20% in replicated field trials (Vasantha-Srinivasan et al. 2025).
- Soybean–*Heterodera glycines*: Foliar application of *B. velezensis* formulations decreased cyst populations by 60%, with concomitant increases in nodulation and nitrogen fixation (Vasantha-Srinivasan et al. 2025).
- Rice–*Pratylenchus zeae*: Soil amendment with *Streptomyces* spp. and *Bacillus* spp. synergistically suppressed lesion nematodes; improving root health and grain yield (Antil et al. 2023).

4.4.4 Specific Examples and Efficacy

- *B. subtilis* OKB105 and *B. cereus* 09B18 show up to 95% nematicidal activity against J2s of *M. javanica* and *H. filipjevi.*
- *P. fluorescens* F113 affects *G. rostochiensis* J2 mobility and egg hatchability.
- *B. firmus* GB-126 controls *R. reniformis* by reducing egg hatching and alleviating infestation in cotton (Castillo et al. 2013).

Table 4.3 Key bacterial groups, their target nematodes, and mechanisms

Bacterial group	Genus/Species	Target nematodes	Pathogenic effects	Action mode
Parasitic bacteria	*Pasteuria penetrans*, *P. thornei*, *P. nishizawae*, *Candidatus P. usgae*	323 nematode species of 116 genera, including major economic PPNs	Parasitize major economic PPNs	Parasitism
Opportunistic parasitic bacteria	*Bacillus nematocida* (B16), *Brevibacillus laterosporus*, *Bacillus* sp. *RH219*	*Panagrellus redivius*, *Bursaphelenchus xylophilus*	*Br. laterosporus* strain G4 penetrates cuticles and digests target organisms in lab	Parasitism, production of enzymes and toxins
Rhizobacteria	>29 genera, *Bacillus* (>15 species), *Pseudomonas* (>11 species)	Reduce nematode populations in soil	Different suppression degrees; three commercial bionematicides (deny, bio yield™, BioNem)	Interfering with recognition, toxin production, nutrient competition, plant-growth promotion, ISR
Parasporal crystal-forming bacteria	*Bacillus thuringiensis* (Cry5, Cry6, Cry12, Cry13, Cry14, Cry21)	*Trichostrongylus colubriformis*, *Caenorhabditis elegans*, *Nippostrongylus brasiliensis*	Cry proteins toxic to larval stages, damaging intestines	Cry proteins cause intestinal damage
Endophytic bacteria	Majority overlap with rhizobacteria	Root-knot nematode, root-lesion nematode, etc.	Suppress root-knot and root-lesion nematodes	Similar mechanisms to rhizobacteria (competition, inhibitory chemicals, ISR)
Symbionts of Entomopatho-genic nematodes	*Xenorhabdus*, *Photorhabdus*	*B. xylophilus*, *M. incognita* and eggs	Toxic to juveniles, inhibit egg hatch	Toxin production (ammonia, indole, stilbene derivative)

4.4.5 Nematicidal Compounds from Bacteria

The following Table 4.5 lists key nematicidal compounds, their bacterial sources, target nematodes, and mechanisms:

Table 4.4 Modes of action of bacterial bioprotectants against phytonematodes (Askary and Martinelli 2015)

Bioprotectants	Mode of action
Pasteuria penetrans	Bacterial spores are attached to the nematode's body and germinate forming a germ tube that penetrates the body cuticle. Vegetative mycelial colonies eventually fill the body with a large number of endospores
Pseudomonas fluorescens	Produce antibiotics, viz., phenazines, tropolone, pyrrolnitrin, pyocyanin, and 2, 4-diacetylphloroglucinol, which have suppressive effect on plant-parasitic nematodes
Bacillus firmus	Enzymatic action, degradation of root exudates, root-protection, and the production of a phytohormone
B. thuringiensis	Nematicidal toxins found in families of *B. thuringiensis* proteins
B. subtilis	The genes are encoding surfactin and iturin synthesis as antibiotics

Table 4.5 Key nematicidal compounds, their bacterial sources, target nematodes, and mechanisms of action

Compound	Source	Target nematode	Activity/Mechanism
Chitinases, proteases	Rhizobacteria (*Bacillus*)	Various PPNs	Hydrolyze exoskeleton, eggshell rupture
Cry proteins (Cry5, etc.)	*B. thuringiensis*	Various PPNs	Damage internal organs
Sphingosine	*B. cereus*	*M. incognita*	Causes oxidative damage
Thuringiensin	*B. thuringiensis*	*H. glycines*	Inhibits RNA production, like aldicarb
2-Undecanone, 2-Nonanone	*Bacillus* spp.	*B. xylophilus*	Destroys pharynx and intestine integrity
2,4-DAPG	*P. fluorescens*	*G. rostochiensis*	Reduces motility, inhibits egg hatching
Hydrogen cyanide	*Pseudomonas* spp.	*M. javanica*	Inhibits mitochondrial cytochrome oxidase
Avermectin B1 (Abamectin)	*Streptomyces avermitilis*	*B. xylophilus*, *M. incognita*	Disrupts glutamate-gated chloride channels
Milbemectin	*S. hygroscopicus*	Similar to Avermectin	Comparable mode of action
Spectinabilin	*S. spectinabilis*	*B. xylophilus*	LC50 0.84 mg/L
Spinosad	*Saccharopolyspora spinosa*	*M. incognita*	Disrupts GABA and nicotinic acetylcholine receptors

4.5 Predatory Nematodes

Biological control of plant-parasitic nematodes (PPNs) using predatory nematodes is an eco-friendly and sustainable approach to manage nematode-related crop damage, which cause significant crop losses worldwide. Among these, free-living predatory nematodes offer direct antagonism of PPNs in the soil food web. Predatory nematodes, which feed on other nematodes, offer a promising alternative due to their natural predatory behavior, environmental safety, and potential for integration into sustainable agriculture.

4.5.1 Taxonomy and Diversity of Predatory Nematodes

Predatory nematodes occur chiefly in the orders Mononchida, Dorylaimida, and Diplogasterida, classified by mouthpart morphology and prey-capture strategy:

- *Mononchida*: Genera *Mononchoides*, *Clarkus*, and *Fictor* possess well-developed, claw-like teeth for piercing PPN cuticles (en.wikipedia.org).
- *Dorylaimida*: Species such as *Dorylaimus* exhibit broad odontostyles enabling suction feeding on juvenile nematodes.
- *Diplogasterida*: *Fictor composticola* demonstrates both fungal and nematode predation, with opportunistic feeding on root-knot juveniles.

Common predatory nematode genera include *Mononchus, Clarkus, Mylonchulus*, and *Diplogaster*. These nematodes target PPNs like *Meloidogyne* (root-knot nematodes), *Pratylenchus*, and *Heterodera*.

4.5.2 Feeding Mechanisms

Predatory nematodes employ diverse strategies:

- *Piercing-sucking*: Sharp stylets or teeth to pierce PPN second-stage juveniles (J2), followed by extraction of cellular contents (Fig. 4.14).

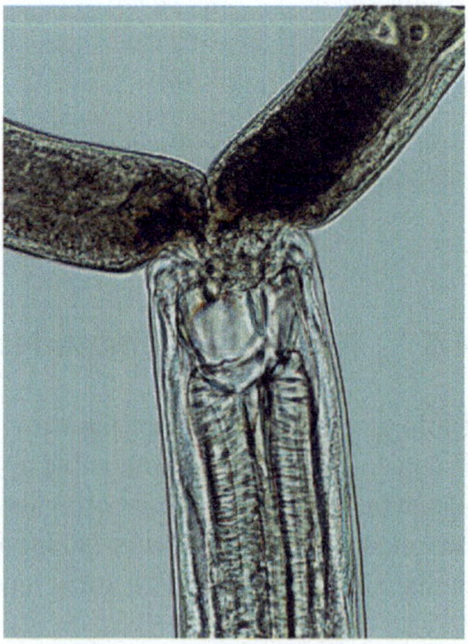

Fig. 4.14 Predatory nematode, *Mononchus* (Mononchida—Adenophorea)

- *Swallowing whole*: Smaller predators may ingest entire J2 or eggs.
- *Aggregation behavior*: Some mononchus aggregate around dense PPN populations, enhancing local predation pressure.

The efficacy of predation is influenced by environmental factors such as soil temperature, moisture, and organic matter content.

4.5.3 Nematode Management

Efficacy Studies demonstrate that predatory nematodes can significantly reduce PPN populations. Field and greenhouse experiments show reductions in PPN densities by 30–70% when predatory nematodes are applied, depending on species and environmental conditions. Specific examples include *Mononchus aquaticus* reducing *M. incognita* populations in tomato crops and *Odontopharynx longicaudata* suppressing *Tylenchorhynchus* species. Efficacy is enhanced when predatory nematodes are combined with organic amendments (e.g., compost, manure), which increase soil microbial activity and support predator populations.

4.5.4 Case Studies of Efficacy

- Mononchoides fortidens *against* Meloidogyne *spp.*: Pot trials demonstrated that pre-plant inoculation with *M. fortidens* at 1×10^4 nematodes/kg soil reduced gall indices by over 50% and enhanced plant biomass in tomato, with greater suppression at warmer temperatures (30–35 °C) (en.wikipedia.org).
- Fictor composticola *integrated with organic amendments*: In cucumber greenhouse trials, co-application of *F. composticola* with chicken manure, neem cake, and neem leaves lowered *M. incognita* populations by ~70%, outperforming amendment or predator alone (Sidhu and Kanwar 2020).
- *Other Predators*: Genera *Clarkus* and *Dorylaimus* spp. have shown moderate suppression of lesion nematodes in field micro-plots when combined with cover cropping, indicating potential for integrated approaches.

4.5.5 Integration with Agronomic Practices

Predatory nematodes perform best within integrated pest management (IPM) frameworks:

- *Organic amendments*: They supply microbial prey and habitat structure, boosting predator establishment.
- *Cover crops*: They increase soil moisture and organic matter, indirectly supporting predator populations.

- *Reduced tillage*: It preserves soil structure, minimizing disruption of nematode communities.

Such synergies can deliver durable PPN suppression while improving soil health.

4.5.6 Advantages and Limitations

Advantages

- Target specificity minimizes non-target effects.
- Compatibility with organic and low-input farming systems.
- Potential for self-sustaining populations under conducive conditions.
- Predatory nematodes are nontoxic, biodegradable, and do not leave harmful residues, unlike chemical nematicides.
- They primarily target PPNs, minimizing harm to non-target organisms.
- They can persist in soil ecosystems, providing long-term control when conditions are favorable.
- Compatible with integrated pest management (IPM) strategies, including crop rotation and biological amendments.

Limitations

- High initial production costs for mass culture.
- Sensitivity to abiotic stress (temperature extremes, desiccation).
- Variable field performance due to soil heterogeneity.

Addressing these limitations requires formulation improvements and ecological studies.

4.5.7 Mechanisms of Action

Predatory nematodes belong to various families, including Mononchidae, Dorylaimidae, and Diplogasteridae. They control PPN populations through direct predation or by altering soil microbial dynamics. Key mechanisms include:

- *Predation*: Predatory nematodes actively hunt PPNs, using specialized mouthparts (e.g., stylets or teeth) to puncture and consume their prey. For example, *Mononchus* species can consume multiple PPNs daily.
- *Competition*: Predatory nematodes compete with PPNs for resources, reducing their population density.
- *Indirect Effects*: Their presence may enhance soil microbial activity, indirectly suppressing PPNs by fostering antagonistic microorganisms.

4.6 Integrated Nematode Management (INM)

Combining biocontrol with cultural practices (e.g., crop rotation, organic amendments) enhances efficacy. For instance, incorporating compost with *Trichoderma* boosts suppression by improving soil microbial diversity.

In vegetables (e.g., tomato, cucumber), *P. chlamydosporia* and *B. subtilis* are effective against root-knot nematodes. In cereals, *P. penetrans* targets cyst nematodes like *H. avenae*. In perennials (e.g., banana), endophytic fungi like *F. oxysporum* reduce *Radopholus similis* damage.

4.7 Commercial Products

Pasteuria spp. and *Purpureocillium lilacinum* have been intensively researched as potential bionematicides for seed bio-priming. *B. firmus* and *Myrothecium verrucaria* are also the most widely used bionematicides (Table 4.6).

4.8 Challenges and Limitations

4.8.1 Environmental Conditions

Effectiveness varies with soil type, moisture, temperature, and nematode species. For example, *P. penetrans* requires sandy soils for optimal spore attachment. Efficacy is influenced by temperature, moisture, light, soil pH, nutrient content, and texture. High temperatures and low moisture can reduce fungal viability (Ur Rahman et al. 2024). Efficacy is inconsistent across soil types, pH ranges, and moisture levels, impacting fungal survival and virulence (Luo et al. 2004).

4.8.2 Slow Action

Biocontrol agents take longer to suppress nematodes compared to chemical nematicides, which can be a limitation in acute infestations. Fungi establish gradually, which may not align with the rapid control demands of high-value crops (Ur Rahman et al. 2023).

4.8.3 Application Barriers

High costs, limited shelf life, and need for precise application timing hinder commercial adoption.

Table 4.6 Key commercially available bioprotectants, target nematodes, crops, and applications rates

Active ingredient/Product name	Crop	Target nematodes/Application rate (spores/ha)
Bacillus firmus/VOTIVO	Maize	*Heterodera* spp. (1.5×10^{11})
	Cotton	*Hoplolaimus* spp. (2×10^{11})
	Sorghum	*Criconema* spp. (3.7×10^{10})
	Soybean	*Pratylenchus* spp. (1.24×10^{11})
	Sugar beet	*Meloidogyne* spp. (2×10^{10})
B. firmus/Nortica	Turf grass	*Belonolaimus* spp. (1.33×10^{14})
		Hoplolaimus spp.
Myrothecium verrucaria/DiTera WP	Bedding plants	*Heterodera* spp.
	Ornamentals	*Globodera* spp.
	Leafy vegetables	*Pratylenchus* spp.
	Cole crops	*Tylenchorhynchus* spp.
	Pome fruits	*Tylenchulus* semipenetrans
	Stone fruits	*Trichodorus* spp.
	Tree nuts	*Longidorus* spp.
	Grapevine	*Paratylenchus* spp.
	Pineapple	*Rotylenchulus* spp.
		Xiphinema spp.
		Belonolaimus spp.
		Criconemoides spp.
		Criconemella spp.
		Tylenchorhynchus spp.
		Hoplolaimus spp.
		Rotylenchus spp.
		Helicotylenchus spp.
		Radopholus spp.
Pastueria usage/Econem	Turf grass	*Belonolaimus longicaudatus* (9.8×10^{8}–9.8×10^{9})
Purpureocillium lilacinum/BioAct WG	Strawberries	*Radopholus similis* (2×10^{14}–1×10^{15})
P. lilacinum/MeloCon	Pineapple	*Heterodera* spp.
P. lilacinum/NemOut	Ornamentals	*Globodera* spp.
	Tobacco	*Rotylenchulus reniformis*
	Citrus	*Nacobbus* spp.
	Tree nuts	*Helicotylenchus* spp.
	Peaches	*Belonolaimus* spp.
	Grapevine	*Pratylenchus* spp.
	Banana	*Pratylenchus* spp.

4.8.4 Host Specificity

Some agents (e.g., *Pasteuria*) are species-specific, limiting their broad-spectrum use.

4.8.5 Nematode Diversity and Abundance

High diversity and abundance can challenge control, with variable results across different nematode species.

4.8.6 Crop Management Practices

Tillage, fertilization, and pesticide use impact fungal effectiveness, with heavy pesticide application reducing efficacy.

4.8.7 Production and Formulation Costs

Scale-up of fungal biopesticides involves complex fermentation and drying processes, leading to higher costs compared to chemical nematicides (Luo et al. 2004).

4.8.8 Regulatory Hurdles

Registration of microbial agents is complex and varies by region. Registration protocols for microbial agents are stringent and vary by region, delaying market approval (Luo et al. 2004).

4.9 Future Directions

Recent research highlights several advancements and areas for improvement:

4.9.1 Synergistic Effects

Combining multiple agents, such as *Trichoderma* and *Bacillus*, can achieve up to 80% nematode suppression in greenhouse trials, suggesting a synergistic approach for enhanced efficacy (Ayaz et al. 2024).

4.9.2 Organic Amendments

Incorporating organic matter like neem cake and compost enhances biocontrol efficacy by providing nutrients for antagonistic microbes, improving soil health (Kerry 1997).

4.9.3 Genomics and Omics

Advances in genomics are improving strain selection, with genetically enhanced *Bacillus* strains showing 20–30% higher efficacy. Metagenomics and transcriptomics are being used to discover novel nematicides, with ongoing genome sequencing of *Pasteuria penetrans* for understanding nematicidal activity (Kerry 1997).

4.9.4 Farmer Adoption

Farmers from India and Brazil report success with *Trichoderma*-based products, indicating growing interest, though scalability remains a concern.

4.9.5 Identification and Compatibility

Identifying effective bioagent and ensuring compatibility with other control strategies.

4.9.6 Scalable Production

Developing technologies for mass production remains urgent for commercial application (Siddiqui and Mahmood 1996).

4.9.7 Environmental Impact

Understanding long-term effects on soil health and ecosystems.

4.9.8 Quantification and Applicability

- Developing new methods for quantifying efficacy and applying findings across diverse agro-ecological systems.
- Advances in genomics, transcriptomics, and metabolomics engineering hold promise for strain improvement, targeting enhanced enzyme production, faster trap formation, and broader environmental tolerance.

- The integration of nematophagous fungi with other biocontrol agents (e.g., *Bacillus* spp.), plant elicitors (chitosan), or nanocarriers may yield synergistic effects and more consistent field performance (de Freitas Soares et al. 2023; Al-Ani et al. 2022).
- There is a need for integrated pest management and molecular studies to address these gaps (Ur Rahman et al. 2024).

4.10 Recommendations

4.10.1 Soil Health Management

Incorporate organic matter to support biocontrol agents, enhancing their activity, and persistence.

4.10.2 Site-Specific Strategies

Select agents based on nematode species, crop, and soil conditions. For example, use *P. lilacinum* in tropical soils for *Meloidogyne* control.

4.10.3 Integrated Approaches

Combine biocontrol with resistant cultivars and crop rotation for optimal results, leveraging multiple control methods.

4.10.4 Research Needs

Focus on improving formulation stability, reducing costs, and developing broad-spectrum agents to overcome current limitations.

This comprehensive analysis underscores the potential of biological control as a sustainable alternative to chemical nematicides, with ongoing research poised to address existing challenges and enhance adoption in agricultural systems.

4.11 Conclusion

Biological control of plant-parasitic nematodes using antagonistic fungi, bacteria, and nematodes offers a sustainable and environmentally friendly alternative to chemical nematicides, with significant potential demonstrated in recent studies. However, challenges in efficacy and scalability highlight the need for continued research to realize its full potential in agricultural systems.

Biological control agents represent a cornerstone of sustainable PPN management, offering diverse modes of action and environmental safety. While challenges in field consistency, cost, and regulation remain, ongoing research in formulation technology and fungal biology is poised to unlock their full potential, reducing reliance on chemical nematicides and promoting agricultural resilience.

4.12 Suggested Readings

- Askary TH, Martinelli PRP (2015) Biocontrol agents of phytonematodes. CAB International, Wallingford, 470 p.
- Stirling GR (2014) *Biological control of plant-parasitic nematodes: soil ecosystem management in sustainable agriculture* (2nd edn). CABI. (A standard reference focusing on biocontrol principles, mechanisms, and practices).
- Davies KG, Spiegel Y (2011) *Biological control of plant-parasitic nematodes: building coherence between microbial ecology and molecular mechanisms.* Springer. (Explores ecological and molecular aspects of nematode biocontrol).
- Kerry BR (2000) Rhizosphere interactions and the exploitation of microbial agents for the biological control of plant-parasitic nematodes. Ann Rev Phytopathol 38:423–441. (Landmark review on rhizosphere-mediated biological control).
- Abd-Elgawad MMM (2021) Biological control agents of plant-parasitic nematodes: a review. Egypt J Biol Pest Control 31:1–15. (Summarizes recent microbial biocontrol strategies with special focus on eco-friendly applications).
- Timper P (2014) Conserving and enhancing biological control of nematodes. J Nematol 46(2):75–89. (Modern perspectives on biological control within integrated nematode management).
- Mhatre PH, Karthik C, Kadirvelu K et al (2019) Plant growth promoting rhizobacteria (PGPR): a potential alternative tool for nematodes bio-control. Biocatal Agric Biotechnol 17:119–128.

4.13 Review Questions

- Define biological control in the context of plant-parasitic nematodes.
- What are the advantages of biological control over chemical nematicides?
- List important groups of microorganisms used in nematode biocontrol.
- How do nematophagous fungi suppress nematode populations?
- Explain the role of *Pasteuria penetrans* in nematode management.
- What is the significance of endophytic fungi and bacteria in nematode suppression?
- How do predatory nematodes contribute to nematode biocontrol?
- Explain the mechanism of induced systemic resistance (ISR) triggered by biocontrol agents.

- Give examples of commercially available biocontrol formulations against nematodes.
- How do biocontrol strategies fit into Integrated Nematode Management (INM)?
- Discuss case studies where biological control has successfully managed plant-parasitic nematodes.

References

Al-Ani LKT, Soares FEF, Sharma A et al (2022) Strategy of nematophagous fungi in determining the activity of plant parasitic nematodes and their prospective role in sustainable agriculture. Front. Fungal Biol. 3:863198. https://doi.org/10.3389/ffunb.2022.863198

Antil S, Kumar R, Pathak DV, Kumari A (2023) Recent advances in utilizing bacteria as bio-control agents against plant parasitic nematodes emphasizing *Meloidogyne* spp. Biol Control 183:105244

Askary TH, Martinelli PRP (2015) Biocontrol agents of phytonematodes. CAB International, Wallingford, p 470

Ayaz M, Zhao W, Chi YK et al (2024) Biocontrol of plant parasitic nematodes by bacteria and fungi: a multi-omics approach for the exploration of novel nematicides in sustainable agriculture. Front Microbiol 15:1433716. https://doi.org/10.3389/fmicb.2024.1433716

Castillo JD, Lawrence KS, Kloepper J (2013) Biocontrol of the reniform nematode by *Bacillus firmus* GB-126 and *Paecilomyces lilacinus* 251 on cotton. Plant Dis 97(7):967–976

de Freitas Soares FE, Ferreira JM, Genier HLA et al (2023) Biological control 2.0: use of nematophagous fungi enzymes for nematode control. J Nat Pestic Res 4:100025

Javeed MT, Al-Hazmi A (2015) Effect of *Trichoderma harzianum* on *Meloidogyne javanica* in tomatoes as influenced by time of the fungus introduction into soil. J Pure Appl Microbiol 9(1):535–539

Jones DL, Akhtar M (2019) Field evaluation of *Purpureocillium lilacinum* granular formulation against soybean cyst nematode. J Agric Sci 11(3):45–52

Kerry (1997) Biological control of nematodes: prospects and opportunities. In: Maqbool MA, Kerry B (eds) Plant nematode problems and their control in the near east region. FAO, Rome

Li X, Li Y, Li H (2018) Seed-coating with *Trichoderma harzianum* enhances cucumber growth and nematode resistance. Biol Control 126:80–87

Luo H, Mo MH, Huang XW et al (2004) *Coprinus comatus*: a basidiomycete fungus forms novel spiny structures and infects nematodes. Mycologia 96:1218–1225

Mani A (1988) Studies on the bacterial parasite *Pasteuria penetrans:* I. Spore viability after storage. II. Culture on citrus nematode *Tylenchulus semipenetrans*. Int Nematol Netw Newsl 5(2):24–25

Nemec S, Meridith FI (1981) Amino acid content of leaves in mycorrhizal and non-mycorrhizal citrus rootstocks. Ann Bot 47:351–358

Niknam GR, Dhawan SC (2001) Induction of systemic resistance by *Bacillus subtilis* isolate Bs1 against *Rotylenchulus reniformis* in tomato. In: National congress on centenary of nematology in India—appraisal & future plans. Indian Agricultural Research Institute, New Delhi, pp 143–144

Poveda J, Abril-Urias P, Escobar C (2020) Biological control of plant-parasitic nematodes by filamentous fungi inducers of resistance: *Trichoderma*, mycorrhizal and endophytic fungi. Front Microbiol 11:992

Prasad R, Sharma S, Singh U (2020) Efficacy of *Pochonia chlamydosporia* for managing root-knot nematode in tomato under Indian conditions. J Nematol 52:e2020

Rao MS, Gowen SR, Pembroke B, Parvatha Reddy P (1997) Relationship of *Pasteuria penetrans* spore encumbrance on juveniles of *Meloidogyne incognita* and their infection in adults. Nematol Medit 25:129–131

Roberts DP, Scott M, Lohrke S et al (2005) Biocontrol agents applied individually and in combination for suppression of soil-borne diseases of cucumber. Crop Prot 24:141–155

Shanthi A, Rajeswari-Sundarababu, Sivakumar CV, Mehta UK (1998) Soil application of Pseudomonas fluorescens (Migula) for the control of root knot nematode (Meloidogyne incognita) on grapevine (Vitis vinifera L.). Nematology: Challenges and opportunities in 21st Century, pp 203–206. Sugarcane Breeding Institute (ICAR), Coimbatore, India

Schouteden N, De Waele D, Panis B, Vos CM (2015) Arbuscular mycorrhizal fungi for the biocontrol of plant-parasitic nematodes: a review of the mechanisms involved. Front Microbiol 6:1280

Siddiqui ZA, Mahmood I (1996) Biological control of plant parasitic nematodes by fungi: a review. Bioresour Technol 58:229–239

Sidhu HS, Kanwar RS (2020) Integration of some organic amendments and the predatory nematode, *Fictor composticola*, for the management of *Meloidogyne incognita* in cucumber. Egypt J Biol Pest Control 30:146

Sosamma VK, Koshy PK (1995) Effect of *Pasteuria penetrans* and *Paecilomyces lilacinus* on population buildup of root-knot nematode, *Meloidogyne incognita* on black pepper. In: National symposium on nematode problems of India—An appraisal of the nematode management with eco-friendly approaches and biocomponents. Indian Agricultural Research Institute, New Delhi, p 47

Ur Rahman M, Chen P, Zhang X, Fan B (2023) Predacious strategies of nematophagous fungi as bio-control agents. Agronomy 13(11):2685. https://doi.org/10.3390/agronomy13112685

Ur Rahman M, Zhong X, Uzair M, Ben Fan B (2024) Application of fungi as biological control strategies for nematode management in horticultural crops. Phytopathol Res 6:38. https://doi.org/10.1186/s42483-024-00257-6

Vasantha-Srinivasan P, Park KB, Kim KY et al (2025) The role of *Bacillus* species in the management of plant-parasitic nematodes. Front Microbiol 15:1510036

Weibelzahl-Fulton E, Dickson DW, Whitty EB (1996) Suppression of *Meloidogyne incognita* and *M. javanica* by *Pasteuria penetrans* in field soil. J Nematol 28:43–49

Bio-intensive Integrated Nematode Management: An Eco-Friendly Strategy

5

Abstract

Bio-intensive integrated management of plant parasitic nematodes is a sustainable approach that combines multiple strategies to reduce reliance on harmful chemicals, focusing on biological, cultural, and physical methods. It employs nematophagous fungi, antagonistic bacteria, predatory nematodes, and resistant crops to reduce nematode damage while minimizing chemical use. The evidence leans toward integrating these methods for effective, environmentally friendly nematode management, though approaches may vary by crop and region.

Keywords

Ecofriendly nematode management · Physical methods · Cultural practices · Biological control · Resistant varieties · Proactive options · Reactive options

5.1 Learning Objectives

- Define the concept of bio-intensive integrated nematode management (BINM) and explain how it differs from conventional nematode management approaches.
- Understand the ecological and biological principles underlying BINM.
- Identify and describe the key components of BINM including cultural, biological, physical, botanical, and organic amendments.
- Explain the role of bio-control agents (fungi, bacteria, predatory nematodes) in nematode suppression.
- Evaluate the role of host resistance, crop rotation, organic amendments, cover crops, and trap crops in nematode management.
- Explain how enhancing soil health, biodiversity, and microbial activity contributes to nematode suppression.
- Understand how different management practices can be integrated for a synergistic effect.
- Demonstrate knowledge of successful case studies and field applications of BINM in major crops.

R. P. Parvatha, *Plant Parasitic Nematodes*,
https://doi.org/10.1007/978-981-95-9306-4_5

- Analyze nematode management problems in different cropping systems and propose BINM-based solutions.
- Evaluate constraints, challenges, and opportunities for wider adoption of BINM in integrated pest management (IPM) programs.
- Recognize the importance of BINM in climate-resilient and eco-friendly agriculture.
- Explore future trends, innovations, and research needs in the development of bio-intensive nematode management strategies.

5.2 Introduction

Plant parasitic nematodes (PPNs), such as *Meloidogyne* spp. (root-knot nematodes) and *Heterodera* spp. (cyst nematodes), pose significant threats to global agriculture, causing substantial economic losses. Research indicates an average worldwide crop loss of 12.3%, equating to $157 billion annually for the top 20 life-sustaining crops, based on 2010–2013 figures (Singh et al. 2015). To address this, bio-intensive integrated nematode management (BINM) offers a sustainable, environmentally friendly approach by integrating biological, cultural, physical, and, when necessary, chemical methods (Fig. 5.1). It combines multiple strategies to reduce reliance on harmful chemicals, focusing on biological, cultural, and physical methods. This chapter synthesizes findings from recent studies to provide a detailed overview of BINM strategies, their applications, and effectiveness.

Bio-intensive integrated nematode management (BINM) focuses on sustainable, eco-friendly strategies to reduce nematode populations and crop damage while minimizing reliance on chemical nematicides. BINM is a holistic, multi-tactic approach aiming to keep nematode damage below economic thresholds while minimizing environmental impact. It involves careful monitoring of nematode populations (soil assays, root health scouting) and using multiple compatible tactics synergistically. In practice, bio-intensive INM for PPNs emphasizes biological and cultural controls first, reserving chemicals only as a last resort.

An important difference between conventional and bio-intensive INM is that the emphasis of the latter is on proactive measures to redesign the agricultural ecosystem to the disadvantage of a nematode and to the advantage of its parasite and predator complex (Fig. 5.2).

5.3 Bio-intensive INM Principles

5.3.1 Objectives

The objectives of bio-intensive integrated nematode management (BINM) include:

- To minimize environmental and health hazards.
- Utilization of several compatible control measures.

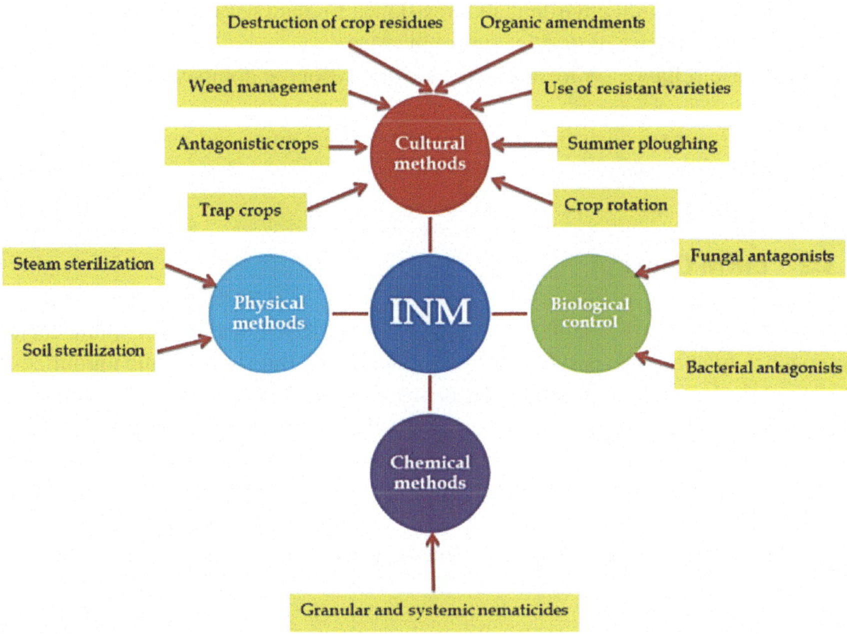

Fig. 5.1 Components of integrated nematode management

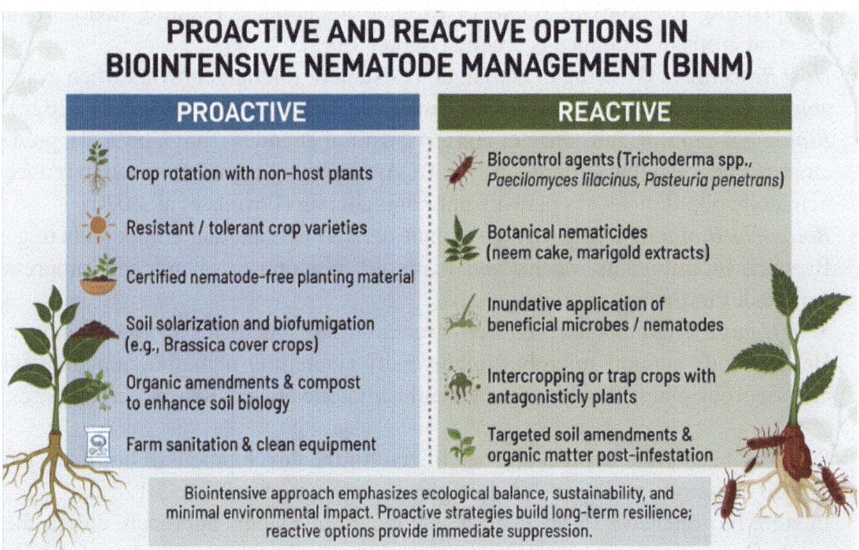

Fig. 5.2 Proactive and reactive options of bio-intensive integrated nematode management

- To maximize natural environmental resistance to plant parasitic nematodes.
- To minimize the use of drastic control measures.
- To increase reliance on location-specific and resource compatible management strategy.
- To minimize input costs in harmony with potential gains and maximize profit to the concerned grower.

5.3.2 BINM Concept

Bio-intensive Integrated Nematode Management (BINM) is a holistic, multi-tactic approach aiming to keep pest damage below economic thresholds while minimizing environmental impact (Barker 1992). It involves careful monitoring of nematode populations (soil assays, root health scouting) and using multiple compatible tactics synergistically (Barker 1992). As one classic statement put it, "a well-planned combination of practices will go much further for control of nematodes than any…treatment alone" (Barker 1992). In practice, bio-intensive INM emphasizes biological and cultural controls first, reserving chemicals only as needed. Key principles include:

- *Monitoring and Decision-Making*: Use of soil and root sampling to identify PPN species and densities, and apply control only when populations exceed damage thresholds.
- *Cultural Manipulations*: Crop rotation, fallow, and sanitation (clean equipment and planting material) to starve or exclude nematodes. Planting non-host or resistant crops in sequence is crucial (Barker 1992).
- *Host Resistance*: Growing of nematode-resistant or tolerant cultivars when available; these reduce nematode reproduction and protect yield (Krueger et al. 2022).
- *Biological Control*: Introduce or conserve natural enemies (fungi, bacteria, predatory nematodes) that suppress PPNs. BCAs are living organisms used to reduce nematode populations via parasitism or antagonism (Poveda et al. 2020).
- *Botanical/Biofumigant Agents*: Use plant-derived nematicidal compounds (e.g., Brassica biofumigants, neem, and marigold extracts) to chemically suppress nematodes in the soil.
- *Soil Health*: Maintain fertile, well-structured soils with high organic matter. Healthy soils support microbial biodiversity (including nematode antagonists) and vigorous plants that better withstand nematode attack (Krueger et al. 2022).

Some of the approaches that can be utilized for the development of bio-intensive integrated nematode management strategies are presented in Table 5.1.

In sum, bio-intensive INM for PPNs builds on a systems approach: enrich the soil and plant environment (via organic amendments, cover crops, etc.) so that PPN predators and antagonists thrive and nematode populations stay in check, combined with resistant varieties and targeted use of safer controls (Fig. 5.3). This maximizes crop health and yields sustainably (Poveda et al. 2020).

Table 5.1 Approaches that can be utilized for the development of bio-intensive integrated nematode management

Approaches	Bio-intensive INM practices
Regulatory methods	Seed certification (disease-free seeds and plants)
Physical methods	Hot water treatment of planting material, soil solarization
Cultural methods	Crop rotation, intercropping, strip cropping, fallowing, flooding, summer plowing, green manuring, trap crops, organic soil amendments, field sanitation, spacing of plants, altered planting dates, organic mulches
Chemical methods	Reduced-risk pesticides, site-specific treatment, nanotechnology
Biological methods	Nematophagous fungi, arbuscular mycorrhizal fungi, antagonistic bacteria, predatory nematodes
Host plant resistance	Resistant varieties, genetically modified crops, genome edited crops

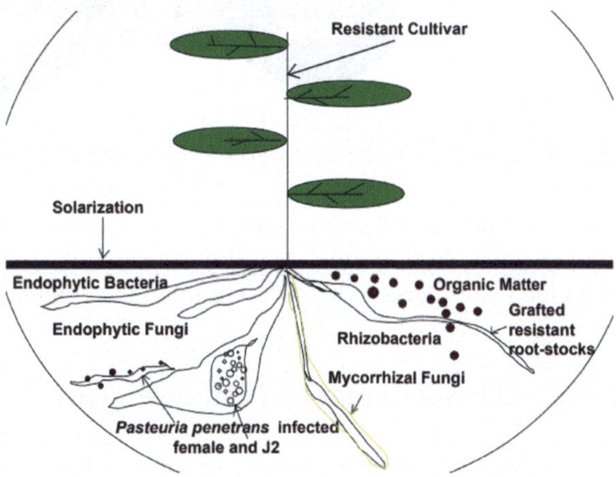

Fig. 5.3 BINM methods that can be used to manage plant nematodes

5.4 Pillars of Integrated Nematode Management

The five major pillars of INM used here are: crop rotation, cultivar choice and establishment, soil management, targeted control, and monitoring and evaluation (Fig. 5.4). Each pillar is constructed of numerous control tools that prevent, reduce populations, or enhance plant tolerance to nematode damage as well as provide tactical tools to make INM more effective.

5.4.1 Crop Rotation

Nematode problems are preventable by expanding the length of rotations with non-host crops. This is true only for nematode species with narrow host ranges such as

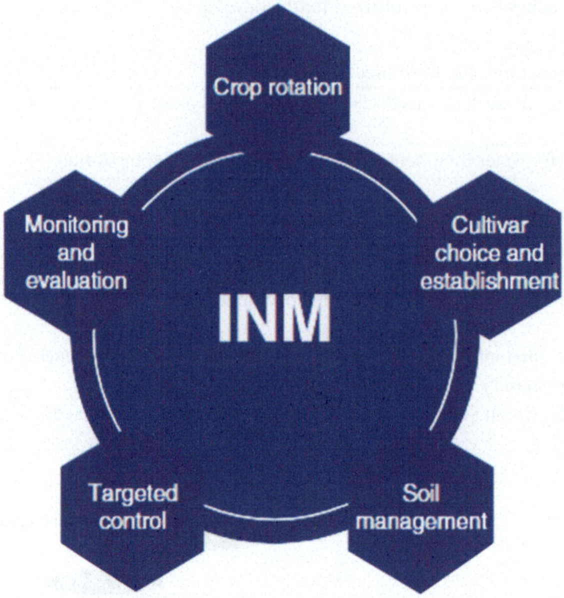

Fig. 5.4 Five pillars of integrated nematode management

cyst nematode species. For polyphagous species such as those of the genera *Meloidogyne, Rotylenchulus*, and *Pratylenchus*, it is not the frequency of cultivation of a crop that is decisive but the makeup of the cropping sequence.

5.4.2 Cultivar Choice and Establishment

Crop cultivars with resistance or tolerance traits are available and offer a good option to build an effective INM system. However, most breeders search for resistance in crops in a rotation with known nematode problems. They usually overlook potential resistance in other crops in the rotation sequence planted prior to the main susceptible crop.

5.4.3 Soil Management

Optimal growing conditions help crops to endure nematode damage. Therefore, proper soil management is an important pillar of INM. Nutrient and water management, proper tillage, and organic matter content are all basic conditions for a healthy crop. A strong plant is more resilient to nematode impacts.

5.4.4 Targeted Control

In the past, synthetic nematicides were the starting point of targeted control in INM. This view is no longer tenable when we aim to build a sustainable agricultural system. Targeted variable rate application within the field where necessary and use of nano-formulations are increasingly becoming the favored approach. Physical methods such as solarization, inundation, and anaerobic soil disinfestation can also be utilized to manage nematodes.

5.4.5 Monitoring and Evaluation

The monitoring and evaluation pillars are the first and final step in an INM system. Knowing what species are present at what infestation levels make it possible to begin to develop an INM plan and to check whether the results of INM are satisfactory or need restructuring. Historical information of a field is a good starting point. Once a nematode species has been introduced into a field, it will typically stay there below or above detection levels.

5.5 BINM Options

BINM options may be considered as proactive or reactive.

5.5.1 Proactive Options

BINM emphasizes proactive and sustainable methods to prevent nematode damage, integrating biological, cultural, and ecological approaches. Proactive options, such as crop rotations and creation of habitat for beneficial organisms, permanently lower the carrying capacity of the farm for the nematode. The carrying capacity is determined by the factors like food, shelter, natural enemy complex, and weather, which affect the reproduction and survival of a nematode species. Cultural control practices are generally considered to be proactive strategies. Proactive practices include crop rotation, resistant crop cultivars including transgenic plants, disease-free seeds and plants, crop sanitation, spacing of plants, altering planting dates, mulches, etc.

The proactive strategies (cultural controls) include:

- Healthy, biologically active soils (increasing below-ground diversity).
- Crop diversity with differences in nematode resistance (increasing above-ground biodiversity).
- Resistant crop cultivars.

5.5.2 Reactive Options

The reactive options mean that the grower responds to a situation, such as an economically damaging population of nematodes, with some type of short-term suppressive action. Reactive methods generally include inundate release of biological control agents, mechanical and physical controls, botanical pesticides, and chemical controls.

5.6 Proactive Options

The key proactive components form the backbone of a bio-intensive, integrated approach for managing plant-parasitic nematodes. Each element is designed to anticipate and prevent nematode outbreaks—rather than reacting once populations have already built up. BINM for PPNs is characterized by its preventive focus, aiming to reduce nematode populations and mitigate damage before significant infestations occur. The following components have been identified as central to this approach:

5.6.1 Preventive Application of BCAs

Applying BCAs early in the crop cycle or before nematode populations build up is a key proactive strategy:

- Early inoculation with fungi like *Esteya vermicola* and *E. floridanum* for pinewood nematode (*Bursaphelenchus xylophilus*) control has increased pine survival rates by 30–50% over 3–6 years.
- Combining BCAs like *Purpureocillium lilacinum* with *Bacillus subtilis* reduced *Pratylenchus penetrans* populations more effectively (reproduction factor Rf = 0.4) than single applications, outperforming some chemical controls.

This preventive approach ensures nematodes are managed before they cause significant economic damage.

5.6.2 Biofumigation

Biofumigation involves incorporating plant material into the soil to release nematicidal compounds, offering a nonchemical, proactive method. Recent studies emphasize:

- *Brassica Plants*: Species like mustard release isothiocyanates from glucosinolate hydrolysis, toxic to nematodes. This method is particularly effective in integrated nematode management (INM) practices, reducing PPN populations before planting.

- *Non-Brassica Plants*: Graminaceous plants like sorghum and Sudan grass produce nematicidal cyanides via enzymatic hydrolysis of cyanogenic glycosides/dhurrin. Marigold (*Tagetes* spp.) produces α-terthienyl, showing biofumigation potential against PPNs.
- *Application*: Mechanically chopped plant material is incorporated into the soil, creating an unfavorable environment for nematodes, aligning with BINM's preventive focus.

5.6.3 Cultural Practices

Cultural practices form another vital component, focusing on crop rotation, resistant cultivars, and cover crops. Replacing susceptible cash crops with resistant varieties, such as soybean cultivars resistant to both *Meloidogyne incognita* and *Heterodera glycines*, has been tested in projects involving eight cultivars at early and late planting dates in infested soil, and two in fumigated plots with 1, 3-dichloropropene (Westphal et al. 2006). Cover crops like "Boss" oilseed radish (*Raphanus sativus conv. oleiformis*) and canola (*Brassica napus*) with biofumigation potential have shown lower *M. incognita* juveniles in soil compared to susceptible "Siletina" and rye by May, with similar numbers at harvest (Westphal et al. 2006).

Cultural practices are proactive by disrupting nematode life cycles and reducing their habitat suitability:

- *Crop Rotation*: Alternating susceptible crops with non-hosts or resistant varieties disrupts nematode reproduction, reducing population densities over time. Rotation of susceptible crops with non-host or resistant crops helps to drive down nematode populations (Barker 1992). For instance, rotating potato after a grain or clover crop reduces potato cyst nematode build-up; rotating soybean with corn or oats helps manage soybean cyst nematode. Rotation length should be long enough for nematode decline (often >2 years for strong suppression).
- *Cover Crops and Green Manures*: Small grains can be replaced with cover crops resistant to nematodes or those with biofumigation potential, further enhancing prevention.
- Planting cover crops that suppress nematodes (e.g., marigold, sun hemp, Crotalaria, certain grasses/legumes) can reduce PPN densities. Some cover crops act as "trap crops" inducing nematode hatch without sustaining them, or release nematicidal compounds on decomposition. For example, Tagetes (marigold) roots produce thiophenes that are toxic to many nematodes, and brassica cover crops (e.g., mustard, radish, rapeseed) release isothiocyanates upon incorporation that fumigate the soil (Krueger et al. 2022).
- *Use of Resistant Varieties*: Planting crops genetically resistant to specific nematodes (e.g., resistant tomato varieties against *M. incognita*) prevents damage proactively.

- Crop varieties bred for nematode resistance or tolerance should be used. Resistant plants allow little nematode reproduction, so populations decline; tolerant plants sustain yield despite nematode presence. Many vegetable, fruit, and grain cultivars have resistance to particular PPNs (e.g., *Mi*-resistance in tomato and pepper for root-knot nematode, *Hs1pro-1* in sugar beet for cyst nematode, various cereals resistant to lesion nematodes). Integrating resistant varieties is a cornerstone of PPN IPM (Krueger et al. 2022).
- *Soil Solarization*: Heating soil with plastic mulch kills nematodes and other pathogens, preparing fields for planting with reduced nematode pressure.

5.6.4 Soil Health Management

Many IPM recommendations for PPNs focus on general soil health. Practices such as reduced tillage, cover cropping, balanced nutrient management, and proper irrigation improve soil structure and biota. This fosters free-living nematodes (bacterial- and fungal-feeders) and other beneficial organisms that compete with or prey on PPNs (Krueger et al. 2022). Well-drained, oxygenated soil also limits nematode survival (e.g., severe frosts kill surface nematodes). Overall, promoting healthy soil ecosystem—through organic amendments, crop diversity, and avoidance of toxic chemicals—naturally suppresses plant parasites and strengthens crops against nematode stress.

5.6.5 Integrated Use of BCAs with Other Inputs

No single tactic is sufficient on its own. By layering these proactive measures—rotations, resistant varieties, biofumigation, biological controls, cultural sanitation, and decision-support monitoring—a soil environment that is inhospitable to nematodes, limits their reproduction, and sustains crop health over multiple seasons can be created.

Proactive management often involves combining BCAs with other agricultural inputs to enhance efficacy, ensuring sustainability:

- *Compatibility with Chemicals*: BCAs like endospore-forming bacteria (*Bacillus*, *Clostridium*, *Pasteuria*) are tolerant to agrochemicals, with shelf lives exceeding 1 year, allowing tank-mixing or rotation with chemical nematicides. For example, *VOTiVO*™ (containing *B. firmus*) is combined with *PONCHO*® (a synthetic insecticide) as a seed treatment, delaying nematode resistance.
- *Synergy with Organic Amendments*: Combining *Pseudomonas fluorescens* GRP3 with organic manure increased tomato shoot dry weight ($P \leq 0.05$) more than standalone application (Siddiqui et al. 2001).
- *Combination with Cultural Practices*: Using BCAs alongside neem cake (e.g., *Purpureocillium lilacinum* with neem cake for *M. incognita* in chrysanthemum) enhances control, leveraging additive or synergistic effects.

Table 5.2 Key examples from recent research, highlighting proactive components and their impacts

Crop/Nematode	Proactive component	Application method	Impact	Reference
Tomato/*Meloidogyne* spp.	Bacterial BCA (*Bacillus velezensis* Bv-25)	Soil application	100% J2 mortality in 12 h, 98.6% infection rate reduction, 73.8% root-knots reduced	Pires et al. (2022)
Tomato/*Meloidogyne incognita*	Fungal BCA (*Trichoderma asperellum* T34)	Soil application	71% egg masses reduced, 54% eggs per plant reduced	Pires et al. (2022)
General/PPNs	Biofumigation (*brassica* sp.)	Incorporation of mustard residues	Releases isothiocyanates, reduces PPN populations	Dutta et al. (2019)
Tomato/*Meloidogyne* spp.	Combined BCA + organic amendment	*Pseudomonas fluorescens* with organic manure	Increased tomato shoot dry weight ($P \leq 0.05$) more than standalone	Siddiqui et al. (2001)
General/PPNs	Crop rotation	Rotating with non-host crops	Disrupts nematode life cycles, reduces population densities	Westphal (2011)

This integration ensures proactive management by creating a multifaceted defense against nematodes.

5.6.6 Quantitative Insights and Examples

To illustrate the effectiveness, the following Table 5.2 summarizes key examples from recent research, highlighting proactive components and their impacts:

5.6.7 Ecological and Practical Considerations

The effectiveness of these proactive components varies by soil type, climate, and crop system. For instance, bacterial BCAs like *Pasteuria* spp. show high host specificity, requiring careful selection for target nematodes. Fungal BCAs like *Trichoderma* spp. perform better in soils with adequate moisture, while biofumigation requires precise timing to maximize compound release. Integration with cultural practices enhances sustainability, but grower acceptance depends on research into biology, ecology, and interactions with other inputs, as highlighted in recent studies.

5.6.8 Conclusion

The proactive components of bio-intensive integrated management of PPNs encompass a range of biological and cultural strategies, including biocontrol agents (bacteria and fungi), biofumigation, integrated use with other inputs, cultural practices, and preventive applications. These methods, supported by recent research, offer a sustainable, environmentally friendly approach to nematode management, with specific examples demonstrating significant reductions in nematode damage.

5.7 Reactive Options

5.7.1 Targeted Biocontrol Agents (BCAs)

Biological control is a cornerstone of BINM, leveraging microorganisms antagonistic to nematodes. This involves using microorganisms like fungi (*Pochonia chlamydosporia, P. lilacinum*) and bacteria (*P. fluorescens, B. firmus*) that naturally combat nematodes. These bio-nematicides can be used alone or with other methods for better results and are highlighted for their efficacy. For instance, *B. firmus* is a component of commercial products like VOTiVO™ and PONCHO®/VOTiVO™ mix, used as seed treatments (BASF). These bio-nematicides act through mechanisms such as nematode-trapping, endoparasitism, and toxin production, with endospore-forming bacteria offering advantages like heat resistance and prolonged shelf life (>1 year), enabling tank-mixing with fertilizers and pesticides.

Research priorities include understanding the ecology, biology, and mode of action of these agents, as well as their interactions with other agricultural inputs. These bio-nematicides can be used alone or with other methods for better results. For example, *P. fluorescens* GRP3 combined with organic manure increased tomato shoot dry weight ($P \leq 0.05$) for *M. incognita* management (Westphal 2011). Compatibility with chemical pesticides, surfactants, and soil amendments is crucial for integration into INM programs, as outlined in detailed studies (Eissa and Abd-Elgawad 2015).

A wide array of natural enemies can suppress PPNs (Poveda et al. 2020). Common BCAs include:

- *Nematophagous fungi*: *P. chlamydosporia, P. lilacinum* (formerly *Paecilomyces*), *Arthrobotrys* spp., *Trichoderma harzianum*, mycorrhizal, and endophytic fungi. These attach to or parasitize nematode eggs and juveniles, produce lytic enzymes or toxins, compete for space/nutrients, and can induce plant defenses (Poveda et al. 2020). For example, filamentous fungi not only kill nematodes directly (parasitism, antibiosis) but also improve plant nutrition and root morphology, bolstering plant resistance (Poveda et al. 2020). Examples include the following:
  - *Pochonia chlamydosporia* reduces *Meloidogyne* spp. infection by 32–43%, reproduction by 44–59%, and female fecundity by 14.7–27.6%, while inducing salicylic acid (SA) and jasmonic acid (JA) pathways.

  - *Trichoderma asperellum* T34 reduces egg masses by 71% and eggs per plant by 54% in tomato.
  - *Metarrhizium anisopliae* ITCC9014 parasitizes 97 ± 2% of J2 nematodes, reducing the nematode multiplication factor by 82%.
  - *Glomus etunicatum* decreases *H. glycines* females by 28%, enhancing plant tolerance.
- *Bacterial bioagents*: Beneficial bacteria include *Pasteuria penetrans* (an obligate parasite of PPNs), *Bacillus* spp. (e.g., *B. thuringiensis*, *B. amyloliquefaciens*), *P. fluorescens*, *Xenorhabdus*, *Serratia*, *Streptomyces*, etc. These can attach to nematode cuticles, produce nematicidal toxins, or colonize roots to stimulate induced resistance. Many successful BCAs (e.g. *Bacillus*, *Pseudomonas*, *Purpureocillium*) have been shown to reduce egg hatch and kill juvenile nematodes (Ayaz et al. 2024). Examples include the following:
  - *Bacillus firmus* I-1582 degrades nematode eggshells and induces systemic resistance in crops like tomato, protecting against *Meloidogyne* spp.
  - *B. velezensis* Bv-25 achieves 100% mortality of second-stage juveniles (J2) within 12 h, reducing infection rates by 98.6%, root-knots by 73.8%, and disease index by 61.6%, while increasing plant height by 14.4% and yield by 36.5%.
  - *Pasteuria penetrans* shows host specificity changes, reducing root-knot nematode populations by 96–99%.
  - *Pseudomonas fluorescens* CHA0 reduces disease severity more effectively in tomato than cucumber, highlighting host-specific efficacy.
- *Predatory nematodes*: Certain free-living nematodes (order Mononchida and others) feed on plant-parasitic species and help regulate their numbers. Though less commonly applied, they can play a role in suppressing PPNs and enhancing soil nutrient cycling (Poveda et al. 2020).
- *Examples*: Commercial bioproducts based on BCAs (e.g. *Bacillus subtilis*, *Pochonia*, *Trichoderma* formulations) are increasingly used in organic and IPM systems to reduce nematode populations with minimal residues.

These BCAs act through parasitism, nematicidal compound production (e.g., volatile organic compounds, toxins), and systemic resistance induction, making them proactive by preempting nematode establishment.

5.7.2 Physical Methods

Physical methods such as soil solarization help suppress nematodes. Soil solarization by covering infested beds with transparent polyethylene during hottest months, inactivate nematodes in detected hotspots. Hot water or steam treatment of transplant media or tools is recommended when nematodes are confirmed in potting mixes or equipment.

5.7.3 Cultural Sanitation

Reduced tillage has decreased *H. glycines* population densities in corn-soybean rotation, reducing foliar sudden death syndrome (SDS) symptoms, unlike in soybean monoculture where densities remained high (Barker 1992). Postharvest plowing or disking can result in a 90% or greater reduction of nematode populations, particularly effective in regions with high temperatures like the Near East (FAO). Heavily infested crop residues (roots and above-ground parts) should be removed and destroyed rather than incorporated into the soil.

5.7.4 Organic Amendments

Incorporating organic matter (composts, manures, crop residues) improves soil microbial activity and fertility. Decomposition products (e.g., ammonia, organic acids) can directly harm nematodes. More importantly, organics feed the soil food web, boosting populations of nematode predators and antagonists (Krueger et al. 2022). Well-amended soils also support healthy plant roots that better tolerate nematode feeding. Organic mulches and fallow periods can also reduce PPNs.

5.7.5 Botanical Nematicides

Plant-derived biochemicals can suppress nematodes. Broad examples include neem (azadirachtin), neem cake, castor bean derivatives, and essential oils (e.g., from thyme, garlic). A well-studied approach is biofumigation: growing brassica (mustard, canola, arugula) or certain weeds, then incorporating the biomass into the soil. The breakdown of glucosinolates in brassicas produces isothiocyanates (natural fumigants) that can markedly reduce nematode populations. For example, mustard (*Brassica juncea*) shoots turned under before planting have been shown to release nematicidal compounds suppressing potato cyst nematode (Abd-Elgawad 2021). Marigold root extracts and ground sun hemp or velvet bean residues are also used as natural nematicides. These botanical amendments are compatible with organic farming and can be integrated with other tactics.

5.7.6 Less Toxic Chemical Treatments (as Last Resort)

Use of low-toxicity or "bio-nematicides" (e.g., fluensulfone, oxamyl) applied as in-furrow or soil drench only when monitoring indicates severe infestations and other measures are insufficient. Precise spot treatments can be used to minimize off-target impacts.

5.7.7 Integrated Practices

While BINM aims to minimize chemical use, environmentally friendly nematicides may be integrated, especially non-fumigants like fluazaindolizine, showing promise in field trials against root-knot nematodes (Paez 2023). Future research is needed on the biology, ecology, and interaction of bio-nematicides with other inputs, using genes and metabolites to reduce PPN effects and promote plant responses (Wilson and Jackson 2013).

Combining these methods, like using *Pasteuria penetrans*, *Purpureocillium lilacinum*, and neem cake for tomato crops enhances effectiveness. Arbuscular mycorrhizal fungi in transplant potting mixes can improve plant growth in nematode-infested soils, supporting sustainable agriculture.

The BINM practices against plant-parasitic nematodes are presented in Table 5.3.

5.8 Case Studies and Specific Applications

Several studies demonstrate how integrated, bio-intensive tactics can markedly improve nematode management in real crops. For instance, integrating *Pasteuria penetrans*, *Purpureocillium lilacinum*, and neem cake managed root-knot nematodes in tomato (Parvatha Reddy et al. 1997). These studies underscore the versatility of BINM across crops like tomato, bell pepper, eggplant, and medicinal Coleus (Rao et al. 2004; Seenivasan and Devrajan 2008; Singh 2013).

5.8.1 Potato (Egypt)

A field study on nematode-prone potato combined a *Bacillus subtilis*–based bionematicide (Stanes® Sting) with biofertilizers (e.g., phosphorus and potassium solubilizers). This integrated treatment suppressed root-lesion nematodes (*Pratylenchus* spp.) and increased tuber yield by ~30% over the untreated control (Abd-Elgawad 2021).

5.8.2 Tomato/Potato (GH/In-Field)

In greenhouse trials (Siddiqui et al. 2001), tomato plants infested with root-knot nematode (*Meloidogyne*) were grown with organic manure alone, a *Pseudomonas fluorescens* inoculant alone, or both together. The combined treatment produced significantly higher shoot biomass than either treatment alone, indicating a synergistic effect of PGPR with organic amendment (Abd-Elgawad 2021).

Table 5.3 Best bio-intensive IPM practices against plant-parasitic nematodes, with examples and outcomes

Strategy	Tactics/Examples	Effect on PPNs/Plants	References
Biological control	Use of BCAs such as *Pochonia chlamydosporia*, *Trichoderma*, *Bacillus*, *Pasteuria*, predatory nematodes, etc.	Directly parasitize or kill nematodes; reduce egg hatch and juvenile survival; induce plant resistance; enhance root growth. Improves PPN suppression with minimal residue	Poveda et al. (2020)
Crop rotation	Rotate to non-host/resistant crops (e.g., cereals or clover after potato; corn/rye after soybean)	Starves specific nematodes by depriving them of hosts, causing population crash. Can reduce PPNs below damage thresholds	Barker (1992)
Cover crops	Nematode-suppressive plants (e.g., marigold, sun hemp, velvet bean, mustard greens)	Trap or kill PPNs via bioactive compounds (thiophenes, cyanides, etc.); adds organic matter. Reduces RKN and lesion nematodes significantly in experiments	Krueger et al. (2022)
Organic amendments	Compost, manure, biochar, green manure incorporation	Stimulates soil food web and nematode-antagonistic microbes; may release nematicidal compounds on decomposition. Often improves crop vigor and indirectly reduces PPN damage	Krueger et al. (2022)
Resistant cultivars	Planting nematode-resistant/tolerant varieties (e.g., RKN-resistant tomato, SCN-resistant soybean)	Limits nematode reproduction and disease; allows near-normal yields despite nematode presence. Essential for managing high-PPN fields	Krueger et al. (2022)
Botanical nematicides + biofumigation	Use of neem products, essential oils, and brassica cover crops with glucosinolate content	Releases natural nematicides (azadirachtin, isothiocyanates) in the soil that kill or repel PPNs. Shown to significantly lower nematode counts in trials (e.g., mustard biofumigation against cyst nematodes)	Abd-Elgawad (2021)
Soil health management	Maintain organic matter, balanced fertilization, proper pH, moisture; avoid soil compaction	Promotes beneficial microbes and free-living nematodes that compete with PPNs; improves plant root growth. Healthier soil reduces PPN impacts over time	Krueger et al. (2022)

5.8.3 Potato (India)

A field trial integrated multiple biocontrol agents with a reduced-rate chemical nematicide. Seed potatoes were treated with *Pochonia chlamydosporia*, *Pseudomonas fluorescens*, and *Trichoderma viride*, combined with a low dose of carbofuran. This consortium outperformed any single treatment, giving the highest tuber number and weight and the greatest reduction in potato cyst nematode eggs and juveniles (Abd-Elgawad 2021). This exemplifies "stacked" IPM: biologicals + minimal pesticide yield maximal control.

5.8.4 Rice White Tip Nematode

Using farmer-saved seeds, brine treatment (20% NaCl), hot water treatment (55 °C for 10 minutes), and application of carbofuran (Furadan 3G) in the main field during final land preparation can greatly minimize yield losses caused by the seed-borne nematode Aphelenchoides besseyi in the commonly cultivated rice cultivars BR11 and Nizersial (Islam et al. 2015).

5.8.5 Black Pepper Foot Rot and Nematodes Disease Complex

Integrated management of foot rot (*Phytophthora capsicii*) and nematodes (*Meloidogyne incognita* and *Radopholus similis*) on black pepper was achieved by:

- Mixing AMF and *Trichoderma harzianum* in solarized nursery mixture to raise healthy and robust seedlings.
- Application of *T. harzianum* and FYM in planting pit.
- Field application of Neem cake at 1 kg/vine mixed with 50 g of *T. harzianum* during August (Sarma 2003).

5.8.6 Banana Burrowing Nematode

Integration of neem cake at 200 g/plant with *Glomus mosseae* at 100 g/plant (containing 25–30 chlamydospores/g of inoculum) was most effective in reducing the *Radopholus similis* population both in soil and roots, while karanj cake with *G. mosseae* gave maximum increase in fruit yield of banana. Mycorrhizal root colonization and number of chlamydospores of *G. mosseae* were highest in neem cake–amended soil (Parvatha Reddy et al. 2002).

5.8.7 Potato Rot Nematode

The control of potato rot nematode, *Ditylenchus destructor* was achieved by the combination of disease escape; hygiene, in the form of seed certification and crop rotation (Winslow and Willis 1968).

- Healthy seed, planted late, harvested early and stored as cool and dry as possible.
- Proper rotation-potatoes not more frequently than once in 3 or 4 years and avoid other susceptible crops.
- Field hygiene (removal of old infested tubers and weed control).

5.8.8 Tomato Root-Knot Nematode

Proper site selection and preparation for transplant productions are essential components of INM. Fallowing, crop rotation, cover crops, and nematicides can be used. The transplants are inspected at the site of origin for root-knot nematode galls. Certification and inspection should be followed. *Meloidogyne* spp. infections that avoid detection are usually below the damage thresholds (Bird 1981).

New variations on these involve combinations of cultural (deep plowing to a depth of 20 cm) and chemical methods (nursery bed treatment with Carbofuran at 0.4 g a.i. per m^2 and main field treatment with Carbofuran at 1 kg a.i. per ha) for the control of root-knot nematodes in tomato, which also registered maximum yield.

Red gram (*Cajanus cajan*) and *Crotalaria spectabilis* were used in conjunction with clean fallow and a nematicide (Fensulfothion) for managing root-knot nematode populations in tomato transplants.

In nursery, integration of *P. penetrans* (at 28×10^4 spores/m^2), *P. lilacinum* (at 10 g/m^2 with 19×10^9 spores/g) and neem cake (at 0.5 kg/m^2) gave maximum increase in plant growth and number of seedlings/bed. Parasitization of *M. incognita* females was highest when neem cake was integrated with *P. penetrans*, while parasitization of eggs was highest when neem cake was integrated with *P. lilacinum*. In field, planting of tomato seedlings (raised in nursery beds amended with neem cake + *P. penetrans*) in pits incorporated with *P. lilacinum* (at 0.5 g/plant) gave least root galling and nematode multiplication rate and increased fruit weight and yield of tomato (Parvatha Reddy et al. 1997).

5.8.9 Brinjal Root-Knot Nematode

Deep plowing (20 cm) together with spot application of Carbofuran at 1.0 kg a.i. per ha increased growth of brinjal plants and significantly reduced root-knot nematodes.

Nursery treatment with Metham sodium at 25 ml per sq. m. along with deep plowing (20 cm) in the main field gave significant reduction in root-knot nematode population and better brinjal yields.

5.8.10 Okra Root-Knot and Reniform Nematodes

Deep plowing (20 cm) together with seed treatment (Carbofuran at 3% w/w) was found to be effective for the control of root-knot nematodes in okra.

Deep plowing (20 cm) followed by fallowing for 1 month, fallowing for 1 month after weeding, integration of Carbofuran application at 1 kg a. i. per ha at sowing after either of the cultural practices or deep plowing (20 cm) together with Carbofuran seed treatment resulted in the control of the reniform nematode (*Rotylenchulus reniformis*) and better yield of okra (Lakshmanan and Sivakumar 1981).

5.8.11 Sweet Potato Root-Knot Nematode

Control of root-knot damage to sweet potatoes involves integration of at least three methods:

- The selection of seed roots, which are either free of nematodes or freed of nematodes by hot water treatment.
- Planting of the seed roots into beds of sand or coarse textured soil that is either nematode—free or, if infested, is pre-plant fumigated.
- Transplanting of the clean "slips" into nematode-free soil or pre-plant fumigated soil.

5.8.12 Carrot Root-Knot Nematode

Meloidogyne hapla is a severe pathogen of carrots. The damage threshold is very low and infection by one *M. hapla* can make a carrot unsuitable for the fresh market. A highly commercial INM system has been evolved using a 5-year rotation in conjunction with a cover crop, soil fumigation, and tolerant cultivars. A soil fumigant is applied the fall before the first of two consecutive carrot crops, followed by 2 years of onion production and 1 year of the cover crop Sudex. Onions are relatively poor hosts for *M. hapla* and have a high economic threshold. In fields where nematode problems are detected after the first growing season, a low rate of a soil fumigant may be applied before the second carrot crop, or the rotation design may be altered to delete the second carrot crop. Tolerance to *M. hapla* exists in commercial cultivars of carrots (Bird 1981).

5.8.13 Tobacco Root-Knot Nematode

The development of an integrated program for the control of *Meloidogyne incognita* in tobacco has been of immense value to the state of North Carolina (Taylor and Sasser 1978). On problem fields where the average yearly income without control is \$1047 per hectare:

- Sanitary procedures, such as plowing up tobacco roots at the end of the season, add $711.
- Rotation with resistant crops, such as groundnut or cotton, adds $1104.
- Use of resistant tobacco varieties, such as NC 95, adds $939.
- Use of nematicides adds another $1378.

The total increase is $5179 or almost five times as much as without nematode and disease control.

5.8.14 Strawberry Lesion Nematode

When a soil is heavily infested with *Pratylenchus penetrans,* intelligent combination of nematicides and crop rotation is desirable. Soil application of nematicides may reduce the nematode population to a very low level. An advisable cropping sequence could then be:

- Strawberry (which is susceptible but does not cause a large increase of the residual nematode population).
- Potato (moderately susceptible and causing moderate nematode reproduction).
- Oats or barley (not susceptible but causing nematode reproduction).
- Beet (not susceptible and suppressing the population).

After the above cropping sequence, moderately susceptible crops can be grown again successfully.

5.8.15 Additional Examples

In a US study, integrating cover crops and organic mulches in strawberry reduced root-knot and sting nematodes, improving yields compared to monoculture (Krueger et al. 2022). In Florida tomato fields, marigold cover crops cut root-knot nematode populations by ~60% (detailed field data vary by region and crop, but generally support that well-designed bio-intensive IPM boosts yields and suppresses nematodes).

It is seen from the above discussion that the integration of the different methods offers effective control possibilities which are of great practical value in the suppression of harmful nematode population densities as well in preventing the build-up of measurable populations of nematodes.

The following Table 5.4 summarizes key studies on BINM, highlighting methods, crops, and references:

This table encapsulates specific applications, demonstrating the adaptability of BINM across various crops and conditions.

Table 5.4 Integrated management of root-knot nematodes

Crop	INM components	Outcome/References
Potato (Egypt)	*Bacillus subtilis* biopesticide (Stanes® sting) + biofertilizers	+30% tuber yield vs. untreated; strong nematode suppression (Abd-Elgawad 2021)
Tomato (India)	*Pseudomonas fluorescens* PGPR + organic manure (vs. either alone)	Significantly greater plant biomass and root gall reduction (Abd-Elgawad 2021)
Potato (India)	Combined BCAs (*Pochonia*, *Pseudomonas*, *Trichoderma*) + low-dose carbofuran	Highest tuber number/weight and nematode reduction vs. single tactics (Abd-Elgawad 2021)
Strawberry (USA)	Legume cover crop + compost mulch	Markedly lower sting and root-knot nematodes; increased yield
Natural field soil	Agrochemicals, organic matter, *Purpureocillium lilacinum*	Oduor-Owino (2003)
Tomato	Integration of *Pasteuria penetrans*, *P. lilacinum*, neem cake	Parvatha Reddy et al. (1997)
Bell pepper	*Pochonia chlamydosporia*, *Pseudomonas fluorescens*	Rao et al. (2004)
Medicinal coleus	Not specified in detail	Seenivasan and Devrajan (2008)
Eggplant	Not specified in detail	Singh (2013)

5.9 Challenges and Gaps

5.9.1 Field Efficacy Variability

Many promising BCAs perform well under controlled conditions but show inconsistent results in diverse soils and climates (Ayaz et al. 2024). Soil pH, temperature, moisture, and texture can affect survival and activity of biocontrol agents (Ayaz et al. 2024). Finding BCAs that reliably work across environments and multiple PPN species is difficult.

5.9.2 Formulation and Adoption Issues

Biocontrol products often face practical hurdles: high production costs, short shelf life, and lack of efficient delivery systems limit their use (Ayaz et al. 2024). Regulatory approval and registration for microbial nematicides can be costly and time-consuming (Ayaz et al. 2024). Farmers accustomed to fast-acting nematicides may hesitate to adopt slower-acting BCAs; lack of awareness and extension support further hampers uptake.

5.9.3 Knowledge Gaps

PPN agro-ecology is complex. Our understanding of nematode–plant–microbe interactions (especially in the rhizosphere microbiome) remains incomplete. Advanced "omics" (genomics, metabolomics, microbiome analysis) are needed to unravel how BCAs work and to discover novel bioactive compounds (Ayaz et al. 2024). There is also a need for more field trials in diverse cropping systems (tropical, organic farms, etc.) and for defining economic thresholds of damage for mixed-species nematode complexes.

5.9.4 Climate Change and Emerging Pests

Changing climates may shift nematode distributions and pressure. Management strategies must adapt to such shifts (e.g., develop heat-tolerant BCAs, revised crop rotations) (Ayaz et al. 2024). New PPNs may invade areas (e.g., subtropical nematodes moving poleward), requiring vigilant monitoring.

5.10 Future Directions

Research is pushing toward:

5.10.1 Next-Generation BCAs

Using biotechnology to enhance biocontrol strains (engineered bacteria or fungi with stronger nematicidal traits) (Ayaz et al. 2024).

5.10.2 Microbiome Engineering

Shaping soil microbial communities to favor nematode suppression (e.g., custom probiotic consortia) (Ayaz et al. 2024).

5.10.3 Precision Agriculture Tools

Applying drones and soil sensors to map nematode hotspots and deploy site-specific controls (Ayaz et al. 2024).

5.10.4 Farmer Outreach and Policy

Educate stakeholders on safe, bio-intensive INM, and incentivizing adoption. Overall, combining traditional agronomy with cutting-edge science (genomics, big data), offers a path for more effective and sustainable nematode management.

5.11 Conclusion

Bio-intensive integrated management of plant parasitic nematodes is a multifaceted strategy that integrates biological control, cultural practices, and physical methods to manage nematode damage sustainably. It addresses significant economic losses while promoting environmentally friendly agriculture, with on-going research enhancing its efficacy and applicability.

Research opportunities remain in characterizing nematode communities and in the assessment of some management tactics as well as in developing related damage functions/thresholds and the integration of compatible strategies/tactics. Components for this integration include crop rotation/cropping systems, resistant cultivars, fallow/destruction of residual infected roots, solarization, less toxic nematicides, and soil amendments/biological control. Carefully developed research-extension priorities, and the use of some type of data/decision-support system, should facilitate the development and deployment of effective INM programs. Long-term INM goals and priorities might focus on interdisciplinary IPM (all pathogen and pest groups) and sustainable crop-production systems with critical inputs from experts in technology development/transfer, institutional resources/training, and economic/social/cultural components.

Promising new technologies for nematode diagnosis and identification, critical to IPM, are on the horizon. In addition, novel management weapons, including transgenic plants with nematode resistance and/or biological control genes, new safer nematicides, and more effective biological control agents should be forthcoming. Improved data management and decision-support systems should facilitate the integration of new and traditional INM strategies and tactics.

5.12 Suggested Readings

- Parvatha Reddy P (2008) Integrated nematode management in horticultural crops. Scientific Publishers (India), Jodhpur.
- Parvatha Reddy P (2012) Recent advances in crop protection. Springer, New Delhi. (Chapters on bio-intensive nematode management).
- Singh AK, Thakur SK, Keshari N et al (2022) Integrated nematode management. Parmar Publication.
- Narayana Swamy G, Tirupal D, Manisha M et al (2022) Integrated nematode management in vegetables: review article. Pharma Innov J 11(7): 839–844.

- Singh AU (2025) Integrated nematode management. In: Maurya AK, Pant H, John V, Srivastava DK (eds) Recent advances in plant nematology. Kripa-Drishti Publications, Pune, pp 80–82.
- Abd-Elgawad MMM (2021) Optimizing safe approaches to manage plant-parasitic nematodes. Plants 10(9):1911. https://doi.org/10.3390/plants10091911.
- Westphal A (2011) Sustainable approaches to the management of plant-parasitic nematodes and disease complexes. J Nematol 43(2):122–125.

5.13 Review Questions

- Define Bio-intensive Integrated Nematode Management (BINM). How does it differ from conventional nematode management approaches?
- What are the core principles of BINM in sustainable agriculture?
- List and briefly describe the biological control agents used in BINM against plant-parasitic nematodes.
- How do cultural practices (crop rotation, organic amendments, trap crops, soil solarization) contribute to nematode suppression?
- Discuss the role of resistant cultivars in BINM strategies.
- What is the importance of organic inputs (farmyard manure, compost, oil cakes, green manures) in BINM?
- How does biodiversity in the soil ecosystem support BINM?
- Give examples of successful case studies where BINM has been effectively applied.
- What are the future research priorities to improve the efficiency of BINM practices?
- Evaluate the role of policy support and farmer training in successful adoption of BINM.

References

Abd-Elgawad MMM (2021) Optimizing safe approaches to manage plant-parasitic nematodes. Plants 10(9):1911. https://doi.org/10.3390/plants10091911

Ayaz M, Zhao JT, Zhao W et al (2024) Biocontrol of plant parasitic nematodes by bacteria and fungi: a multi-omics approach for the exploration of novel nematicides in sustainable agriculture. Front Microbiol 15:1433716. https://doi.org/10.3389/fmicb.2024.1433716

Barker KR (1992) Opportunities for integrated management of plant-parasitic nematodes in the near east. In: Maqbool MA, Kerry B (eds) Plant nematode problems and their control in the near east region. FAO, Rome

Bird GW (1981) Integrated nematode management for plant protection. In: Zuckerman BM, Rohde RA (eds) Plant parasitic nematodes, vol 3. Academic, New York, pp 355–375

Dutta TK, Khan MR, Phani V (2019) Plant-parasitic nematode management via biofumigation using brassica and non-brassica plants: current status and future prospects. Curr Plant Biol 17:17–32

Eissa MFM, Abd-Elgawad MMM (2015) Nematophagous bacteria as biocontrol agents of phytonematodes. In: Askary TH, Martinelli PRP (eds) Biocontrol agents of Phytonematodes. CAB International, Wallingford, pp 217–243

Islam MS, Rahman MH, Farazi MM et al (2015) Integrated management of seed-borne nematode (*Aphelenchoides besseyi*) in T. Aman rice (*Oryza sativa* L). Agriculturists 13(1):79–86

Krueger R, McSorley R, Grabau ZJ (2022) Nematode management in organic agriculture. ENY-058, Department of Entomology and Nematology, UF/IFAS Extension

Lakshmanan PL, Sivakumar CV (1981) Control of Rotylenchulus reniformis in bhendi by integrated method (Abstr.) Nematol Soc India Symp., Tamil Nadu Agricultural University, Coimbatore, p 54

Oduor-Owino P (2003) Integrated management of root-knot nematodes using agrochemicals, organic matter and the antagonistic fungus, *Paecilomyces lilacinus* in natural field soil. Nematol Medit 31:121–123

Paez AR (2023) Plant-parasitic nematodes and their management: a focus on new nematicides. InTech Open. https://doi.org/10.5772/intechopen.1002237

Parvatha Reddy P, Nagesh M, Devappa V (1997) Effect of integration of *Pasteuria penetrans*, *Paecilomyces lilacinus* and neem cake for the management of root-knot nematodes infecting tomato. Pest Manag Hortic Ecosyst 3(2):100–104

Parvatha Reddy P, Rao MS, Nagesh M (2002) Integrated management of burrowing nematode (*Radopholus similis)* using endomycorrhiza (*glomus mosseae*) and oil cakes. In: Singh HP, Chadha KL (eds) Banana—improvement, production and utilization. AIPUB, Trichy, pp 344–348

Pires D, Vicente CSL, Menéndez E et al (2022) The fight against plant-parasitic nematodes: current status of bacterial and fungal biocontrol agents. Pathogens 11(10):1178. https://doi.org/10.3390/pathogens11101178

Poveda J, Abril-Urias P, Escobar C (2020) Biological control of plant-parasitic nematodes by filamentous fungi inducers of resistance: *Trichoderma*, mycorrhizal and endophytic fungi. Front Microbiol 11:992. https://doi.org/10.3389/fmicb.2020.00992

Rao MS, Naik D, Shylaja M (2004) Bio intensive management of root-knot nematodes on bell pepper using *Pochonia chlamydosporia* and *Pseudomonas fluorescens*. Nematol Mediterr 32(2):159–163

Sarma YR (2003) Recent trends in the use of antagonistic organisms for the disease management in spice crops. In: Ramanujam B, Rabindra RJ (eds) Current status of biological control of plant diseases using antagonistic organisms in India. Project Directorate of Biological Control, Bangalore, pp 49–73

Seenivasan N, Devrajan K (2008) Integrated approach for the management of root-knot nematode, *Meloidogyne incognita* in medicinal coleus. Ind J Nematol 38(2):154–158

Siddiqui ZA, Iqbal A, Mahmood I (2001) Effects of *Pseudomonas fluorescens* and fertilizers on the reproduction of *Meloidogyne incognita* and growth of tomato. Appl Soil Ecol 16:179–185

Singh S (2013) Integrated approach for the management of the root-knot nematode, *Meloidogyne incognita*, on eggplant under field conditions. Nematology 15(6):747–757

Singh S, Singh B, Singh AP (2015) Nematodes: a threat to sustainability of agriculture. Procedia Environ Sci 29:215–216

Taylor AL, Sasser JN (1978) Biology, identification and control of root-knot nematodes (Meloidogyne spp.). North Carolina State University Graphics, Raleigh, p 111

Westphal A (2011) Sustainable approaches to the management of plant-parasitic nematodes and disease complexes. J Nematol 43(2):122–125

Westphal A, Xing LJ, Egel DS (2006) Use of cover crops for management of root knot nematodes in cucurbits (Abstr.). Annual international research conference on methyl bromide alternatives emission reduction, Orlando, FL, Abstract 22-1-3

Wilson MJ, Jackson TA (2013) Progress in the commercialisation of bionematicides. BioCont 58:715–722

Winslow RD, Willis RJ (1968) Nematode diseases of potatoes. In: Webster JM (ed) Economic nematology. Academic, New York, pp 17–48

Part III

Novel Less Toxic Nematicides

Next-Generation Nematicides

6

Abstract

Nematicides like fluensulfone, fluopyram, and fluazaindolizine are novel, less toxic nematicides for managing plant nematodes. These nematicides, registered since 2014, offer lower toxicity and environmental impact compared to older options. Biological options, like Arino (powered by RinoTec™ Technology), offer sustainable, low-toxicity alternatives. The evidence leans toward ongoing development, but specific 2025 innovations are not widely documented yet. This chapter provides a comprehensive overview of novel, less toxic nematicides for managing plant-parasitic nematodes. It includes detailed characteristics, efficacy, and market trends, ensuring a thorough understanding for agricultural professionals and researchers.

Keywords

Fluensulfone · Fluopyram · Fluazaindolizine · Sulfonamide nematicides

6.1 Learning Objectives

- Understand the concept and need for new generation nematicides in sustainable nematode management
- Explain the mechanisms of action of new generation nematicides compared to conventional chemical nematicides.
- Evaluate the advantages and limitations of new generation nematicides in terms of efficacy, specificity, environmental safety, and resistance management.
- Discuss regulatory aspects, commercialization challenges, and future prospects for the adoption of novel new nematicides in integrated nematode management (INM)

R. P. Parvatha, *Plant Parasitic Nematodes*,
https://doi.org/10.1007/978-981-95-9306-4_6

6.2 Introduction

Plant-parasitic nematodes, such as root-knot (*Meloidogyne* spp.) and cyst nematodes, cause significant economic losses in agriculture, estimated at 12.6% of global crop yield for top crops, equating to $215.77 billion annually (Abd-Elgawad and Askary 2018). Traditional nematicides, including fumigants like methyl bromide and organophosphates/carbamates like aldicarb, have been effective but pose high toxicity risks and environmental concerns, leading to bans or restrictions in many regions. For instance, methyl bromide is banned due to ozone depletion, and aldicarb use is prohibited in the EU due to health risks (Fairfield, Nematicide Market).

Novel, less toxic nematicides are crucial for sustainable agriculture, targeting plant-parasitic nematodes while minimizing harm to humans and the environment. These chemicals, developed in recent years, aim to replace older, more toxic fumigants and organophosphates. Under the pressures of climate change—including elevated temperatures, altered precipitation patterns, and increased CO_2 concentrations—PPN populations are expected to expand their geographic ranges, increase in abundance and reproductive rates, and potentially overcome host resistance more readily (Dutta and Phani 2023). Consequently, there is an urgent need to develop novel, less toxic chemical controls that remain effective under shifting climatic conditions.

The demand for novel less toxic alternatives has surged, driven by regulatory pressures, environmental awareness, and the need for sustainable agriculture. The global nematicides market, valued at $1.82 billion in 2024, is projected to reach $2.92 billion by 2033, growing at a CAGR of 5.35% during 2025–2033 (Nematicides Market Includes Latest Trends, Share, and Forecast 2033). This growth is fueled by the development of bio-nematicides and chemical nematicides with lower toxicity profiles.

6.3 Novel Less Toxic Nematicides

With old toxic nematicides (e.g., fumigants, organophosphates, and carbamates) phased out, several new synthetic nematicides have been introduced or are in development. Many use novel chemistries and safer toxicity profiles. Based on recent research, the following nematicides are identified as new generation, novel and less toxic, offering improved safety and efficacy for nematode management: Fluopyram (Velum®), Fluensulfone (Nimitz), Fluazaindolizine (Salibro®), Spirotetramat, Tioxazafen (NemaStrike® ST), Cyclobutrifluram (Tymirium®), and Imicyafos (Imicyafos®) (Fig. 6.1).

6.3.1 Fluensulfone (Nimitz)

ADAMA's *Nimitz* (480 g/L EC) contains fluensulfone, the first new non-fumigant nematicide in decades. It was registered by the US EPA in 2014 for cucurbits (e.g.,

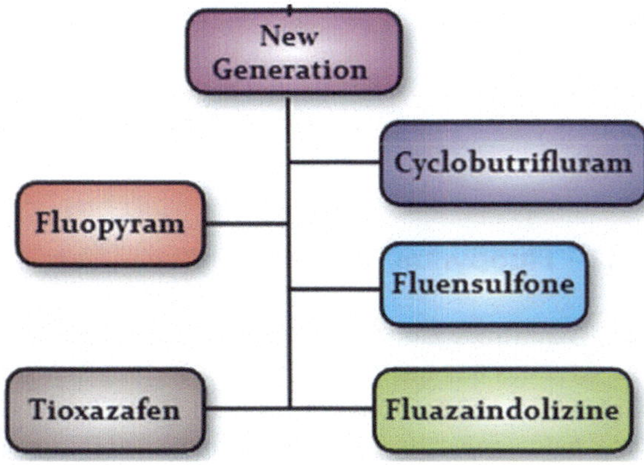

Fig. 6.1 New generation less toxic nematicides

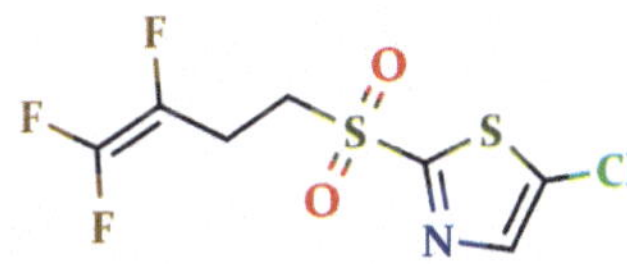

Fig. 6.2 Chemical structure of fluensulfone

cucumber, watermelon, squash) and fruiting vegetables (tomato, pepper, eggplant, okra). Subsequent approvals include California (and other US states), as well as Australia in 2015; filings are underway in Asia (Japan, S. Korea, Taiwan) and other regions. Fluensulfone belongs to a novel chemical class (Fig. 6.2). It acts as a direct-contact nematicide that causes rapid, irreversible paralysis and death of nematodes. Unlike older OP or carbamate nematicides, *Nimitz* carries only a "Caution" label (indicating lower toxicity). In efficacy trials, fluensulfone has shown broad-spectrum control—especially strong against root-knot nematodes (*Meloidogyne* spp.)—with performance comparable to conventional fumigants.

- *Registration*: First registered in 2014 in the USA, commercial name Nimitz® by Adama.
- *Chemical Class*: 1, 3-Thiazoles.
- *Mode of Action*: Acts as a beta oxidation inhibitor, disrupting nematode energy metabolism.
- *Toxicity*: LD_{50} of 671 mg/kg for rats classified with a signal word "Caution," indicating lower toxicity compared to older nematicides.
- *Efficacy*: Highly effective against *Meloidogyne* spp., reducing nematode mobility at 25 ppm and egg hatch at 95 ppm (Desaeger et al. 2020).
- *Environmental Profile*: Soil half-life of 7–17 days, water solubility of 545 ppm, making it relatively short-lived and less persistent.
- *Application*: Doses range from 1.9–3.3 kg/ha, lower than traditional fumigants, reducing environmental impact.

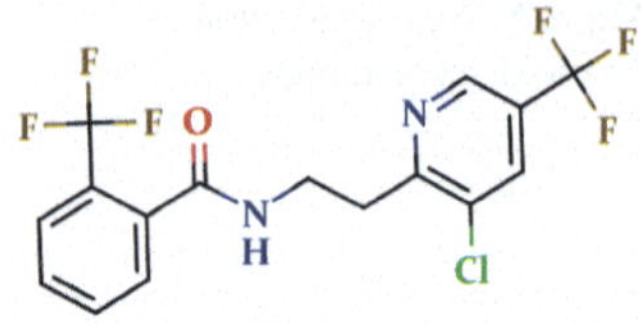

Fig. 6.3 Chemical structure of Fluopyram

6.3.2 Fluopyram (Velum®)

Fluopyram (trade name *Velum® Prime*, Bayer) is primarily an SDHI fungicide that also serves as a nematicide (Fig. 6.3). It is registered globally (USA, EU, etc.) for many crops; Velum Prime SC is labeled for nematode control on vegetables and some field crops. Its mode of action is inhibition of succinate dehydrogenase (Complex II) in nematode mitochondria, causing energy disruption. Fluopyram immobilizes nematodes rapidly and irreversibly. It has demonstrated efficacy against a broad range of PPNs: for example, field trials in Michigan showed that fluopyram seed treatment significantly suppresses soybean cyst nematode (*Heterodera glycines*) populations. It also controls root-knot nematodes (*Meloidogyne* spp.) and lesion nematodes (*Pratylenchus* spp.), often outperforming older non-fumigants. For instance, 24-h EC_{50} values against *M. incognita* are in the low μg/mL range, far more toxic than many predecessors).

- *Registration*: First registered in 2013 in Honduras and 2015 in the USA, commercial names include Velum®, Verango® by Bayer CropScience.
- *Chemical Class*: Benzamides.
- *Mode of Action*: Succinate dehydrogenase inhibitor (SDHI), affecting nematode energy production.
- *Toxicity*: LD_{50} > 2000 mg/kg for rats, signal word "Caution," indicating very low acute toxicity.
- *Efficacy*: Fast-acting, with 80% reduced mobility at 5.58 ppm for 2 h, but poor ovicidal activity, limiting its use against nematode eggs (Desaeger et al. 2020).
- *Environmental Profile*: Soil half-life > 200 days, water solubility 10 ppm, suggesting longer persistence but low leaching potential.
- *Application*: Doses range from 0.197 to 0.207 kg/ha, with a maximum annual rate of 0.494 kg/ha, enhancing safety.

6.3.3 Fluazaindolizine (Salibro®)

Fluazaindolizine (Corteva's technical name Reklemel) is a new sulfonamide nematicide sold as *Salibro®* (Fig. 6.4). It received US EPA registration in September 2023 for a range of horticultural crops (root/tuber crops, cucurbits, fruiting vegetables, nuts, nonbearing trees/vines, etc.). Like fluensulfone and fluopyram, Salibro is not a restricted-use pesticide and carries only a "Caution" label, reflecting its safer profile. Its exact mode of action is still unknown, but it is chemically distinct from all

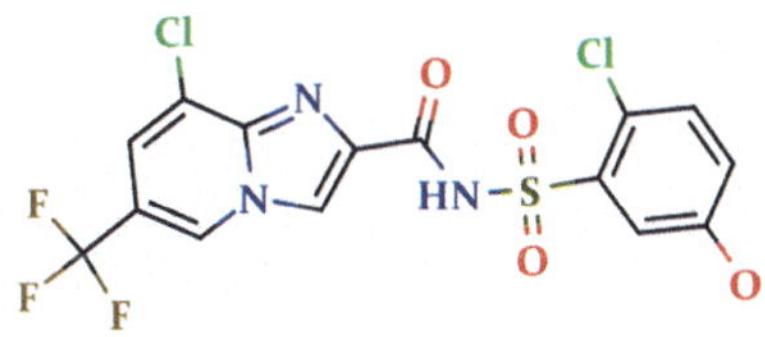

Fig. 6.4 Chemical structure of fluazaindolizine

older nematicides. University of Florida trials show that Salibro has strong activity against root-knot nematodes. In sweet potato plots, Salibro reduced *Meloidogyne* root galling as well or better than other new nematicides (Velum and Nimitz). However, as with other non-fumigants, it is generally less potent than fumigants in severely infested fields.

- *Registration*: Expected in 2021, commercial name Salibro™ by Corteva Agriscience, likely available by 2025.
- *Chemical Class*: Carboxamides.
- *Mode of Action*: Mode of action is distinct and not fully elucidated, potentially offering novel resistance management.
- *Toxicity*: LD_{50} > 1187 mg/kg for rats, signal word "Caution," indicating low toxicity.
- *Efficacy*: Slower-acting, with EC_{50} > 2x that of fluensulfone, and poor ovicidal activity at >250 ppm for 7 days, suitable for systemic control (Oka 2020).
- *Environmental Profile*: Soil half-life ~35 days, water solubility 2000 ppm, balancing persistence and mobility.
- *Application*: Doses range from 0.25–2.0 kg/ha, aligning with modern low-dose strategies.

6.3.4 Spirotetramat

- *Registration*: First registered in 2008 in the USA, commercial applications include Movento® by Bayer.
- *Chemical Class*: Tetramic acid (Fig. 6.5).
- *Mode of Action*: Acts as an ACC inhibitor, leading to lipid depletion in nematodes.
- *Toxicity*: LD_{50} > 2000 mg/kg for rats, but classified with a signal word "Warning," indicating higher caution compared to others.
- *Efficacy*: Reduces *Meloidogyne incognita* by 54% at 0.017 kg/ha, but efficacy varies across nematode species, limiting broad-spectrum use (Desaeger et al. 2020).
- *Environmental Profile*: Soil half-life <1 day, water solubility 30 ppm, indicating rapid degradation and low environmental persistence.
- *Application*: Typically used at low rates, suitable for integrated pest management.

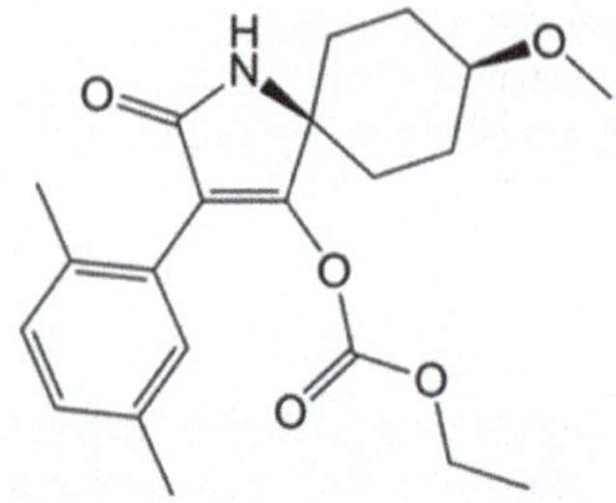

Fig. 6.5 Chemical structure of spirotetramat

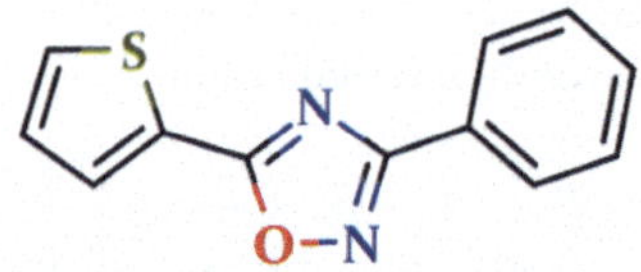

Fig. 6.6 Chemical structure of tioxazafen

6.3.5 Tioxazafen (NemaStrike® ST)

Tioxazafen (3-phenyl-5-(2-thienyl)-1, 2, 4-oxadiazole) is a seed-treatment nematicide developed by Monsanto (now Bayer) and sold as *Acceleron® NemaStrike™ ST* (Fig. 6.6). It was registered by the US EPA (and Health Canada) in 2017 for corn, soybean and cotton seeds. Its mode of action is unique: tioxazafen disrupts ribosomal protein synthesis in nematodes, a mechanism distinct from other nematicide classes. In vitro, tioxazafen causes nematode paralysis: for example, 24-h EC_{50} values are on the order of 50–60 μg/mL for *M. incognita* and *Rotylenchulus reniformis*. This paralysis appears largely irreversible after exposure. However, field trials have shown only modest nematode suppression and yield benefit with tioxazafen seed treatments. For instance, studies on cotton and soybean report limited reduction of *M. incognita* infection and yield improvement, suggesting it may need combination or higher rates in heavy infestations.

- *Registration*: First registered in 2017 in the USA as a seed treatment, but discontinued in 2020 due to safety concerns.
- *Chemical Class*: Oxadiazole.
- *Mode of Action*: Disrupts ribosomal activity, affecting nematode protein synthesis.
- *Toxicity*: Initially classified with "Caution," but withdrawn due to safety issues, highlighting the importance of ongoing evaluations.
- *Efficacy*: Effective as a seed treatment, but limited data due to discontinuation.
- *Environmental Profile*: Soil half-life 48–303 days, water solubility 1.24 ppm, suggesting potential for long-term residue.
- *Application*: No longer available, but noted for historical context.

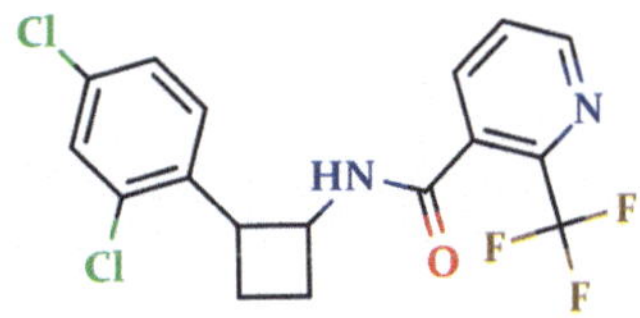

Fig. 6.7 Chemical structure of cyclobutrifluram

6.3.6 Cyclobutrifluram (Tymirium®)

Cyclobutrifluram is an emerging nematicide from Syngenta (trademark *Tymirium*™) (Fig. 6.7). It is an enantio-enriched cyclopropyl carboxamide and a potent SDH inhibitor. Syngenta has announced filing for registration (e.g., China ICAMA in 2023) and patent grants in the USA, EU, Japan. The US EPA issued a proposed registration notice in April 2025. Cyclobutrifluram targets the mitochondrial succinate dehydrogenase (Complex II) of nematodes, disrupting the TCA cycle and energy production. Pre-release trials indicate it has very broad activity: it controls root-knot nematodes on tomatoes, cucumbers, corn and other crops; the sugar beet cyst nematode (*Heterodera schachtii*); and lesion nematode (*Pratylenchus zeae*) on corn. Mode-of-action studies show rapid nematode paralysis and reduced egg hatching. If approved, Cyclobutrifluram is expected to offer long-lasting, multi-crop nematode control with a novel MOA.

6.3.7 Imicyafos (Imicyafos®)

Imicyafos is a synthetic phosphonothioate nematicide (an organophosphate analogue) with high soil activity (Fig. 6.8). It has been registered in the EU (in many member states) since the mid-2000s and introduced in South Korea around 2012. Its mode of action is as an acetyl-cholinesterase inhibitor (similar to fosthiazate), causing contact paralysis of nematodes. Imicyafos controls a broad range of nematodes: in greenhouse tests, it achieved ~91–95% mortality of *M. incognita* and *M. hapla* (60 days post-application) and ~ 90% reduction in root galling, outperforming a standard (fosthiazate) in those trials. In field studies on radish, imicyafos greatly reduced populations of the root-lesion nematode *P. penetrans* and improved yield, while having minimal impact on non-parasitic soil nematodes. Formulated products (granular or soluble) of imicyafos are used on vegetables and other crops in markets where it is registered.

6.4 Comparative Analysis

The following Table 6.1 summarizes the main novel, less toxic nematicides identified:

These nematicides are applied at lower rates (0.5–2 kg ai/ha) compared to traditional fumigants (200–300 kg a.i./ha), enhancing their environmental profile.

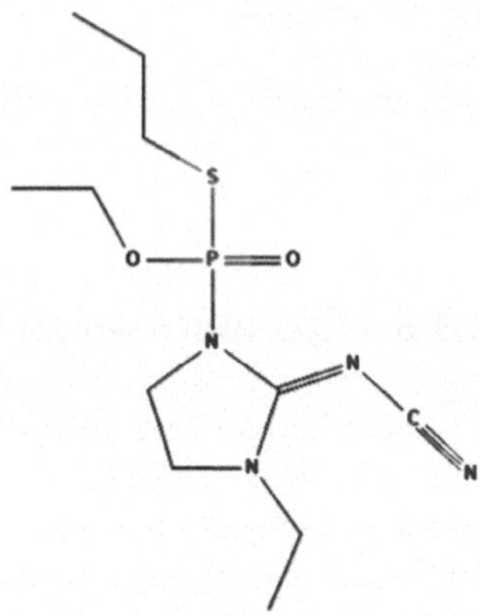

Fig. 6.8 Chemical structure of Imicyafos

Table 6.1 Characteristics of selected next-generation nematicides

Aspect	Fluensulfone	Fluopyram	Fluazaindolizine	Cyclobutrifluram
Chemical group	Fluoroalkenyl-sulfone	Pyridinyl-ethyl-benzamide	Imidazopyridine	Pyridine-Carboxamide
Commercial name	Nimitz	Velum, Verango, ILeVo, indemnify	Salibro	Tymirium
Developer	Adama	Bayer CropScience	Corteva Agriscience	Syngenta
Discovery (year)	2004	2009 (as nematicide)	2017 (reported)	2013
Release (year)	2014	2014 (as nematicide)	2020 (expected)	2020
LD 50 (mg/kg)[a]	671 (APVMA 2015)[b]	>2000 (PPDB 2019)	>1187 (CIB 2018)	>2000
DT 50 (day)[c]	7–17 (APVMA 2015)	4–25 (PPDB 2019; Chawla et al. 2018)	About 35 (Lahm et al. 2017; Chen et al. 2018)	30
Log P[d]	1.96 (APVMA 2015)	3.3 (PPDB 2019)	Not available	5.01
Mode of action	Unknown	Succinate-dehydrogenase inhibitor (Burns et al. 2015; Faske and Hurd 2015)	Unknown	Mitochondrial Complex II electron transport inhibitors. Succinate coenzyme Q reductase

[a]Median lethal dose in rats by oral administration
[b]Reference number
[c]Half-life in soil
[d]n-octanol–water partition coefficient

The following Table 6.2 compares key attributes of these nematicides, aiding in selection based on crop type, nematode species, and environmental considerations:

These nematicides are applied at significantly lower rates (0.5–2 kg ai/ha) compared to traditional fumigants (200–300 kg ai/ha), reducing environmental impact and aligning with sustainable agriculture practices (Hajihassani 2017).

Table 6.2 Summary of emerging synthetic nematicides (currently marketed or in development)

Product (AI)	Developer	Regulatory status (Region/Year)	Mode of action	Efficacy (Target nematodes)
Nimitz (Fluensulfone)	ADAMA (Advanta)	US EPA (2014, cucurbits/ fruiting veg); approved in CA, FL, AZ, etc., Australia (2015); Japan, S. Korea, Taiwan filings; Israel, Mexico, others	Novel non-fumigant; contact agent causing rapid, irreversible paralysis and death	Broad-spectrum; excellent control of root-knot (*Meloidogyne* spp.) and other PPNs; trials report efficacy comparable to standard fumigants
Velum® (Fluopyram)	Bayer CropScience	Registered globally (USA, EU, etc.) for multi-crops (fungicide with nematicidal label)	Succinate dehydrogenase (SDH) inhibitor; disrupts nematode respiration	Rapid immobilization (often irreversible) of nematodes. Broad efficacy against *Meloidogyne, Heterodera*, etc.; e.g., significantly suppresses *H. Glycines* in field and greenhouse
Salibro® (Fluazaindolizine)	Corteva Agriscience	US EPA (2023) for root/tuber, cucurbit, fruiting vegetables, nuts, nonbearing vines; in use in FL and other states; EU/other pending	Sulfonamide class; novel/unknown target	Effective on root-knot nematodes (*Meloidogyne* spp.); university trials show control similar or superior to velum (fluopyram) and Nimitz. Less potent than fumigants but safer for workers/ beneficials

(continued)

Table 6.2 (continued)

Product (AI)	Developer	Regulatory status (Region/Year)	Mode of action	Efficacy (Target nematodes)
NemaStrike® ST (Tioxazafen)	Monsanto/ Bayer (Acceleron)	US EPA (2017) seed treatment on corn, soybean, cotton (also Canada)	Disrupts nematode ribosomal protein synthesis (novel MOA)	Broad-spectrum seed-applied nematicide. Laboratory EC_{50} ~47–60 µg/mL for *M. incognita* and *R. reniformis*. Causes irreversible paralysis; however, field trials have shown limited yield protection in cotton/soy
Tymirium® (Cyclobutriflu-ram)	Syngenta	EPA proposed (2025); China ICAMA filed (2023); patents granted (USA, EU, JP, etc.)	SDH (complex II) inhibitor; disrupts nematode mitochondrial respiration, leading to paralysis	Potent on *Meloidogyne* species (e.g., cucumber, tomato, corn), *Heterodera* spp. (beet cyst nematode), and lesion nematode *Pratylenchus zeae*. Demonstrates high mortality and broad nematode spectrum in trials
Imicyafos GR/ SL (Imicyafos)		Registered in EU (member states; since ~2006); used in S. Korea (from 2012)	Organophosphate phosphonothioate; AChE. High soil contact activity, fast diffusion	Broad nematicidal action. In greenhouse tests, gave ~90–95% mortality of *Meloidogyne incognita/hapla* (60 days) and ~90% control of root galling; also suppresses *P. penetrans* with minimal impact on beneficial nematodes. Improves radish yield by reducing lesion nematode damage

AI active ingredient, *RUP* restricted-use pesticide

6.5 Nematode Management

The 3-F nematicides (fluensulfone, fluopyram, fluazaindolizine) have been evaluated on a variety of crops for their ability to suppress a diversity of plant-parasitic nematodes (Table 6.3).

Table 6.3 Evaluation of new reduced-risk agricultural nematicides against plant nematodes

Nematicide	Nematode	Greenhouse	Field	Reference
Fluopyram	*Belonolaimus longicaudatus*	–	Strawberry	Watson and Desaeger (2019)
	Heterodera glycines	Soybean	–	Heiken (2017)
	Meloidogyne hapla	–	Strawberry	Watson and Desaeger (2019)
	M. incognita	Tomato	–	Wram and Zasada (2019), Faske and Hurd (2015), and Silva et al. (2019)
	M. incognita	Lima bean	Lima bean	Jones et al. (2017)
	M. incognita	–	Cucumber	Hajihassani et al. (2019)
	M. incognita	–	Carrot	Becker et al. (2019)
	M. javanica	–	Tomato	Desaeger and Watson (2019)
	Pratylenchus penetrans	–	Strawberry	Watson and Desaeger (2019)
	Rotylenchulus reniformis	Tomato	–	Faske and Hurd (2015)
Fluensulfone	*B. longicaudatus*	–	Strawberry	Watson and Desaeger (2019)
	B. longicaudatus	–	Potato	Grabau et al. (2019)
	Globodera pallida	–	Potato	Norshie et al. (2016)
	Longidorus vineacola	–	Bell pepper	Oka (2019)
	M. hapla	–	Strawberry	Watson and Desaeger (2019)
	M. incognita	Tomato	–	Wram and Zasada (2019) and Silva et al. (2019)
	M. incognita	–	Carrot	Becker et al. (2019)
	M. incognita	Lima bean	Lima bean	Jones et al. (2017)
	M. incognita	–	Cucumber	Hajihassani et al. (2019) and Morris et al. (2016)
	M. incognita	–	Sweet potato	Ploeg et al. (2019)
	M. javanica	–	Tomato	Desaeger and Watson (2019)

(continued)

Table 6.3 (continued)

Nematicide	Nematode	Greenhouse	Field	Reference
	Paratrichodorus sp.	–	Potato	Grabau et al. (2019)
	P. penetrans	–	Lettuce	Oka (2019)
	P. penetrans	–	Strawberry	Watson and Desaeger (2019)
	P. thornei	–	Chickpea	Oka (2019)
	Pratylenchus sp.	–	Potato	Grabau et al. (2019)
	Xiphinema index	–	Fig	Oka (2019)
Fluazaindolizine	*B. longicaudatus*	–	Strawberry	Watson and Desaeger (2019)
	M. hapla	–	Strawberry	Watson and Desaeger (2019)
	M. incognita	Tomato	–	Wram and Zasada (2019) and Silva et al. (2019)
	M. incognita	–	Carrot	Becker et al. (2019)
	M. incognita	–	Cucumber	Hajihassani et al. (2019)
	M. javanica	–	Tomato	Desaeger and Watson (2019)
	P. penetrans	–	Strawberry	Watson and Desaeger (2019)
Spirotetramat	*Aphelenchoides ritzemabosi*	Nursery plant	–	Chalanska et al. (2017)
	Heterodera avenae	–	Wheat	Smiley et al. (2011, 2012)
	M. incognita	Lima bean	Lima bean	Jones et al. (2017)
	M. incognita	Peach		Shirley et al. (2019)
	M. incognita	Nursery plant	Nursery plant	Baidoo et al. (2017)
	Meloidogyne spp.	–	Multiple perennials	McKenry et al. (2009)
	Mesocriconema xenoplax	Peach	–	Shirley et al. (2019)
	M. xenoplax	–	Multiple perennials	McKenry et al. (2009)
	P. neglectus	–	Wheat	Smiley et al. (2012)
	P. penetrans	Raspberry	–	Zasada et al. (2010)
	P. vulnus	–	Multiple perennials	McKenry et al. (2009)
	Tylenchulus semipenetrans	–	Multiple perennials	McKenry et al. (2009)
	X. americanum	–	Multiple perennials	McKenry et al. (2009)
	X. index	–	Multiple perennials	McKenry et al. (2009)

Note: Empty fields within an experimental venue indicate that there is no data available

6.6 Approaches to Reduce the Cost of Nematicides Treatment

6.6.1 Seed Treatment

Seed treatment with nematicides is being used increasingly to cut down the cost of chemical without any adverse effect on the plant growth (Table 6.4). Seed treatment gives adequate initial protection from the nematode larvae to seedlings leading to better plant growth and robust nematode-free seedlings.

Seed treatment with carbofuran, fenamiphos, and phorate gave effective control of root-knot nematodes (*Meloidogyn incognita*/*M. javanica*) infecting French bean, cowpea and peas (Parvatha Reddy 1984), okra (Sivakumar et al. 1976; Jain and Bhatti 1981), chili and bottle gourd (Kandasamy and Sivakumar 1981) and increased fruit yields. Seed treatment with carbofuran and phorate resulted in significant reduction in population of *Tylenchorhynchus brassicae* and increased cauliflower yield (Singh et al. 1984).

6.6.2 Nursery Bed Treatment

Considerable loss occurs from nematode infestation in plant nursery beds. An effective and economical control can often be achieved by elimination of nematodes in nursery beds by treatment with nematicides.

Nursery bed treatment with carbofuran @ 0.2 to 0.4 g a.i./m^2 in infected areas gave effective management of citrus nematode (Parvatha Reddy 1986), root-knot nematodes on tomato, brinjal, and chili (Jain and Bhatti 1983; Ramakrishnan and Balasubramanian 1981), and *P. delatteri* on crossandra (Vadivelu and Muthukrishnan 1979). Nursery bed treatment with metham sodium at 250 liters/ha gave effective control of *R. reniformis* in tomato (Sivakumar et al. 1977).

Table 6.4 Seed treatment chemicals for nematode management

Active ingredient	Brand name	Organization	Target crop	Target nematode	Mode of action
Abamectin (Ivermectins)	Avicta Complete	Syngenta	Cotton, corn, soybean	All plant Parasitic Nematodes	Glutamate-gated chloride channel (GluCl) allosteric modulators
Fluopyram	ILoVO	Bayer crop Science	Soybean	Soybean cyst nematode, root-knot nematode	SDHI enzyme inhibitor
Tioxazafen	Nemastrike	Bayer crop Science	Cotton, corn Soybean	Soybean cyst, root-knot, and Reniform nematodes	Mitochondrial Translation inhibitor

6.6.3 Bare Root Dip Treatment

Nematode damage is more harmful to seedlings than older plants. Hence, nematicides treatment applied only to seedling roots may lessen nematode damage and be more economical. Seedling bare-root dip treatment of transplantable seedlings in systemic nematicides like thionazin, carbofuran, and oxamyl at 500 ppm for 15 min completely eliminates root-knot infection and protects the seedlings for 6 to 8 weeks from nematode infestation after planting.

Carbofuran and oxamyl at 1000 ppm for 15–30 min.; fenamiphos at 250–750 ppm for 30 min. (Thakar and Patel 1984); dimethoate at 500 ppm for 6 h, phosphamidon and dichlofenthion at 1000 ppm for 8 h (Jain and Bhatti 1987); and thionazin 500 ppm for 15 min. (Reddy and Seshadri 1975) gave effective control of root-knot nematodes in tomato when used as bare root dips. Alam et al. (1973) reported that oxamyl was effective for the control of root-knot nematodes in brinjal and okra. VC-13 was found effective against *M. incognita* in chilies, brinjal, and okra (Saxena et al. 1974; Alam et al. 1973).

Bare root dip treatment of brinjal seedlings with carbofuran and turbophos at 500–1000 ppm was effective in reducing *R. reniformis* population (Prasad and Krishnappa 1981).

Root dip treatment in 0.05–0.1% of fensulfothion, fenamiphos, and phorate protected crossandra seedlings from *P. delatteri* for 30 days after transplantation (Vadivelu and Muthukrishnan 1979).

6.6.4 Row Treatment

An important way to reduce the cost on nematicides is by row rather than overall application. For crops grown in rows more than 60 cm apart, row application is widely practiced. The nematicide is applied only to the rows where the plants are to grow, e.g., tobacco, cotton, vegetables. Depending on the spacing, the cost of the chemical is from one half to one tenth of that for broadcast application.

6.6.5 Strip Treatment

A strip preplant treatment can also be used for some perennial crops, such as grape and citrus, saving as much as 50 percent of the chemical as compared with overall application. In this method only planting strip is treated with the nematicide. The disadvantage of this method is that recontamination from cultivation or irrigation is almost certain.

6.6.6 Spot Treatment

For annual crops grown in widely spaced hills, spot application is practiced, e.g., tobacco, cotton, vegetables. This type of application is more economical than row application. The spot application is commonly used for replant problems in established orchards or vineyards. Experimentally, it has been demonstrated to be feasible with wide-spaced annual crops such as melons. For melons, a treating site of 60 sq. cm is sufficient, while tree crops require a minimum area of 2–2.5 m. The planting site method reduces the amount of chemical used considerably.

6.6.7 Precision Nematicides Application

The site-specific nematicide placement is a means of limiting nematicide applications where they are needed within a field. This system can reduce nematicide usage more than 75% compared to uniform-rate applications while still enhancing crop yields. The success of site-specific nematode management depends on an affordable map of the nematode distribution within a field as the basis for making management decisions. The clustered distribution of nematode pests makes them suitable for management through precision nematicide application. It enables to take more informed management decisions by using GPS, GIS, and RS to revolutionize the way data are collected (at resolutions of 1–5 m) and analyzed.

Application of the Flevoland simulation model (based on intensive sampling in selected portions of potato-cyst infested fields that facilitated the identification of scattered infestation foci) could decrease nematicide usage by 86% in fields with highly contagious dispersal patterns of cyst nematodes in the Netherlands (Schomaker and Been 1999). Nematicide treatment by variable rate application resulted in 36–42% reduction in the amount of nematicide applied relative to whole-field application to manage root-knot nematodes in cotton. Cotton lint yields were increased by 84 kg/ha. Site-specific application of nematicide based on root-knot nematode population density in over 1200 ha of potato resulted in 30% reduction in chemical usage and production cost savings of US$ 180 ha^{-1} (King and Taberna Jr. 2013). Takács-György et al. (2014) estimated that the amount of pesticides saved on the level of EU-25 countries is 31.7–84.5 thousand tons in case 15% of farms apply precision farming, 63.4–169.1 thousand tons in case 25% of them introduce it, while in the most favorable case (40%) it is 126.8–338.1 thousand tons.

6.6.8 Nanoformulations of Nematicides

Nanotechnology holds the promise of controlled delivery of agrochemicals to improve pest management. Nanomaterials have a tremendous potential in making crop protection methodologies cost-effective and environmentally friendly. Productivity enhancement through nanotechnology-driven precision farming and maximization of output and minimization of inputs through better monitoring and targeted action is desirable. Nanotechnology enables plants to use nematicides more efficiently.

Nanoscale systems, such as encapsulation and entrapment of nematicides by using polymers, dendrimers, surface ionic attachments, and other mechanisms, may be used in controlled and slow release of nematicides, which allows the slow uptake of active ingredients and, in turn, reduces the amount of nematicide application by minimizing the input and waste. The nanoscale delivery vehicles increase effectiveness by binding firmly to the plant surface and reducing the amount of nematicides by preventing run-off into the environment (Chen and Yada 2011).

Silver NP treated rice plants were robust with the 75 and 50 g treatment, which resulted in a considerable reduction in numbers of rice cyst nematode (*Heterodera sacchari*) at harvest. The phytosynthesized AgNPs at 0.0005 percent was highly effective to reduce the *M. incognita* population in soil by 88.16%, in roots by 92.79%, number of egg masses/g root by 79.80%, eggs/egg mass by 65.14%, and gall index (0.2 as compared to 3.0 in carbofuran) in tomato (Surega 2015). Similarly, the biweekly application of 90.4 mg/m^2 of AgNPs caused reduction in gall formation in root and improved quality of turf grass for 2 years without phytotoxicity (Cromwell et al. 2014).

6.6.9 Other Methods

Indigenous manufacture of nematicides, reduction of customs duty on imported nematicides, and sale of nematicides at subsidized price by Government agencies can also reduce the cost of nematicides.

6.7 Market Trends and Future Developments

The nematicides market is witnessing increased investment in sustainable and environmentally friendly products, with a focus on bio-nematicides and chemical innovations. Companies like Bayer, Corteva, and Adama are leading in developing these products, with significant R and D spending noted in market reports (Nematicides Market - Forecasts from 2025 to 2030). The growing adoption of integrated pest management (IPM) strategies, including biological controls, complements chemical nematicides, enhancing overall efficacy (Abd-Elgawad and Askary 2018).

Despite these advancements, challenges remain, including limited awareness of bio-nematicides and stringent regulations, particularly in Europe, which has phased out several hazardous chemicals (Nematicide Market Size, Share, Trends, Insights, Growth 2030). Emerging markets in Asia-Pacific and Latin America offer growth opportunities, driven by increasing crop losses due to nematodes and the need for higher yields (Nematicides Market Size, Share, Growth Trends, and Forecast to 2027 | MarketsandMarkets).

6.8 Research Gaps and Limitations

While the identified nematicides are promising, specific developments in 2025 are not extensively documented in accessible sources, suggesting a potential gap in recent publications or industry announcements. Future research should focus on field efficacy, resistance management, and integration into INM programs, especially as fumigant regulations tighten (Desaeger et al. 2020). The discontinuation of tioxazafen underscores the need for continuous safety evaluations, ensuring long-term viability of new products.

6.9 Conclusion

Fluensulfone, fluopyram, fluazaindolizine, spirotetramat, and the historically noted tioxazafen represent the forefront of novel, less toxic nematicides for plant nematode management. These options offer improved safety profiles and efficacy, aligning with sustainable agricultural practices. Ongoing research and market trends suggest further innovations, but current evidence highlights these as the most significant developments to date.

6.10 Suggested Readings

- Oka, Y. 2020. From old-generation to next-generation nematicides. *Agronomy 10*(9): 1387. https://doi.org/10.3390/agronomy10091387.(Comprehensivereview of recently introduced chemical nematicides, their properties, and field use).
- Matthew A.B., Martin C.H. and Tim C.T. 2024. Chemical control of nematodes. In: R.N. Perry, M. Moens and J.T Jones (eds.) *Plant Nematology*, third Edition, pp. 514–538. CABI. (Explains the transition from conventional fumigants to safer, targeted nematicides).
- Alam, M.S., Khanal, C., Rutter, W. and Roberts, J. 2022. Non-fumigant nematicides are promising alternatives to fumigants for the management of *Meloidogyne Enterolobii* in tobacco. *J. Nematol.* 54(1): e2022–1. doi: 10.2478/jofnem-2022-0045. (Summarizes new molecules, formulations, and environmental considerations).
- Desaeger, J., Wram, C. and Zasada, I. 2020. New reduced-risk agricultural nematicides - rationale and review. *J Nematol* 3; 52:e2020–2091.
- Paez, A.R. 2023. Plant-parasitic nematodes and their management: A focus on new nematicides. *InTech Open*. https://doi.org/10.5772/intechopen.1002237.

6.11 Review Questions

- What are "new generation nematicides," and how do they differ from conventional nematicides in terms of chemistry and mode of action?
- Explain the importance of developing safer and environmentally sustainable nematicides.
- List the major classes of new generation nematicides.
- Discuss the unique chemical features of at least two new generation nematicides.
- Describe the mechanism of action of fluensulfone and how it differs from organophosphates and carbamates.
- Why are new generation nematicides considered more eco-friendly than conventional fumigants like methyl bromide?
- Suggest research gaps and limitations of these novel nematicides.

References

Abd-Elgawad MM, Askary TH (2018) Fungal and bacterial nematicides in integrated nematode management strategies. Egypt J Biol Pest Control 28:74. https://doi.org/10.1186/s41938-018-0080-x

Alam MM, Khan AM, Saxena SK (1973) Some new host records of root-knot nematode, *Meloidogyne incognita* (Kofoid & White, 1919) Chitwood, 1949. Indian J Nematol 3:159–160

APVMA (2015) *Public Release Summery on the Evaluation of the New Active Fluensulfone in the Product Nimitz 480 EC Nematicide*; Australian Pesticides and Veterinary Medicines Authority: Kingstone, Australia, 47 pp

Baidoo R, Yan GP, Nelson B et al (2017) Use of chemical flocculation and nested PCR for *Heterodera glycines* detection in DNA extracts from field soils with low population densities. Plant Dis 101:1153–1161

Becker JO, Ploeg A, Nuñez JJ (2019) Multiyear field evaluation of fluorinated nematicides against *Meloidogyne incognita* in carrots. Plant Dis 103:2392–2396

Burns AR, Luciani GM, Musso G et al (2015) *Caenorhabditis elegans* is a useful model for anthelmintic discovery. Nat Commun 6:74–85

Chalanska A, Bogumil A, Labanowski G (2017) Management of foliar nematode *Aphelenchoides ritzemabosi* on *Anemone hupehensis* using plant extracts and pesticides. J Plant Dis Prot 124:437–443

Chawla S, Patel DJ, Patel SH et al (2018) Behaviour and risk assessment of fluopyram and its metabolite in cucumber (*Cucumis sativus*) fruit and in soil. Environ Sci Pollut Res 25:11626–11634

Chen H, Yada R (2011) Nanotechnologies in agriculture: new tools for sustainable development. Trends Food Sci Technol 22(11):585–594

Chen X, Li X, Pang K et al (2018) Dissipation behavior and residue distribution of fluazaindolizine and its seven metabolites in tomato ecosystem based on SAX SPE procedure using HPLC-QqQ-MS/MS technique. J Hazard Mater 342:698–704

CIB (2018) Minutes of the 54th meeting of the Central Insecticides Board, New Delhi. Available online: http://ppqs.gov.in/sites/default/files/minutes_of_54th_central_insecticides_board_cib_meeting.pdf. Accessed 2 June 2020

Cromwell WA, Yang J, Starr JL, Young-Ki J (2014) Nematicidal effects of silver nano particles on root-knot nematode in Bermuda grass. J Nematol 46(3):261–266

Desaeger JA, Watson TT (2019) Evaluation of new chemical and biological nematicides for managing *Meloidogyne javanica* in tomato production and associated double-crops in Florida. Pest Manag Sci 75:3363–3370

Desaeger J, Wram C, Zasada I (2020) New reduced-risk agricultural nematicides–rationale and review. J Nematol 3(52):e2020–e2091

Dutta TK, Phani V (2023) The pervasive impact of global climate change on plant nematode interaction continuum. Front Plant Sci 14:1143889

Faske TR, Hurd K (2015) Sensitivity of *Meloidogyne incognita* and *Rotylenchulus reniformis* to fluopyram. J Nematol 47:316–321

Grabau ZJ, Noling JW, Navia PAG (2019) Fluensulfone and 1, 3-dichloroprene for plant-parasitic nematode management in potato production. J Nematol 51:1–12

Hajihassani A (2017) Chemical Nematicides for control of plant-parasitic nematodes in Georgia vegetable crops. University of Georgea Cooperative Extension, Bull, p 1502

Hajihassani A, Davis RF, Timper P (2019) Evaluation of selected nonfumigant nematicides on increasing inoculation densities of *Meloidogyne incognita* on cucumber. Plant Dis 103:3161–3165

Heiken JA (2017) The effects of Fluopyram on nematodes. Master's thesis,. North Carolina State University, Raleigh, NC, p 75

Jain RK, Bhatti DS (1981) Systemics as seed treatment for the control of *Meloidogyne javanica* in okra. In: Second nematology symposium. Tamil Nadu Agricultural University, Coimbatore, p 44

Jain RK, Bhatti DS (1983) Chemical control of *Meloidogyne javanica* in tomato through nursery bed treatment. Himachal Pradesh Agricultural University, Solan, Third Nematology Symposium, pp 37–38

Jain RK, Bhatti DS (1987) Population development of root-knot nematode (*Meloidogyne javanica*) and tomato yield as influenced by summer ploughing. Trop Pest Manag 33:122–125

Jones JG, Kleczewski NM, Desaeger J et al (2017) Evaluation of nematicides for southern root-knot nematode management in lima bean. Crop Prot 96:151–157

Kandasamy K, Sivakumar CV (1981) Effect of *Meloidogyne arenaria* on the germination of chilli (*Capsicum annum* Linn.) seeds. In: Second Nematology Symposium. Tamil Nadu Agricultural University, Coimbatore, p 16

King BA, Taberna JP Jr (2013) Site-specific management of *Meloidogyne chitwoodi* in Idaho potatoes using 1, 3-dichlorpropene; approach, experiences, and economics. J Nematol 45:202–213

Lahm GP, Desaeger J, Smith BK et al (2017) The discovery of fluazaindolizine: a new product for the control of plant-parasitic nematodes. Bioorg Med Chem Lett 7:1572–1575

McKenry M, Kaku S, Buzo T (2009) Evaluation of Movento (Spirotetramat) for efficacy against nematodes infesting perennial crops. J Nematol 41:355

Morris KA, Langston DB, Davis RF et al (2016) Efficacy of various application methods of fluensulfone for managing root-knot nematodes in vegetables. J Nematol 48:65–71

Norshie PM, Grove IG, Back MA (2016) Field evaluation of the nematicide fluensulfone for control of the potato cyst nematode *Globodera pallida*. Pest Manag Sci 72:2001–2007

Oka Y (2019) Nematicidal activity of fluensulfone compared to that of organophosphate and carbamate nematicides against *Xiphinema index* and *Longidorus vineacola*. Eur J Plant Pathol 154:565–574

Oka Y (2020) From old-generation to next-generation nematicides. *Agronomy* 10:1387. https://doi.org/10.3390/agronomy10091387

Parvatha Reddy P (1984) Efficacy of seed treatment with three nematicides for the control of *Meloidogyne incognita* infecting cowpea, French bean and peas. Indian J Nematol 14:39–40

Parvatha Reddy P (1986) Chemical control of plant parasitic nematodes infecting citrus and grapevine. Indian J Nematol 43:303–305

Ploeg A, Stoddard S, Becker JO (2019) Control of *Meloidogyne incognita* in sweet potato with fluensulfone. J Nematol 51:1–8

PPDB (2019) Fluopyram. Pesticide properties data base. University of Hertfordshire, Hertfordshire

Prasad KSK, Krishnappa K (1981) Effect of bare-root dip treatment with pesticides on *Rotylenchulus reniformis* affecting brinjal. Indian J Nematol 11:154–158

Ramakrishnan C, Balasubramanian P (1981) Control of Meloidogyne incognita in chilli (Capsicum annum L.) nursery with nematicides. Second Nematology Symposium, Tamil Nadu Agricultural University, Coimbatore, p 54

Reddy DDR, Seshadri AR (1975) Elimination of root-knot nematode infestation from tomato seedlings by chemical bare-root dips or soil application. Indian J Nematol 5:170–175

Saxena SK, Alam MM, Khan AM (1974) Chemotherapeutic control of nematodes with Vydate oxamyl, VC-13 and Dazomet on chilli plants. Indian J Nematol 4:235–238

Schomaker CH, Been TH (1999) A model for infestation foci of potato cyst nematodes, *Globodera rostochiensis* and *G pallida*. *Phytopathology* 89:583–590

Shirley A, Noe JP, Nyczepir AP et al (2019) Effect of spirotetramat and fluensulfone on population densities of *Mesocriconema xenoplax* and *Meloidogyne incognita* on peach. J Nematol 51:1–10

Silva J, de O, Loffredo A et al (2019) Efficacy of new nematicides for managing *Meloidogyne incognita* in tomato crop. J Phytopathol 167:295–298

Singh SP, Pant V, Khan AM, Saxena SK (1984) Efficacy of certain nematicides and oil cakes for the control of stunt nematode, *Tylenchorhynchus brassicae* and parasitic fungi in the rhizosphere of cauliflower. Indian Phytopathol 37:647–651

Sivakumar CV, Palanisamy S, Balasubramanain M (1977) Control of *Rotylenchulus reniformis* in tomato nursery. Indian J Nematol 7:78–80

Sivakumar CV, Palanisamy S, Naganathan TG (1976) Persistence of nematicidal activity in seed treatment of okra with carbofuran and aldicarb sulfone. Indian J Nematol 6:106–108

Smiley RW, Marshall JM, Yan GP (2011) Effect of foliarly applied spirotetramat on reproduction of *Heterodera avenae* on wheat roots. Plant Dis 95:983–989

Smiley RW, Gourlie JA, Rhinhart KE et al (2012) Influence of nematicides and fungicides on spring wheat in fields infested with soil-borne pathogens. Plant Dis 96:1537–1547

Surega R (2015) Green synthesis of bioactive silver nanoparticles using plant extracts and their Antinemic properties. Ph. D. thesis,. Department of Nematology, Tamil Nadu Agriculture. University, Coimbatore

Takács-György K, Turek R, Maria M, Takács I (2014) Sustainable new agricultural technology—economic aspects of precision crop protection. In: 1st international conference on 'economic scientific research–theoretical, empirical and practical approaches', ESPERA 2013. www.sciencedirect.com

Thakar NA, Patel CC (1984) Screening of few cowpea varieties against reniform nematode, *Rotylenchulus reniformis* Linford and Oliveira, 1940. Indian J Nematol 14:204

Vadivelu RS, Muthukrishnan TS (1979) Control of root lesion nematode, *Pratylenchus delattrei* on crossandra. Indian J Nematol 9:86

Watson TT, Desaeger JA (2019) Evaluation of non-fumigant chemical and biological nematicides for strawberry production in Florida. Crop Prot 117:100–107

Wram CL, Zasada IA (2019) Short-term effects of sublethal doses of nematicides on *Meloidogyne incognita*. Phytopathology 109:1605–1613

Zasada IA, Walters TW, Pinkerton JN (2010) Post-plant nematicides for the control of root lesion nematode in red raspberry. Hort Technol 20:856–862

Part IV

Novel Approaches

Precision/Site-Specific Nematode Management

7

Abstract

Precision farming or precision agriculture leverages advanced technologies to tailor crop management practices to subfield variability, enhancing input use efficiency and sustainability. In the context of plant parasitic nematodes (PPNs), which exhibit highly heterogeneous spatial distributions and cause substantial yield losses, precision techniques, enable site-specific detection, mapping, and control, reducing nematicide overuse and environmental impact. This chapter synthesizes recent advances in precision tools for PPN management, including remote sensing, soil and molecular diagnostics, geospatial sampling and mapping, variable-rate nematicide applications, decision support systems (DSS), and emerging robotic and automation technologies. Challenges, integration with biological control, and future research directions are also discussed.

Keywords

Global positioning system (GPS) · Geographic information system (GIS) · Remote sensing (RS) · Variable rate technologies (VRT) · Yield monitors (YM)

7.1 Learning Objectives

- Define *precision nematode management (PNM)* and explain its role within precision agriculture.
- Understand the limitations of conventional nematode management practices and the need for site-specific approaches.
- Describe advanced diagnostic tools like molecular markers, imaging, Artificial Intelligence, and remote sensing used for early detection and quantification of nematodes.

R. P. Parvatha, *Plant Parasitic Nematodes*,
https://doi.org/10.1007/978-981-95-9306-4_7

- Interpret spatial and temporal variability of nematode populations within fields using geospatial and GIS-based data.
- Explain how precision agriculture technologies (e.g., drones, sensors, IoT devices, variable-rate applicators) can be integrated into nematode management.
- Analyze the role of decision support systems and predictive models in forecasting nematode outbreaks.
- Develop integrated management recommendations based on multisource data (soil, climate, nematode population dynamics).
- Evaluate emerging tools and strategies for precision nematode control, such as variable-rate nematicide applications and biological controls, to reduce pesticide use and enhance crop yields.
- Assess the economic, environmental, and ecological benefits of precision nematode management compared to blanket approaches.
- Identify challenges (cost, technology adoption, training needs) and propose future research directions in site-specific nematode management.

7.2 Introduction

Plant-parasitic nematodes, microscopic worms that infest plant roots, such as root-knot (*Meloidogyne* spp.), cyst (*Heterodera* and *Globodera* spp.), and lesion nematodes (*Pratylenchus* spp.), are associated with nearly every major crop, leading to global economic losses exceeding US$150 billion annually. These organisms thrive in soil, feeding on roots and causing stunting, wilting, and reduced yields, posing a threat to food security. The increasing demand for sustainable agriculture, driven by environmental concerns and restrictions on synthetic pesticides, has intensified research into precision management strategies.

Precision plant nematode management is a sophisticated approach that focuses on using advanced technology for controlling plant-parasitic nematodes. This approach aims to apply treatments exactly where needed, reducing waste and environmental impact. GPS and variable-rate technology allow farmers to apply nematicides only to areas with high nematode populations, guided by detailed soil maps. For example, fields are often pre-mapped for soil texture or nematode hotspots (via soil sampling or sensors), then prescription maps guide GPS-guided sprayers or planters that vary nematicide rate by location (Lawrance and Lawrance 2020). This chapter explores the technologies, strategies, and recent advancements shaping this field, providing a comprehensive overview for researchers, farmers, and policymakers.

Besides tractors equipped with computers, unmanned aerial vehicles (UAVs) (drones) are used to carry precision application systems or multispectral cameras. Drones can rapidly treat or scout fields, greatly reducing time and inputs. In one report, a drone covered a 6-acre paddy field in 2–3 h (versus 3–4 days manually) and cut water use (from 50 L/acre to 10 L/acre activity meters), and created near-continuous maps of soil texture or salinity (Lawrance and Lawrance 2020). Because nematode risk often correlates with soil texture, these maps become proxies for infestation risk. Yield monitors and aerial imagery further refine targeting: growers overlay nematode maps with low-yield zones to create as-applied maps (Lawrance and Lawrance 2020).

7.3 Advanced Detection and Diagnostics

Accurate nematode detection underpins any precision approach. Conventional soil sampling and microscopic counts are laborious and time-consuming. Recent innovations include:

7.3.1 Remote Sensing and Imagery

UAV-mounted multispectral and thermal cameras detect plant stress signatures—such as canopy temperature anomalies and vegetation indices—associated with nematode damage, flagging infestation zones for follow-up sampling.

7.3.2 Soil Sensor Networks

In-field probes measure soil moisture, temperature, electrical conductivity, and other parameters that correlate with nematode activity, enabling near-real-time monitoring of conditions favoring nematode proliferation.

7.3.3 Molecular Diagnostics

PCR-based assays and loop-mediated isothermal amplification (LAMP) detect nematode DNA directly from soil samples at low population densities, shortening diagnosis time from days to hours and improving species-level identification.

These tools reduce the sampling lag inherent in traditional methods, allowing for earlier, more precise interventions that curb population buildup.

7.4 Geospatial Sampling and Mapping

Because PPNs are spatially clustered, geospatial sampling, whether systematic grid or zone-based, enables creation of infestation maps to guide management.

7.4.1 Grid Sampling

Fields are divided into uniform grids (e.g., 0.2–1.0 ha cells), with soil cores collected from each cell. Veterinary geostatistics then interpolate nematode densities across the field.

7.4.2 Zone Sampling

Management zones are delineated based on soil texture, topography, yield history, or remote-sensing data. Sampling intensity is concentrated where infestation risk is highest (Lawrance and Lawrance 2020).

Nematicide rate prescription maps derived from these approaches tailor chemical applications to the spatial distribution of nematode populations, maximizing efficacy where needed and reducing waste elsewhere (Lawrance and Lawrance 2020).

Overall, *geospatial and sensor technologies* enable site-specific nematicide prescriptions, avoiding uniform blanket applications. As one extension summary notes, nematicide VRT applies "only to the areas where the nematode population has reached the economic threshold" (Bergtold and Sailus 2020). Equipment uses embedded GPS so that "the specified amount of nematicide is applied to the selected areas as the equipment moves across the field," with an as-applied map logged in real time (Bergtold and Sailus 2020).

7.5 Key Technologies in Precision Farming

Precision plant nematode management leverages advanced technologies to target treatments effectively, minimizing resource use and environmental impact. Key technologies include global positioning system (GPS), geographic information system (GIS), remote sensing (RS), variable rate technologies (VRT), and yield monitors (YM) (Fig. 7.1).

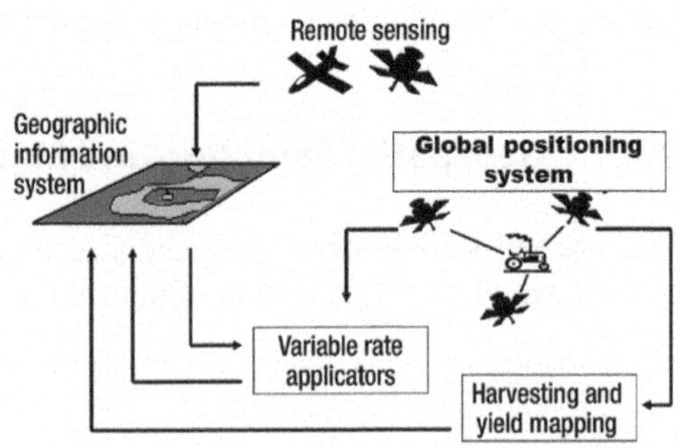

Fig. 7.1 Components of map-based precision nematode management systems

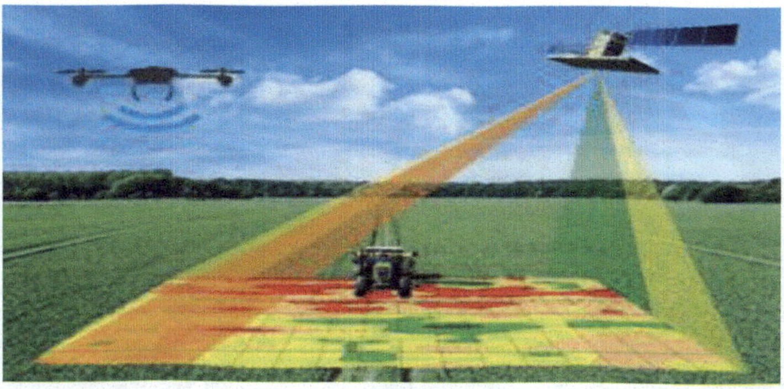

Fig. 7.2 Geographic information system (GIS)

7.5.1 Geographic Information System (GIS)

GIS is a specifically designed data-management system to store spatial data in order to create variable-intensity maps (Fig. 7.2). GISs are computer hardware and software that use feature attributes and location data to produce maps. Information on the current crop is collected to assess treatments and to potentially generate projected harvest maps, along with input data on weather, nematode densities, seed varieties, and planting populations.

7.5.2 Global Positioning System (GPS) and Variable-Rate Technology (VRT)

GPS enables site-specific application of nematicides, focusing on areas with high nematode populations (Fig. 7.3). Research from the US Cotton Belt highlights losses of at least 200 pounds per acre due to nematodes, with GPS technology offering a breakthrough by adapting variable-rate applications (Fig. 7.4). Soil sampling on uniform grids (0.025–0.5 acre) or zones, geo-referenced with GPS, creates contour maps overlaid with yield maps to identify "hot spots" for treatment.

7.5.3 Remote Sensing and Soil Electrical Conductivity

Remote sensing detects nematode distribution, aiding in creating prescription maps for targeted applications (Fig. 7.5). Soil electrical conductivity, measured at depths like 0–12 inches and 0–36 inches using systems like Veris with GPS, helps adjust nematicide rates based on soil texture, reducing application in high-clay areas where nematodes are less prevalent.

Satellite or drone multispectral imagery can highlight stunted "nematode-stress" patches using vegetation indices (Trevisan 2017). Hyperspectral sensors and machine learning are emerging to differentiate nematode damage from other

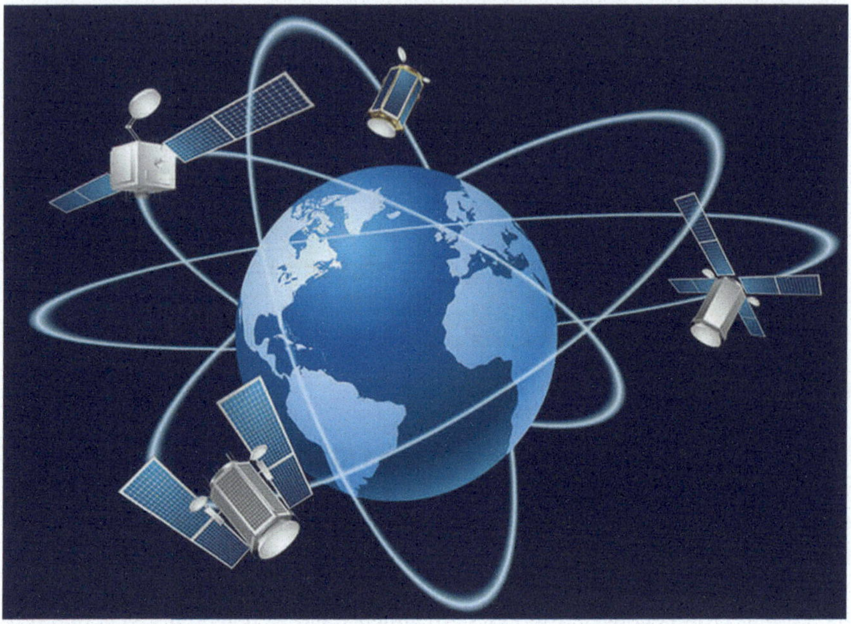

Fig. 7.3 Global positioning system (GPS)

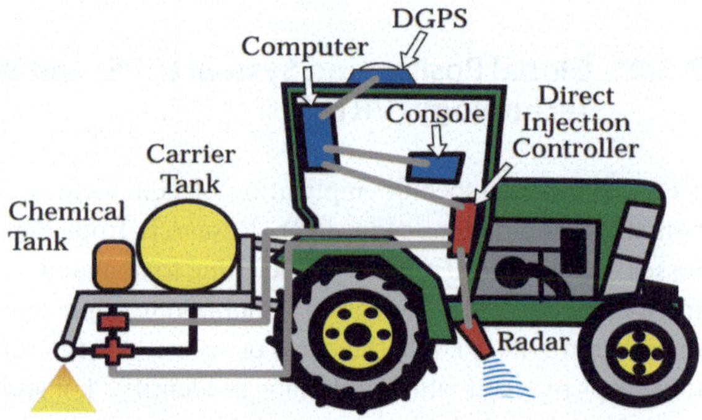

Fig. 7.4 Variable rate applicator

stressors (Trevisan 2017). Small drones spray designated areas or gather images. They significantly cut labor and inputs: e.g., one farmer's drone operation reduced spray time from days to hours, halved water uses per acre, and promised 30–35% higher yields (Shankar 2024).

Overall, geospatial and sensor technologies enable site-specific nematicide prescriptions, avoiding uniform blanket applications. As one extension summary notes, nematicide VRT applies "only to the areas where the nematode population has reached the economic threshold" (Lawrance and Lawrance 2020). Equipment uses embedded GPS so that "the specified amount of nematicide is applied to the selected areas as the equipment moves across the field," with an as-applied map logged in real time (Lawrance and Lawrance 2020).

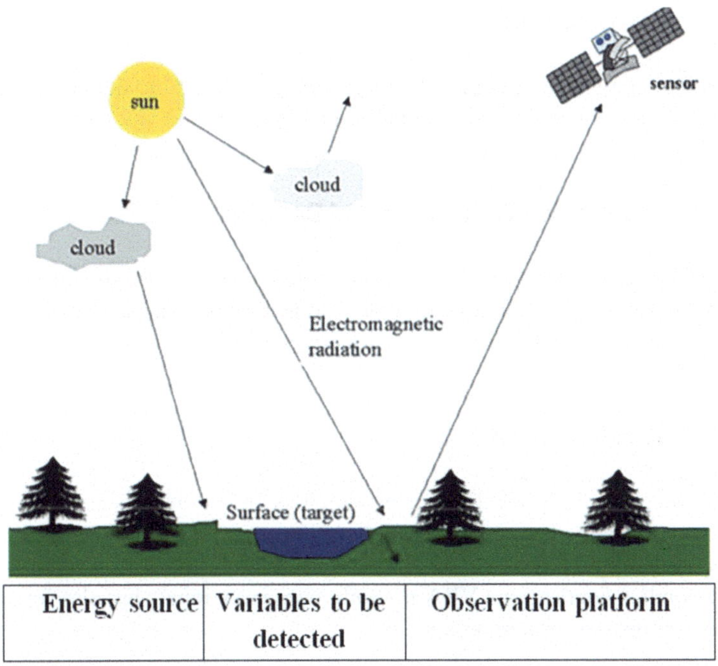

Fig. 7.5 Remote sensing (RS)

7.6 Artificial Intelligence (AI) and Automation for Nematode Counting

A significant recent advancement is the use of AI, particularly Convolutional Neural Networks (CNNs) like YOLOv8x, for automating nematode counting. A 2024 study published in Frontiers in Plant Science achieved 94% accuracy for egg detection and 93% for *Meloidogyne* juveniles, processing images in 4–5 seconds compared to 2.5 hours manually, integrates these models, enhancing precision in population assessments critical for management decisions (Saikai et al. 2024).

7.7 Variable-Rate Nematicidal Application

Variable-rate application (VRA) systems adjust nematicide doses on-the-fly according to prescription maps:

7.7.1 Implementation

GPS-enabled applicators modulate hoppers or injection rates to deliver more active ingredient in hotspot zones and minimal or no application in low-risk areas.

7.7.2 Economic and Environmental Benefits

Field trials on cotton in the south eastern United States demonstrated up to 40% reduction in total nematicide use without yield penalties—translating to significant cost savings and reduced soil residues (Roberson 2006).

7.7.3 Global Adoption

In Brazil, site-specific applications of nematicides on cotton reduced product usage by 40% and maintained or improved yields, highlighting applicability across diverse cropping systems (Ortiz et al. 2012).

These studies confirm that VRA not only conserves inputs but also lessens off-target environmental contamination.

7.8 Decision Support Systems (DSS)

Integration of detection, mapping, and VRA is facilitated by DSS platforms that:

- Aggregate real-time sensor data, historical infestation maps, and weather forecasts.
- Model nematode population dynamics under varying soil and climatic conditions.
- Recommend optimal timing and dosage of nematicides or nonchemical interventions (e.g., resistant cultivars, biological agents).

Although specialized PPN-focused DSS are still emerging, general precision-agriculture platforms increasingly incorporate nematode modules, aiding farmers in evidence-based decision making.

7.9 Automation and Robotics

Automation is poised to further refine precision PPN management:

7.9.1 Robotic Soil Processing

Automated instruments extract nematode cysts and eggs from soil samples, grind cyst walls, and count eggs with minimal human oversight, speeding diagnostics and standardizing results across laboratories (Legner et al. 2021).

7.9.2 Autonomous Vehicles

Ground robots and UAVs equipped with soil-sampling probes or spectral sensors can traverse fields systematically, reducing labor costs and enabling more frequent monitoring cycles.

These technologies promise to integrate seamlessly with DSS and VRA systems, closing the loop from detection to action.

7.10 Benefits

7.10.1 Primary Benefits

Economically, precision management optimizes inputs by applying treatments only where needed, potentially cutting costs by 36–42% in fumigant use while yielding comparable or better results than blanket methods. Environmentally, reduced chemical loads lessen risks to ecosystems and support integrated nematode management (INM) practices. Agronomically, it improves root health and crop performance in infested zones, with examples showing yield boosts of 75–120 pounds per acre in cotton (Overstreet et al. 2014).

7.10.2 Reduced Input Costs and Higher Profitability

Traditional nematode management often applies treatments uniformly, even where nematodes are absent or below economic thresholds. Precision management enables variable-rate nematicide application, optimized seed treatment and bioagent deployment, reduced labor and chemical costs, and better return on investment (ROI).

7.10.3 Early Detection and Preventive Management

Precision systems incorporate molecular diagnostics (PCR/QPCR-based detection) and remote sensing, allowing detection of nematodes before visible symptoms appear, timely intervention at sub-economic population levels, and prevention of epidemic population buildup, resulting in shift from reactive control to proactive nematode management.

7.10.4 Enhanced Crop Yield and Quality

Plant-parasitic nematodes reduce nutrient uptake, root efficiency, and water absorption. Precision management minimizes root damage through timely suppression. Benefits to crop performance include improved root architecture and nutrient absorption, better plant vigor and uniform growth, increased tolerance to abiotic stresses (drought, salinity), and improved marketable yield and product quality.

7.10.5 Improved Soil Health and Microbial Balance

Targeted interventions prevent excessive disruption of soil biota. Advantages include conservation of beneficial fungi and bacteria, enhanced soil biodiversity, improved nutrient cycling, and long-term soil productivity maintenance. Healthy soil ecosystems naturally suppress nematode resurgence.

7.10.6 Data-Driven Decision Making

Precision systems generate long-term datasets including nematode population dynamics, crop response patterns, soil variability maps, and seasonal risk forecasts. These datasets allow evidence-based management planning, optimization of crop rotation strategies, and development of predictive nematode risk models.

7.10.7 Compatibility with Modern Smart Agriculture

Precision nematode management integrates seamlessly with GIS-based farm management systems, remote sensing platforms, AI-assisted disease diagnostics, and automated farm machinery. This makes nematode control part of a broader digital agriculture ecosystem.

7.10.8 Reduced Resistance Development

Repeated blanket chemical use promotes resistance or reduced sensitivity in nematode populations. Precision approaches minimize selection pressure, enable rotation of management tactics, and encourage biological and cultural controls. This results in sustained long-term effectiveness of nematicides.

7.10.9 Support for Integrated Pest and Crop Management

Precision nematode management enhances coordination with nutrient management, irrigation scheduling, disease and weed control, and resistant cultivar deployment. This holistic integration improves overall farm system resilience.

7.10.10 Economic Benefits

One of the most compelling advantages is cost reduction through minimized inputs. Farmers typically achieve lower input expenditure per hectare while maintaining or increasing yield. The site-specific nematicide placement (SNP) can slash nematicide usage by more than 75% compared to uniform rates, without sacrificing yields. In Arkansas trials, site-specific fumigant applications reduced rates by 36–42%, saving costs while boosting lint yields by 75 pounds per acre over untreated areas. Economic analyses from multistate projects confirm higher profitability, as targeted treatments avoid wasteful applications in low-risk zones, potentially covering setup costs through improved returns. For high-value crops, integration with precision tools like drip irrigation enhances uptake and flexibility, further driving market growth in nematicides (Table 7.1) (Erdoğan et al. 2023; Afzal and Mukhtar 2024).

7.10.11 Environmental and Sustainability Benefits

By limiting chemicals to hotspots, precision management reduces environmental footprints, including lower risks of groundwater pollution and biodiversity loss. It supports soil health by preserving beneficial nematodes that drive nutrient cycling and microbial activity. Bionematicides and EPNs, when applied precisely, enhance rhizosphere balance, outcompete pathogens, and align with organic farming trends. Market outlooks predict growth in biologicals due to their compatibility with IPM, reducing reliance on synthetics and addressing resistance issues. This aligns strongly with INM and climate-smart agriculture principles. Overall, it contributes to global food security by mitigating nematode-induced losses while promoting resilient ecosystems (Table 7.2).

In conclusion, the key benefits are presented in Table 7.3.

Table 7.1 Economic benefits

Category	Quantified benefit	Example context
Nematicide reduction	>75% vs. uniform	Cotton SNP systems
Cost savings	36–42% in fumigant rates (~$26/acre)	Arkansas cotton fields
Yield improvement	75–120 lbs./acre increase	Targeted sandy zones in cotton
Economic return	Slight advantage over blanket; no yield loss	Site-specific vs. uniform

Table 7.2 Environmental and sustainability benefits

Environmental aspect	Benefit	Supporting practices
Reduced chemical use	Minimizes pollution and residues	Variable-rate applications
Soil health enhancement	Boosts beneficial microbes and nutrient cycling	Integration with cover crops
Biodiversity protection	Safe for non-target organisms	Biological controls like EPNs
Sustainability	Supports IPM and organic systems	Site-specific zoning

Table 7.3 Summary of benefits

Category	Major benefits
Agronomic	Higher yield, healthier roots, uniform crop growth
Economic	Reduced input costs, improved profitability
Environmental	Lower chemical usage, improved sustainability
Technological	Data-driven and predictive decision making
Soil health	Enhanced microbial balance and long-term productivity

7.11 Nematode Management

Precision farming optimizes nematode management by using tools like GPS-guided mapping, variable-rate applications, and data-driven zoning to apply treatments only where needed to optimize inputs. This approach minimizes environmental impact and costs, as nematodes often cluster unevenly in fields. Common methods include electrical conductivity (EC) soil mapping to correlate soil texture with nematode hotspots, grid or zone soil sampling for population thresholds, and geo-referenced prescription maps for nematicides. The SNP system exemplifies this, applying nematicides only where populations exceed economic thresholds, based on mapped distributions. Key technologies include:

Soil Mapping and Zoning: Electrical conductivity (EC) mapping with tools like Veris rigs identifies soil texture variations correlated with nematode hotspots. Fields are divided into management zones (e.g., high-risk sandy areas), guiding targeted sampling and applications (Greer et al. 2009).

GPS and Variable Rate Technology (VRT): GPS-equipped applicators deliver variable rates of nematicides (e.g., Telone at 1 gal/acre average vs. 3 gal/acre uniform), automated via maps or manually monitored.

Table 7.4 The technology, its application, benefits derived, and examples of precision farming

Technology	Application	Benefits	Examples
Soil EC mapping (Veris rig)	Identifies texture-based zones	Precise targeting; reduces sampling needs	Used in Arkansas for zoning (Greer et al. 2009)
GPS-VRT applicators	Variable nematicide delivery	50–75% input reduction; cost savings	Telone application in high-risk areas (Robinson 2015)
Geospatial modeling	Maps variability using imagery/ semi-variograms	Improved zone delineation	Reduces variability for site-specific management (Oritz et al. 2006)
Precision tillage	Deep ripping for root channels	Yield increases; stress mitigation	Winter application in California

Geospatial Analysis: Tools like aerial imagery, remote sensing, and semi-variograms model spatial variability, delineating zones to reduce field-scale inconsistencies and improve site-specific decisions (Starr et al. 2007; Oritz et al. 2006).

Precision Tillage: Deep ripping (18+ inches) along bed centers creates channels for root escape from infested upper soil, increasing yields without reducing nematode counts.

In practice, farmers in Arkansas use EC mapping to treat only half their fields, achieving 50% cost savings and integrating with resistant varieties for over 100-pound yield boosts.

The technology, its application, benefits derived, and examples of precision farming are presented in Table 7.4.

Precision methods optimize profits by limiting applications to infested areas, reducing nematicide use by 75% + and enhancing environmental quality through lower groundwater contamination. They promote sustainability, especially when integrated with rotations and resistance, doubling yields in some infested fields.

7.11.1 Cotton Root-Knot Nematode, *Meloidogyne incognita*

Cotton fields in the US Mid-South and Southeast often exhibit nematode clustering tied to soil sand content (45–75%). University of Arkansas trials (2010 onward) used Veris EC mapping to define zones, correlating higher sand (65%+) with greater damage thresholds (e.g., <2000 nematodes/pint soil tolerated in 45% sand but not >75%). Site-specific Telone applications in targeted hotspots, integrated with GPS-guided ripping (alternating 48-row strips) and 12-row nematicide sections. This method reduced nematicide by 36%, yielding 75 lbs./acre more than untreated; ripped/treated plots gained 120 lbs./acre by improving root penetration (Increased profitability ~$38/acre return improvement) (Fig. 7.6) (Overstreet et al. 2014).

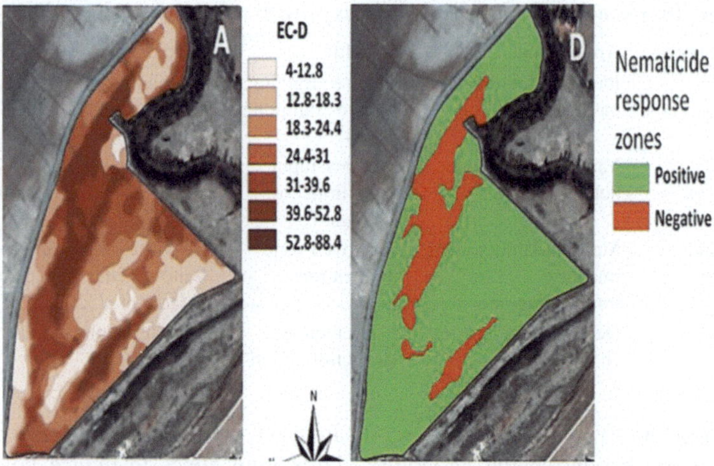

Fig. 7.6 *Left*—EC values of cotton soil. *Right*—Nematicide response zones of cotton soil

Hence, nematode damage strongly correlates with soil texture variability, making them ideal targets for precision agriculture.

7.11.2 Soybean Cyst Nematode (SCN), *Heterodera glycines*

Michigan and Minnesota trials integrated cover crops in rotations under tillage/no-till, assessing SCN (*Heterodera glycines*) in sandy loam soils using metabarcoding for community analysis; site-specific via soil type zoning (e.g., sandy loam favored fast-reproducers). More than 3-year rotations suppress more nematodes than short rotations. No-till lowered SCN versus tilled; rotations with resistant varieties (Peking/PI 88788) reduced populations and boosted yields. Cover crops like camelina are ideal for nematode-resistant zones (Melakeberhan and Kakaire 2021).

7.11.3 Potato (Columbia Root-Knot, Stubby-Root, and Cyst Nematodes)

In Idaho and the Columbia Basin, site-specific fumigation addresses *Meloidogyne chitwoodi* (CRKN) and stubby-root (SRN). In 2007, 640 ha was treated variably with 1, 3-Dichloropropene (1,3-D), using grid sampling (e.g., 261–291 m range variograms) and block kriging to map densities against thresholds (50 J2/500 cm^3 soil). In clustered fields, 1, 3-D applied only to hotspots (e.g., northwest quadrants). Reduced rates (10–15 gal./acre Telone) combined with metham sodium (30–38 gal./

acre) cut culls from 88% to 1–2%. Culls dropped from 57–94% to 0–8% in trials (1993–2000); Mocap (6–12 lb. a.i./acre) enhanced low-rate Telone efficacy by targeting surface nematodes. For cyst nematodes (*Globodera* spp.), trap crops like *Solanum sisymbriifolium* integrated with zoning, reduced quarantine risks. In Minnesota, cover crops (e.g., resistant oilseed radish) variably affected communities, favoring no-till for lower establishment (King and Taberna Jr. 2013; Ingham 2017).

7.11.4 Watermelon Root-Knot Nematode, *Meloidogyne incognita*

A 2-year Georgia field study targeted *M. incognita* with drip-applied novel nematicides (fluensulfone, fluazaindolizine, fluopyram) at varying timings/rates, including oxamyl combinations. Drip irrigation ensured targeted delivery, reducing galls and soil populations versus controls. Drip systems enable variable-rate via zoning, aligning with precision principles for horticultural crops. All 3-F treatments lowered infestations profitably; e.g., multi-timing reduced populations more than untreated. This method complements cover crops and profitable for high-value fields (Abd-Elgawad 2024).

7.11.5 Carrot Root, Lesion, and Stunt Nematodes

Michigan studies in Grattan sand and sandy clay loam tested fall cover crops (oats, Defender radish, Dwarf Essex rape) preceding carrots for the management of carrot nematodes. Bacterivores increased post-cover crops like oats/radish (Melakeberhan and Kakaire 2021).

7.11.6 Tomato Root-Knot Nematodes, *Meloidogyne incognita*, *M. javanica*

Florida and California trials focused on resistant grafting/rootstocks (e.g., interspecific hybrids) for *Meloidogyne* spp., with variable rate of drip nematicides (metham sodium). Grafting reduced galling, while resistant varieties (e.g., Sanibel) were effective against tropical RKN resulting in lower nematode density at harvest. Metam sodium penetrated 18–24 inches via irrigation (Barrett et al. 2012; Hajihassani 2022).

The summary of nematode management studies is presented in Table 7.5.

Table 7.5 Comparison of nematode management studies

Crop/ Nematode	Precision method	Key outcome (Yield/Cost)	Source/Location
Cotton, root-knot	Veris EC soil mapping, variable rate Telone appln.	−36% nematicide, +75 lbs. yield /acre	U. Arkansas, Mid-South
Potato, Columbia root-knot	Grid kriging, site-specific 1,3-D application	Culls from 88% to 1–2%	Idaho, Columbia Basin (King and Taberna Jr. 2013; Ingham 2017)
Watermelon root-knot (*M. incognita*)	Drip-targeted 3-F nematicides application	Reduced galls/ populations, profitably	Georgia (Abd-Elgawad 2024)
Soybean cyst (*H. glycines*)	Rotation zoning, cover crop selection	Lower nematode establishment in no-till	Michigan, Minnesota
Carrot, root-lesion (*P. penetrans*)	Timed sampling, indices	Increased bacterivores, variable densities	Michigan (Melakeberhan and Kakaire 2021)
Tomato, root-knot	Grafting, drip nematicides	Lower density, no yield loss	Florida, California (Barrett et al. 2012; Hajihassani 2022)

7.12 Case Studies and Implementation Examples

7.12.1 USA—Cotton (Southeast)

Clemson University and USDA projects have pioneered nematicide VRT. In one case, a South Carolina farm was mapped by soil conductivity into three zones; only the high-risk sandy zone received full-rate nematicide via a Gandy variable-rate hopper. This saved ~25–34% of nematicide costs and boosted yields by ~5% (Khalilian et al. 2003). In adjacent states (Georgia), field trials compared uniform versus VR aldicarb and Telone® treatments. Results showed that VR application of Telone (1, 3-D) gave the highest yield and profit, with substantial reduction in total fumigant applied (Baird et al. 2001). A multiyear Georgia study also found VR aldicarb side-dress improved profits over the standard single rate (Baird et al. 2001). The key was using GPS-linked soil maps and nematode assays to prescribe rates.

7.12.2 USA—Technology Development

A USDA-funded project (Clemson/Arkansas) developed practical tools for growers: high-density soil EC mapping, remote yield and imagery analysis, and on-the-go VR sprayers for nematicides (USDA 2005). They showed how Veris soil scanners and zone-based sampling could create nematode-density distribution maps, which in turn drove VR applications (USDA 2005). This integrated mapping approach is now part of extension training for cotton growers.

7.12.3 Brazil—Cotton

Brazilian researchers emphasize that nematodes form patchy foci and advocate precision tactics. One outcome was a prescription system for planting-time nematicides: maps from high-resolution remote imagery and selective sampling identify infestation spots, and only those zones are treated (Trevisan 2017). Trials found this site-specific planting used ~40% less nematicide overall while cotton yields remained equal or higher than flat-rate fields (Trevisan 2017). The cost-benefit was large: saving ~$60/ha in nematicide for a $10 mapping investment (Trevisan 2017). The Brazilian team also noted innovative ideas like switching plant varieties by zone (resistant in nematode areas, susceptible elsewhere) if multi-hybrid planters are available (Trevisan 2017).

7.12.4 Global—Biological Nematicides Research

Studies in Europe (Slovenia) have quantified biocontrol efficacy. For example, *Bacillus firmus* I-1582 (marketed, e.g., as VOTiVO®) was tested on tomatoes: hyperspectral sensors measured that this bacterium reduced root-knot nematode counts by ~50% relative to untreated controls (Susič et al. 2020). In contrast, the chemical standard (fluopyram) treatment completely eliminated nematodes in greenhouse trials (Susič et al. 2020). Such work confirms that while bionematicides help, they often need to be combined with precision placement to approach the efficacy of synthetics.

7.12.5 Asia and Others—Emerging Use

Although detailed field studies are fewer, precision tools are rapidly adopted in Asia for general pest control. In India, farmers widely use drones for precision spraying of fertilizers and pesticides to save labor and inputs (Shankar 2024). This infrastructure could extend to nematicides as specialized nozzles and sensors develop. Similarly, digital agriculture platforms (IoT sensors, mobile apps) are beginning to include nematode risk data; for example, hyperspectral drones are being tried to detect nematode stress early, enabling spot spraying.

The following Table 7.6 provides information on precision approaches for nematode management.

The following Table 7.7 compares traditional and precision approaches for nematode management:

Table 7.6 Case studies on precision nematode management

Location/ Crop	Nematicide	Precision tech	Key findings	Reference
USA (Georgia)/ cotton	Aldicarb, 1,3-D (Telone)	Soil zones (EC-based); VRT sprayer	Site-specific zones (from soil texture) were used. 1, 3-D gave better control in coarse (sandy) zones, while aldicarb sufficed in finer zones. VR application boosted yields and reduced fumigant use in low-risk areas	Ortiz et al. (2012)
USA (Georgia)/ cotton	Aldicarb, 1,3-D	Grid mapping; VR sprayers	Three field trials compared uniform vs. VR rates. All 1, 3-D treatments out-yielded an aldicarb-only control. VR 1, 3-D achieved the highest economic return and cut total fumigant used. VR aldicarb (side-dress) had higher yield and return than the standard uniform rate	Baird et al. (2001)
USA (SC)/ cotton	Aldicarb, 1,3-D (Telone)	Veris EC mapping; GPS; Gandy applicator	A farmer's 3-zone map led to VR nematicide placement. This cut nematicide costs by ~25–34% (e.g., Telone cost fell to $8–10/acre) and raised yields ~5%. The system "enables applying fewer chemicals…to control nematodes," saving money and protecting water quality	Khalilian et al. (2003)
Brazil/ cotton	Fumigant (planted in furrow)	Remote-sensing + GIS zoning; VR planting	Mapping via drone/satellite images and targeted soil sampling defined nematode hotspots. VR seed-applied nematicide used ~40% less pesticide (especially omitting nematode-free zones) with equal or higher yields. A $10/ha mapping cost saved ~$60/ha in chemical	Trevisan et al. (2018)
Slovenia/ tomato	*B. Firmus* (bioagent) vs. Fluopyram (chemical)	Hyperspectral imaging	In glasshouse/microplot trials on RKN-infested tomatoes, *B. firmus* I-1582 reduced nematode counts by ~51–53%. By contrast, the chemical standard (fluopyram) nearly eliminated nematodes in pots and suppressed them similarly in micro-plots. This illustrates that under controlled conditions chemicals can outperform biologicals, though both had similar (50%) control in field-like tests	Susič et al. (2020)

Table 7.7 Comparison of traditional and precision approaches for nematode management

Method	Traditional approach	Precision farming approach	Advantages of precision
Nematicide application	Uniform rate across the field	Site-specific, variable rate based on nematode density	Reduces chemical use, lowers costs, less environmental impact
Biological control agents	Broad application, often less effective	Targeted application using GIS/GPS, higher efficacy	Improved efficacy, reduced waste, sustainable
Soil amendments	Applied uniformly, variable effectiveness	Applied to high-density areas, timed for optimal impact	Enhanced effectiveness, resource efficiency
Crop rotation	General rotation, less targeted	Precision-guided rotation based on nematode maps	Better nematode suppression, improved yields

7.13 Chemical Versus Biological Nematicides: Efficacy

Chemical fumigants and nematicides (e.g., 1, 3-dichloropropene, metham-sodium, carbamates like aldicarb) typically kill nematodes rapidly and give strong short-term control. In field trials on cotton, treatments with 1, 3-D or other chemicals significantly raised yields over untreated controls (Khalilian et al. 2003; Susič et al. 2020). For example, Georgia trials found that *all plots treated with 1, 3-D had significantly higher yields than untreated checks with only aldicarb* (Susič et al. 2020). In the Clemson/Arkansas study, variable-rate applications of 1, 3-D produced the highest economic returns, substantially cutting the total fumigant applied (Susič et al. 2020). In contrast, biological nematicides (bionematicides)—microbes or their metabolites—usually act more slowly or modestly. *Bacillus firmus* strain I-1582, a widely used biocontrol agent, was shown to cut root-knot nematode (*Meloidogyne*) counts by about 50% in pot and microplot trials (Susič et al. 2020). By comparison, a standard chemical (fluopyram) completely eliminated nematodes in a controlled pot experiment (Susič et al. 2020). The same study reported that in field-like microplots, fluopyram and *B. firmus* achieved similar suppression (both about 50%), but the chemical was inconsistent under heavy infestation conditions (Susič et al. 2020).

In summary, *chemical* nematicides often achieve higher immediate knockdown of nematodes. *Biological* nematicides are generally safer but give partial control; for example, one trial found *B. firmus* halved nematode loads whereas the chemical completely knocked them out (Susič et al. 2020). Efficacy also depends on placement: precision delivery may boost both types. Site-specific high doses of biologicals (where populations are worst) or combined biological + chemical regimens are emerging as strategies (reviewed in Egypt, noting variable-rate use of bionematicides) (Abd-Elgawad 2024). In practice, growers may use both: for instance, chemical in severe hotspots and bionematicides elsewhere as part of an integrated plan.

7.14 Integration with IPM and Other Systems

Precision nematicide use fits within larger INM and precision agriculture frameworks. Data from nematode maps can be layered with other field information (weed maps, disease patches, and irrigation zones) to optimize all inputs. For instance, nematode-infested spots often coincide with moisture stress or nutrient deficiencies, so combined site-specific fertilizer, irrigation, and nematicide application may be coordinated. Extension guides note that nematicide prescriptions can be "overlaid with yield maps to determine problem areas," and the VR applicator runs whenever moving across any infested zone (Lawrance and Lawrance 2020).

Decision-support systems (e.g., NemaDecide for potato) are beginning to incorporate spatial nematode data to refine timing and rate decisions. Remote-sensing insights (NDVI or hyperspectral) used for disease or fertility stress can be cross-referenced to distinguish nematode damage. Cover cropping or resistant varieties can also be applied in a site-specific way: e.g., sowing trap crops or tolerant cultivars only in hotspot subfields (Trevisan 2017).

Finally, precision nematicide tools often share platforms with other sprayers. A GPS-guided tractor outfitted with AI-driven weed sprayers could carry a nematicide tank and switch to apply only when in a mapped nematode zone. Similarly, robotics and automated seeders are being developed to change seed varieties or place biocontrol products on-the-go. As one review noted, recognizing nematode spatial patterns "enables best seed-location matching, leverage variable rates of bionematicides, and grasp relationships between beneficial/harmful organisms in space and time for IPM" (Abd-Elgawad 2024).

Integrated nematode management (INM) has been the approach of choice for nematode management for the last three decades, and several principles of INM are highly compatible with the ideology behind precision farming, which includes:

- Off-the-farm inputs like nematicide applications to be reduced.
- Use of action thresholds, i.e., taking corrective measures based on economic and ecological criteria.
- Use of cultural and biological control measures, and resistant varieties, to enhance sustainability.

The optimization of inputs such as nematicides and minimization of economic and environmental damage provided in precision agriculture are similar to INM. However, the spatial component so central to precision agriculture is lacking in INM.

Beyond technology, methods like crop rotation, using resistant crop varieties, and cover crops help manage nematodes sustainably. Biological controls, such as beneficial microbes, are also gaining traction, especially with restrictions on chemical nematicides.

Precision management extends beyond technology to include INM practices, ensuring sustainable control:

7.14.1 Crop Rotation and Resistant Varieties

Rotating crops with non-host plants reduces nematode populations, while using resistant or tolerant varieties, such as those for *M. incognita*, can lower densities. For example, in southern Indiana, watermelon systems with soybean and corn rotations are modified to include resistant cash crops.

7.14.2 Cover Crops and Biological Controls

Cover crops with resistance or biofumigation potential, like certain small grains, suppress nematodes. Biological controls, such as beneficial nematodes and microbial biopesticides, are gaining traction, with recent research focusing on microbial formulations for sustainable crop protection.

Precision tools can also optimize application of biocontrol agents (e.g., *Bacillus*, *Pochonia*, and *Pasteuria*):

- Site-specific delivery of microbial formulations to infested zones maximizes contact with nematodes while sparing non-target soil biota.
- Variable-rate irrigation and fertigation systems can co-apply biocontrol organisms with minimal disruption of their viability.

Although biocontrol in precision frameworks is less studied than chemical VRA, initial trials indicate enhanced efficacy and sustainability when spatial targeting is employed.

7.14.3 Chemical Nematicides

While chemical options like fumigants (e.g., Telone II) and non-fumigants are effective, environmental concerns have led to restrictions, with new nematicides emerging over the last decade offering safer mechanisms. Precision application reduces reliance, aligning with INM.

7.15 Comparative Analysis of Methods

The following Table 7.8 summarizes key precision management methods, their applications, and benefits:

This table illustrates the multifaceted nature of precision management, combining technology with sustainable practices.

Table 7.8 Key precision management methods, their applications, and benefits

Method	Application	Benefits
GPS and variable-rate tech	Site-specific nematicide application	Reduces costs, minimizes environmental impact
Remote sensing	Mapping nematode distribution	Enhances targeting accuracy
AI automation (e.g., SEGNEMA)	Rapid nematode counting (eggs, juveniles)	Saves time, improves accuracy (up to 94%)
Crop rotation	Alternating non-host crops	Reduces nematode populations sustainably
Resistant varieties	Planting tolerant/resistant crops	Lowers nematode damage, enhances yield
Biological controls	Using beneficial microbes/ nematodes	Eco-friendly, reduces chemical reliance

7.16 Challenges and Considerations

Managing nematodes is inherently complex due to their microscopic size, wide host range (over 2500 plant species), and ability to survive in soil for years. Environmental concerns limit chemical options, pushing for integrated approaches, though debate continues on balancing these methods.

AI tools face difficulties with overlapped objects and organic debris, with improvements suggested through expanded image datasets and global collaboration. The debate continues on balancing chemical and biological methods, with biopesticides offering promise but requiring further validation for efficacy across diverse conditions.

Key challenges remain:

7.16.1 Cost and Accessibility

High upfront investment in sensors, GPS guidance, and robotics can be prohibitive for smallholder farmers, especially in developing regions.

7.16.2 Standardization

Calibration protocols for remote sensing, soil sensors, and molecular assays need harmonization to ensure cross-field comparability.

7.16.3 Data Integration

Seamless interoperability among detection tools, mapping software, DSS, and application machinery is critical but often lacking.

7.16.4 Environmental Variability

Soil heterogeneity and rapid nematode movement can challenge predictive models, necessitating adaptive algorithms and machine-learning approaches.

7.17 Environmental and Economic Impacts

Precision nematicide application reduces chemical use and cost. By not treating low-risk zones, growers save money on expensive nematicides and lower environmental loading. For example, Clemson Extension reported a South Carolina cotton farm cut its nematicide bill by at least 25–34% using site-specific sprayers. In one case, a grower's fumigant costs fell from ~$36–39/acre to $8–10/acre with VRT, while yield rose ~5%. Similarly, Brazilian studies found variable-rate planting-time application of nematicides used about 40% less total pesticide (mainly by omitting areas without nematodes) while maintaining equal or higher cotton yields (Trevisan 2017). This analysis projected that a $10/ha investment in soil/nematode mapping could save ~$60/ha in nematicide costs (Trevisan 2017).

Economic analyses show mixed effects: in some trials, the cost of additional soil sampling offsets gains. For example, Georgia plots treated with VR aldicarb had similar yields to uniform treatments, but higher sampling and chemical costs "further lowered the total economic return." However, the VR fumigant trials showed higher profit: VR 1, 3-D gave the largest return versus uniform rates. Overall, precision targeting tends to cut unnecessary input. As one report concluded: "enables farmers to apply fewer chemicals to control nematodes in cotton, save money and protect water quality."

Environmental benefits are also clear. Bionematicides (e.g., *Pasteuria*, *Trichoderma*, *Bacillus* spp.) are inherently less toxic: they "suppress nematodes naturally without harming beneficial soil organisms" (Susič et al. 2020). By limiting applications, precision methods reduce runoff and exposure. For instance, sites with heavy clay (high EC) often receive little or no nematicide, reflecting studies showing traditional nematicides work poorly there (Lawrance and Lawrance 2020). Thus, precision maps "apply less nematicide to zones with higher conductivity" (Lawrance and Lawrance 2020). In Brazil, roughly 40% fewer nematicides were needed overall, cutting contamination risk (Trevisan 2017). In short, the eco-impact of nematicides is mitigated by site-specific use: lower total dose and nonuniform coverage mean fewer residues leach into groundwater or harm non-targets.

7.18 Recent Advancements and Future Directions

AI-powered automation for nematode counting represents a transformative step, enhancing precision in population assessments and supporting real-time decision-making. The call for global collaboration to develop universal automated systems, leveraging machine learning interest among nematologists, underscores the field's dynamic evolution.

Future research should focus on scaling precision farming technologies for smallholder farmers, improving data accessibility, developing cost-effective sensors for nematode detection, open-source DSS platforms for nematodes, integration of AI/ML for real-time prediction, and socioeconomic analyses to foster adoption. The integration of artificial intelligence and machine learning could further enhance predictive models for nematode outbreaks, as suggested in recent reviews. Additionally, climate-smart packages are discussed in a ScienceDirect article on PPNs in protected agriculture (Phani et al. 2021).

7.19 Conclusion

Precision plant nematode management integrates GPS, remote sensing, and AI with INM strategies like crop rotation and biological controls. Recent advancements, particularly in AI automation, enhance efficiency and accuracy, addressing the complex challenge of nematode control while supporting sustainable agriculture. Continued research and collaboration are essential to refine these methods, ensuring food security and environmental stewardship.

Precision farming offers transformative potential for sustainable PPN management. By combining rapid detection, geospatial mapping, variable-rate applications, and automation, farmers can achieve high control efficacy with markedly reduced chemical inputs. Continued innovation, coupled with efforts to lower technology barriers and standardize methodologies, will be pivotal in mainstreaming precision nematode management across diverse agricultural landscapes.

7.20 Suggested Readings

- Gebbers, R. and Adamchuk, V.I. 2010. *Precision agriculture and food security. Science* 327(5967): 828–831. (Overview of precision agriculture concepts applicable to nematode management).
- Bulacio Fischer, P.T., Carella, A., Massenti, R. et al. 2025. Advances in monitoring crop and soil nutrient status: Proximal and remote sensing techniques. *Horticulturae 11*(2): 182. https://doi.org/10.3390/horticulturae11020182. (Describes tools like remote sensing, GIS, and proximal sensors for site-specific nematode detection).
- Plumblee, M.T. and Mueller, J.D. Implementing precision agriculture concepts and technologies into crop production and site-specific management of nematodes. In: R.A. Sikora et al. (eds.) Integrated nematode management: state-of-the-art and visions for the future, pp. 421–427. doi:10.1079/9781789247541.0059. (Key paper linking nematode ecology to precision agriculture methods).

- King, B.A. and Taberna, J.P. 2013. *Site-specific nematode management of Meloidogyne chitwoodi in Idaho potatoes using 1, 3-Dichloropropene; Approach, experiences, and economics. J. Nematol.* 45(3): 202–213. (Case study applying spatial mapping to nematode management).
- Dhanasri, V. and Roy, R.A. (2025). 2020. *Artificial intelligence in nematology: A step towards precision agriculture. Vigyan Varta* 6(2): 186–188. (Latest developments and practical challenges in implementing SSNM).
- Ortiz, B.V., Perry, C., Sullivan, D. et al. 2012. Variable rate application of nematicides on cotton fields: a promising site-specific management strategy. *J. Nematol.* 44(1): 31–39.
- Overstreet, C., McGawley, E.C., Khalilian, A. et al. 2014. Site specific nematode management - Development and success in cotton production in the United States. *Journal of Nematology* 46(4): 309–320.
- Trevisan, R. 2017. Using precision agriculture to manage nematodes in Brazil. *Global AgTech.* globalagtechinitiative.comindia.mongabay.com.
- USDA. 2005. *Site Specific Detection and Control of Crop Nematodes.* Clemson University, Clemson, SC 29634.

7.21 Review Questions

- Define precision nematode management (PNM). How does it differ from conventional nematode management strategies?
- Explain the key technologies of site-specific nematode management. Why is spatial variability of nematode populations important?
- Describe the different sampling and diagnostic methods used in precision nematode management.
- Discuss the role of GPS and GIS in mapping nematode infestations.
- Evaluate how remote sensing tools (e.g., drones, multispectral imaging) can be applied for nematode damage detection.
- Discuss the role of molecular and AI-based diagnostic tools in enhancing the precision of nematode management.
- What are the potential economic and environmental benefits of adopting site-specific nematode management practices?
- How can variable-rate technology (VRT) be utilized for nematicide application in precision nematode management?
- Examine case studies where site-specific nematode management has been successfully implemented in different crops.
- What are the future prospects and research priorities for precision/site-specific nematode management?

References

Abd-Elgawad MMM (2024) Nematode spatial distribution in the service of biological pest control. Egypt J Biol Pest Control 34:3. https://doi.org/10.1186/s41938-024-00768-6

Afzal A, Mukhtar T (2024) Revolutionizing nematode management to achieve global food security goals—an overview. Heliyon 10(3):e25325. https://doi.org/10.1016/j.heliyon

Baird RE, Rich JR, Waters D (2001) Evaluation of variable rate nematicide applications using precision farming methods to manage *Meloidogyne incognita* on cotton. Nematol Medit 29:247–254

Barrett CE, Zhao X, McSorley R (2012) Grafting for root-knot nematode control and yield improvement in organic heirloom tomato production. HortScience 47(5):614–620

Bergtold J, Sailus M (2020) Conservation tillage systems in the Southeast: production, profitability and stewardship. In: SARE handbook series 15, Sustainable Agriculture Research and Education (SARE) program

Erdoğan H, Halil Ünal I, Susurluk A, Lewis EE (2023) Precision application of the entomopathogenic nematode *Heterorhabditis bacteriophora* as a biological control agent through the Nemabot. Crop Prot 174:106429. https://doi.org/10.1016/j.cropro.2023.106429

Greer A, Wilson G, Kirkpatrick T (2009) Management of economically important nematodes of arkansas cotton—FSA7567. University of Arkansas, United States Department of Agriculture, and County Governments Cooperating, Fayetteville

Hajihassani A (2022) Grafting in tomato: a practical tool for root-knot nematode control. University of Georgia Co-op, Extn

Ingham R (2017) Nematode management in the face of short supply of Telone and Vydate. In: Proceedings of the Washington—Oregon potato conference, Kennewick, pp 33–38. www.potatoconference.com

Khalilian A, Mueller JD, Han YJ (2003) Performance of variable rate nematicide application systems. In: *Proceedings 2003 Beltwide cotton conference. National Cotton Council*, Memphis, pp 578–582

King BA, Taberna JP Jr (2013) Site-specific management of *Meloidogyne chitwoodi* in Idaho potatoes using 1, 3-dichlorpropene; approach, experiences, and economics. J Nematol 45:202–213

Lawrance KS, Lawrance GW (2020) Plant-parasitic nematode management. In: Bergtold J, Sailus M (eds) Conservation tillage systems in the southeast production, profitability and stewardship, pp 164–180. SARE handbook series 15. www.sare.org

Legner CM, Tylka GL, Pandey S (2021) Robotic agricultural instrument for automated extraction of nematode cysts and eggs from soil to improve integrated pest management. Sci Rep 11(1):3212. https://doi.org/10.1038/s41598-021-82261-w

Melakeberhan H, Kakaire S (2021) Managing nematodes, cover crops, and soil health in diverse cropping systems. Michigan State Univ Extn, East Lansing

Oritz B, Sullivan DG, Perry C, Vellidis G (2006) Geospatial solutions for precision management of cotton root-knot nematodes. In: ASAE annual meeting, Portland. https://doi.org/10.13031/2013.20639

Ortiz BV, Perry C, Sullivan D et al (2012) Variable rate application of nematicides on cotton fields: a promising site-specific management strategy. J Nematol 44(1):31–39

Overstreet C, McGawley EC, Khalilian A et al (2014) Site specific nematode management - development and success in cotton production in the United States. J Nematol 46(4):309–320

Phani V, Khan MR, Dutta TK (2021) Plant-parasitic nematodes as a potential threat to protected agriculture: current status and management options. Crop Prot 144:105573. https://doi.org/10.1016/j.cropro.2021.105573

Roberson R (2006) Variable rate application controls cotton nematodes, reduces cost. In: Farm progress, 19 July 2006

Robinson E (ed) (2015) Greg Lyerly: precision application for nematode control in cotton. Delta Farm Press, New York

Saikai KK, Bresilla T, Kool J et al (2024) Counting nematodes made easy: leveraging AI-powered automation for enhanced efficiency and precision. Front Plant Sci 15:1349209. https://doi.org/10.3389/fpls.2024.1349209

Shankar P (2024) Farming with AI and drones to increase yields, manage resources and reduce pests. Mongabay. india.mongabay.comindia.mongabay.com

Starr JL, Koenning SR, Kirkpatrick TL et al (2007) The future of nematode management in cotton. J Nematol 39(4):283–294

Susič N, Žibrat U, Sinkovič L et al (2020) From genome to field—observation of the multimodal nematicidal and plant growth-promoting effects of *Bacillus firmus* I-1582 on tomatoes using hyperspectral remote sensing. Plants 9(5):592. https://doi.org/10.3390/plants9050592

Trevisan R (2017) Using precision agriculture to manage nematodes in Brazil. Global AgTech. globalagtechinitiative.comindia.mongabay.com

Trevisan RG, Vilanova Jr. NS, Eitelwein MT, Molin JP (2018) Management of plant growth regulators in cotton using active crop canopy sensors. Agriculture 8(7):101. https://doi.org/10.3390/agriculture8070101

USDA (2005) Site specific detection and control of crop nematodes. Clemson University, Clemson, p 29634

Nanotechnology: State-of-the-Art Nematode Management

8

Abstract

Nanotechnology provides promising tools for plant nematode management through precise delivery of nematicides, direct nematicidal effects, enhanced biocontrol, early detection, improved plant resilience, RNA interference, and nematicidal nanoparticles. Nanomaterials like silver and plant-derived particles offer sustainable nematode management. Nanotechnology reduces environmental impact compared to traditional chemical nematicides. While promising, its large-scale application requires addressing safety, cost, and regulatory challenges. There is a debate about the safety of nanomaterials, with studies showing mixed results on non-target organisms. Ongoing research and field trials will further refine its role in sustainable agriculture.

Keywords

Nanomaterials · Nanosensors · Targeted nematicide delivery · Enhanced plant resistance · RNA interference · Improved soil health

8.1 Learning Objectives

- Define nano-formulations and explain their relevance in crop protection and nematode management.
- Differentiate between conventional nematicides and nano-based delivery systems.
- Describe different nano-formulations (e.g., nano-emulsions, nano-capsules, nano-clays, polymeric nanoparticles, metallic nanoparticles).
- Explain the mechanisms by which these nano-formulations enhance nematicidal activity.

R. P. Parvatha, *Plant Parasitic Nematodes*,
https://doi.org/10.1007/978-981-95-9306-4_8

- Understand how nano-formulations improve solubility, stability, and controlled release of active ingredients.
- Analyze case studies where nano-formulations have been successfully used against root-knot, cyst, and other plant-parasitic nematodes.
- Compare their efficacy with traditional chemical nematicides.
- Discuss the potential of nano-formulations in reducing chemical load and promoting sustainable agriculture.
- Identify research gaps and future directions for the development of smart nano-delivery systems.
- Assess the role of nano-biotechnology in integrated nematode management strategies.

8.2 Introduction

Nanotechnology is emerging as a promising tool in plant nematology, offering innovative solutions for the detection and management of plant-parasitic nematodes, potentially reducing reliance on harmful chemicals. These microscopic worms such as root-knot (*Meloidogyne* spp.) and cyst (*Heterodera* and *Globodera* spp.) can cause significant damage to crops, leading to reduced yields and economic losses estimated at US$157 billion annually. Traditional control methods rely on chemical nematicides, which are often toxic, persistent, environmentally harmful, and face regulatory restrictions due to their impact on non-target organisms. Nanotechnology offers innovative, sustainable alternatives, leveraging nanoscale materials (1–100 nm) to enhance more targeted and efficient pest management strategies. Nano-formulations (e.g., metal or polymer nanoparticles, nano-emulsions, virus-based carriers) can increase efficacy and stability of nematicidal agents while potentially reducing environmental impact. This chapter explores the key roles played by nanotechnology in controlling plant-parasitic nematodes, along with the current research and applications, challenges, and future considerations.

8.3 Types of Nanomaterials Used

Various types of nano-formulated pesticides have been developed, including nano-encapsulations, nanosuspensions (Fig. 8.1), lipid-based, polymer-based, clay-based, and porous silica-based formulations, as well as metal nanoparticles (NPs). A wide variety of nanomaterials have been explored for nematode control. Key categories include:

8.3.1 Metal and Metal-Oxide Nanoparticles

Silver (AgNP) and copper NPs are most studied. Biogenic AgNPs (from plant extracts or bacteria) have broad-spectrum nematicidal action (e.g., against

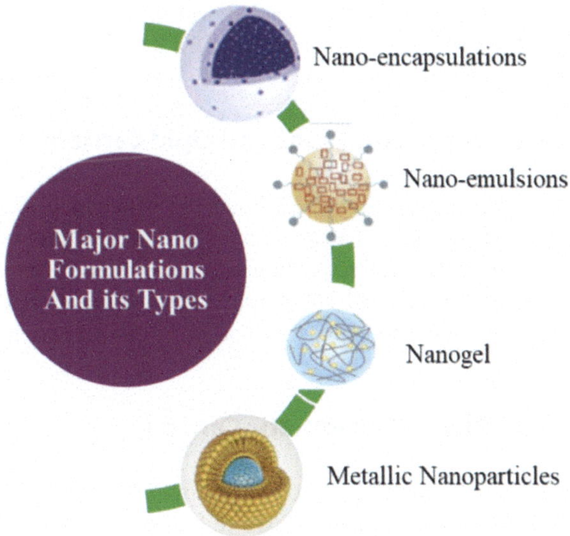

Fig. 8.1 Different types of nano-formulations available for crop pest management

Meloidogyne spp.) (Cromwell et al. 2014). Zinc oxide and titanium dioxide NPs also show nematode toxicity in vitro. For instance, green-synthesized ZnO and CuO NPs caused high mortality of *Heterodera* and *Meloidogyne* eggs and J2 in laboratory tests. Many metal NPs can release ions or generate ROS that kill nematodes; EPA notes nano-silver/nano-copper are common active ingredients in pesticide products (Wang et al. 2022).

8.3.2 Silica and Silicon-Based Nanoparticles

Amorphous SiO_2 NPs and nano-chelated silicon fertilizers have been applied as nematicides. These not only directly reduce egg hatching and juvenile survival, but also deposit silicon in roots, bolstering cell walls and defense responses (Yu et al. 2022). Unlike heavy metals, silicon is generally considered environmentally benign; many Si-containing treatments are labeled eco-friendly (Yu et al. 2022).

8.3.3 Polymeric Nanocarriers

Biodegradable polymers like chitosan, polylactic-co-glycolic acid (PLGA), polylactic acid (PLA), polycaprolactone (PCL), and lignin are used to make nanoparticles or nano-capsules. Chitosan NPs (often formed by ionic gelation) are intrinsically toxic to nematodes and also serve as carriers for botanical extracts or synthetics. For example, chitosan nanospheres (≈2000 ppm) reduced *M. incognita* egg hatching by 95% and juvenile survival by 97%, while reducing soil/root nematode populations by ~78–98% and boosting tomato growth (Mouniga et al. 2023). Nano-capsules of

PLA-PEG and PCL-PEG loaded with abamectin achieved superior control compared to free abamectin (Fu et al. 2018).

8.3.4 Nano-Emulsions and Lipid Carriers

Oil-in-water nanodroplets of botanical nematicides (e.g., garlic oil, essential oils, plant extracts) have enhanced bioactivity. Field studies show that nano-emulsified garlic oil or bacterial metabolites control root-knot nematodes more effectively than their nonencapsulated counterparts (Osman et al. 2025). The ultrafine emulsions adhere to roots and nematodes and often improve plant health.

8.3.5 Nano-Biosynthetic Carriers

Engineered virus-like particles and protein cages can serve as precise delivery vehicles. For instance, plant viruses (e.g., tobacco mosaic virus, RCNMV) have been loaded with RNA or chemicals for nematode targeting. These biologically derived NPs are stable in soil and release payloads on nematode encounter (Opdensteinen et al. 2024; Cao et al. 2015).

8.3.6 Hybrid/Composite Materials

Some formulations combine metals with biopolymers or other nanostructures (e.g., metal NPs embedded in chitosan, layered silicates loaded with botanicals) to leverage multiple modes of action. Research is ongoing on nano-chelated nutrients (e.g., nano-Fe, nano-Mg) that may have by-product nematicidal effects when delivered to crops.

8.4 Key Roles of Nanotechnology

Plants are one of the important sources of nanoparticle-based nematicides, and it contains different types of nanoparticles as a nematicide component in agriculture applications (Fig. 8.2). Researchers have focused their attention on understanding the mechanism of synthesis of botanical-based nanoparticles for controlling nematodes, to prevent diseases caused by them and to protect our environment instead of chemical nematicides.

8.4.1 Nano Sensors for Early Detection

Early detection of nematode infestations is critical to prevent widespread crop damage. Nanosensors, capable of detecting specific biomolecules or physiological

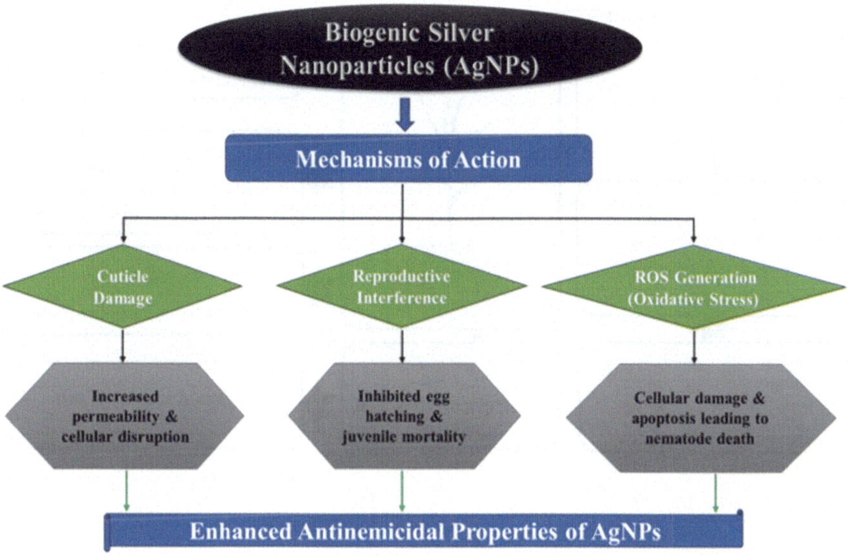

Fig. 8.2 Benefits of association between nematicides and nanotechnology. It improves the properties of the bioactive compounds and increases their effectiveness in nematode control

changes in plants, enable farmers to identify nematode presence before visible symptoms emerge. Examples include:

- *Carbon Nanotube-Based Sensors*: These can detect volatile organic compounds (VOCs) released by plants under nematode stress, providing an early warning system.
- *Quantum Dot-Based Sensors*: These monitor changes in plant root exudates, which often signal nematode activity, offering a highly sensitive detection method.

8.4.2 Applications in Nematode Control

Nanotechnology aids in controlling plant nematodes primarily through two methods: enhancing RNA interference (RNAi) for targeted gene silencing and using nanoparticles with direct nematicidal effects. For RNAi, nanomaterials deliver double-stranded RNA (dsRNA) to nematodes, protecting it from degradation and ensuring it reaches the target, which can disrupt nematode reproduction and survival. Nanoparticles, such as silver, titanium, and silicon, have shown potential to kill nematodes directly, offering an alternative to toxic chemical nematicides.

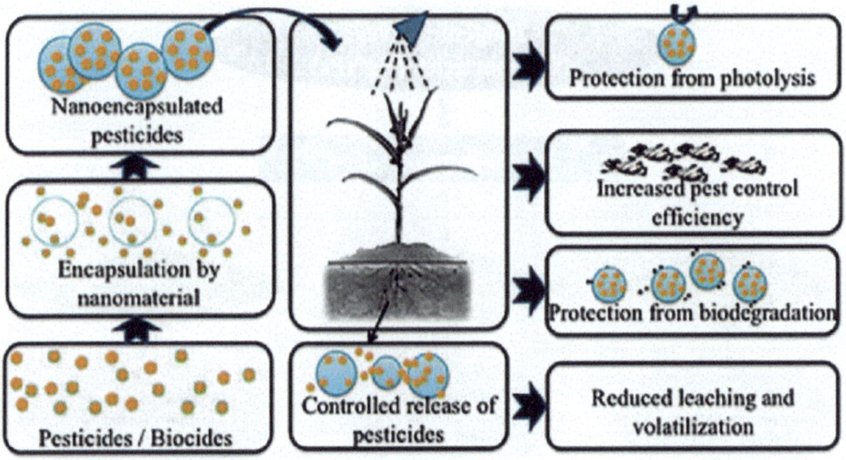

Fig. 8.3 Encapsulation of nematicides

8.4.3 Targeted Delivery of Nematicides

Nanoparticles can be engineered to deliver nematicides directly to nematodes or their infection sites, minimizing the impact on non-target organisms and the environment. This precision reduces the amount of chemical needed and enhances sustainability. Examples include:

- *Silver Nanoparticles (AgNPs)*: Known for their antimicrobial properties, AgNPs have shown potential in disrupting nematode metabolism or damaging their cuticles, thereby reducing their ability to infect plants. Application of Ag NPs improved tomato plant growth and reduced root-knot nematode (*M. incognita*) infection in comparison to silver nitrate and control treatments.
- *Lipid-Based Nanoparticles*: These can encapsulate nematicides (Fig. 8.3), allowing for controlled release over time (Fig. 8.4), which decreases the frequency of applications and improves efficiency.

8.4.4 Enhancing Plant Resistance

Nanotechnology can contribute to developing plants that are more resilient to nematode attacks, reducing reliance on external control measures. NPs can improve the efficacy of biocontrol agents like *Trichoderma* or *Bacillus* species by protecting them from environmental stress or enhancing their delivery to nematode-infested roots. Some NPs, such as chitosan NPs, stimulate plant defense mechanisms, upregulating the genes involved in systemic acquired resistance (SAR) against nematodes. Potential applications include:

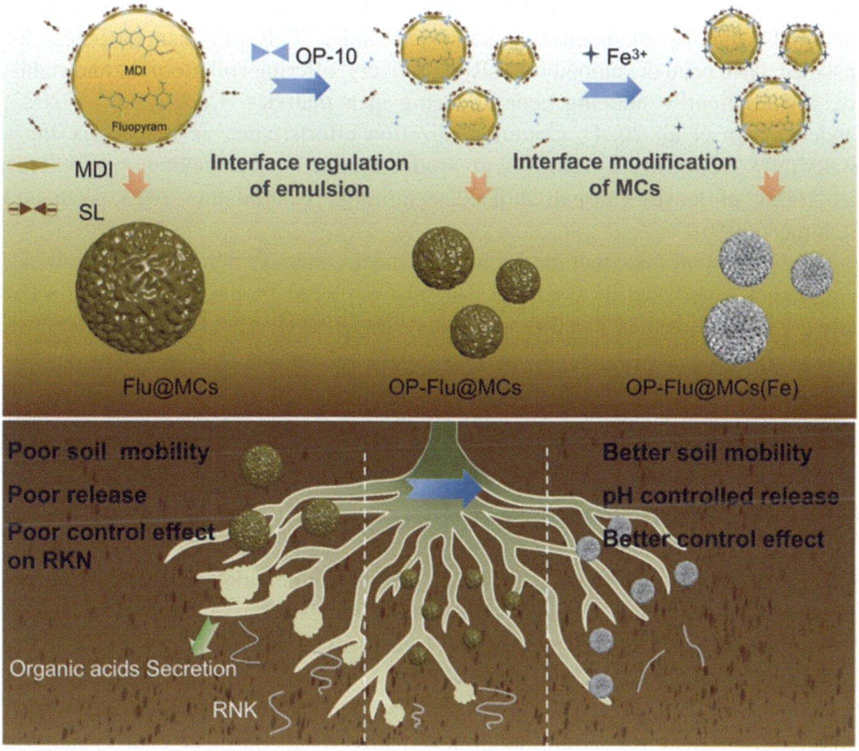

Fig. 8.4 Controlled release of nematicides from nano-capsules

- *Nano-Enabled Gene Delivery Systems*: These systems can introduce nematode-resistant genes into plants more efficiently than traditional genetic engineering methods.
- *Nanoparticle Coatings on Seeds*: Such coatings may trigger systemic resistance in plants, enhancing their natural defenses against nematodes.

8.4.5 Enhanced RNA Interference (RNAi) with Nanomaterials

One of the primary applications is enhancing RNA interference, a biological process where dsRNA silences specific nematode genes to inhibit their development, reproduction, and parasitism. Research has identified over 75 genes in root-knot and cyst nematodes suitable for RNAi targeting, but soil delivery of dsRNA is challenging due to rapid degradation, often within 48 h (Dubelman et al. 2014). Nanotechnology addresses this by using nanomaterials to protect and deliver dsRNA effectively.

Plant virus-derived nanoparticles, such as tobacco mild green mosaic virus (TMGMV) and red clover necrotic mottle virus (RCNMV), are particularly promising. These nanomaterials are biodegradable, soil-stable, and noninfectious to mammals, making them environmentally friendly. TMGMV nanoparticles, for instance, have been developed for dsRNA delivery, offering soil mobility and stability, and efficiently inducing gene silencing in nematodes (Caparco et al. 2023; Opdensteinen et al. 2024). Commercialization efforts, such as BioProdex Inc.'s SOLVINIX LC, utilize TMGMV, with production methods achieving up to 500 mg per 100 g of leaves through molecular farming (Charudattan 2023; Wen and Steinmetz 2016).

Other nanomaterials, including clay nanosheets, chitosan nanoparticles, polymer nanoparticles, and peptide assemblies, are also explored, though primarily for foliar delivery to plants rather than direct nematode targeting (Liu et al. 2020; Mitter et al. 2017; Zhang et al. 2010; Christiaens et al. 2018; Numata et al. 2014). These materials enhance RNAi efficacy against pests, potentially extending to nematodes, but further research is needed for soil applications.

8.4.6 Nanoparticles as Nematicides

Beyond RNAi, nanoparticles themselves exhibit nematicidal properties, offering direct control over plant parasitic nematodes. Studies have shown that silver (Ag NPs), titanium, and silicon nanoparticles can reduce nematode populations. For example, Cromwell et al. (2014) demonstrated nematicidal effects of silver nanoparticles on root-knot nematodes in Bermuda grass. Similarly, Ardakani (2013) explored the toxicity of silver, titanium, and silicon nanoparticles on *Meloidogyne incognita* (root-knot nematode) and tomato growth parameters. Jo et al. (2013) also used silver nanoparticles for nematode control on Bermuda grass putting greens. These studies highlight nanoparticles' potential as effective nematicides, with multisite activity against nematodes without causing phytotoxicity, making them a suitable alternative to traditional chemical nematicides.

8.4.7 Soil Health and Nematode Suppression

Certain nanomaterials can improve soil conditions, indirectly suppressing nematode populations and supporting plant health. Examples include:

- *Nano-Clays*: These improve soil structure and water retention, creating an environment less favorable for nematode survival and reproduction.
- *Nano-Fertilizers*: By enhancing plant vigor, these fertilizers help crops withstand nematode damage more effectively.

Table 8.1 summarizes the nanomaterials discussed, their applications, and specific examples, highlighting both direct and potential uses in nematode control.

Table 8.1 Key nanomaterials and their applications in nematode control

Nanomaterial	Application	Specific example	Reference
TMGMV-derived nanoparticles	dsRNA delivery to nematodes	Soil mobility and stability for gene silencing	Caparco et al. 2023
RCNMV-derived nanoparticles	Encapsulation for nematicide delivery	Biodegradable, soil-stable delivery system	Cao et al. 2015
Silver nanoparticles (ag NPs)	Direct nematicidal effect	Control of root-knot nematodes in Bermuda grass	Cromwell et al. 2014
Titanium and silicon NPs	Nematicidal toxicity	Toxicity to *Meloidogyne incognita* in tomatoes	Ardakani 2013
Clay nanosheets	Potential RNAi delivery (analogous use)	Delivery for plant DNA viruses, potential for nematodes	Liu et al. 2020

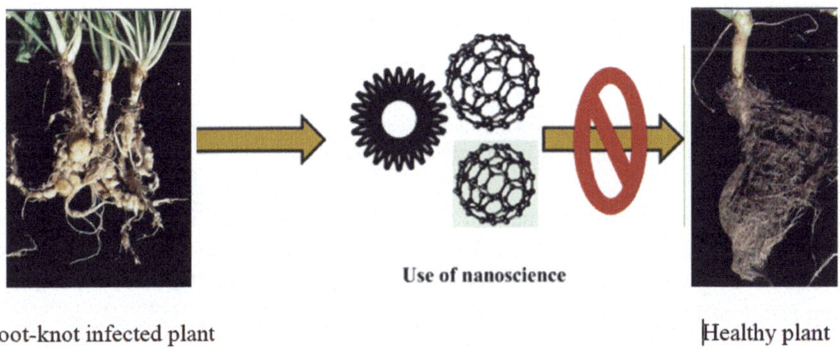

Fig. 8.5 Management of root-knot nematodes on vegetables using nano-nematicides

8.5 Case Studies

Researchers have tested nano-nematicides against a wide range of plant-parasitic nematodes, mainly focusing on economically important pests:

8.5.1 Root-Knot Nematodes (*Meloidogyne* spp.)

It is by far the most studied group. *M. incognita, M. javanica,* and *M. graminicola* have been targeted on many crops. Studies report effective control on vegetables (tomato, eggplant, cucumber) (Fig. 8.5) (Mohammad et al. 2022; Osman et al. 2025), fruits (citrus, etc.), field crops (rice, cotton), and even turf grass (Cromwell et al. 2014). For example, silver NP treatments improved Bermuda grass growth and reduced *M. graminicola* root galls over 2 years (Cromwell et al. 2014).

Application of Ag NPs improved tomato plant growth and reduced root-knot nematode (*M. incognita*) infection in comparison to silver nitrate and control

Table 8.2 Effect of soil treatment with AgNPs on galls and egg-masses of tomato (index of 0 to 9) caused by *M. incognita* under greenhouse conditions

Treatments	Conc. (mM)	Fresh plant wt. (g)	No. of galls	RGI	No. of egg-masses	Egg-mass index
Biological AgNP	0.25	48.2 c	4.7 cd	2 ef	0 f	1 d
	0.50	52.0 b	2.0 d	2.0 ef	0 f	1 d
	1.00	55.4 a	1.3 d	1.3 f	0 f	1 d
Chemical AgNP	0.25	41.1 d	6.0 cd	2.7 de	1 ef	1.7 cd
	0.50	37.9 de	10.3 c	3.3 d	3.7 de	2 c
	1.00	23.4 g	20.7 b	4.3 c	5.3 d	2.3 c
$AgNO_3$	0.25	37.5 e	25.0 b	5 c	12 c	4 b
	0.50	33.6 f	28.0 b	5 c	14.7 bc	4 b
	1.00	20.9 g	48.0 a	6 b	6 b	4 b
Control (check)	–	16.3 h	55.0 a	7 a	24	5 a
CD ($P \leq 0.05$)	–	3.2	8.01	0.85	2.9	0.72

Note: Each value represents the mean of four replicates. Within each concentration, values with the same letter do not differ significantly according to CD ($P \leq 0.05$)

treatments. The highest increment of fresh weight as well as the lowest numbers of galls and egg-masses was obtained when tomato plants were treated with Ag NP produced by ginger extract at 1 mM (Table 8.2) (El-Deen and El-Deeb 2018; Ahmed and Bahig 2018).

8.5.2 Cyst Nematodes (*Heterodera, Globodera*)

Some work has explored nano-agents against soybean cyst (*Heterodera glycines*), sugar beet cyst (*H. schachtii*), and potato cyst (*Globodera pallida*) nematodes. The results are promising but less numerous. In lab tests, ZnO and TiO_2 NPs inhibited *H. schachtii* juveniles and egg hatch. Incorporating NPs (Si, Mg, and Zn) into fertilizer has been examined for suppressing *H. schachtii* in sugar beet.

8.5.3 Lesion and Other Sedentary Nematodes

Root lesion nematodes (*Pratylenchus* spp.) and reniform nematodes (*Rotylenchulus* spp.) have been tested with biogenic NPs and chitosan, generally showing reduced populations in treated plants. Migratory nematodes like burrowing nematode (*Radopholus* spp.) and dagger nematodes (*Xiphinema* spp.) have few reports, but silica and copper NPs have shown some activity in vitro.

8.5.4 Special Cases

Pinewood nematode (*Bursaphelenchus xylophilus*) control has been attempted using bacterially produced AgNPs and nano-clays to deliver biocontrol bacteria (Zinovieva et al. 2023). Nematodes that vector diseases (e.g., *Longidorus* spp.) could also be targets in soil sterilization strategies.

8.5.5 Crop Contexts

Nano-nematicides have been evaluated across climates and regions, including greenhouse vegetables (Mediterranean, Asia), tropical fruits, temperate cereals and forage, and managed turf. Many studies originate from Asia and the Middle East, but European and US trials are increasing. There is no inherent climate limitation—nanoscale delivery can work in any system where nematodes persist. The versatility of nanoparticle carriers means they can be formulated for root drench, soil injection, seed coating, or foliar drench as needed.

8.5.6 Laboratory and Field Study Findings

- *Laboratory Assays*: Virtually all publications report potent nematicidal activity in in vitro or greenhouse conditions. For instance, nano-formulated plant extracts consistently caused higher *Meloidogyne* J2 mortality than the raw extracts (Mohammad et al., 2022). In chitosan studies, >95% juvenile kill was typical within days (Mouniga et al. 2023). Polymer and virus-NPs with abamectin produced controlled, long-lasting nematode suppression in pots (Fu et al. 2018; Cao et al. 2015). These lab studies also often show improved plant growth and defense markers in NP-treated plants, suggesting systemic benefits.
- *Greenhouse Trials*: In controlled pot experiments, nano-treatments usually outperform conventional formulations. For example, tomato seedlings drenched with abamectin-loaded viral nanoparticles had 50–80% fewer galls than those given free abamectin (Cao et al. 2015). Nano-silver soil amendments significantly reduced *Meloidogyne* infection in *Cynodon dactylon* (turf grass) seedlings in replicated trials (Cromwell et al., 2014). Greenhouse eggplant trials using nano-emulsions of *Bacillus* metabolites or garlic oil showed higher yield and lower nematode counts than untreated plants (Osman et al. 2025).
- *Field Trials*: Fewer studies have reached field scale, but results are encouraging. In a 2-year Bermuda grass field trial, biweekly application of AgNP ($\approx$90 mg/m^2) improved turf quality and reduced root galling, although soil nematode counts during the season were not significantly lowered (Cromwell et al., 2014). A 2025 open-field trial on eggplant showed that nanoencapsulated garlic oil and *Bacillus* metabolites significantly increased fruit yield (up to 80–130% of control) and decreased root-knot numbers, whereas the same bulk treatments had lesser effect

Table 8.3 Nanoparticle formulated nematicides along with dose and application techniques for management of different nematodes

Crop	Nematode pest	Nanoparticle formulated nematicides application rate	Application techniques
Potato	*Globodera* spp.	Aldicarb at 2.24–3.36	Incorporated in row
		Oxamyl at 4.0–5.5	
		Carbofuran at 4.0–5.5	
Tomato, cucurbits	*Meloidogyne* spp.	Aldicarb at 3.36	Incorporated in 30 cm bands
		Ethoprophos at 0.9–2.9	Incorporated in bands
		Oxamyl at 0.6–1.2	
		Fenamiphos at 1.6–3.3	
		Dazomet at 30–50 g/m^2	Incorporated in bands and irrigated time interval after planting
Citrus	*Tylenchulus semipenetrans*	Fenamiphos at 10.8–21.6	Annual treatment applied along drip-line
		Aldicarb at 5.5–11.0	
Grape	*Xiphinema index*	Aldicarb at 10 ml	In band for nursery use
	Meloidogyne spp.	Fenamiphos (EC formulation) at 10 ml	
Banana	*Radopholus similis*	Carbofuran at 2–4 g a.i./plant	Applied around plant two to three times per year
	Helicotylenchus multicinctus	Ethoprophos at 2–4 g a.i./plant	
	Pratylenchus spp.	Fenamiphos at 2–4 g a.i./plant	
	Meloidogyne spp.	Isazofos at 2–4 g a.i./plant	
		Ebufos at 2–4 g a.i./plant	

(Osman et al. 2025). Similarly, foliar SiNP sprays in eggplant fields combined with reduced-rate chemical nematicide cut nematode reproduction by ~50% (El-Ashry et al. 2022).

These studies illustrate that lab efficacy does translate to some field benefit, especially in improved crop vigor, even if complete nematode control is seldom achieved. In practice, nano-formulations often need repeated applications. Importantly, several trials noted no phytotoxicity at the effective NP doses, and in some cases plant health surpassed that of chemically fumigated plots.

Use of nanoparticles in nematode management is a novel and fancy approach that may prove very effective in future with the progress of application aspect of nanotechnology (Table 8.3) (Ladner et al. 2008).

8.6 Phytosynthesized Nano-Formulations for Nematode Management

8.6.1 Laboratory Studies

- *Inhibition of Egg Hatching of* M. incognita: The inhibition in egg hatching was maximum (91.49%) at the highest concentration of 0.0005% of AgNPs (The AgNPs synthesized using *Tridax procumbens*) at 7 days after inoculation followed by 87.35% in 0.0001% and 82.64% in 0.00005% (Surega 2015).
- *Morphological Changes Caused by the Phytosynthesized AgNPs on Root-Knot Nematode Egg-Masses*: SEM studies on the ultrastructure of root-knot nematode eggs on exposure to AgNPs revealed that the eggs were completely deformed. Healthy eggs were intact whereas the treated eggs had lost their structural integrity (Surega 2015).
- *Juvenile Immobility and Mortality*: Cromwell et al. (2014) observed that more than 99% of J_2 of *M. incognita* became inactive on exposure to AgNPs at 30 to 150 µg/ml of water under laboratory conditions.

Tridax procumbens-mediated synthesized AgNPs, SiO_2, and TiO_2 gave cent percent immobility and mortality of J_2 of *M. incognita* in vitro at 800, 400, and 200 mg ml/lit except SiO_2 (Ardakani 2013).

There was cent percent mortality of juveniles of *M. incognita* at higher concentrations of 0.0005 and 0.0001% of the phytosynthesized AgNPs within the shortest period of exposure of 24 h and the trend was continued at all periods of exposure. Similarly, the cent percent mortality of juveniles of *M. incognita* was also caused by the lower concentration of 0.00005 but after sixth day of exposure (Surega 2015).

- *Morphological Changes Caused by the Green Synthesized AgNPs on Root-Knot Nematode Juveniles*: The dead juveniles of root-knot nematode on exposure to the most effective concentration of 0.0005% of AgNPs appeared to be deformed with rupture of cell wall (Surega 2015).

8.6.2 Glasshouse Studies

- *Influence of Phytosynthesized AgNPs on Attraction of* M. incognita *to Tomato Roots*: The highest reduction (94.67%) in the attraction of J2 of *M. incognita* was recorded 24 h after inoculation by 0.0005% concentration and it was ranging from 59.15% to 91.14% in other concentrations of 0.00001–0.0001 of the Ag NPs irrespective of period of exposure from 1 to 7 days (Surega 2015).
- *Influence of AgNPs on Tomato Root Penetration of* M. incognita: The maximum reduction (79.76–90.16%) of root penetration of *M. incognita* (J2) was recorded by highest concentration of 0.0005% of phytosynthesized AgNPs 1 day after inoculation (Surega 2015).

- *Influence of AgNPs on Plant Growth Parameters of Tomato*: Under pot experimentations, AgNPs and TiO_2 were found to reduce the shoot height and root length of tomato compared to untreated control indicating that NPs are causing toxic effect on tomato (Ardakani 2013).

Application of phytosynthesized AgNPs gave maximum increase in plant growth parameters (shoot height—13.11–26.99%, shoot weight—13.92%, root length—26.57%, root weight—23.45%, and fruit yield—20.54% over control) (Surega 2015).

- *Influence of AgNPs on* M. incognita *Population on Tomato*: The phytosynthesized AgNPs at 0.0005% was highly effective in reducing the *M. incognita* population in soil by 88.16%, in roots by 92.79%, number of egg-masses/g root by 79.80%, eggs/egg-mass by 65.14%, and gall index (0.2 as compared to 3.0 in carbofuran) (Surega 2015).

8.6.3 Field Studies

In turf grass, AgNPs reduced *M. graminis* population in soil by 92 and 82% after 4 and 2 days of exposure, respectively, under field conditions. Similarly, the biweekly application of 90.4 mg/m^2 of AgNPs caused reduction in gall formation in root and improved quality of turf grass for 2 years without phytotoxicity (Cromwell et al. 2014).

8.7 Effectiveness Compared to Conventional Nematicides

Nano formulations often outperform bulk or conventional nematicides in efficacy as enumerated below.

8.7.1 Higher Potency at Low Dose

Laboratory bioassays frequently report near-100% mortality at much lower concentrations than chemical controls. *Meloidogyne* J2 exposed to AgNPs at 30–150 μg/mL showed >99% inactivation in 6 h (Cromwell et al. 2014), whereas typical nematicide dosages are orders of magnitude higher. Chitosan NPs (2000 ppm) killed 97% of *M. incognita* juveniles in 72 h. and reduced soil/root nematode loads by ~80–98% (Mouniga et al. 2023), comparable to systemic nematicides.

8.7.2 Enhanced Plant Benefits

Unlike toxic fumigants, many nano-treatments also stimulate plant growth. In field and greenhouse trials, nano-silica and plant-extract NPs not only suppressed nematodes but also improved plant vigor and yield. For instance, tomato treated with nano-formulated *Artemisia annua* extract had higher fruit yield and enhanced defense enzyme activity than with unformulated extract (Mohammad et al. 2022). Similarly, eggplants treated with garlic oil nano-emulsion showed significantly greater growth and fruit yield while nematode infestation was reduced, outperforming the original extract (Osman et al. 2025).

8.7.3 Broader Spectrum with Biopesticides

Nano-encapsulation can make biological agents more reliable. *Bacillus*-derived nematicidal compounds, when nano-formulated, showed more consistent activity under field conditions than unprotected metabolites (Osman et al. 2025).

8.7.4 Reducing Conventional Dose

Notably, combining NPs with lower-rate chemicals can maintain control. One report showed that adding SiNPs allowed use of only 50% standard nematicide rate while still cutting nematode reproduction and galling on eggplant (El-Ashry et al. 2022). Regulatory agencies note that nano-pesticides generally require less active ingredient and leave less non-target residue (Wang et al. 2022).

Overall, the consensus in recent reviews is that well-designed nano-formulations can exceed or match traditional nematicides' performance, often with lower dosage and collateral benefits (Osman et al. 2025). For example, Zinovieva et al. (2023) report many NPs "effective against nematodes" and sometimes "more effective than known commercial nematicides.". Complete nematode eradication is rarely achieved in the field; instead, nano-formulations tend to mitigate population levels and reduce crop damage comparably to soil fumigants.

8.8 Mechanisms of Action

Nanomaterials attack nematodes through multiple pathways:

8.8.1 Direct Toxicity

Many metal/metal-oxide NPs (e.g., Ag, Cu, ZnO, SiO_2) generate reactive oxygen species (ROS) or metal ions that damage nematode cells. For example, nematode stress-response and DNA-repair genes are upregulated upon exposure to nanoparticles, indicating oxidative damage. High surface-area NPs can penetrate or abrade the cuticle, causing mortality. In vitro assays often show extremely high juvenile death rates—e.g., *Meloidogyne* J2 exposed to 30–150 μg/mL AgNP become >99% inactive within 6 h (Cromwell et al. 2014).

8.8.2 Delivery of Actives

Nanocarriers can encapsulate or bind conventional nematicides (e.g., abamectin) to improve solubility, soil mobility, and controlled release. For instance, abamectin-loaded polymer or virus nanoparticles (e.g., PLGA-PEG or plant-virus capsids) release the pesticide more slowly and expand its zone of protection around roots (Fu et al. 2018; Cao et al. 2015). Spindle-shaped PLGA–PEG NPs with abamectin achieved 98.4% nematode mortality at just 1 ppm, outperforming spherical formulations (Fu et al. 2018). Similarly, abamectin packaged in plant-virus nanoparticles markedly increased soil mobility and reduced *Meloidogyne* galling on tomato (Cao et al. 2015).

8.8.3 Plant Immune Stimulation

Some nanoparticles indirectly protect plants by priming their defenses. Silicon-based nanomaterials (SiO_2 NPs, nano-silicon fertilizers) accumulate in roots and trigger stress-response pathways. Foliar or soil-applied SiNPs are reported to concentrate at nematode infection sites, activating plant ROS and defense enzymes and thereby reducing nematode proliferation (Yu et al. 2022). In one study, SiO_2 NPs caused degeneration of nematode reproductive organs and blackened their tissues, suggesting both direct and plant-mediated effects (Yu et al. 2022). Organic carriers like chitosan can also elicit plant resistance signals (Ca^{++}, nitric oxide, phenolics) while releasing inhibitory amines upon degradation.

8.8.4 Physical and Biochemical Interruption

Nanoparticles may clog nematode feeding structures or disturb host–pathogen signaling. Encapsulated essential oils or microbial metabolites (in nano-emulsions) can interfere with nematode sensory or digestive functions. The nanoscale size allows penetration into soil microenvironments and uptake by nematodes, enhancing contact with the active agent.

In summary, nano-nematicides can act by multiple synergistic routes: direct nematode toxicity, controlled delivery of chemicals, and activation of plant defenses (Yu et al. 2022). These mechanisms often yield higher effectiveness than bulk formulations.

8.9 Sustainability and Environmental Impact

Traditional nematicides, such as carbamates and organophosphates, are often banned due to their undesirable effects on non-target organisms and the environment. Nanotechnology offers a more sustainable approach by reducing reliance on these chemicals. Plant virus-based nanomaterials, being biodegradable and non-toxic to mammals, address public acceptance concerns, especially for non-GMO solutions. Transient dsRNA delivery, favored over transgenic approaches like SmartStax Pro maize (approved in 2017 by the USA, 2021 by China) (De Schutter et al. 2022), enhances eco-friendliness. However, challenges remain, including environmental safety concerns for beneficial insects, livestock, and humans, though lab studies show no adverse effects (Fletcher et al. 2020; Jensen et al. 2013).

Nano-formulations lower the need for broad-spectrum chemical nematicides, reducing soil and water pollution. Targeted delivery and enhanced efficacy reduce application frequency and costs for farmers. Biodegradable NPs (e.g., chitosan or starch-based) minimize long-term environmental impact compared to conventional pesticides. Nanotechnology promises reduced chemical loads, but also raises novel safety questions. Key points include:

8.9.1 Reduced Non-Target Impact

EPA research indicates nano-enabled pesticides tend to adhere better to plant surfaces and have reduced effects on non-target organisms relative to conventional formulations (Wang et al. 2022). Lower application rates of active ingredient mean less residue drift. For example, biocompatible carriers (chitosan, silica) are biodegradable or naturally occurring. Silicon and other mineral NPs are largely benign to mammals and many soil biota, making them relatively safe (Yu et al. 2022). Some nano-bioencapsulated agents (plant extracts, proteins) may also pose minimal risk if naturally degradable.

8.9.2 Potential Hazards

Conversely, metallic nanoparticles (especially nano-Ag, CuO, ZnO) can be toxic to beneficial soil organisms at high concentrations. Laboratory studies (often on *Caenorhabditis elegans*) show that ultrasmall NPs can cause oxidative stress in non-target nematodes and earthworms. Bioaccumulation in the food web is a concern if heavy metals leach into water. Engineering more "green" particles (e.g., carbon-based, lignin, algae-derived) is an active area to mitigate these risks. Current knowledge gaps include long-term fate of NPs in soil, leaching behavior, and effects on microbial communities.

8.9.3 Regulatory Status

As of 2025, few nano-nematicides have formal regulatory registration. EPA and other agencies are developing guidelines specifically for nano-pesticides. They emphasize thorough risk assessment for any novel carrier; each new nano-formulation must demonstrate safety to pollinators, wildlife, and humans. The US

EPA notes that while nano-pesticides show great promise for sustainable agriculture, additional research on adverse impacts is needed (Wang et al. 2022).

8.9.4 Worker Safety and Acceptance

Manufacturing and handling nano-formulations require safety protocols (gloves, masks, controlled release systems) to avoid inhalation or dermal exposure. Education on nano-pesticide use will be needed for growers. Public acceptance of "nano" in food production may hinge on demonstrating clear environmental benefits and absence of residues.

Overall, the environmental profile of nano-nematicides can be better than or comparable to current nematicides if designed carefully. Many studies explicitly note that the novel formulations are intended as eco-friendly alternatives (Yu et al. 2022). However, caution and monitoring remain essential as this technology is new.

8.10 Challenges and Considerations

While nanotechnology offers significant potential, several challenges must be addressed to ensure its successful application in plant nematology:

8.10.1 Research and Development

The field is rapidly evolving, with limited but encouraging reports on nanotechnology's role in nematode control. Research continues to explore new nanomaterials and their applications, such as clay nanosheets for RNAi delivery, though currently more focused on foliar applications (Liu et al. 2020), focusing on plant DNA viruses but potentially analogous. The growing attention to this area suggests potential for further innovation, addressing challenges like soil delivery efficiency and environmental safety.

8.10.2 Environmental Impact

The long-term effects of nanoparticles on soil ecosystems, beneficial microbes, and non-target organisms remain poorly understood. There is a risk of accumulation or unintended ecological consequences, necessitating further study.

8.10.3 Cost and Scalability

High production costs and limited field-scale studies hinder widespread adoption. Developing and deploying nanotechnology-based solutions can be costly, and adapting them for large-scale agricultural use poses practical challenges.

8.10.4 Regulatory Hurdles

Standardized guidelines for the safe use of nanomaterials in agriculture are still evolving. As a novel technology, nanotechnology in agriculture must comply with strict safety and efficacy regulations, which could delay its widespread adoption.

8.10.5 Toxicity Concerns

The long-term impact of NPs on soil ecosystems, beneficial microbes, and human health requires further study.

8.11 Future Directives

Nanotechnology for nematode control is promising but still emerging. Key areas that require emphasis include:

8.11.1 Scale-Up and Cost

Many studies are lab-scale. Producing nanomaterials at field quantities (tons of crops) must be cost-effective. Methods like green synthesis (using plant waste or microbes) are being optimized to reduce cost. Carrier reuse (e.g., polymers that degrade) is also being investigated.

8.11.2 Consistency and Stability

Ensuring uniform particle size, stability under diverse field conditions (pH, organic matter), and shelf-life of formulations is critical. Humic substances in soil can agglomerate NPs; research is focusing on coatings (e.g., PEG, chitosan) to enhance dispersal.

8.11.3 Regulatory Hurdles

As mentioned, regulatory pathways for nano-pesticides are still under development. Comprehensive ecotoxicology data will be needed to gain approval. Collaboration between researchers, industry, and agencies is underway to define standards.

8.11.4 Targeted Delivery

Future work aims to increase specificity. For example, nanoparticle carriers could be engineered to bind nematode-specific attractants or release actives only when nematode enzymes (e.g., cellulases) are present. Research into nanotechnology-mediated RNA interference is also advancing: combining dsRNA with clay nanosheets or plant-virus particles could silence essential nematode genes in situ (Opdensteinen et al. 2024). This approach could complement chemical control and slow resistance.

8.11.5 Field Validation

More multi-season field trials in diverse agro-ecosystems are needed. Integration into pest management (rotations, biocontrol, soil health) will determine real-world viability. Several studies have shown synergy between NPs and reduced chemical use (El-Ashry et al. 2022), indicating that nano-formulations may first be adopted as part of integrated strategies.

8.11.6 Long-Term Effects

As the technology matures, long-term field studies (multi-year monitoring of soil health, nematode populations, and crop yields) will be crucial to confirming sustainability. It will also be important to engineer nanoparticles that degrade harmlessly after action.

Despite these challenges, recent advances suggest a bright future. The ability to tailor nanoparticle composition, size, and payload for specific nematode targets offers unprecedented precision. With ongoing research, nano-nematicides could become an integral tool in sustainable nematode management worldwide (Wang et al. 2022).

8.12 Conclusion

Nanotechnology provides innovative approaches to address the challenges posed by plant-parasitic nematodes, including targeted nematicide delivery, early detection through nanosensors, enhanced plant resistance, and improved soil health. These advancements have the potential to revolutionize nematode management, making it more effective and sustainable than traditional methods. However, realizing this potential requires ongoing research to fully understand environmental impacts, reduce costs, and meet regulatory standards. As the field progresses, nanotechnology could become a cornerstone of integrated pest management strategies in plant nematology, benefiting agriculture worldwide.

In conclusion, nanotechnology is transforming plant nematology by offering innovative, sustainable solutions for controlling phytoparasitic nematodes, with ongoing research poised to address current limitations and enhance agricultural sustainability.

8.13 Suggested Readings

- Jena, R., Das, M., Chattopadhyay, A. et al. 2024. Nanotechnology. In: S. Roy and A. Hossain (eds.) *An Exciting Perspective for Nematode Management*, 277–298. CRC Press, Boca Raton. [Discusses different nanomaterials (metal nanoparticles, nano emulsions, nano-capsules)] and their role in nematode suppression.
- Abd-Elgawad, M.M.M. 2024. Revolutionizing nematode management—Nanomaterials as a promising approach for managing economically important plant-parasitic nematodes: Current knowledge and future challenges. In: I. Mahmood, R.A. and Ansari, R. Rizvi (eds.) *Nanotechnology and Plant Disease Management*. CRC Press, Boca Raton. [comprehensive review of nano-bioformulations (nano-encapsulated biocontrol agents, essential oils, etc.)].
- Devi, G. 2024. Utilization of Nanotechnology for the Management of Plant-parasitic Nematodes. *Archives of Current Research International* 24(12): 347–357.
- Opdensteinen, P., Charudattan, R., Hong, J.C. et al. 2024. Biochemical and nanotechnological approaches to combat phytoparasitic nematodes. *Plant Biotechnol J.* 22(9): 2444–2460.
- Sabry, K.H. 2019. Role of Nanotechnology Applications in Plant-Parasitic Nematode Control., pp. 223–240. In: K. Abd-Elsalam and R. Prasad (eds.) *Nanobiotechnology Applications in Plant Protection*, pp. 223–240. Springer, Cham.
- Wang, D., Saleh, N.B., Byro, A. et al. 2022. Nano-enabled pesticides for sustainable agriculture and global food security. *Nat. Nanotechnol.* 17: 347–360.
- Zinovieva S.V., Udalova Z.V. and Khasanova O.S. 2023. Nanomaterials in plant protection against parasitic nematodes. *Uspehi sovremennoj biologii* 143(3): 278–299.

8.14 Review Questions

- Explain the role of nanotechnology in revolutionizing plant nematode diagnosis and management. Highlight its advantages over conventional methods.
- Critically analyze the mechanisms through which nano-pesticides or nano formulations act against plant-parasitic nematodes.
- What are the challenges in commercialization and field application of nanoformulations for nematode control?
- Write a critical review on the role of green-synthesized nanoparticles in sustainable nematode management.

- What are the limitations of current diagnostic tools for plant-parasitic nematodes, and how can nanotechnology overcome them?
- Discuss future prospects: Could nanotechnology pave the way for precision.

References

Ahmed HNE, Bahig AE (2018) Effectiveness of silver nanoparticles against root-knot nematode, *Meloidogyne incognita* infecting tomato under greenhouse conditions. J Agric Sci 10(2):148–156

Ardakani AS (2013) Toxicity of silver, titanium and silicon nanoparticles on the root-knot nematode, *Meloidogyne incognita* and growth parameters of tomato. Nematol 15(6):671–677

Cao J, Guenther RH, Sit TL et al (2015) Development of abamectin loaded plant virus nanoparticles for efficacious plant parasitic nematode control. ACS Appl Mater Interfaces 7(18):9546–9553

Caparco AA, González-Gamboa I, Hays SS et al (2023) Delivery of nematicides using TMGMV-derived spherical nanoparticles. Nano Lett 23(12):5785–5793

Charudattan R (2023) Use of plant viruses as bioherbicides: the first virus-based bioherbicide and future opportunities. Pest Manag Sci 80:103–114

Christiaens O, Tardajos MG, Martinez Reyna ZL et al (2018) Increased RNAi efficacy in *Spodoptera exigua* via the formulation of dsRNA with guanylated polymers. Front Physiol 9:316

Cromwell WA, Yang J, Starr JL, Jo YK (2014) Nematicidal effects of silver nanoparticles on root-knot nematode in Bermuda grass. J Nematol 46(3):261–266

De Schutter K, Taning CNT, Daele LV (2022) RNAi-based biocontrol products: market status, regulatory aspects, and risk assessment. Front Insect Sci 1:818037. https://doi.org/10.3389/finsc.2021.818037

Dubelman S, Fischer J, Zapata F et al (2014) Environmental fate of double-stranded RNA in agricultural soils. PLoSONE 9(3):e93155. https://doi.org/10.1371/journal.pone.0093155

El-Ashry RM, El-Saadony MT, El-Sobki AEA et al (2022) Biological silicon nanoparticles maximize the efficiency of nematicides against biotic stress induced by *Meloidogyne incognita* in eggplant. Saudi J Biol Sci 29(2):920–932

El-Deen AHN, El-Deeb BA (2018) Effectiveness of silver nanoparticles against root-knot nematode, *Meloidogyne incognita* infecting tomato under greenhouse conditions. J Agric Sci 10(2):148–156

Fletcher SJ, Reeves PT, Hoang BT, Mitter N (2020) A perspective on RNAi-based biopesticides. Front Plant Sci 11:51. https://doi.org/10.3389/fpls.2020.00051

Fu Z, Chen K, Li L et al (2018) Spherical and spindle-like abamectin-loaded nanoparticles by flash nanoprecipitation for southern root-knot nematode control: preparation and characterization. Nano 8(6):449. https://doi.org/10.3390/nano8060449

Jensen PD, Zhang Y, Wiggins BE et al (2013) Computational sequence analysis of predicted long dsRNA transcriptomes of major crops reveals sequence complementarity with human genes. GM Crops Food 4:90–97

Jo YK, Starr JL, Deng Y (2013) Use of silver nanoparticles for nematode control on Bermuda grass putting green. Turf Grass Environ Res 12(2):22–24

Ladner DC, Tchounwou PB, Lawrence GW (2008) Evaluation of the effect of ecologic on root knot nematode, *Meloidogyne incognita*, and tomato plant, *Lycopersicon esculenum*. Int J Environ Res Public Health 5:104–110

Liu Q, Li Y, Xu K et al (2020) Clay nanosheet - mediated delivery of recombinant plasmids expressing artificial miRNAs via leaf spray to prevent infection by plant DNA viruses. Hortic Res 7(1):179

Mitter N, Worrall EA, Robinson KE et al (2017) Clay nanosheets for topical delivery of RNAi for sustained protection against plant viruses. Nat Plants 3:16207

Mohammad AA, Amer HM, El-Sawy SM et al (2022) Nematicidal activity of sweet annie and garden cress nano-formulations and their impact on the vegetative growth and fruit quality of tomato plants. Sci Rep 12:22302. https://doi.org/10.1038/s41598-022-26819-2

Mouniga R, Anita B, Lakshmanan A et al (2023) Nematicidal properties of chitosan nano formulation. J Nematol 55. https://doi.org/10.2478/jofnem-2023-0033e2023-1

Numata K, Ohtani M, Yoshizumi T et al (2014) Local gene silencing in plants via synthetic dsRNA and carrier peptide. Plant Biotechnol J 12(8):1027–1034

Opdensteinen P, Charudattan R, Hong JC et al (2024) Biochemical and nanotechnological approaches to combat phytoparasitic nematodes. Plant Biotechnol J 22(9):2444–2460

Osman HA, Ameen HH, Abd El-Aziz ME et al (2025) Comparative field study of silver nanoparticles and garlic oil nano emulsion for nematode control and yield enhancement in eggplant. Sci Rep 15:20220. https://doi.org/10.1038/s41598-025-06697-0

Surega R (2015) Green synthesis of bioactive silver nanoparticles using plant extracts and their Antinemic properties. Ph. D. thesis, Department of Nematology, Tamil Nadu Agriculture University, Coimbatore

Wang D, Saleh NB, Byro A et al (2022) Nano-enabled pesticides for sustainable agriculture and global food security. Nat Nanotechnol 17:347–360

Wen AM, Steinmetz NF (2016) Design of virus-based nanomaterials for medicine, biotechnology, and energy. Chem Soc Rev 45(15):4074–4126

Yu J, Yu X, Li C et al (2022) Silicon mediated plant immunity against nematodes: summarizing the underline defence mechanisms in plant nematodes interaction. Int J Mol Sci 23(22):14026. https://doi.org/10.3390/ijms232214026

Zhang X, Zhang J, Zhu KY (2010) Chitosan/double- stranded RNA nanoparticle-mediated RNA interference to silence chitin synthase genes through larval feeding in the African malaria mosquito (*Anopheles gambiae*). Insect Mol Biol 19:683–693

Zinovieva SV, Udalova ZV, Khasanova OS (2023) Nanomaterials in plant protection against parasitic nematodes. Uspehi Sovremennoj Biologii 143(3):278–299

Remote Sensing/Drone Technology: Innovation to Nematode Detection and Monitoring

9

Abstract

Remote sensing offers a noninvasive approach to diagnose and manage plant-parasitic nematodes (PPNs) in agriculture, addressing the significant economic losses they cause, estimated at $80–157 billion annually. PPNs, such as root-knot (*Meloidogyne* spp.) and cyst nematodes (*Heterodera* and *Globodera* spp.), are challenging to detect due to their microscopic size and often nonspecific symptoms like stunted growth or chlorosis. Traditional methods rely on labor-intensive soil sampling and microscopic analysis, but remote sensing provides a scalable, efficient alternative by detecting plant stress and nematode infestations through spectral reflectance and imaging technologies.

Keywords

Remote sensing · Drone technology · Nematode detection · Hyperspectral imaging · Spectral signatures · Thermal imaging

9.1 Learning Objectives

- Understand the principles of remote sensing and drone-based imaging in agriculture, with special reference to nematode-infested crop systems.
- Explain the types of sensors (multispectral, hyperspectral, thermal, LiDAR) used in drones for detecting nematode-induced stress in plants.
- Assess the advantages, challenges, and limitations of drone technology for nematode diagnosis and management under field conditions.
- Learn how drones and remote sensing help detect nematode damage in crops.
- Understand the sensors and vegetation indices used for diagnosing nematode stress.

R. P. Parvatha, *Plant Parasitic Nematodes*,
https://doi.org/10.1007/978-981-95-9306-4_9

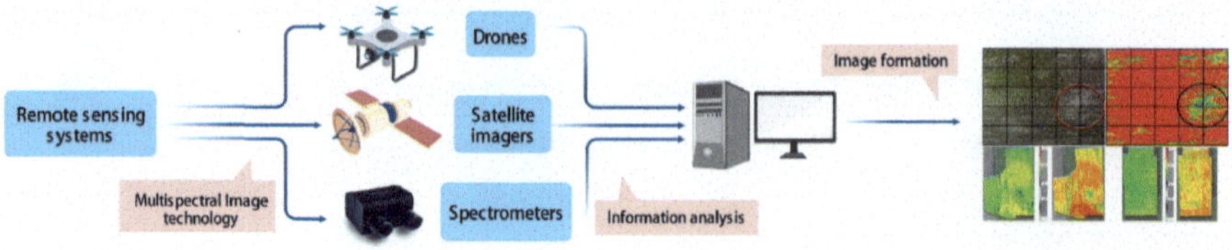

Fig. 9.1 Demonstration of the principles and working processes involved in remote sensing technologies

- Explore how drones enable mapping and precision management of nematode hotspots.
- Recognize benefits, challenges, and future prospects of drone-based nematode management.

9.2 Introduction

Remote sensing and unmanned aerial vehicle (UAV) (drone) platforms are emerging as powerful tools to detect plant-parasitic nematodes and guide their management in crops. Nematode infestations often induce subtle stress in plants (chlorosis, reduced vigor) before symptoms are visible; these stresses alter canopy reflectance and temperature. By capturing spectral data from above—via satellites, aircraft, or drones—researchers can identify *"hotspots"* of nematode damage across fields (Fig. 9.1). For example, pine trees dying from pine wilt disease showed distinct hyperspectral signatures; a study using a DJI Matrice-600 UAV and hyperspectral camera achieved tree-level classification accuracies of 75–95% by late summer (Yu et al. 2022). Similarly, in soybean and potato studies, vegetation indices (VIs) such as NDVI, NDRE, or chlorophyll index derived from multispectral imagery correlated strongly with nematode populations and yield loss (Chandel et al. 2024; Jindo et al. 2023). In one soybean case, spatial maps of NDRE (Normalized Difference Red-Edge) from drone flights clearly delineated areas of high nematode pressure (Fig. 9.2). These imagery-based stress signals can then be integrated into decision models or used to target ground sampling.

Drones can help diagnose plant nematodes by detecting stress symptoms early. They aid nematode management by monitoring treatment effectiveness and soil health. The evidence leans toward drones being useful for large-scale agriculture, but direct nematode targeting is challenging. This chapter provides a comprehensive overview, drawing on recent research to explore how remote sensing technologies are utilized, their effectiveness, and future potential, with a focus on plant parasitic nematodes.

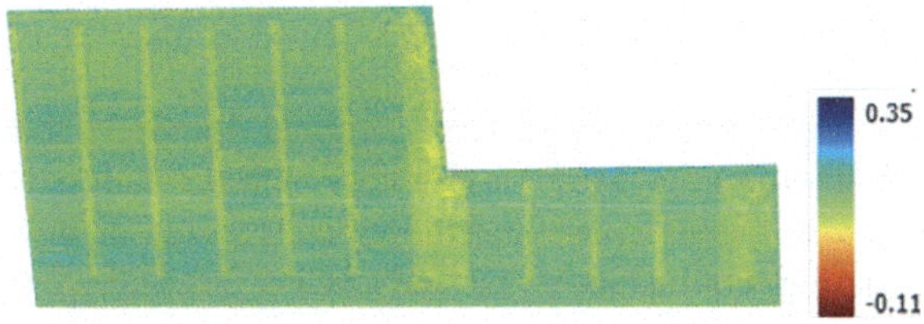

Fig. 9.2 NDRE vegetation index map of the soybean trials derived from aerial multispectral imagery. Such maps allow identification of nematode hotspots for targeted management

9.3 Diagnosis

Key detection methods include:

9.3.1 Vegetation Indices and Spectral Signatures

Plant stress from nematodes alters reflectance, especially in the visible/NIR bands. Indices like NDVI, NDRE, or green-NDVI can highlight areas of reduced vigor (Chandel et al. 2024). (e.g., *Fusarium* or nematode damage often increases red/NIR reflectance).

9.3.2 Multispectral and Hyperspectral Imaging

Multispectral drones (RGB + NIR/red-edge) give quick survey data. Hyperspectral sensors capture hundreds of narrow bands, allowing identification of specific "nematode stress" spectral signatures. Hyperspectral imaging has been used to detect reniform nematodes (*Rotylenchulus reniformis*) in cotton by measuring reflectances at 451–949 nm.

9.3.3 Thermal Imaging

Heavy nematode infestation can cause stomatal closure and reduced transpiration, raising canopy temperature. Drone thermal cameras have been shown to detect such differences. In a Dutch study of potato cyst nematode (PCN), canopy temperature from a UAV thermal camera correlated positively with nematode density, with infested areas noticeably warmer (Jindo et al. 2023).

9.3.4 Multi-platform Sensing

Satellites provide broad area coverage (e.g., Sentinel, Landsat), though at coarser resolution. Drones fly low (e.g., 10–50 m) for ultra high-resolution maps. By choosing flight altitudes and sensors, drones can tailor pixel resolution (down to cm-scale) for field-scale scouting. For instance, one study flew a DJI Matrice 600 at 10–50 m and showed that lower altitude (10 m) gave finer detail but many more images to stitch (Arjoune et al. 2022). Drones also avoid cloud-cover issues and can be flown on demand (Chandel et al. 2024).

9.4 Vegetation Indices and Spectral Signatures

Vegetation indices (VIs) and spectral signatures are increasingly utilized for the noninvasive detection of plant-parasitic nematodes (PPNs). These methods leverage hyperspectral and multispectral remote sensing to detect subtle changes in plant physiology and biochemistry caused by nematode infestations, often before visible symptoms appear (Žibrat et al. 2020).

9.4.1 Vegetation Indices for PPN Detection

Vegetation indices are mathematical transformations of spectral reflectance data typically derived from specific wavelengths, to highlight plant health, stress, or disease. They are calculated using reflectance in visible (400–700 nm), near-infrared (NIR, 700–1100 nm), and sometimes shortwave infrared (SWIR) regions, which are sensitive to changes in chlorophyll, water content, and structural traits affected by PPNs.

- *Normalized Difference Vegetation Index (NDVI)*:
  - *Description*: NDVI = (NIR—Red) / (NIR + Red), where NIR is reflectance in the near-infrared band and Red is the red band (around 620–700 nm). It is widely used to assess plant vigor and biomass.
  - *Application in PPN Detection*: NDVI is effective in detecting nematode-induced stress by capturing reductions in chlorophyll content and leaf area. For example, a study on coffee crops infected with root-knot nematodes (*Meloidogyne* spp.) achieved a classification accuracy of 78% (kappa coefficient 0.71) using NDVI to differentiate healthy from early and severely infected plants (Pun et al. 2024).
  - *Limitations*: NDVI alone may not distinguish nematode damage from other stressors (e.g., drought, fungal infections) due to its lack of disease specificity. It also showed limited ability to differentiate healthy from infected coffee plants using leaf area index (LAI) or chlorophyll content (SPAD) (Pun et al. 2024).

- *Other Vegetation Indices.*
  - *Water Index (WI)*: Sensitive to leaf water content, WI correlates with nematode infection in wet and mild conditions, as seen in sugar beet fields infected with *Heterodera schachtii* (Pun et al. 2024).
  - *Simple Ratio Pigment Index (SRPI) and Structural Independent Pigment Index (SIPI)*: These indices target pigment changes and structural alterations, showing high correlation with nematode population density in sugar beets (Pun et al. 2024).
  - *Photochemical Reflectance Index (PRI) and Carotenoid Reflectance Index (CRI)*: Used in maize disease studies, these indices detect changes in photosynthetic efficiency and carotenoid content, with CRI showing significant potential ($P < 0.000$) across growth stages (Nkuna et al. 2024).
  - *Anthocyanin Reflectance Index (ARI)*: Useful for detecting stress-related pigment changes, though less effective than NDVI or CRI in some contexts (Nkuna et al. 2024).
- *Spectral Disease Indices (SDIs):*
  - SDIs are tailored indices designed for specific diseases, including nematode infestations. For example, in sugar beet studies, SDIs were developed using hyperspectral signatures (450–950 nm) to differentiate healthy leaves from those affected by Cercospora leaf spot, powdery mildew, and sugar beet rust, with potential applicability to nematode damage (Nkuna et al. 2024).
  - SDIs rely on algorithms like RELIEF-F to identify significant wavelengths, enhancing disease-specific detection (Nkuna et al. 2024).
- *Spectral Signatures for PPN Detection*: Spectral signatures refer to the unique reflectance patterns of plants across a range of wavelengths, captured using hyperspectral sensors (e.g., 350–2500 nm). These signatures are altered by PPNs due to changes in leaf pigments, water content, and cellular structure.
- *Key Spectral Regions*:
  - *Visible Region (400–700 nm)*: Reflectance in the green (500–570 nm) and red (620–700 nm) regions is highly sensitive to chlorophyll loss caused by nematode feeding, which reduces photosynthetic capacity. For example, studies on sugar beet and maize noted increased reflectance in the red region due to nematode-induced stress (Pun et al. 2024; Nkuna et al. 2024).
  - *Red-Edge (680–750 nm)*: The red-edge, a sharp transition between red and NIR, is critical for detecting early stress. It was effective in distinguishing healthy from nematode-infected coffee plants (Pun et al. 2024).
  - *Near-Infrared (NIR, 700–1100 nm)*: NIR reflectance decreases with reduced leaf area and structural damage caused by nematodes, as seen in coffee and sugar beet studies (Pun et al. 2024; Nkuna et al. 2024).
  - *Shortwave Infrared (SWIR, 1100–2500 nm)*: Sensitive to water content, SWIR is useful in wet conditions but less reliable in hot, dry seasons (Pun et al. 2024).

9.4.2 Hyperspectral Sensors and Platforms

- *Ground-Based Sensors*: Spectro-radiometers like the ASD FieldSpec Pro (350–2500 nm) provide high-resolution data for precise spectral analysis, used in sugar beet and maize studies (Pun et al. 2024; Nkuna et al. 2024).
- *Airborne Sensors*: Hyperspectral mappers (e.g., 126 bands, 450–2500 nm) and RapidEye sensors capture canopy-level data, as used in sugar beet and coffee studies (Pun et al. 2024).
- *Drones (UAVs)*: UAVs equipped with multispectral sensors (e.g., Phantom 4, Sequoia, Sentera RedEdge) enable rapid, high-resolution imaging. In soybean fields, red and NIR bands detected *H. glycines* and *Pratylenchus brachyurus* at the flowering stage (Arantes et al., 2023).
- *Satellite-Based Sensors*: RapidEye sensors provided radiometric data for coffee crop monitoring, though spatial resolution can limit accuracy (Pun et al. 2024).

9.4.3 Analytical Methods

- *Spectral Angle Mapper (SAM)*: Used to classify hyperspectral images, SAM achieved 72% accuracy (kappa 0.65) in sugar beet fields for *H. schachtii* detection (Pun et al. 2024).
- *First Derivative Reflectance*: Peaks at 715–722 nm in maize studies helped discriminate growth stages and disease severity (Nkuna et al. 2024).
- *Generalized Linear Models (GLM)*: GLMs with binomial link functions identified NDVI and CRI as robust indicators across maize growth stages (Nkuna et al. 2024).
- *Inverse Distance Weighting*: This interpolation method enhances spatial distribution maps of nematode infestations (Pun et al. 2024).

9.4.4 Case Studies

- *Coffee* (Meloidogyne *spp.*): Hyperspectral data from RapidEye sensors showed that red, NIR, and red-edge spectra outperformed LAI and SPAD in detecting *Meloidogyne* spp. infections, with NDVI achieving 78% accuracy (Pun et al. 2024).
- *Sugar Beet* (H. schachtii): Hyperspectral data (450–2500 nm) correlated NDVI, WI, SRPI, and SIPI with nematode density, achieving 72% classification accuracy in wet conditions (Pun et al. 2024).
- *Soybean* (H. glycines, P. brachyurus): UAV-based multispectral sensors identified red and NIR bands as key for detecting nematode-induced stress at the flowering stage, with multiple linear regression improving R^2 over simple models (Arantes et al., 2023).

- *Pine Trees* (*Bursaphelenchus xylophilus*): Drone-based imaging and GPS data detected pine wilt disease, with spectral changes in green and NIR regions indicating infection (Pun et al. 2024; Cotrozzi 2022).

9.4.5 Advantages

- *Noninvasive and Presymptomatic Detection*: Hyperspectral sensing detects stress before visible symptoms, enabling early intervention (Žibrat et al. 2020; Lowe et al. 2017).
- *Scalability*: UAVs and satellites allow monitoring of large areas, unlike traditional microscopy-based methods (Arantes et al., 2023).
- *High Sensitivity*: Narrow-band hyperspectral data capture subtle physiological changes, improving accuracy over multispectral systems (Žibrat et al. 2020).

9.4.6 Limitations

- *Environmental Variability*: Hot and dry conditions reduce the reliability of spectral signatures, particularly in SWIR bands (Pun et al. 2024).
- *Disease Specificity*: Most VIs lack specificity to differentiate nematode damage from other stressors (e.g., fungal diseases, drought). SDIs are a promising solution but require further development (Pun et al. 2024).
- *Cost and Complexity*: Hyperspectral sensors and data processing are expensive and require expertise, limiting adoption in resource-constrained regions (Pun et al. 2024).
- *Spatial Resolution*: Satellite data may lack the resolution needed for small-scale infestations, though UAVs mitigate this (Pun et al. 2024).

9.4.7 Future Directions

- *Development of SDIs*: Tailored indices for specific nematode species could improve detection accuracy (Mahlein et al. 2013).
- *Integration with Machine Learning*: Deep learning models, like YOLOv5 for nematode detection in soil samples, could enhance spectral data analysis (Agarwal et al. 2023; Yuan et al. 2024).
- *Improved Sensors*: Higher-resolution hyperspectral sensors and lower-cost multispectral alternatives could broaden accessibility (Pereira et al. 2024).
- *Long-Term Monitoring*: Continuous monitoring using drones and satellites could track nematode population dynamics over time (Cotrozzi 2022).

9.4.8 Conclusion

Vegetation indices like NDVI, WI, SRPI, SIPI, PRI, and CRI, combined with hyperspectral signatures in the visible, red-edge, and NIR regions, offer a powerful approach for detecting PPNs in crops like coffee, sugar beet, soybean, and pine trees. While NDVI is the most widely used, achieving up to 78% accuracy in some studies, tailored SDIs and advanced analytical methods like SAM and GLMs show promise for improving specificity. However, challenges like environmental variability, cost, and lack of disease specificity require further research to make these tools practical for widespread agricultural use. Integrating remote sensing with machine learning and developing affordable sensors will be critical for sustainable PPN management (Pun et al. 2024; Mahlein et al. 2013; Žibrat et al. 2020).

9.5 Multispectral and Hyperspectral Imaging

These techniques leverage remote sensing to analyze spectral reflectance, enabling noninvasive detection of nematode-induced plant stress, often before visible symptoms appear. This is crucial for early intervention, supporting sustainable agriculture and precision farming.

Multispectral and hyperspectral imaging are advanced tools used to detect plant parasitic nematodes. These methods analyze how plants reflect light across different wavelengths to spot stress or damage caused by nematodes, often before visible symptoms appear. This early detection is vital for protecting crop yields and managing agricultural health effectively. Multispectral imaging uses a few specific light wavelengths to assess plant health, making it cost-effective and suitable for large areas, like fields monitored by drones or satellites. Hyperspectral imaging captures many detailed light wavelengths, offering a deeper look at plant conditions, ideal for precise, research-focused applications.

9.5.1 Background and Economic Impact

Plant parasitic nematodes, including key genera like *Globodera*, *Heterodera* (cyst nematodes), *Meloidogyne* (root-knot nematodes), and the pinewood nematode *Bursaphelenchus xylophilus* in forestry, pose substantial threats to crop production. Traditional detection methods rely on visual inspection based on characteristic symptoms, which are labor-intensive, subject to human bias, and often detect infestations only after significant damage has occurred. Remote sensing, including multispectral and hyperspectral imaging, offers a promising alternative by providing accurate, scalable detection without the need for individual plant checks or manual uprooting.

9.5.2 Multispectral Imaging: Overview and Applications

Multispectral imaging captures images in a few specific wavelength bands, typically in the visible, near-infrared (NIR), and sometimes short-wave infrared (SWIR) ranges. It is more cost-effective and practical for large-scale applications, often deployed via drones or satellites. Studies have demonstrated its utility in detecting nematode impacts indirectly through vegetation indices (VIs) that reflect plant health and stress.

In coffee fields infected with root-knot nematode, hyperspectral leaf scans pinpointed NIR bands sensitive to nematode damage (Martins et al. 2017); when simulated to a 5-band satellite sensor (RapidEye), red, red-edge, and NIR bands (plus NDVI) enabled classifying healthy versus infected trees at ~78% accuracy (Chandel et al. 2024). Hyperspectral imaging has been used to detect reniform nematodes (*R. reniformis*) in cotton by measuring reflectances at 451–949 nm.

- *Case Study—Coffee*: Orlando and Vieira (2024) evaluated orbital multispectral imaging using the Planet sensor to discriminate effective strategies for reducing nematode populations in coffee. Coffee plants were treated with various biological and chemical nematicides, and multispectral data classified treatments with 80% global accuracy. For instance, *Bacillus amyloliquefaciens* isolate B266 reduced nematode populations by 35.90%, and *B. subtilis* isolate B33 by 55.13%, highlighting multispectral imaging's role in assessing management efficacy.
- *Case Study—Soybeans*: Chandel et al. (2024) used drone-based multispectral imaging (DJI Phantom 4 Multispectral) to evaluate nematode impacts on soybeans. Vegetation indices like GNDVI ($r = -0.6$), NDRE ($r = -0.5$), and MSAVI2 ($r = -0.37$) correlated negatively with nematode populations, indicating plant stress. Machine learning models, such as Random Forest ($r = 0.93$, rRMSE = 13.32%), enabled preharvest yield estimation, demonstrating multispectral imaging's potential for large-scale monitoring.

Multispectral imaging is particularly suited for monitoring large areas, as it can detect general stress indicators but may miss subtle spectral changes. Its feasibility is enhanced by lower costs, making it accessible for farmers, especially when integrated with drone technology.

9.5.3 Hyperspectral Imaging: Overview and Applications

Hyperspectral imaging captures images across many contiguous wavelength bands, providing a continuous spectrum for each pixel. This high spectral resolution allows for detailed analysis of plant biochemical changes, making it ideal for precise detection and research applications.

- *Case Study—Potato Tubers*: Žibrat et al. (2020) used hyperspectral imaging to detect root-knot nematode (*M. luci*) infestations in potato tubers. Reflectance spectra were collected in 448 bands (VNIR: 400–988 nm, 160 bands; SWIR: 950–2500 nm, 288 bands) using Norsk Elektro Optikk AS push-broom cameras. Dimensionality reduction methods, including Linear Discriminant Analysis (LDA), achieved 100% accuracy for both inside and outside tuber images, with all methods exceeding 60% accuracy. Relevant spectral regions included 1500–1600 nm, 1850–2000 nm, and 2300–2450 nm, linked to hydrocarbons, polysaccharides, and glucose, indicating nematode-induced changes.

- *Early Detection Potential*: Hyperspectral imaging has shown promise for presymptomatic detection, as evidenced by studies on other plant diseases, such as bacterial leaf spot in tomatoes (Zhang et al. 2023). This capability extends to nematodes, with research suggesting detection before visible symptoms, crucial for timely management.

Hyperspectral imaging's detailed spectral data enable it to identify specific signatures of nematode damage, but its complexity and cost limit widespread field use. It is often employed in research settings or for high-value crops where precision is paramount.

9.5.4 Comparative Analysis

Both techniques detect nematode impacts by analyzing plant health rather than directly imaging the nematodes, given their microscopic size. Hyperspectral imaging offers higher precision due to its detailed spectral resolution, while multispectral imaging is more practical for large-scale and cost-effective monitoring. A key finding from the potato study suggests that hyperspectral data can inform the design of multispectral systems with selected bands (e.g., ~5 bands) for similar classification accuracy, potentially reducing costs while maintaining effectiveness (Table 9.1).

Table 9.1 Comparative analysis of multispectral and hyperspectral imaging for detection of nematodes

Aspect	Multispectral imaging	Hyperspectral imaging
Spectral resolution	Few specific bands (e.g., visible, NIR, SWIR)	Many contiguous bands (e.g., 448 bands in study)
Cost and complexity	Lower cost, more practical for large-scale use	Higher cost, more complex, research-focused
Detection method	Indirect, via vegetation indices (e.g., NDVI, GNDVI)	Direct analysis of spectral signatures, higher accuracy
Accuracy examples	80% for coffee treatment discrimination, r = −0.6 for soybean GNDVI	Up to 100% for potato tuber classification (LDA)
Scalability	High, suitable for drones/satellites	Lower, often lab-based or small-scale
Early detection	Possible, but less detailed than hyperspectral	Strong, presymptomatic detection demonstrated

9.5.5 Advantages and Limitations

- *Advantages*:
  - Both methods enable noninvasive, remote detection, eliminating labor-intensive manual inspections.
  - They support early detection, critical for reducing crop losses, with hyperspectral offering presymptomatic capabilities and multispectral providing scalability.
  - Integration with machine learning enhances accuracy, as seen in soybean yield estimation and coffee treatment classification.
- *Limitations*:
  - Multispectral imaging may miss subtle spectral changes detectable by hyperspectral systems, potentially reducing sensitivity for early detection.
  - Hyperspectral systems are less practical for widespread field use due to higher costs and complexity.
  - Both techniques primarily detect nematode effects on plants (e.g., stress, biochemical changes) rather than the nematodes themselves, which may limit direct quantification.

9.5.6 Future Directions

Ongoing research aims to optimize these techniques for real-time, field-based applications. For instance, developing dedicated multispectral systems based on hyperspectral data could balance cost and accuracy. Integration with weather and soil factors, as explored in soybean studies, may further enhance detection capabilities. Additionally, machine learning and AI are increasingly used to automate classification, improving efficiency and precision in nematode detection.

9.5.7 Conclusion

Multispectral and hyperspectral imagings are valuable tools for detecting plant parasitic nematodes, with hyperspectral offering higher precision for research and early detection, and multispectral providing practical scalability for large-scale monitoring. Their noninvasive nature and ability to detect infestations before visible symptoms make them essential for modern agriculture, supporting precision farming and sustainable crop management. Continued advancements may further bridge the gap between cost, accuracy, and scalability, enhancing their adoption in agricultural practices.

9.6 Thermal Imaging

Thermal imaging, or infrared thermography, shows promise as a nondestructive method for detecting plant-parasitic nematodes (PPNs) by monitoring changes in plant canopy temperature caused by nematode infestations. PPNs, such as root-knot nematodes (*Meloidogyne* spp.) and cyst nematodes (*Heterodera* spp.), damage plant roots, reducing water uptake and altering transpiration, which can lead to detectable temperature differences in leaves or tubers.

9.6.1 How Thermal Imaging Works for PPN Detection

- *Principle*: Thermal imaging detects infrared radiation to measure surface temperature variations. Nematode infestations disrupt plant water status, often increasing canopy temperature due to reduced evaporation from damaged roots (Schmitz et al. 2004).
- *Application*: It captures temperature differences in crop canopies, which may indicate stress from PPNs before visible symptoms appear. For example, in sugar beets, canopy temperature correlated significantly with *H. schachtii* density in susceptible cultivars, detectable via high-resolution digital thermography (Schmitz et al. 2004).

9.6.2 Advantages and Limitations

- *Advantages*:
  - Noninvasive and rapid, allowing real-time monitoring.
  - Can detect presymptomatic stress, enabling early intervention (Pineda et al. 2021).
  - Reduces reliance on chemical treatments by pinpointing affected areas.
- *Limitations*:
  - Environmental factors (e.g., air temperature, humidity, canopy architecture) can interfere with thermal data, requiring corrections for accurate interpretation (Pineda et al. 2021).
  - Less effective for resistant cultivars where temperature changes are minimal (Pineda et al. 2021).
  - Requires training to interpret thermal patterns, as misinterpretation is common.

9.6.3 Key Studies and Findings

- *Sugar Beets and* H. schachtii:
  - Research showed that thermal imaging could distinguish between low and high nematode infestations in susceptible sugar beet cultivars, with canopy temperature serving as an early stress indicator. It was less effective for toler-

  ant cultivars like BlueFox, where no significant correlation was found (Schmitz et al. 2004).
  - Visible imaging outperformed thermography for early detection (15 days after sowing), but thermography was useful once the canopy reached full coverage (Joalland et al. 2017).
- *Potato Tubers and* Meloidogyne luci:
  - Hyperspectral imaging, often combined with thermal methods, detected latent infestations in potato tubers without visible symptoms. While not directly thermal imaging, this suggests potential for integrating thermal data with other imaging techniques for enhanced detection (Schmitz et al. 2004; Žibrat et al. 2020).
- *General Plant Stress Detection*:
  - Thermal imaging has been used to detect biotic stress from other pathogens, like tomato powdery mildew, by measuring maximum temperature differences (MTD) ranging from 0.2 C to 2 C during incubation. Similar principles could apply to PPNs, though specific MTD ranges for nematodes are less documented (Zhu et al. 2018).
  - Studies on wheat and tomato plants showed thermal imaging's ability to detect disease-induced temperature changes before visible symptoms, suggesting its potential for PPNs (Raza et al. 2015).

9.6.4 Challenges and Considerations

- *Environmental Interference*: Factors like leaf angles, sunlit/shaded areas, and ambient conditions affect thermal readings, necessitating calibration (e.g., correcting for emissivity, humidity) (Panda and Rathore 2019).
- *Resolution Needs*: Detecting subtle temperature changes requires high-resolution cameras (e.g., FLIR Exx-Series for precise measurements vs. entry-level FLIR ONE for broader hotspots).
- *Integration with Other Methods*: Combining thermal imaging with hyperspectral imaging, spectrometry, or molecular diagnostics (e.g., real-time PCR) improves accuracy, especially for latent infestations (Žibrat et al., 2020; Pun et al. 2024).
- *Field Application*: Natural temperature variations in crop canopies can obscure PPN-related signals, making stand-alone thermal imaging less reliable outdoors compared to controlled environments (Hoffmann et al. 2013).

9.6.5 Practical Recommendations

- *Equipment*: Use high-sensitivity thermal cameras (e.g., FLIR Exx-Series) for detecting subtle temperature changes in plants.
- *Calibration*: Adjust for environmental factors like ambient temperature and humidity using tools like portable Dew Point Meters (Panda and Rathore 2019).

- *Complementary Techniques*: Pair thermal imaging with hyperspectral imaging or deep learning (e.g., CNN for automated PPN detection) to enhance accuracy and handle complex cases (Agarwal et al. 2023; Pun et al. 2024).
- *Training*: Operators need training to interpret thermal patterns accurately, as misinterpretation can lead to false positives.

9.6.6 Future Directions

- *Improved Sensors*: Advances in thermal camera resolution and affordability could make field-scale PPN detection more feasible (Pineda et al. 2021).
- *Machine Learning Integration*: Combining thermal data with AI-driven image analysis could automate detection and improve precision, especially for complex nematode infestations (Agarwal et al. 2023).
- *Standardized Protocols*: Developing standardized thermal thresholds (e.g., MTD values specific to PPNs) could enhance reliability across crops and environments (Hashim et al. 2020).

9.6.7 Conclusion

Thermal imaging is a promising tool for detecting PPN infestations by identifying temperature changes caused by root damage, particularly in susceptible crops like sugar beets and potatoes. However, its effectiveness is limited by environmental variables and the need for high-resolution equipment. For optimal results, it should be integrated with hyperspectral imaging, molecular diagnostics, or deep learning methods. While not yet a stand-alone solution for PPN detection, ongoing advancements in technology and data analysis could make it a valuable part of integrated pest management in agriculture.

9.7 Multi-platform Sensing

Multi-platform sensing involves using different technologies, like satellites for broad views, drones for detailed imaging, and ground sensors for close-up data, to detect plant parasitic nematodes. This method helps to detect nematodes early without digging up plants.

9.7.1 Conceptual Framework

Multi-platform sensing refers to the integration of data collected from multiple sensing platforms, such as satellites, UAVs, and ground-based systems, each equipped with diverse sensors like hyperspectral, thermal, and optical devices. This approach leverages the strengths of each platform to achieve a holistic understanding of PPN infestations. The concept aligns with precision agriculture, where advanced technologies are employed to optimize crop management and mitigate nematode-related losses.

The term "multi-platform" is understood through its application in related fields, such as geospatial data collection, where it involves using land vehicles, aircraft, and water-based vessels with sensors like GPS and cameras. In the context of PPN detection, it extends to combining satellite imagery for broad-scale monitoring, UAV-based hyperspectral imaging for detailed analysis, and ground sensors for in situ validation, ensuring a comprehensive dataset for analysis.

9.7.2 Historical Context and Research Evolution

Research on remote sensing for PPN detection began in the 1960s, with significant advancements noted in recent decades. Early studies focused on visual detection based on characteristic symptoms, which were labor-intensive and subject to human bias. The shift to remote sensing methods, particularly with the advent of hyperspectral and multispectral imaging, has enabled noninvasive, presymptomatic detection, addressing the limitations of traditional methods.

9.7.3 Applications in PPN Detection

Multi-platform sensing has been applied across various crops, with specific studies highlighting its efficacy. Below is a Table 9.2 summarizing key applications and accuracies from recent research.

These studies demonstrate the versatility of multi-platform sensing, with hyperspectral imaging particularly noted for its ability to detect early infestations by analyzing plant responses at the molecular level. For instance, machine learning classifiers like Linear Discriminant Analysis (LDA) and Principal Component Analysis (PCA) have been used to achieve high accuracies in cotton, while Partial

Table 9.2 Key applications of multi-platform sensing for detection of nematodes (Pun et al. 2024)

Nematode species	Host crop	Detection method	Accuracy
Meloidogyne exigua, M. paranaesis, M. incognita	Coffee (*Coffea arabica*)	NDVI (remote sensing)	78%
Heterodera schachtii, Rhizoctonia solani	Sugar beet (*Beta vulgaris*)	Spectral angle mapper (SAM)	72%
Meloidogyne incognita	Cotton (*Gossypium* spp.)	Hyperspectral imaging (350–1000 nm)	95%
Meloidogyne spp.	Tomato (*Solanum lycopersicum*)	Hyperspectral imaging (400–2500 nm, PLS-SVM)	90–100%
Bursaphelenchus xylophilus	Pine (genus *Pinus*)	Remote sensing	Not specified

Least Squares-Support Vector Machines (PLS-SVM) have differentiated nematode infestations from water stress in tomatoes.

9.7.4 Technological Components

The effectiveness of multi-platform sensing relies on the integration of various sensors and platforms:

- *Remote Sensing*: Utilizes satellites and UAVs to capture images at different bandwidths of the electromagnetic spectrum, facilitating the identification of properties invisible to specific wavelengths. The Normalized Difference Vegetation Index (NDVI) and Spectral Angle Mapper (SAM) are common metrics, with UAVs mitigating challenges like cloudy weather and nonhomogeneous illumination.
- *Hyperspectral Imaging*: Employs spectroscopy and radiometric methods to observe biological sample responses, generating high-dimensional data that requires efficient algorithms for analysis. It is particularly effective for early detection, though it faces challenges like longer image acquisition times due to indirect measurement of chemical composition.
- *Ground-Based Sensors*: Complement aerial and satellite data by providing in situ measurements, such as soil moisture and nutrient levels, which can correlate with nematode presence.

9.7.5 Challenges and Limitations

Despite its potential, multi-platform sensing faces several challenges:

- *Resolution and Weather Conditions*: Inadequate temporal and spatial resolution, as well as cloudy weather, can affect data quality. UAVs partially address this by offering higher resolution and flexibility, but their deployment is weather-dependent.
- *Data Integration*: The complexity of fusing data from multiple platforms and sensors requires sophisticated algorithms and computational resources, which can be prohibitive for small-scale farmers.
- *Cost and Accessibility*: Advanced sensors and platforms, such as hyperspectral cameras and UAVs, involve significant investment, limiting accessibility in developing regions.

9.7.6 Future Directions

Future research areas identified include improving sensor resolution, developing cost-effective solutions, and enhancing data fusion algorithms to make multi-platform sensing widely adopted. The integration with AI and machine learning, as seen in recent studies on deep learning for nematode identification, offers promising avenues for automation and precision.

9.7.7 Broader Context and Complementary Technologies

While this note focuses on sensing technologies, it is worth noting that multi-platform sensing complements other emerging methods for PPN detection, such as molecular diagnostics [e.g., Polymerase Chain Reaction (PCR), DNA metabarcoding] and AI-powered image analysis. These molecular techniques, while not part of sensing platforms, can be integrated into a holistic detection strategy, enhancing overall efficacy.

9.7.8 Conclusion

Multi-platform sensing represents a cutting-edge approach for the noninvasive detection of plant parasitic nematodes, leveraging the strengths of diverse platforms and sensors to achieve early, accurate identification. With applications across various crops and accuracies up to 95% in some cases, it supports precision agriculture by reducing economic losses and improving crop management. However, challenges like resolution, weather, and cost need addressing to ensure widespread adoption, particularly in resource-limited settings. This comprehensive analysis underscores the potential of multi-platform sensing as a vital tool in modern agriculture, with on-going research poised to enhance its effectiveness.

9.8 Management Applications

Remote sensing data can directly inform precision control strategies. For example, researchers have used drone-derived nematode maps to implement variable-rate nematicide applications. By mapping hot and cold spots of infection, farmers can apply soil fumigants or biological nematicides only where needed. In a soybean study, multiple regression models using UAV multispectral bands predicted soil cyst counts; the resulting maps were used to create a prescription for variable-rate nematicide application, reducing unnecessary treatments (Arantes et al. 2023). Similarly, a cotton field trial established nematode thresholds from soil maps and applied fumigant (Telone®) and metam sodium (Vapam®) at variable rates accordingly. Other precision strategies include site-specific soil sampling (targeting suspected hotspots rather than random grids) and integrating imagery into yield-monitoring workflows. In Virginia Tech field trials, for instance, drone-derived NDRE and green-NDVI were used in machine learning models to estimate soybean yield pre-harvest (Chandel et al. 2024). These yield estimates and stress maps can trigger timely interventions (e.g., foliar sprays, irrigation) and help optimize fertilizer or nematicide use (Chandel et al. 2024).

9.8.1 Representative Case Studies

Remote sensing and UAVs have been tested in diverse crop–nematode systems worldwide. For example:

- *Pine (Japan/China)*: Hyperspectral UAV imaging was used to detect pine wilt nematode (*B. xylophilus*) infection in pine stands. Multi-temporal hyperspectral data enabled high classification accuracy (75–95%) of infected versus healthy trees by late summer (Yu et al. 2022).
- UAV helps identify pine wilt disease symptoms based on hyperspectral drone images and assesses the potential for early detection of nematode infection (Syifa et al. 2020; Tian et al. 2022).
- *Soybean (Brazil, USA)*: In Brazil, drone-based multispectral surveys of SCN-infested soybeans revealed specific bands (e.g., 586 nm) that correlate with cyst counts (Arantes et al. 2023). Combining red, green, and NIR bands in a regression model improved detection of soybean cyst nematode (*H. glycines*) better than single bands (Arantes et al. 2023). In Virginia (USA), a UAV (DJI Phantom 4) collected multispectral data over resistant versus susceptible soybean plots. Vegetation indices (NDRE, green-NDVI) showed strong negative correlations ($r \approx -0.5$ to -0.6) with SCN population differences and high correlation with end-of-season yield (Chandel et al. 2024). These drone maps highlighted nematode-induced stress zones that matched higher cyst counts, and machine learning models (e.g., Random Forest) using the imagery predicted yield with $r \approx 0.93$ (Chandel et al. 2024).
- *Potato (Netherlands)*: A study of potato cyst nematode (*Globodera pallida*) used a UAV equipped with optical and thermal sensors ~2.5 months before harvest. The authors found that all vegetation indices (NDRE, NDVI, chlorophyll index) responded to initial nematode density (P < sub>i</sub>) in a manner analogous to yield loss curves (Jindo et al. 2023). Crucially, canopy temperature from the thermal sensor rose significantly in plots with >10 eggs·g < sup> − 1</sup>, confirming that thermal imagery can serve as an early nematode stress signal (Jindo et al. 2023). They concluded UAV indices could achieve ~90% of the detection accuracy of intensive soil sampling, and that the chlorophyll index may be the most reliable predictor of nematode-induced yield loss (Jindo et al. 2023).
- *Coffee* (*Brazil*): In perennial crops like coffee, root-knot nematodes cause patchy declines. A multispectral satellite study (RapidEye) in Minas Gerais, Brazil, classified coffee fields into healthy versus moderately/severely RKN-infected. Leaf-level hyperspectral scans had identified key NIR wavelengths; using those bands (red, red-edge, NIR) plus NDVI, the RapidEye image was classified with 78% overall accuracy (Martins et al. 2017). This mapped nematode infestation clusters across the plantation, information critical for isolating and managing foci (Martins et al. 2017).
- *Sugar Beet and Cotton*: Earlier research has shown remote sensing can also detect beet cyst nematode (*H. schachtii*) in sugar beet fields and reniform nema-

todes in cotton (Sikora et al. 2023). While not all have published quantitative results, these examples illustrate broad applicability across climates and cropping systems. In several studies, physiological leaf symptoms of different sugar beet cultivars were detected using hyperspectral signatures (HSS) and the data was correlated to initial population densities (*Pi*) of beet cyst nematode (BCN) after processing via the Nemaplot-population model (Schmidt 2015).

- *Green Gram*: The effect of *M. incognita* on the root system of green gram can be noninvasively diagnosed by using hyperspectral data at the early growth stage of aerial plant canopy (Waghmare 2024).

The Table 9.3 below summarizes representative applications.

9.8.2 Precision Management Strategies

Once detection maps are generated, precision agriculture tactics can optimize nematode management:

- *Variable-Rate Nematicide Application*: By correlating imagery or yield maps with soil nematode counts, practitioners can apply nematicides only in infested zones. This reduces chemical use and cost. For instance, cotton trials in Mississippi developed soil-sample-based nematode maps to guide variable-rate Telone® and Vapam® application, with thresholds indicating appropriate rates for each zone. In soybean, UAV-mapped cyst densities were turned into a prescription map for site-specific nematicide treatment (Arantes et al. 2023). Such maps typically combine remote-sensing hotspots with GIS-soil interpolation and are fed into tractor/VRA (variable-rate applicator) control systems.
- *Site-Specific Sampling and Zoning*: Remote-sensed maps help target field scouting. Instead of random grids, growers can sample within identified hotspot boundaries, improving diagnosis accuracy. For example, researchers highlight that drones can prescreen fields so that only areas showing stress are intensively sampled for nematodes (Arantes et al. 2023). This hotspot sampling streamlines lab diagnostics and speeds decision-making.
- *Decision Support Systems*: Imagery and derived indices can be fed into advisory tools. Machine learning models ingest drone data to predict nematode pressure and yield outcomes. In the Virginia Tech study, a Random Forest model using NDRE and other VIs predicted final soybean yield with $r \approx 0.93$ (Chandel et al. 2024). Such models give growers early warnings to adjust fertilization, irrigation, or select resistant varieties before harvest. Integrating remote-sensed data with environmental sensors (soil moisture, weather) is an emerging direction for even richer decision support (Jindo et al. 2023).

In summary, remote sensing enables site-specific operations; only infested zones receive control inputs (chemical, biological, or cultural), maximizing efficiency. This fits into integrated nematode management; for instance, mapping

Table 9.3 Application of remote sensing/drone technologies for nematode management

Crop	Nematode	Sensing method	Key outcome/use	References
Pine	Pine wilt (*B. xylophilus*)	UAV hyperspectral	Early detection of wilt; 75–95% accuracy distinguishing healthy vs. infested trees	Yu et al. 2022
Soybean	Soybean cyst (*H. glycines*)	UAV multispectral (NDRE)	Identified SCN "hotspots"; enabled design of variable-rate nematicide prescription	Arantes et al. 2023
Soybean	Soybean cyst (*H. glycines*)	UAV multispectral + ML	Vegetation indices (NDRE, GNDVI) correlated with SCN counts ($r \approx 0.5$–0.6) and yield ($r \approx 0.75$)	Chandel et al. 2024
Potato	Potato cyst (*G. pallida*)	UAV multispectral + thermal	UAV NDVI/NDRE indices (and canopy temperature) detected PCN presence; thermal index rose sharply in high-density plots	Jindo et al. 2023
Coffee	Root-knot (*Meloidogyne* spp.)	Satellite multispectral (RapidEye)	Mapped RKN infection with red/red-edge/NIR bands; classified healthy vs. infected with ~78% accuracy	Martins et al. 2017
Cotton	Reniform (*R. reniformis*)	UAV hyperspectral + soil mapping	Soil and hyperspectral scans used to create infestation maps guiding variable-rate fumigant applications	Sikora et al. 2023

nematode clusters in a field informs crop rotation plans and prevents moving infested soil into clean areas (Martins et al. 2017). Ultimately, combining maps of nematode pressure with other precision data (yield, soil) can reduce production risks and input costs.

9.9 Emerging Trends, Challenges, and Opportunities

9.9.1 Trends

Emerging research is rapidly advancing this field. Trends include the integration of machine learning/deep learning with high-dimensional image data and the fusion of multiple sensors. For example, new studies advocate combining UAV imagery with soil and climate data and even molecular (DNA) analyses to improve nematode diagnosis (Jindo et al. 2023). Hyperspectral sensors are shrinking in size and cost, allowing more frequent drone surveys. There is also growing use of time-series remote sensing: tracking vegetation indices throughout the season to spot nematode effects as they emerge.

9.9.2 Challenges

However, significant challenges remain. Nematode symptoms often mimic other stresses (drought, nutrient deficiency, disease) (Shao et al. 2023), so false positives can occur. Remote detection requires careful calibration: variable lighting, soil background, and canopy differences all add noise. High-resolution images generate huge datasets, demanding substantial processing and domain-specific training data for AI models. Moreover, adopting these technologies can be costly for individual farmers (drones, sensors, software) and may require expertise. Regulatory issues (e.g., drone flight approvals) and the need for rapid ground realities are additional barriers.

9.9.3 Opportunities

Opportunities abound as well. The recent commercial launch of Syngenta's nematode detection service (in partnership with AgTech startup Gamaya) shows industry momentum. This satellite-based tool (part of Syngenta's *Cropwise* platform) analyzes field images over hundreds of thousands of hectares, identifying high-risk soybean zones and estimating yield loss (Marges 2022). Such services can be scaled globally—Agri-tech vendors like Gamaya report that their algorithms now cover millions of hectares with "instantaneous and reliable nematode damage detection" (Marges 2022). On the horizon are likely UAV-based commercial scouting packages and software-as-a-service platforms that integrate drone flights with nematode-specific analytics.

In research, promising directions include hyperspectral imaging from unmanned aerial systems (UAS) for even finer detail (Martins et al. 2017), and low-altitude thermal photogrammetry to catch early stomatal responses. Likewise, increased use of affordable multispectral cameras (e.g., 5–12 band sensors) and smartphone-based spectrometers may democratize on-farm monitoring. Finally, these remote tools dovetail with new control methods (targeted biological nematicides, resistant cultivars); knowing when and where nematodes are active enables smarter use of these tactics.

In sum, remote sensing and drones offer a data-driven revolution in nematode management. By enabling early, field-wide diagnosis and precision interventions, these technologies can reduce yield losses and chemical usage. Ongoing advances in sensor technology, computing, and agronomic modeling—coupled with practical field trials—are steadily closing the gap between research and on-farm application. The coming years should see more integrated solutions (images + AI + decision support) that make nematode monitoring as routine as weather forecasts, further enhancing sustainable crop protection.

9.10 Conclusion

Remote sensing offers a fast, cost-effective, and scalable method for PPN diagnosis and management by enabling noninvasive, large-scale monitoring of plant stress and nematode distribution. It complements traditional methods like soil sampling and molecular diagnostics and supports precision agriculture, reducing reliance on chemical controls and promoting sustainable practices. Continued research into hyperspectral imaging, UAV optimization, and integrative approaches will enhance its efficacy, making it a cornerstone of modern nematode management. For practical implementation, growers should consult local Extension services or nematology labs to integrate remote sensing with soil sampling and tailored IPM strategies. While challenges like resolution and weather dependency persist, ongoing advancements in sensor technology and data analytics promise broader adoption in precision agriculture, reducing crop losses and promoting sustainable nematode management.

9.11 Suggested Readings

- Jindo, K., Teklu, M.G., van Boheeman, K. et al. 2023. Unmanned aerial vehicle (UAV) for detection and prediction of damage caused by potato cyst nematode *G. pallida* on selected potato cultivars. *Remote Sens.* 15: 1429.
- Sun, Z., Ibrayim, M. and Hamdulla, A. 2022. Detection of pine wilt nematode from drone images using UAV. *Sensors* 22, 4704. https://doi.org/10.3390/s22134704.
- Gano, B., Bhadra, S., Vilbig, J.M. et al. 2024. Drone-based imaging sensors, techniques, and applications in plant phenotyping for crop breeding: A compre-

hensive review. *The Plant Phenome Journal* 7: e20100. https://doi.org/10.1002/ppj2.20100.

- Žibrat, U., Širca, S., Susič, N. et al. 2020. Noninvasive detection of plant parasitic nematodes using hyperspectral and other remote sensing systems. In: P.C. Pandey, P.K. Srivastava, H. Balzter et al. (eds.) *Earth Observation, Hyperspectral Remote Sensing*, pp. 357–375. Elsevier. https://doi.org/10.1016/B978-0-08-102894-0.00015-2.
- Kuska, M.T., Daub, M. and Mahlein, A.K. 2022. Emerging technologies for integrated nematode management: remote sensing or proximal sensing as a potential tool to detect and identify nematode infestation. In: R.A. Sikora et al. (eds.) *Integrated Nematode Management: State-of-the-Art and Visions for the Future*, pp. 414–420. CAB International.

9.12 Review Questions

- What is remote sensing, and how is it applied in agriculture for nematode management?
- Define UAVs/drones. Why are they considered a valuable tool in plant health monitoring?
- What types of sensors (e.g., multispectral, hyperspectral, thermal) are used in drone-based nematode diagnosis, and what information do they provide?
- Explain how spectral signatures of nematode-infested plants differ from healthy plants.
- What are the advantages of using drones over traditional field scouting methods for nematode detection?
- How can drone-based imagery help in the early detection of nematode infestations?
- Give an example of how vegetation indices (e.g., NDVI, NDRE) can be used to assess plant stress caused by nematodes.
- In what ways can drone technology reduce chemical nematicide usage and support eco-friendly management strategies?
- What are the limitations and challenges in using drones for nematode diagnosis and management?

References

Agarwal S, Curran ZC, Yu G et al (2023) Plant parasitic nematode identification in complex samples with deep learning. J Nematol 55(1):20230045

Arantes BHT, Moraes VH, Geraldine AM et al (2023) Detection of nematodes in soybean crop by drone. Rev Ciênc Agron 54. https://doi.org/10.5935/1806-6690.20230038

Arjoune Y, Sugunaraj N, Peri S et al (2022) Soybean cyst nematode detection and management: a review. Plant Methods 18:110

Chandel A, Jjagwe P, Langston D (2024). Drone imaging to evaluate impact of nematodes on soybean yield. Virginia Cooperative Extension VCE Publications https://www.pubs.ext.vt.edu/BSE/bse-362/bse-362.html

Cotrozzi L (2022) Spectroscopic detection of forest diseases: a review (1970–2020). J For Res 33:21–38

Hashim IC, Shariff ARM, Bejo SK (2020) Application of thermal imaging for plant disease detection. In: IOP conference series: earth and environmental science, vol 540. IOP publishing, p 012052. https://doi.org/10.1088/1755-1315/540/1/012052

Hoffmann N, Schröder T, Schlüter F, Meinlschmidt P (2013) Potential of infrared thermography to detect insect stages and defects in young trees. J fur Kulturpflanzen 65(9):337–346

Jindo K, Teklu MG, van Boheeman K et al (2023) Unmanned aerial vehicle (UAV) for detection and prediction of damage caused by potato cyst nematode *G. Pallida* on selected potato cultivars. Remote Sens 15:1429

Joalland S, Screpanti C, Liebisch F et al (2017) Comparison of visible imaging, thermography and spectrometry methods to evaluate the effect of *Heterodera schachtii* inoculation on sugar beets. Plant Methods 13:73

Lowe A, Harrison N, French AP (2017) Hyperspectral image analysis techniques for the detection and classification of the early onset of plant disease and stress. Plant Methods 13:80

Mahlein AK, Rumpf T, Welke P et al (2013) Development of spectral indices for detecting and identifying plant diseases. Remote Sens Environ 128:21–30

Marges (2022) The unique digital nematode diagnostics tool co-developed by Gamaya and Syngenta is being launched for commercial use. Gamaya. 27 Sep 2022, Syngenta

Martins GD, Galo M d LTB, Vieira BS (2017) Detecting and mapping root-knot nematode infection in coffee crop using remote sensing measurements. IEEE J of Selected Topics in Applied Earth Observations and Remote Sensing 10(12):5395–5403

Nkuna BL, Chirima JG, Newete SW et al (2024) Developing models to detect maize diseases using spectral vegetation indices derived from spectral signatures. The Egyptian Journal of Remote Sensing and Space Sciences 27:597–603

Orlando VSW, Vieira BS (2024) Orbital multispectral imaging: a tool for discriminating management strategies for nematodes in coffee. Precis Agric 25(5):573–2588

Panda S, Rathore R (2019) Thermal imaging as a potential element of integrated pest management in Indian agriculture. Int Arch App Sci Technol 10(3):11–15

Pereira MR, Verrelst J, Tosin R et al (2024) Plant disease diagnosis based on hyperspectral sensing: comparative analysis of parametric spectral vegetation indices and nonparametric Gaussian process classification approaches. Agronomy 14(3):493

Pineda M, Barón M, Pérez-Bueno ML (2021) Thermal imaging for plant stress detection and phenotyping. Remote Sens 13(1):68

Pun TB, Thapa Magar R, Koech R et al (2024) Emerging trends and technologies used for the identification, detection, and characterisation of plant-parasitic nematode infestation in crops. Plants 13(21):3041. https://doi.org/10.3390/plants13213041

Raza SE, Prince G, Clarkson JP, Rajpoot NM (2015) Automatic detection of diseased tomato plants using thermal and stereo visible light images. PLoS One 10(4):e0123262

Schmidt, K. 2015. Density estimation of *Heterodera schachtii* populations by hyperspectral sensors on a biological time scale. Available at: http://nemaplot.com/BCN_Spectral_with_Leslie.html. Accessed 2 July 2020

Schmitz A, Kiewnick S, Schlang J, Sikora RA (2004) Use of high resolution digital thermography to detect *Heterodera schachtii* infestation in sugar beets. Commun Agric Appl Biol Sci 69(3):359–363

Shao H, Zhang P, Peng D et al (2023) Current advances in the identification of plant nematode diseases: from lab assays to in-field diagnostics. Front Plant Sci 14:1106784

Sikora RA, Helder J, Molendijk LPG et al (2023) Integrated nematode management in a world in transition: constraints, policy, processes, and technologies for the future. Annu Rev Phytopathol 61:201–230

Syifa M, Park S, Chang-Wook L (2020) Detection of the pine wilt disease tree candidates for drone remote sensing using artificial intelligence techniques. Engineering 6(8):919–926

Tian H, Koski TM, Zhao L et al (2022) Invasion history of the pinewood nematode *Bursaphelenchus xylophilus* influences the abundance of *Serratia* sp. in pupal chambers and tracheae of insect-vector *Monochamus alternatus*. Front Plant Sci 13:856841. https://doi.org/10.3389/fpls.2022.85684/full

Waghmare CD (2024) Hyperspectral remote sensing detection of root-knot nematode infestation in mungbean under various management strategies. J Sci Ind Res 83:309–314

Yu R, Huo L, Huang H et al (2022) Early detection of pine wilt disease tree candidates using time-series of spectral signatures. Front Plant Sci 13:1000093

Yuan Z, Musa N, Dybal K et al (2024) Quantifying nematodes through images: datasets, models, and baselines of deep learning. arXiv:240419748v1 [csCV]. 30 Apr 2024

Zhang X, Vinatzer A, Li S (2023) Hyperspectral imaging analysis for early detection of tomato bacterial leaf spot disease. Res Square. https://doi.org/10.21203/rs.3.rs-3778695/v1

Zhu W, Chen H, Ciechanowska I, Spaner D (2018) Application of infrared thermal imaging for the rapid diagnosis of crop disease. IFAC-Papers Online 51(17):424–430

Žibrat R, Širca S, Susič N et al (2020) Noninvasive detection of plant parasitic nematodes using hyperspectral and other remote sensing systems. In: Pandey PC, Srivastava PK, Balzter H, Bhattacharya B, Petropoulos GP (eds) Earth observation, hyperspectral remote sensing. Elsevier, pp 357–375

Decision Support Systems for Nematode Management

10

Abstract

Decision support systems (DSS) for plant-parasitic nematodes are an active area of development. DSS integrate data acquisition, modeling, and recommendation modules to aid farmers in timely and precise interventions. Current tools combine agronomic models (rotations, damage thresholds) with growing computational support (machine learning, simulation). Real-world deployments are most advanced in high-value crops like potato and sugar beet in Europe, but new sensing and AI technologies promise broader applicability. Continued integration of diverse data (soil, remote sensing, genomics, and climate) and user-friendly platforms (web/mobile apps) are driving DSS toward making nematode management both more precise and more practical for growers. This chapter surveys the development of DSS for PPN management, categorizes major system architectures, highlights recent AI- and simulation-based tools, and discusses case studies such as NemDST, NemaDecide, and SBN-Watch. Key challenges and future research directions are identified to guide the deployment of efficient and scalable DSS solutions.

Keywords

Decision support systems • Nematode management • NemDST • NemaDecide • SBN-Watch • AI technologies

10.1 Learning Objectives

- Understand the concept and components of Decision Support Systems (DSS) and their relevance in nematode management.

R. P. Parvatha, *Plant Parasitic Nematodes*,
https://doi.org/10.1007/978-981-95-9306-4_10

- Explain the role of DSS in nematology, including integration of diagnostic tools, data management, predictive modeling, and expert knowledge for decision making.
- Analyze different types of DSS approaches (rule-based systems, expert systems, AI/ML-based models, GIS-enabled DSS) used for nematode detection, forecasting, and management.
- Evaluate the advantages and limitations of DSS in comparison to conventional nematode management practices.
- Demonstrate how DSS supports precision nematode management, including site-specific interventions, optimal use of nematicides, and sustainable approaches.
- Interpret case studies or examples where DSS has been successfully applied in nematode diagnosis and management at farm or regional levels.

10.2 Introduction

Integrated management of plant-parasitic nematodes (PPN) increasingly relies on computerized DSS that combine biological models, farm data, and expert rules. Several platforms have been developed recently to guide decisions on crop rotation, resistant varieties, soil sampling, and control measures for nematodes in field crops.

Decision support systems (DSS) play a crucial role in plant nematode management by providing farmers and agricultural professionals with data-driven insights to make informed decisions. A DSS is a computer-based tool that integrates various data sources, models, and analytical techniques to assist in complex decision-making processes. These systems leverage technology to enhance precision, efficiency, and sustainability in managing plant-parasitic nematodes, which can cause substantial crop losses. In the context of plant nematode management, these systems help optimize strategies for controlling nematodes by offering tailored recommendations based on specific field conditions.

10.3 Importance of DSS in Integrated Nematode Management

The integration of DSS into nematode management is crucial for several reasons. First, PPNs often interact with other biotic and abiotic factors, such as soil-borne pathogens and climate variability, making management complex. DSS provide a holistic approach by synthesizing data from various sources, enabling farmers to make decisions that balance efficacy, sustainability, and economic viability. For instance, they can help determine the optimal timing for interventions, assess the effectiveness of crop rotations, and predict the outcomes of using biological controls like cover crops with resistance to specific nematodes.

A review article from Annual Reviews, titled "Integrated Nematode Management in a World in Transition: Constraints, Policy, Processes, and Technologies for the

Future," underscores the need for such holistic approaches (Sikora et al. 2023). It references specific DSS tools, such as NemaDecide, which is designed for managing potato cyst nematodes, demonstrating practical applications in agriculture. Other references include site-specific detection and management of nematodes, using geostatistical analysis to delineate management zones, and predicting nematode hotspots using soil EC data, all of which enhance the precision and effectiveness of INM strategies.

10.4 Role of DSS in Nematode Management

DSS are computer-based tools designed to assist decision making by providing relevant information and analysis. In the context of PPN management, they help decision makers understand nematode biology, the specific crop systems affected, and the potential environmental impacts of management practices. According to the Food and Agriculture Organization (FAO), a successful integrated nematode management (INM) system for nematodes must include a DSS to support decision making (Barker 1997). This involves innovative research programs, effective information delivery and addressing socioeconomic issues, nematode damage functions and thresholds, sampling techniques, management options, education, and grower inputs.

The FAO document highlights that computer communications and simulation models are becoming increasingly important in INM and nematology. These models help in characterizing damage functions, thresholds for major nematode-crop combinations, and general population dynamics (Barker et al. 1985; Duncan 1991; Ferris and Noling 1987). Improved data management and DSS are expected to facilitate the integration of new and traditional IPM strategies, including transgenic plants, new nematicides, and biological control agents.

10.4.1 Data Integration and Analysis

DSS aggregates data from various sources, such as soil sampling, weather conditions, crop type, nematode population levels, and historical farm data, to provide a comprehensive view of the nematode problem. By analyzing this data, DSS helps identify nematode species, assess infestation severity, and predict potential crop damage, enabling targeted management strategies. Tools like the Nemabase system integrate soil temperature, moisture, and nematode life cycle data to predict population dynamics and recommend control measures.

10.4.2 Risk Assessment and Prediction

DSS uses predictive models to assess the risk of nematode outbreaks based on environmental factors, crop susceptibility, and nematode biology. Farmers can anticipate high-risk periods or areas and implement preventive measures, such as crop rotation or resistant varieties, before significant damage occurs. Models within DSS can predict the hatching of root-knot nematodes (*Meloidogyne* spp.) based on degree-day accumulation, guiding the timing of interventions.

10.4.3 Decision-Making Support for Control Strategies

DSS provides evidence-based recommendations for nematode management, including chemical control (nematicides), biological control (e.g., beneficial microorganisms), cultural practices (e.g., crop rotation, cover crops), and resistant cultivars. By evaluating the efficacy, cost, and environmental impact of each option, DSS helps select the most sustainable and cost-effective strategy. A DSS might recommend planting a resistant soybean variety in fields with high cyst nematode (*Heterodera glycines*) populations, reducing reliance on chemical treatments.

10.4.4 Precision Agriculture and Site-Specific Management

DSS integrates with precision agriculture technologies, such as GPS and remote sensing, to map nematode distribution across fields and recommend site-specific treatments. This minimizes unnecessary pesticide applications, reduces costs, and mitigates environmental impacts by targeting only infested areas. Variable-rate nematicide application guided by DSS maps of nematode hotspots improves efficiency and reduces chemical runoff.

10.4.5 Monitoring and Evaluation

DSS facilitates ongoing monitoring of nematode populations and the effectiveness of management practices through real-time data collection and analysis. Farmers can adjust strategies dynamically, ensuring long-term control and preventing resistance development in nematodes. A DSS might track nematode population trends posttreatment and suggest follow-up actions if thresholds are exceeded.

10.4.6 Education and Knowledge Dissemination

DSS serves as an educational tool, providing farmers and extension agents with accessible information on nematode biology, identification, and management practices. It enhances grower awareness and adoption of integrated pest management (IPM) principles, leading to more sustainable practices. Web-based DSS platforms, like those developed for potato cyst nematode management, offer user-friendly interfaces with actionable advice tailored to local conditions.

10.4.7 Sustainability and Economic Benefits

By optimizing resource use and reducing reliance on chemical controls, DSS promotes environmentally friendly and economically viable nematode management. It lowers input costs, improves crop yields, and reduces the ecological footprint of farming practices. A DSS might recommend bio-fumigation with mustard cover crops to suppress nematodes, reducing the need for expensive synthetic nematicides.

10.5 DSS Tools and Platforms

10.5.1 FarmMaps (Formerly Akkerweb)

An open-access precision-farming platform (Wageningen University) integrates weather services, soil maps, satellite imagery and multiple agro-model "apps." FarmMaps (launched 2021) includes over 30 modules (apps) for water balance, crop growth, disease risk, variable-rate applications, and a nematode module (NemaDecide) (Molendijk and Sikora 2021). Users can upload field data (maps of sampled strips, infestation levels, etc.) and run models of nematode population dynamics, yield loss and cost–benefit of nematicides (Molendijk and Sikora 2021). Farmers can compare "what-if" scenarios (variety choice, crop sequence, nematicide use) to optimize returns. FarmMaps' dashboard provides a single interface for all field data, enabling real-time decision support in potato and other crops (Molendijk and Sikora 2021).

10.5.2 NemaDecide

A potato-specific DSS was originally developed in the Netherlands (2005) for managing cyst (*Globodera* spp.), lesion (*Pratylenchus* spp.) and root-knot (*Meloidogyne* spp.) nematodes (Been et al. 2005). NemaDecide uses population-growth models (modified Seinhorst equations), crop tolerance, and yield-loss functions to predict nematode levels and potato yields under different rotations, resistances, and control measures (Been et al. 2005). It estimates probabilities of detection by soil sampling and economic outcomes of management strategies, helping growers choose optimal nematode control and planting schedules (Been et al. 2005). NemaDecide is now integrated into FarmMaps as a GIS-enabled "apps" (NemaDecide GEO) that visualizes field infestations and scenario comparisons (Been et al. 2005).

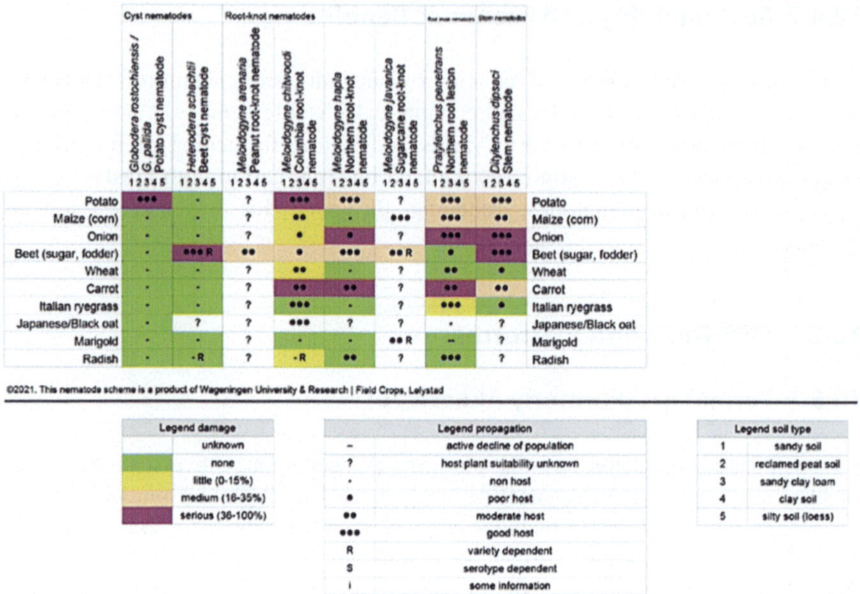

Fig. 10.1 Nematode scheme of the Best4Soil Decision Support System. The colors express the sensitivity for damage, the dots the host status of the crop

10.5.3 Best4Soil

A European collaborative DSS/network (launched ~2020) aggregates expert and farmer knowledge on rotations to suppress soil-borne pests, including nematodes. Best4Soil provides interactive web tools (in 22 languages) for designing "clever crop rotations" that minimize nematode and disease damage (Molendijk and Sikora 2021). Its design is based on the Dutch *aaltjesschema* tradition (dating from 1968). Users select crops and years in the rotation, and the system highlights host status, sensitivity, and damage thresholds (Fig. 10.1). Best4Soil essentially functions as a decision aid for rotation planning across European cropping systems.

10.5.4 SBN-Watch

A decision-support program developed in Sweden for sugar beet fields infested with the sugar beet cyst nematode (*H. schachtii*) (Omer et al. 2019). SBN-Watch allows growers to simulate rotations and beet variety choices. Its database includes parameters (yield loss limits, reproduction factors) for tolerant and standard varieties, and intercrop options (e.g., resistant mustard/radish) (Omer et al. 2019). Based on Seinhorst-type damage models, the system calculates the final nematode population (Pf) and sugar yield for a given initial population (Pi), crop sequence, and varietal tolerance (Omer et al. 2019). In practice, Swedish growers have used SBN-Watch to set appropriate waiting periods between sugar beet crops (ensuring Pf stays below damage thresholds) (Omer et al. 2019).

10.5.5 Other Tools

A variety of niche or general DSS incorporate nematode modules. For example, the Lucid Crop Rotation Key (Australia) offers an online decision aid for choosing crops resistant to common nematodes (Molendijk and Sikora 2021). General "soil navigator" tools (e.g., if-then expert systems for soil health) are being adapted to include nematode-suppression factors (Pun et al. 2023). Some precision-ag mobile platforms (IPM Decisions, AgroClimate, etc.) allow plug-in models for nematode risk.

In summary, current DSS platforms for PPN management range from specialized regional tools (NemaDecide, SBN-Watch) to broader farm-management suites (FarmMaps, Best4Soil) that now include nematode modules (Table 10.1) (Molendijk and Sikora 2021).

These examples illustrate how DSS can be tailored to specific nematode species and crop systems, enhancing decision making at both farm and regional levels. For instance, NemaDecide provides farmers with tools to assess risk levels and select appropriate management tactics, while site-specific detection allows for precision agriculture practices that target nematode hotspots, reducing unnecessary interventions

10.6 How DSS Aid in Nematode Management

A DSS for nematode management typically incorporates multiple factors to provide actionable insights:

10.6.1 Nematode Population Levels

Data from soil sampling or remote sensing can estimate the density and distribution of nematodes in a field.

10.6.2 Environmental Conditions

Weather data (e.g., temperature, moisture) and soil characteristics (e.g., type, pH) influence nematode activity and crop susceptibility.

Table 10.1 Comparison of selected DSS for nematode management

Tool/Platform	Target crops/ Nematodes	Key functionality and models	Region/Use-case/ Source(s)
FarmMaps (Akkerweb)	Potato (and general)	Integrated farm data (weather, soil, satellite) with >30 model "apps." Includes NemaDecide module for nematodes, water-balance, blight, etc. (Molendijk and Sikora 2021)	The Netherlands/EU; open-access platform for advisers and growers (Molendijk and Sikora 2021)
NemaDecide (standalone)	Potato: *Globodera, Pratylenchus, Meloidogyne*	Population dynamics, yield-loss, and resistance models (Seinhorst functions); soil-sampling risk, nematicide cost–benefit analysis (Molendijk and Sikora 2021). Allows "what-if" rotation/variety scenarios	Origin, the Netherlands (WUR), used by potato advisers. Modules also in FarmMaps (Been et al. 2005)
SBN-watch	Sugar beet: *Heterodera schachtii*	Rotation and variety planning (standard vs. tolerant varieties), two Seinhorst equations to link initial Pi to yield and Pf (Omer et al. 2019). Intercrop options (mustard, radish) included	Sweden/Europe; aids beet growers in scheduling plantings to keep Pi below damage thresholds (Omer et al. 2019)
Best4Soil	Any; multiple crops	Web-based rotation planner; shows host/non-host status and sensitivity of crops to PPN and soil diseases (Molendijk and Sikora 2021). Based on accumulated growers' data	Europe (22-language network); used by practitioners and extension for IPM rotation design (Molendijk and Sikora 2021)
Lucid crop rotation key	Any; multiple crops	Online diagnostic keys to select rotation crops based on nematode resistance traits (Molendijk and Sikora 2021). Not a full DSS but a rule-based decision aid	Australia (global access); used by advisers for rotation planning
Nemaplot	Sugar beet: *H. schachtii*	Simulation of nematode population dynamics under multiyear rotations, crop sequences and climate effects. Graphical output of pi/pf over time	Germany (Bonn); academic model illustrating effects of intercropping and temperature on cyst nematodes
Custom DSS (research)	Various; e.g.. walnuts, tomatoes	Experimental mobile or cloud tools using ML for detection or recommendation	Research-stage: e.g., UC-Merced/Sheffield projects (2021–2023) employing sensors and AI for PPN screening (Niu 2021)

10.6.3 Crop Growth Stages

The vulnerability of crops to nematodes varies depending on their development phase, making timing a critical factor in management.

10.6.4 Historical Data

Information on past nematode outbreaks, control measures, and crop performance helps refine future strategies.

Using this integrated data, a DSS can perform several key functions:

10.6.5 Predict Nematode Risks

By analyzing trends and conditions, the system can forecast potential nematode outbreaks, enabling proactive interventions.

10.6.6 Recommend Control Measures

Based on the severity of the infestation and environmental factors, the DSS can suggest the best timing and methods for control, such as:

- *Application of Nematicides:* Chemical controls can be recommended with precise dosages and timing to maximize effectiveness.
  - *Cultural Practices:* Techniques like crop rotation or cover cropping can disrupt nematode life cycles.
  - *Resistant Varieties:* The system can identify nematode-resistant crop varieties suited to the field conditions.

10.6.7 Optimize Resource Use

By providing precise recommendations, the DSS helps avoid overuse of chemicals, reducing costs and environmental impact. For instance, if nematode populations are detected early in the growing season, the DSS might recommend applying a nematicide at a specific time. Alternatively, if conditions are unfavorable for nematode development, it might suggest delaying treatment or relying on non-chemical methods.

10.7 Decision-Making Models and Computational Techniques

DSS for nematodes use a variety of models and algorithms:

10.7.1 Expert Rules/Knowledge Bases

Many systems incorporate *if–then* rules based on damage thresholds, host status, and agronomic guidelines. For example, general soil-health DSS use rule sets to interpret soil data and suggest management options (Pun et al. 2023). Best4Soil's rotation adviser relies on an expert-compiled table of crop sensitivities (Molendijk and Sikora 2021). These rule bases effectively encode decades of nematology knowledge into DSS logic.

10.7.2 Population and Growth Models

Core components include nematode population dynamics models (often linear or exponential growth with host-related decline). For instance, SBN-Watch and NemaDecide use *Seinhorst* functions (classic damage-response models) to relate initial density (Pi) to final density (Pf) and yield loss (Omer et al. 2019). Simulation models like Nemaplot predict yearly fluctuations of *H. schachtii* under specific crop rotations and temperature regimes. These models are usually parameterized from field trials (e.g., tolerance limits, reproduction rates) and allow forecasting of nematode pressure year-to-year.

10.7.3 Simulation and Optimization

Some DSS embed simulation loops (e.g., running a multiyear rotation) and optimization algorithms. NemaDecide, for example, runs multiple scenarios (rotation length, variety, nematicide use) and computes economic outcomes. Growers can query "what-if" scenarios to optimize financial returns (Molendijk and Sikora 2021). More formally, DSS may use linear programing or heuristic search to suggest optimal rotation sequences or control schedules, though few published examples exist specifically for nematodes.

10.7.4 Machine Learning and AI

Emerging DSS increasingly incorporate ML for nematode detection and decision support. Recent tools employ convolutional neural networks (CNNs) and computer-vision methods to automate nematode counting/identification. For example, Pun et al. (2023) developed NemDST, which integrates a YOLOv5 deep-learning detector to identify nematode juveniles and eggs in soil images. Similarly, Saikai et al. (2024) used YOLOv8 CNNs within a Graphical User Interface (GUI) to count *Meloidogyne* eggs and juveniles, achieving ~94% accuracy. Another CNN-based mobile app classified five nematode genera from microscope images with >98% accuracy (Singh et al. 2023). These AI modules primarily address diagnostic steps (image-based detection and identification), which feed into DSS decisions. In principle, ML could also be used for predictive models (e.g., repressors' predicting nematode population from soil/climate data) but such applications are only beginning.

10.7.5 Hybrid and Other Techniques

Some DSS use Bayesian or fuzzy logic to handle uncertainty. For instance, agricultural DSS in general (though not PPN-specific) have used Bayesian networks to assess pest risk from soil/weather inputs, and fuzzy inference to time irrigation. AgroDSS frameworks incorporate data mining and time-series clustering to reveal patterns from historical farm data (Pun et al. 2023). As nematode DSS mature, similar advanced analytics (e.g., ensemble learning, GANs for image synthesis) are likely to appear, as in projects like the UK *NemaRecognition* pipeline (Yong 2023). However, rule- and model-based approaches (Seinhorst models, crop rotation tables) still predominate in current nematode DSS.

In summary, current DSS draw on a mix of classical models (damage thresholds, population dynamics) and modern AI (deep learning for image analysis). The trend is toward integrating more data-driven techniques to improve speed and accuracy of nematode risk assessment (Saikai et al. 2024).

10.8 Websites and Databases

There are a large number of websites and databases developed by universities and extension services around the world. They are in most cases updated regularly by experts of the organizations involved and are therefore important as a first-line DST in obtaining information on INM. A few examples of websites that cover a broad array of crops and nematode problems are presented in Table 10.2.

10.9 Expansion of Effective Decision Support Tools and DSS

Upgrading PPN control strategies is knowledge-intensive and should combine all modern and traditional components related to effective and sustainable PPN management worldwide. These comprise nematode population level and distribution, crop sensitivity, host status, and damage thresholds with known efficacy of individual management measures (Molendijk and Sikora 2021). Therefore, modern structures of agricultural plans for pest management usually have both decision

Table 10.2 A list of websites available from universities, extension services, and government agencies offering integrated nematode management information on a wide array of nematodes and crops worldwide

Institution	Website[a]	Information coverage
Sociedade Brasileira de Nematologia/ Brazilian Society of Nematology	https://nematologia.com.br/ (in Portuguese)	General nematology, diagnostics, and management of nematodes among others: Soybean, cotton, maize, potato, rice, wheat, sugarcane, coffee, vegetables, and fruit crops
NemaDecide, Wageningen University	http://www.nemadecide.com/	Gives real-time results of soil sampling for specific nematodes on potato along with GPS coordinates to aid in site-specific management. Mainly potato cyst nematodes but also lesion and root-knot
University of Florida Cooperative extension	https://sfyl.ifas.ufl.edu/ search-result/?q=nematods	Covers all major fruits, vegetables, and field crops grown in Florida. Gives pertinent information on nematodes, symptoms, control tools, and management recommendations. Updated regularly by Florida nematologists
University of Florida Entomology and Nematology Department	https://edis.ifas.ufl.edu/ department_entomology_and_ nematology	Provides in-depth profiles of insects, nematodes, arachnids and other organisms, and integrated management tools
Nemaplex, University of California Davis	http://nemaplex.ucdavis.edu/ Uppermnus/topmnu.htm#	Extensive information on diagnostics, management, methods, ecology, including spread sheet tools for INM and making economic threshold decisions
University of California Agricultural & Natural Resources—IPM	https://www2.ipm.ucanr.edu/ agriculture/	Pest management information on 44 crops or groups of crops including nematodes. Chapters updated every 3–5 years by faculty members
Australia Department of Primary Industries and Regions	https://pir.sa.gov.au/ search?collection=PIRSA-- web&query=nematodes	Used in Western and South Australia to determine the presence and levels of *Pratylenchus thornei, P. neglectus, P. penetrans* and *P. quasitereoides*
UK, Agriculture and Horticulture Development Board	https://ahdb.org.uk/ (A('xTcrINcvSX))/ Search?q=nematodes	Covers management of a number of nematode problems including migratory, cyst and leaf nematodes on a number of crops in the UK

(continued)

Table 10.2 (continued)

Institution	Website[a]	Information coverage
Indian Agricultural Research Institute, Nematology Divn., ICAR	https://iari.res.in/index.php?Option=com_content&view=article&id=72&Itemid=161	All India coordinated research project on plant parasitic nematodes with integrated approach for their control
Louisiana State University Cooperative Extension	https://www.lsuagcenter.com/portals/our_offices/departments/plant-pathology-crop-physiology/nematode-advisory-service	Coverage of nematodes on cotton, turf, vegetables, soybean, and others
University of Georgia	https://plantpath.caes.Uga.edu/search.html?q=nematodes&cx=00898429120070070881 7%3Abqxfpp1iz4w&searchScope=siteOnly	Disease clinic provides diagnostic analysis or pests and diseases including nematodes and recommends appropriate management strategies
Lucid, Australia	https://keys.lucidcentral.org/keys/v3/crop_rotation_plant_parasitic_nematodes/	Crop to be used in rotations and their resistance to plant parasitic nematodes
Best4Soil	https://www.best4soil.eu/	A network of practitioners, for sharing knowledge on prevention and reduction of soil-borne diseases in 22 European languages. Offers a tool to develop unique crop rotation sequences to take into account nematodes and soil-borne diseases

[a]Websites accessed 1–15 November 2020

support tools (DSTs) and decision support systems (DSSs). Such tools and systems are established or being evolved mostly in developed countries via scientists working in both governmental and private sectors. These tools can produce information on certain abiotic factors that impact PPN behavior in the field. Consequently, they can be used to optimize INM. Molendijk and Sikora (2021) reported tools such as soil sensors (e.g., Nematool and LoRa Soil sensor), remote sensing, and predictive models. Also, DSS can assist farmers and stakeholders in making real-time decisions for appropriate farm management, which may involve PPN problems. Related programs can merge information on numerous different features necessary for improved farm management. Being holistic, these DSSs are not necessarily centered on INM. For instance, some of these DSSs have been developed by the European Union in order to improve similar plans for other PPNs in adequate cropping systems worldwide (Molendijk and Sikora 2021). Among them, Akkerweb/FarmMaps offers the information needed for practical smart farming, such as immediate data about weather services, soil maps, satellite crop images, and crop polygons (Van Evert et al. 2018). Its related software programs also provide an array of models, a DSS digital store, visualization tools, and a task map generator. Other DSSs for managing PPNs are included in NemaDecide (Been et al. 2005). Various relevant models/equations for population dynamics of *Globodera* spp., *Meloidogyne chitwoodi*, and/or *Pratylenchus penetrans* on potatoes, soil sampling, yield loss, partial resistance, and control measures are provided by NemaDecide. NemaDecide integrates data models/sources to serve strategic decisions at the farm/field level. Thus, NemaDecide supplies useful estimations of needed variables such as risks for possible potato crop loss, the possibility of PPN population growth, level of probability for detecting PPN foci by soil sampling, and assessing costs/benefits of control measures. Thus, it offers advice for growers and stakeholders to improve financial returns and compare different cropping scenarios. Also, the combinatorial impact of several biotechnological tools for accurate PPN detection as a prerequisite for INM plans should be exploited and expanded. For instance, magnetic biosensors combined with a DNA extraction protocol and loop-mediated isothermal amplification technique can form a novel solution for in-field diagnostics of the serious nematode *Globodera pallida*. This method possesses the capacity to detect as low as only one *G. pallida* juvenile, even when found with other closely related PPN species (Camacho et al. 2024).

10.10 Case Studies and Implementations

10.10.1 Potato (Europe)

The NemaDecide/FarmMaps DSS has been used by Dutch and European potato growers to plan nematode management. In field demonstrations, advisers used NemaDecide to compare scenarios: e.g., switching potato varieties (susceptible vs. partially resistant) and partial-dose nematicide treatments, then viewing projected nematode curves and yield loss bars for each nematode species (Figs. 10.2

Fig. 10.2 Screenshot of NEMADECIDE GEO application on Akkerweb, with on the map of the field with the sampled strips and the red strips in which a nematode infestation has been detected. The data from the infested sample strips (red) can be uploaded via the nematicide button and the comparison of scenarios can start

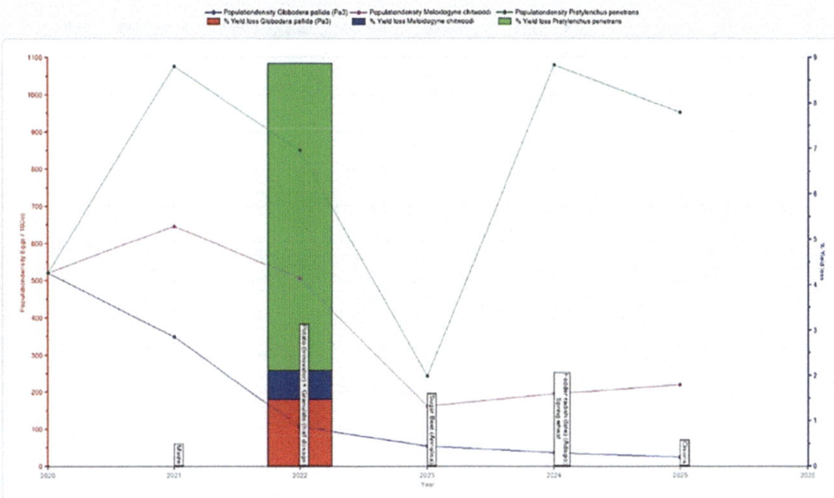

Fig. 10.3 The lines show the population dynamics of *Globodera pallida, Pratylenchus penetrans, and Meloidogyne chitwoodi.* The bar expresses the percentage yield loss in potato (innovator + granular nematicide half dosage) caused by each species

and 10.3). This helped farmers decide whether lower nematicide rates with tolerant varieties could be cost-effective (Molendijk and Sikora 2021). The FarmMaps platform (The Netherlands) aggregates these outputs by field, enabling farm-wide strategy.

10.10.2 Sugar Beet (Sweden/Europe)

SBN-Watch was implemented with Swedish growers facing *H. schachtii* infestations. Users input their soil sample density (Pi) and select rotation sequences and sugar-beet varieties. The DSS then computes expected final population (Pf), sugar yield, and income for each scenario (Fig. 10.4). For example, simulations confirmed that the "Mixer" variety (standard type) maintained high reproduction factor (Rf), whereas tolerant varieties had much lower Rf (Omer et al. 2019). Including sanitation and intercrops (radish) in the rotation sharply reduced nematode resurgence. Growers used these insights to set multiyear intervals between susceptible beet crops (in some cases >4 years as suggested by SBN-Watch) to keep Pf under the economic threshold (Omer et al. 2019).

10.10.3 Multi-crop Rotations (Europe)

Best4Soil is deployed via cooperatives and extension across the EU. For example, a Belgian farmer group used Best4Soil's online tool to test crop sequences that avoid

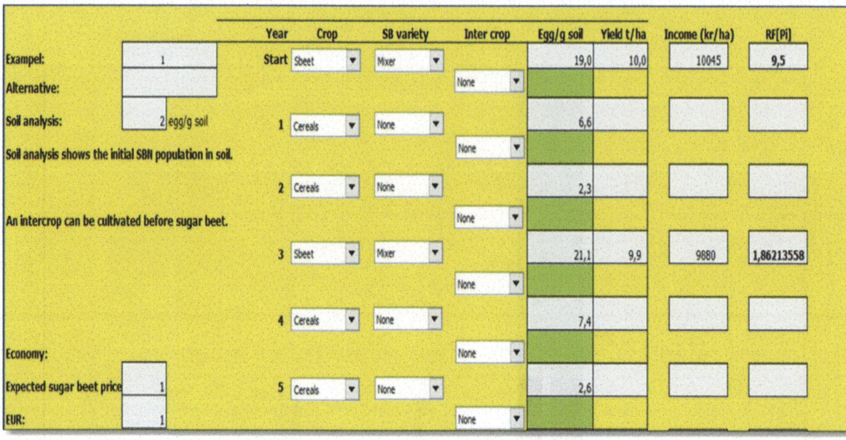

Fig. 10.4 A screenshot of the user interface showing a selected crop rotation and the estimated final SBN population (Pf) values, sugar yield (tonnes/ha), income (SEK/ha), and the reproduction factor (Rf) values

common potato cyst nematodes and root-knot hosts while still controlling soil diseases. Although formal usage stats are unpublished, the fact that Best4Soil offers content in 22 languages indicates broad engagement with growers and consultants (Molendijk and Sikora 2021).

10.10.4 Precision Sensing in Orchards (USA)

A recent project (University of California, Merced, 2021) applied a handheld NIR spectrometer (SCiO device) with machine learning to detect nematode infestation in walnuts (Niu 2021). The Neural Network Model classified nematode infection levels from leaf reflectance with ~72% accuracy in early trials (Niu 2021). This proof-of-concept illustrates how rapid spectral diagnostics could feed into DSS: early nematode alerts would let growers to time treatments or cultivations more sustainably (reducing pesticide use) (Niu 2021).

10.10.5 Remote Sensing Trials

Researchers have correlated field NDVI imagery with nematode damage. For example, in sugar beet research, NDVI maps of canopy stress matched patches of *H. schachtii* infestation (ground-verified by soil counts) (Molendijk and Sikora 2021). Such remote-sensing insights are being incorporated into DSS: coupling drone/satellite NDVI with soil temperature sensors can predict the spatial and seasonal dynamics of nematodes within fields (Molendijk and Sikora 2021).

10.10.6 Nematode Counting in Labs

Automated image-analysis tools from recent research can speed the preparation of DSS inputs. For instance, the GUI tool by Saikai et al. (2024) allows lab technicians to upload microscope images and quickly obtain nematode counts. The counts (e.g., eggs per 250 g soil) can feed directly into damage-threshold calculations in a DSS. Although these are not field-deployed DSS per se, they represent the laboratory "backend" of accurate DSS decision support.

Overall, these cases show that integrated management DSS are being used in real agronomic settings—in Europe for potato and sugar beet, in grower networks for crop rotation planning, and in research trials using novel sensors. The common thread is that DSS outputs (infestation risk, yield loss, break-crop intervals) inform tangible field decisions (variety choice, nematicide application, rotation length) under real-world conditions (Molendijk and Sikora 2021).

10.11 Data Sources and Integration Innovations

Modern nematode DSS leverage diverse data streams:

10.11.1 Soil Diagnostics

Traditional soil-sampling (nematode counts per unit soil) remains fundamental. New technologies are emerging: real-time soil sensors (temperature, moisture) can be networked in fields. For example, recent systems use arrays of soil sensors connected wirelessly to monitor conditions that drive nematode development (Molendijk and Sikora 2021). The data allow dynamic forecasts of nematode life cycle stages (e.g., predicting when Pi will exceed thresholds for the next susceptible crop) (Molendijk and Sikora 2021). In addition, molecular diagnostics (eDNA/metabarcoding) are advancing. Deep sequencing of soil samples can now inventory nematode communities and beneficial microbes, potentially feeding DSS with detailed biodiversity profiles (Molendijk and Sikora 2021). Such genomic data could eventually refine DSS recommendations by quantifying both pest and antagonist species in the soil (Molendijk and Sikora 2021).

10.11.2 Remote Sensing

Satellite, aerial, and drone imagery (multispectral and hyperspectral) are being used to infer nematode presence. The Normalized Difference Vegetation Index (NDVI) is widely used: it highlights zones of reduced vigor that often coincide with nematode hotspots (Molendijk and Sikora 2021). DSS prototypes have shown that combining NDVI maps with soil sensor data can spatially map nematode distribution and inform precision treatments (Molendijk and Sikora 2021). For example, if NDVI shows a persistent low-vigor strip, a DSS could flag that area for targeted nematicide or bio-fumigant application.

10.11.3 Climate and Weather Data

Many DSS incorporate weather forecasts and climate models. Degree-day counters and historic weather trends are used to time interventions (planting, treatments) relative to nematode life cycles. More broadly, climate variability is recognized as affecting nematode dynamics: one review notes that DSS must adapt to "climate volatility," which alters year-to-year population trends. Thus, DSS now often link to meteorological databases or climate-model outputs to adjust risk predictions.

10.11.4 Integrated Farm Data Platforms

Platforms like FarmMaps exemplify data integration: growers upload geo-located soil-sample results and crop fields, and the system automatically fetches local weather forecasts, satellite imagery, and soil-type maps (Molendijk and Sikora 2021). This centralized data repository means that the nematode module can pull all relevant inputs (soil PPN count, field history, weather) to generate recommendations, without the user manually entering each piece.

10.11.5 Decision Support Networks and Knowledge Bases

Online DSS increasingly tap collective knowledge. For instance, Best4Soil's rotation planner is essentially a crowd-sourced database of "clever rotations" contributed by farmers and experts (Molendijk and Sikora 2021). Similarly, global PPN databases (e.g., the Plant-Parasitic Nematodes Database) provide reference genomic and phenotypic data that could be integrated into DSS for identifying local species. Tools like the Lucid rotation key and smartphone apps connect users to expert systems on crop-nematode resistance (Molendijk and Sikora 2021). Even social-data integration (farmer-reported infestation levels via apps) is a potential future stream for DSS knowledge.

10.11.6 Field-Deployable Devices

New sensors are entering the picture. For example, a handheld NIR spectrometer (SCiO) was shown to detect root-knot nematode stress in walnut leaves, analyzed by machine learning (Niu 2021). Such portable devices, combined with cloud-based AI, could allow rapid in-field sampling: a farmer scans leaves, the app predicts nematode infestation level, and the DSS suggests actions. In pest management, similar pocket microscopes with image upload (e.g., smartphone camera + CNN classifier) are also being tried. These innovations move nematode diagnostics out of the lab and into the field, feeding real-time data into DSS.

In summary, data integration in nematode DSS is expanding rapidly. Traditional lab counts and extension bulletins are now supplemented by IoT sensor streams, satellite imagery, genomic surveys, and cloud AI. DSS of the near future will likely combine these: e.g., sensor-validated remote-sensing maps, or ML-enhanced lab counts, to refine nematode risk models (Molendijk and Sikora 2021).

10.12 Benefits of Using Decision Support Systems

The adoption of DSS in plant nematode management offers several significant advantages:

10.12.1 Improved Decision Making

Farmers receive timely, data-backed recommendations, reducing guesswork and enhancing the effectiveness of control measures.

10.12.2 Resource Use Efficiency

Targeted interventions minimize unnecessary pesticide applications, lowering costs and reducing environmental risks.

10.12.3 Increased Crop Yields

By preventing or mitigating nematode damage, DSS helps protect crop health and productivity.

10.12.4 Sustainability

DSS promotes integrated pest management (IPM) practices, balancing economic and ecological considerations for long-term agricultural success.

10.13 Notable Recent Research and Projects (≈2019–2024)

10.13.1 Pun et al. (2023)

The authors integrated a YOLOv5-based image detector into a web tool for PPN management (Pun et al. 2023). NemDST allows uploading nematode images (eggs/juveniles), automatically counts and classifies them, estimates population density against damage thresholds, and recommends control strategies. Although published as a "decision support tool," it chiefly addresses the imaging/counting stage of DSS.

10.13.2 Saikai et al. (2024)

An open-access GUI incorporating YOLOv8 CNN models to identify *Meloidogyne* eggs and juveniles in micrographs has been developed (Saikai et al. 2024). The tool achieved ~94% accuracy and is intended to accelerate nematode quantification for research and extension. This work represents progress toward automated data input for DSS (e.g., replacing manual counts).

10.13.3 Singh et al. (2023)

Singh et al. (2023) presented a lightweight CNN and smartphone app for classifying nematodes from microscope images. The app distinguished five PPN species/groups with ≥98% accuracy, as validated by nematode specialists. The authors suggested such an app could "eliminate the necessity of specialists" for initial ID, feeding into on-farm decision workflows.

10.13.4 Omer et al. (2019)

Provided the scientific basis and validation for the sugar beet DSS described above (Omer et al. 2019). Although slightly beyond 5 years, this work introduced key rotation and variety models that are still used.

10.13.5 Yang Lab (2023–2025)

Seminars and papers (UC Riverside). Focus on integrating AI "co-pilots" and biological control for PPN (Yang 2025). While not yet a concrete DSS, this research indicates direction, combining AI-driven diagnostic aids with nematophagous microbe data. (Yang's ongoing NIH/USDA projects may soon produce new tools.)

10.13.6 NemaRecognition (UK, 2022–2023)

An BBSRC-funded project (U. Sheffield) to develop a machine-learning pipeline for PPN recognition from soil images (Yong 2023). The goal was a real-time image/video system identifying multiple nematode genera, addressing the taxonomist shortage. It represents an investment in AI-based diagnostics that could underpin future DSS.

10.13.7 Halo/Grants4Ag (USA, 2021)

Early-stage research on spectrometer-based nematode sensing in field crops (Niu 2021). Although a blog report, it highlights how novel data sources (handheld NIR spectrometer + ML) can enter the DSS ecosystem.

These examples illustrate that much recent research in nematode DSS is data-driven and AI-enhanced. Studies are pushing boundaries on automated detection (imaging, spectroscopy), integrated sensing, and decision analytics. They complement classic agronomic DSS work by enabling faster, more precise data feeding into the decision models (Pun et al. 2023).

10.14 Challenges and Limitations

Despite their potential, the implementation of DSS in PPN management faces challenges, such as the need for accurate data, user-friendly interfaces, and integration with existing agricultural practices. The FAO notes that progress in the last two decades has included improved simulation models and management tactics, but further advancements are needed to bridge gaps between current and novel strategies. Future research should focus on enhancing DSS to handle the complexities of global agricultural transitions, including climate variability and policy changes affecting access to technology.

10.14.1 Data Requirements

DSS relies on accurate and timely data, which may be difficult to obtain in resource-limited settings. Reliable field-level datasets on nematode counts and environmental variables are limited.

10.14.2 Complexity

Some systems may require technical expertise, limiting adoption among smallholder farmers.

10.14.3 Cost

Developing and maintaining DSS can be expensive, and access may be restricted in some regions. Deep learning models require substantial computation and may not be feasible in low-resource settings.

10.14.4 Adaptability

Systems must be tailored to specific crops, nematode species, and local conditions to be effective. Complexity of model outputs can hinder on-farm uptake without proper training.

10.14.5 Integration with IPM

Bridging DSS outputs with local advisory services remains underdeveloped.

10.15 Future Directions

10.15.1 Hybrid Modeling

Combining mechanistic and data-driven approaches to improve robustness.

10.15.2 Edge Computing

Deploy lightweight inference engines on mobile devices for in-field use.

10.15.3 Crowd Sourced Data Platforms

Engaging farmers in data collection via mobile apps to enrich training datasets.

10.15.4 Decision Personalization

Tailoring recommendations based on crop type, soil health indicators, and farmer preferences.

10.16 Conclusion

Decision support systems are transformative in plant nematode management, enabling data-driven, precise, and sustainable practices. By integrating diverse data sources, predictive models, and user-friendly interfaces, DSS empowers stakeholders to mitigate nematode damage effectively while minimizing environmental and economic costs. However, their success depends on accessibility, local adaptation, and user training to ensure widespread adoption.

In conclusion, decision support systems are invaluable tools in plant nematode management, enabling more precise, efficient, and sustainable agricultural practices. By leveraging data and predictive models, these systems empower farmers to make better decisions, ultimately improving crop outcomes and profitability. Whether through predicting risks, recommending tailored interventions, or optimizing resource use, DSS transforms the way nematode threats are managed in modern agriculture.

10.17 Suggested Readings

- Been, T.H., Schomaker, C.H. and Molendijk, L.P.G. 2005. NemaDecide: A decision support system for the management of potato cyst nematodes. *ResearchGate*. https://www.researchgate.net/publication/40112909.
- Molendijk, L.P.G. and Sikora, R.A. 2021. Decision support systems in integrated nematode management: The need for a holistic approach. In: R.A. Sikora, J. Desaeger and L.P.G. Molendijk (eds.) *Integrated Nematode Management: State-of-the-Art and Visions for the Future*, pp. 428–438. CAB Int., Wallingford, UK.
- Omer, Z.S., Leven-fors, J., Andersson, S. and Wallenhammar, A.C. 2019. Development of a decision support system for managing *Heterodera schahtii* in sugar beet production. *J. Nematol.* 51: e2019–05. DOI: 10.21307/jofnem-2019-005.
- Pun, B., Neupane, A. and Koech, R. 2023. A deep learning-based decision support tool for plant-parasitic nematode management. *J. Imaging* 9(11): 240; https://doi.org/10.3390/jimaging9110240.

10.18 Review Questions

- What are Decision Support Systems (DSS), and why are they important in plant nematode management?
- Differentiate between traditional nematode management approaches and DSS-based management approaches.
- What types of data (e.g., field, climatic, soil, pest population) are essential for effective DSS in nematode management?
- Explain the role of modeling and simulation in nematode DSS development.

- Discuss case studies using DSS for nematode management.
- What are the limitations or challenges in adopting DSS for nematode management at the farm level?

References

Barker KR (1997) Opportunities for integrated management of plant-parasitic nematodes in the near east. In: Maqbool MA, Kerry A (eds) Plant nematode problems and their control in the near east region (FAO plant production and protection paper – 144). FAO, Rome

Barker KR, Schmitt DP, Imbriani JL (1985) Nematode population dynamics with emphasis on determining damage potential on crops. In: Barker KR, Carter CC, Sasser JN (eds) An advanced treatise on Meloidogyne, vol II. North Carolina State University Graphics, Raleigh, NC, USA, pp 135–148

Been TH, Schomaker CH, Molendijk LPG (2005) NemaDecide: a decision support system for the management of potato cyst nematodes. ResearchGate. https://www.researchgate.net/publication/40112909

Camacho MJ, Albuquerque DC, Inácio ML et al (2024) FTA-LAMP based biosensor for a rapid in-field detection of *Globodera pallida*—the pale potato cyst nematode. Front Bioeng Biotechnol 12:1337879

Duncan LW (1991) Current options for nematode management. Annu Rev Phytopathol 29:469–490

Ferris H, Noling JW (1987) Analysis and prediction as a basis for management decisions. In: Brown RH, Kerry BR (eds) Principles and practice of nematode control in crops. Academic Press, Orlando, pp 49–85

Molendijk LPG, Sikora RA (2021) Decision support systems in integrated nematode management: the need for a holistic approach. In: Sikora RA, Desaeger J, Molendijk LPG (eds) Integrated nematode management: state-of-the-art and visions for the future. CAB Int., Wallingford, pp 428–438

Niu H (2021) Detecting nematodes to help growers manage walnut production more sustainably. Halo Blog Newsl. blog.halo.science. gtr.ukri.org

Omer ZS, Leven-fors J, Andersson S, Wallenhammar AC (2019) Development of a decision support system for managing *Heterodera schahtii* in sugar beet production. J Nematol 51:e2019–e2005. https://doi.org/10.21307/jofnem-2019-005

Pun B, Neupane A, Koech R (2023) A deep learning-based decision support tool for plant-parasitic nematode management. *J. Imaging* 9(11):240. https://doi.org/10.3390/jimaging9110240

Saikai KK, Bresilla T, Kool J et al (2024) Counting nematodes made easy: leveraging AI-powered automation for enhanced efficiency and precision. Front Plant Sci 15:1349209. https://doi.org/10.3389/fpls.2024.1349209

Sikora RA, Helder J, Molendijk LPG et al (2023) Integrated nematode management in a world in transition: constraints, policy, processes, and technologies for the future. Annu Rev Phytopathol 61:209–230

Singh N, Singh AK, Dhruw LK et al (2023) Development of a deep learning-assisted mobile application for the identification of nematodes through microscopic images. Mod Agric. ivysci.com. https://doi.org/10.1002/moda.70000

Van Evert FK, Been T, Booij AJ (2018) Akkerweb: a platform for precision farming data, science, and practice. *Proceedings of the 14th International Conference on Precision Agriculture*, June 24 – June 26, 2018, Montreal, Quebec, Canada

Yang J (2025) Artificial intelligence and plant-parasitic nematodes management. Univ of California, Riverside News. 24 Jan 2025. entnem.ucdavis.edu

Yong P (2023) Arable—Nema recognition: an AI-and molecular-driven pipeline for throughput plant parasitic nematode recognition. BBSRC Funded Project Report,. Department of Computer Science, University of Sheffield. gtr.ukri.orggtr.ukri.org

Part V

Biotechnological Approaches

Nematode Molecular Diagnostics: From Bands to Barcodes

11

Abstract

PPNs cause over US$ 157 billion in annual crop losses globally. Accurate diagnosis is crucial for managing these pests effectively. Molecular diagnostics improve nematode identification accuracy, crucial for crop protection. PCR and qPCR are key tools that are widely used for sensitive detection and quantification of plant parasitic nematodes. Isothermal amplification and DNA barcoding are emerging field-friendly techniques. The evidence leans toward combining molecular and morphological methods for best results. There is ongoing debate about the decline in classical taxonomy expertise and integrating multiple methods for accuracy, especially under field conditions.

Keywords

PCR-based methods • Isothermal amplification technologies • DNA sequencing and barcoding • Protein-based methods • Microarrays

11.1 Learning Objectives

- Explain the need for molecular tools in nematode detection and identification compared to traditional methods.
- Describe the basic concepts of DNA, and protein-based diagnostic approaches.
- Understand the principles of molecular diagnostics and how they differ from traditional morphological/biochemical methods in nematode identification.
- Learn about key molecular markers (e.g., rDNA, mtDNA, ITS regions, COI gene) used for species-level identification of plant parasitic nematodes.
- Gain knowledge of major molecular techniques, such as PCR, qPCR, RT-PCR, RFLP, DNA barcoding, and next-generation sequencing, and their role in nematode detection and taxonomy.
- Explore the applications of molecular diagnostics in nematode biodiversity studies, epidemiology, and integrated management strategies.

R. P. Parvatha, *Plant Parasitic Nematodes*,
https://doi.org/10.1007/978-981-95-9306-4_11

- Develop critical skills in evaluating molecular data for accurate nematode diagnosis, including advantages, limitations, and interpretation of results.
- Appreciate recent advances (e.g., CRISPR-based diagnostics, portable DNA sequencers, LAMP assays) and their potential for rapid, field-based nematode detection.

11.2 Introduction

Plant parasitic nematodes are tiny worms that can harm crops by feeding on roots, causing issues like stunted growth and reduced yields. Identifying them accurately is crucial for effective pest management, but traditional methods can be challenging. Plant parasitic nematodes (PPN) control relied on soil fumigants, reducing the need for precise species identification. However, with the prohibition of such chemicals, accurate identification has become crucial for implementing sustainable, nonchemical management strategies, such as crop rotation and resistant plant strains. Molecular diagnostics have evolved to complement and often surpass traditional morphology-based methods, which struggle with the small size and morphological similarity among nematode species, leading to misidentifications and economic losses. Molecular diagnostics offer advanced techniques to pinpoint these pests, enhancing agricultural practices. Molecular diagnostics for PPNs are advanced techniques that help identify and manage these microscopic worms, which cause significant crop losses globally, estimated at over $157 billion annually. These methods are essential because traditional identification based on appearance is often challenging due to the small size and similar looks of nematode species.

11.3 Applications

Molecular diagnostics are crucial for:

11.3.1 Quarantine and Pest Management

Rapid, accurate diagnostics are needed for quarantine, e.g., detecting *Meloidogyne chitwoodi* and *M. fallax* in tubers or *Bursaphelenchus xylophilus* versus *B. mucronatus* in wood support this.

11.3.2 Research and Taxonomy

Case studies like *Xiphinema krugi*, analyzed across 14 populations, identified a species complex with four genotypes/cryptic species using RFLP, ITS-1 sequencing, and morphometrics. This exemplifies integrative taxonomy, combining morphological and molecular data for enhanced resolution.

Effective molecular diagnostics require robust DNA extraction. Methods include:

- Commercial kits using silica columns, offering high-quality DNA but at three times the cost of others.
- Chelex® resin, Phenol/chloroform, and Worm Lysis Buffer (WLB), with chemical lysis using sodium hydroxide requiring no manual disruption.

11.4 Key Molecular Diagnostic Techniques

Molecular diagnostics for PPNs represent a critical advancement in agricultural science, addressing the significant economic impact of these pests. With over 4100 described PPN species, accurate and rapid diagnosis is essential for effective management, especially as chemical nematicides face increasing restrictions, necessitating nonchemical strategies like crop rotation and resistant varieties. These tools are vital for quick and accurate detection, especially in quarantine scenarios to prevent pest spread.

Molecular diagnostics, particularly PCR-based methods, have emerged as a promising solution, offering high sensitivity, specificity, and the ability to handle large sample sizes efficiently. These methods are essential for early detection, species identification, and population quantification, which are crucial for implementing effective pest management strategies.

Molecular diagnostics use DNA and protein analysis to identify nematodes. Key methods include:

- PCR-Based Methods.
- Isothermal Amplification Technologies.
- DNA Sequencing and Barcoding.
- Protein-Based Methods.
- Microarrays.

11.5 PCR-Based Methods

PCR-based methods are molecular techniques that amplify and analyze DNA to diagnose PPNs, which are microscopic worms causing significant crop losses. These methods are vital for identifying and managing nematode infestations, offering precision where traditional methods fall short.

PCR-based methods are foundational in molecular diagnostics, offering high sensitivity and specificity (Fig. 11.1). Conventional PCR uses species-specific primers for identification, while quantitative real-time PCR (qPCR) enables detection and quantification directly from soil or plant tissues. For instance, qPCR has been successfully applied to detect species like *M. enterolobii*, *M. javanica*, *X. elongatum*, *Pratylenchus zeae*, and *Heterodera schachtii*, with studies highlighting its reliability. Real-time PCR is highlighted as the fastest, most sensitive, and 100%

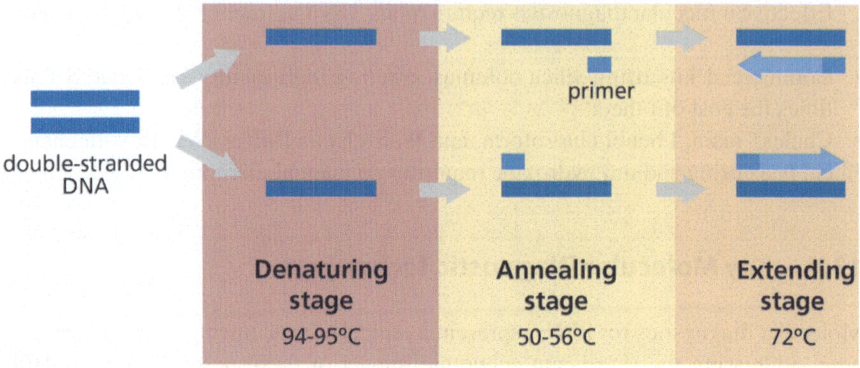

Fig. 11.1 Demonstration of the principles and working processes involved in traditional PCR methods techniques

accurate with Taqman probes. However, challenges include high costs and equipment needs, limiting field applicability. Droplet digital PCR (ddPCR) is also used for precise low-concentration measurements. It amplifies specific DNA for detailed species identification.

11.5.1 Conventional (End-Point) PCR

Amplification of a target DNA region using specific primers, followed by agarose-gel electrophoresis to visualize the presence/absence (and size) of the PCR product.

11.5.2 Multiplex PCR

Simultaneous amplification of two or more targets in one reaction by using multiple primer pairs with distinct amplicon sizes.

11.5.3 PCR-RFLP

PCR amplification of a conserved region followed by digestion with one or more restriction enzymes. Differentiates nematode strains by analyzing DNA fragments after enzyme digestion. Species or pathotypes are discriminated by distinct banding patterns on a gel.

11.5.4 Nested and Semi-nested PCR

Two successive PCRs: the first uses "outer" primers, then a small portion of that reaction is re-amplified with "inner" primers, boosting sensitivity and specificity.

11.5.5 Digital PCR (dPCR)

Partitioning of a PCR reaction into thousands of nanoliter droplets or wells; end-point fluorescence scoring of positive versus negative partitions yields absolute quantification via Poisson statistics.

11.5.6 rDNA-PCR

rDNA-PCR targets ribosomal DNA for species-specific identification, useful for quarantine species like *Globodera*.

11.5.7 Real-Time PCR (qPCR)

qPCR monitors accumulation of fluorescent signal (intercalating dyes or probe-based) during the exponential phase of PCR, allowing both detection and quantification of target DNA. It quantifies nematodes quickly using fluorescent detection, ideal for large-scale screening.

11.5.8 Scar-PCR

Uses species-specific markers for direct detection in soil or plants.

11.5.9 RT-PCR

Detects living nematodes by targeting mRNA, enhancing viability assessments.

The following Table 11.1 summarizes the key PCR-based methods used for the molecular diagnostics of PPNs, based on recent literature.

Each method has been developed to address specific diagnostic needs, with rDNA-PCR and qPCR being particularly prominent due to their versatility and sensitivity. For instance, rDNA-PCR has evolved since the early 1990s to include species-specific primers for quarantine species, while qPCR has become a standard for rapid, high-throughput quantification, as seen in its application for detecting mixed *Meloidogyne* species.

11.5.10 Applications and Case Studies

The applications of these PCR-based methods are diverse, reflecting their adaptability to various agricultural contexts:

Table 11.1 Key PCR-based methods used for the molecular diagnostics of PPNs

Method	Description	Target regions	Key applications	Examples of use
PCR-RFLP	Amplifies DNA and digests with restriction enzymes to analyze fragment patterns for strain differentiation	Total genome, often ITS regions	Differentiating *Meloidogyne* populations at race and strain levels	Used for identifying *Meloidogyne incognita* and *M. Javanica* strains
rDNA-PCR	Targets ribosomal DNA using species-specific primers, including multiplex variants for multiple species	18S, 28S, 5.8S, ITS regions	Identifying quarantine species like *Globodera pallida* and *G. Rostochiensis*, detecting potato cyst nematodes	Multiplex PCR for simultaneous detection of *Globodera* spp. without restriction digestion
Real-time PCR (qPCR)	Combines amplification with real-time fluorescent detection, quantifies DNA using standard curves	Various genomic regions, often ITS and satDNA	Quantifying species like *Meloidogyne incognita, Pratylenchus penetrans*; distinguishing living from dead nematodes with PMA-qPCR	Detection of *Heterodera glycines*, quantification of *Globodera pallida* in soil samples
SCAR-PCR	Uses species-specific markers derived from RAPD or other PCR products for direct detection	Species-specific sequen-ces	Detecting *Heterodera schachtii* in sugar beet, *Meloidogyne arabicida* in coffee, *M. Ethiopica* in kiwi fruit	SCAR markers for direct detection from infected roots and soil, e.g., *Heterodera schachtii*
Reverse transcription-PCR (RT-PCR)	Targets mRNA to detect living nematodes, enhancing viability assessments	mRNA transcri-pts	Detecting living pinewood nematode (*Bursaphelenchus xylophilus*) in wood	Used for distinguishing viable *Bursaphelenchus xylophilus* from dead specimens

- *Species Identification*: PCR-RFLP and rDNA-PCR are crucial for distinguishing between morphologically similar species. For example, multiplex PCR has been used to identify *M. incognita, M. enterolobii*, and *M. javanica* directly from individual galls (Liu et al. 1995). This is particularly important for quarantine and regulatory purposes, ensuring accurate identification of invasive species.
- *Quantification*: Real-time PCR (qPCR) excels in quantifying nematode populations, which is essential for assessing infestation levels. A study by Yerxa (1990) demonstrated the use of multiplex real-time PCR for quantifying *Globodera pallida, G. rostochiensis*, and *G. tabacum*, providing precise data for pest management decisions.

- *Early Detection*: SCAR-PCR and qPCR enable early detection before visible symptoms appear which is critical for timely intervention. For instance, SCAR-PCR has been developed for detecting *H. schachtii* in infected roots and soil (Plesch et al. 2018), enhancing the ability to manage sugar beet cyst nematode infestations.
- *Detection in Complex Samples*: The presence of soil inhibitors, such as clay and humic substances, can challenge PCR efficiency, but methods like digital droplet PCR (ddPCR) offer high precision. For example, ddPCR has been used for *Paratrichodorus allius*, providing direct copy numbers, as noted in recent reviews.
- *Viability Assessment*: RT-PCR, particularly when combined with propidium monoazide (PMA), can distinguish between living and dead nematodes, which is vital for assessing treatment efficacy. A study by Baumann et al. (2013) highlighted its use for detecting living *B. xylophilus* in wood, addressing challenges in posttreatment monitoring.

11.5.11 Advantages and Challenges

- *Advantages*: The advantages of PCR-based methods are significant, making them indispensable in modern nematology:
  - *High Sensitivity and Specificity*: These methods can detect nematodes at low densities and distinguish between closely related species, addressing the limitations of morphological identification.
  - *High-Throughput Capability*: They allow for the processing of large numbers of samples, which is essential for large-scale agricultural surveys and monitoring programs.
  - *Early Intervention*: By enabling detection before visible symptoms, they support proactive management strategies, reducing crop losses.
  - *Phenotypic Variation*: They account for variations across different growth stages, ensuring comprehensive diagnostics.
- *Challenges*: However, challenges remain that require ongoing research and technological advancements:
  - *Cost and Accessibility*: The need for specialized equipment, reagents, and sophisticated facilities can be prohibitive, especially for small-scale farmers or developing regions. For instance, the cost of qPCR reagents and instrumentation is noted as a barrier (Shao et al. 2023a, b).
  - *Soil Inhibitors*: The presence of clay, humic substances, and other soil components can inhibit PCR reactions, affecting accuracy. This is a recurring issue highlighted in multiple studies (Berry et al. 2008).
  - *Optimization Needs*: Each target species may require optimized protocols, adding complexity to implementation. For example, the development of species-specific primers for SCAR-PCR involves significant preliminary work (Yao et al. 2012).

- *False Results*: The presence of dead nematodes or degraded DNA can lead to false positives or negatives, necessitating additional validation steps, such as PMA-qPCR for viability assessment.

11.5.12 Recent Developments and Future Directions

Recent developments include the integration of digital droplet PCR (ddPCR) for enhanced precision and the use of PMA-qPCR for distinguishing living from dead nematodes, addressing some of the challenges mentioned. For instance, a study by Jhaveri et al. (2022) explored the potential of CRISPR-Cas9 delivery via lipofection, which, while primarily for genetic modification, underscores the evolving nature of molecular tools. Future research is likely to focus on reducing costs, improving field applicability, and developing portable diagnostic devices to enhance accessibility, particularly in resource-limited settings.

11.5.13 Conclusion

PCR-based methods for the molecular diagnostics of plant parasitic nematodes represent a cornerstone of modern agricultural science, offering tools that are both sensitive and specific for identifying, quantifying, and managing these pests. From PCR-RFLP for strain differentiation to qPCR for high-throughput quantification, these techniques have transformed nematology, supporting sustainable crop production. However, challenges such as cost, soil inhibitors, and optimization needs highlight the need for continued innovation. As research progresses, these methods are expected to become even more accessible and effective, ensuring robust protection against PPN-induced crop losses.

11.6 Isothermal Amplification Technologies

Isothermal amplification technologies, such as Loop-Mediated Isothermal Amplification (LAMP) and Recombinase Polymerase Amplification (RPA), are field-friendly, requiring no thermal cyclers. They offer high specificity and sensitivity, often combined with lateral flow devices (LFD) for visualization. Recent advancements include integrating CRISPR-Cas12a, enhancing sensitivity to the attomolar level, as demonstrated for *H. avenae* and *H. filipjevi*. These methods are particularly valuable for rapid, on-site diagnostics, with examples including detection of *Bursaphelenchus* spp., *Meloidogyne* spp., and *Radopholus* spp.

11.6.1 Loop-Mediated Isothermal Amplification (LAMP)

Though not PCR per se, LAMP is often lumped under "molecular diagnostics." It amplifies DNA at constant temperature (~ 65 °C) using 4–6 primers.

The use of LAMP method on PPNs has been very popular in recent years. In particular, LAMP technology has been developed for diagnosing many species of PPNs including *Bursaphelenchus* spp., *Meloidogyne* spp., *Anguina* spp., *Radopholus* spp., *Ditylenchus* spp., and *Tylenchulus* spp. (Table 11.2).

Among isothermal amplification techniques, LAMP-LFD and RPA-LFD are quickly evolving in the field of identification, because of their obvious specificity, efficiency, and visualization (Fig. 11.2).

11.6.2 Recombinase Polymerase Amplification (RPA)

Recombinase Polymerase Amplification (RPA) is an isothermal nucleic acid amplification technique that operates optimally between 25 and 42 °C and can generate

Table 11.2 Application of the LAMP technique to plant parasitic nematodes

Nematode	Host	Target region	References
Bursaphelenchus Xylophilus	Pine	ITS-rDNA	Kikuchi et al. (2009)
B. Xylophilus	*Pinus armandii*	ITS-rDNA	Kanetani et al. (2011)
B. Xylophilus	Pine	Pectate lyase-3	Kang et al. (2015)
B. Cocophilus	Coconut and oil palm trees	D2-D3 of rDNA	Ide et al. 2017
Meloidogyne incognita, M. Arenaria, M. Javanica, M. Hapla	Tomato	I ITS of rDNA	Niu et al. (2011)
M. Enterolobii	Tomato	5S rDNA-IGS2	Niu et al. (2012)
M. Mali	Tomato	ITS-5.8S rDNA	Zhou et al. (2017)
M. Chitwoodi, M. Fallax	Tomato	IGS2-18S	Zhang and Gleason (2019)
M. Partityla	Mature pecan trees	ITS-5.8S rDNA	Waliullah et al. (2020)
Anguina wevelli	–	ITS rDNA	Yu et al. 2018
A. Agrostis	–	ITS rDNA	Yu et al. (2020)
Radopholus similis	Anthurium	D2-D3 of rDNA	Peng et al. 2012
Ditylenchus destructor	Potato	28S rRNA	Deng et al. (2019)
Tylenchulus semipenetrans	Citrus orchards	ITS-rDNA	Lin et al. (2016)
T. Semipenetrans	Citrus rhizosphere soil	ITS1	Song et al. (2017)

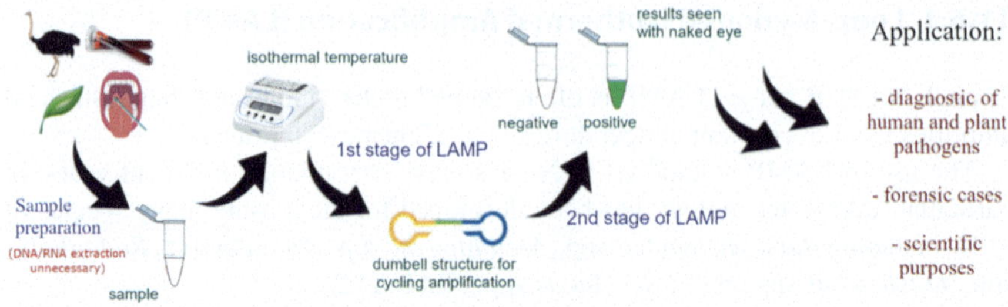

Fig. 11.2 Demonstration of the principles and working processes involved in isothermal amplification techniques

detectable amplicons in as little as 5–20 min. Unlike PCR and RPA, it requires only a simple heat block or even body heat, making it ideally suited for on-site, resource-limited settings.

- Principle of RPA
  - *Recombinase–Primer Complex Formation*: Recombinase proteins load onto outward-facing primers, forming nucleoprotein filaments.
  - *Strand Invasion and Displacement*: These filaments scan double-stranded target DNA and invade at cognate sites, displacing the noncomplementary strand.
  - *Primer Extension*: A strand-displacing polymerase (e.g., Bsu polymerase) extends the 3′ end of the primer, synthesizing new DNA without thermal cycling.
  - *Exponential Amplification*: Repeated rounds of recombinase loading, invasion, and extension yield rapid, exponential amplification of target sequences.
- Key Applications in Nematode Diagnostics
  - Meloidogyne *spp.* (*Root-Knot Nematodes*): RPA assays have been developed to distinguish *M. incognita*, *M. javanica*, *M. arenaria*, and *M. enterolobii* at 38 °C in 20 min, with no cross-reactivity to other nematodes (Ju et al. 2019).
  - Meloidogyne hapla: An RPA-LFD format detected as few as 10^{-3} females or 10^{-2} J2 in 0.5 g of soil, completing the entire workflow—including visualization—in ~1.5 h (Song et al. 2021).
  - Heterodera schachtii (*Sugar Beet Cyst Nematode*): Species-specific RPA primers combined with LFD and Cas12a-mediated fluorescence achieved detection limits down to a single cyst or female (10^{-4}), with results in under 1 h and 100% concordance with conventional PCR on field samples (Yao et al. 2021).
  - Heterodera avenae, H. filipjevi (*Cereal Cyst Nematodes*): RPA coupled with CRISPR-Cas12a detected 10^{-4} J2 and 10^{-1} J2 per 10 g soil within 1 h, showing 10× greater sensitivity than RPA-LFD alone and outperforming conventional PCR in both sensitivity and field-sample performance (Shao et al. 2023a, b).
- *Advantages and Limitations*: Advantages and limitations of isothermal amplification technologies are presented in Table 11.3.

Table 11.3 Advantages and limitations of isothermal amplification technologies

Advantages	Limitations
Speed: Results within 5–20 min (amplification)	Primer design: Requires careful in silico validation to avoid off-targets
Low equipment needs: Single-temperature incubation	Patent restrictions: RPA enzymes are proprietary (e.g., TwistDx kits)
High sensitivity: Down to 10^{-4}–10^{-3} J2 or cyst levels	Multiplexing: More challenging than in PCR
Field Deployability: LFD strips or portable fluorescence readers	Reagent stability: Cold-chain may be required for enzymes in hot climates

- Future Perspectives
  - *Integration with CRISPR/Cas Systems*: Further fusion with Cas13/Cas12 for enhanced specificity and single-molecule sensitivity.
  - *Digital RPA*: Partition-based quantification akin to digital PCR.
  - *Lyophilized, All-in-One Kits*: Cold-chain–free, room-temperature–stable formulations for true point-of-care.
  - *Multiplexed RPA Panels*: Simultaneous detection of multiple nematode species in a single reaction.
  - *Conclusion*: RPA represents a powerful addition to the molecular diagnostics toolkit for plant parasitic nematodes offering rapidity, sensitivity, and portability that are transforming both laboratory and field surveillance efforts.

11.7 DNA Sequencing and Barcoding

DNA sequencing and barcoding are critical for identifying and managing PPNs, which cause significant agricultural losses. These molecular methods address the limitations of traditional morphological identification, which is time-consuming and often unreliable due to morphological similarities among species.

DNA barcoding and sequencing target-specific genetic regions like the Internal Transcribed Spacer (ITS) and Cytochrome c oxidase subunit 1 (COX1) are used for species identification. ITS is particularly useful for closely related species, while COX1 aids in subspecies resolution. Meta-barcoding extends this to analyze nematode communities in environmental samples, offering insights into biodiversity, though challenges like environmental DNA (eDNA) contamination and limited database availability persist. Studies have applied these methods to *Meloidogyne* spp., *Heterodera* spp., and *Bursaphelenchus* spp., with results available in 8–12 h, useful for nematode control and border checks. They help identify cryptic species, especially with next-generation sequencing (NGS).

Below is an overview of key DNA sequencing and barcoding methods used for molecular diagnostics of PPNs, based on current knowledge and advancements.

DNA sequencing enables precise identification by analyzing specific genetic markers. Common techniques include:

11.7.1 Polymerase Chain Reaction (PCR) and Sanger Sequencing

- *Overview*: PCR amplifies target DNA regions, followed by Sanger sequencing to generate readable sequences. This is often used for single-specimen analysis.
- *Target Regions*:
  - *18S and 28S rRNA Genes*: Highly conserved, suitable for genus-level identification due to universal primers (Pun et al. 2024; Powers 2004).
  - *Internal Transcribed Spacer (ITS) Regions (ITS1, ITS2)*: Variable enough for species differentiation, widely used for PPNs like *Meloidogyne* and *Heterodera* (Pun et al. 2024; Shao et al. 2023a, b).
  - *Mitochondrial Cytochrome Oxidase Subunit I (COI)*: Effective for species-level resolution, especially in *Longidoridae*, though less reliable for some *Xiphinema* species (Palomares-Rius et al. 2017; Shao et al. 2023a, b).
  - *Advantages*: High accuracy, established protocols, and extensive reference databases (e.g., GenBank, NEMBASE) (Seesao et al. 2017; Powers 2004).
  - *Limitations*: Time-consuming, requires specialized equipment, and may fail in mixed samples due to nonspecific amplification (Christoforou et al. 2017).

11.7.2 Quantitative PCR (qPCR)

- *Overview*: Real-time PCR detects and quantifies nematode DNA, often using TaqMan probes for specificity. It's the fastest and most sensitive method for routine diagnostics (Christoforou et al. 2017).
- *Applications*: Identifies and quantifies PPNs like *M. incognita* in soil or plant samples (Pun et al. 2024).
- *Advantages*: High sensitivity, suitable for field applications with portable qPCR devices (Christoforou et al. 2017).
- *Limitations*: High reagent and instrumentation costs, requires optimization (Pun et al. 2024).

11.7.3 Next-Generation Sequencing (NGS)

- *Overview*: High-throughput sequencing allows analysis of multiple DNA samples simultaneously, ideal for complex communities.
- *Applications*: Used for *Meloidogyne* and *Heterodera* species, and to study nematode community diversity in agro-ecosystems (Pun et al. 2024; Siddique et al. 2022).
- *Advantages*: High resolution, detects multiple species in mixed samples, and supports meta-barcoding (Pun et al. 2024).
- *Limitations*: Limited primer availability and high costs for some applications (Pun et al. 2024).

11.7.4 MinION Sequencing

- *Overview*: A portable third-generation sequencer by Oxford Nanopore, enabling rapid, in-field sequencing of 18S rRNA genes (Knot et al. 2020).
- *Applications*: Identifies species like *Anisakis simplex* and *Caenorhabditis elegans* with 96–100% accuracy compared to Sanger sequencing, requiring only 10–15 min (Knot et al. 2020).
- *Advantages*: Portability, rapid results, and suitability for field diagnostics with systems like Bentolab (Knot et al. 2020).
- *Limitations*: Requires validation against reference databases and may have lower throughput than NGS.

11.7.5 DNA Barcoding and Meta-Barcoding

DNA barcoding uses standardized genetic markers to identify species by comparing sequences to reference databases. Meta-barcoding extends this to community analysis.

- *DNA Barcoding*:
  - *Process*: Involves DNA extraction, PCR amplification of a barcode region (e.g., COI, 18S rRNA, ITS), sequencing, and comparison with databases like GenBank or SILVA (Shao et al. 2023a, b).
  - *Applications*: Successfully used for species of *Meloidogyne*, *Heterodera*, and *Bursaphelenchus* (Shao et al. 2023a, b).
  - *Advantages*: High primer versatility, stable amplification, and low DNA quality requirements (Shao et al. 2023a, b).
  - *Limitations*: COI is less effective for some PPN groups (e.g., *X. americanum*), and ITS may not resolve all *Heterodera* species (Shao et al. 2023a, b; Palomares-Rius et al. 2017).
- *Meta-barcoding*:
  - *Overview*: Combines barcoding with NGS to identify multiple species in a single sample, using markers like 18S rRNA (V1-V2, V4-V5, V6-V8 regions) or COI (Kawanobe et al. 2021; Sikder et al. 2020).
  - *Applications*: Assesses soil nematode diversity, detecting 97–98% of free-living and 83–92% of PPN species with specific primers (e.g., F548_A/R1912) (Kawanobe et al. 2021).
  - *Advantages*: High throughput, detects diverse taxa, and reveals community structures 3–9% more diverse than morphological methods (Kawanobe et al. 2021).
  - *Limitations*: Requires curated reference databases and may amplify non-nematode DNA without enrichment (Sapkota and Nicolaisen 2015).

11.7.6 Challenges and Future Directions

- *Challenges.*
  - Lack of comprehensive reference databases for some nematode species, particularly for ITS and COI.
  - Nonspecific amplification in mixed samples, requiring nematode enrichment (Sapkota and Nicolaisen 2015).
  - High costs and expertise needed for NGS and qPCR (Pun et al. 2024).
  - Limited COI resolution for certain PPN groups, necessitating multi-marker approaches (e.g., COI + ITS) (Huston et al. 2022).
- *Future Directions.*
  - Development of nematode-specific primers to reduce non-target amplification (Kawanobe et al. 2021).
  - Expansion of curated databases like NEMBASE and SILVA (Kawanobe et al. 2021).
  - Integration of portable sequencing (e.g., MinION) with field-based diagnostics for real-time monitoring (Knot et al. 2020).
  - Combining molecular methods with remote sensing and deep learning for integrative diagnostics (Shao et al. 2023a, b).

11.7.7 Practical Considerations

- *Field Applications*: Portable devices like MinION and mobile qPCR enable in-field diagnostics, reducing turnaround time (Knot et al. 2020; Christoforou et al. 2017).
- *Cost-Effectiveness*: qPCR and LAMP are cost-effective for routine diagnostics, while NGS is better for biodiversity studies (Christoforou et al. 2017).
- *Database Reliance*: Accurate identification depends on robust databases like GenBank, SILVA, or NEMBASE (Seesao et al. 2017; Kenmotsu et al. 2020).

11.7.8 Conclusion

DNA sequencing and barcoding methods, including PCR, qPCR, NGS, and MinION sequencing, have revolutionized PPN diagnostics by offering rapid, accurate, and high-throughput identification. Barcoding with 18S rRNA, ITS, and COI, and metabarcoding with NGS, are particularly effective for species and community analysis. Despite challenges like database limitations and costs, ongoing advancements in portable sequencing and primer design promise to enhance field diagnostics and support sustainable nematode management.

11.8 Protein-Based Methods

Protein-based diagnostics include isozyme analyses, using enzyme phenotypes like esterases and malate dehydrogenase for identification, often supplementing morphological methods. Mass spectrometry, particularly MALDI-ToF-MS, detects diagnostic protein/peptide ions, enabling species-level discrimination, such as between *Anguina tritici*, *A. funesta*, and *M. javanica*. Serological methods use antibodies, with monoclonal antibodies offering higher specificity, as seen in differentiating *G. rostochiensis* from *G. pallida*, detecting protein equivalents of less than one egg. These methods, while effective, face challenges like cross-reactivity and cost, with MALDI-ToF-MS systems costing around 150–200 K euros but low per-sample costs (1-h prep, 3-min analysis).

Protein-based methods for molecular diagnostics of PPNs offer powerful tools for accurate identification, especially when morphological approaches are insufficient due to limited distinguishing features or intraspecific variation. These methods leverage protein profiles or specific protein markers to differentiate nematode species, often providing faster and more reliable results than traditional microscopy. Below is an overview of key protein-based diagnostic methods, their applications, advantages, and limitations, focusing on PPNs.

11.8.1 Matrix-Assisted Laser Desorption/Ionization Time-of-Flight Mass Spectrometry (MALDI-TOF MS)

- *Description*: MALDI-TOF MS generates protein/peptide ion profiles or spectral fingerprints unique to specific nematode taxa. It analyzes intact nematodes or extracted proteins to identify species based on diagnostic peaks in mass spectra.
- *Application*:
  - Used to distinguish species like *A. tritici*, *A. funesta*, and *M. javanica* based on unique spectral peaks.
  - Studies have shown it can identify single *M. incognita* nematodes (adult females or second-stage juveniles, J2s), whether washed, unwashed, crushed, or intact.
  - Protein profiles differ between life stages (e.g., adults vs. J2s), with washed or crushed samples yielding stronger peaks.
- *Advantages*:
  - Rapid analysis (1-h sample preparation, 3-min analysis).
  - Low cost per sample (excluding initial instrumentation cost of ~150–200 K euros in 2017).
  - High sensitivity, detecting subtle differences in protein profiles.
  - Potential for developing commercial databases (e.g., similar to Bruker MALDI Biotyper for microbes) for automated nematode identification.

- *Limitations*:
  - High initial equipment cost.
  - Requires careful selection of solvents and matrix materials to ensure reproducibility and quality of spectra.
  - Limited databases for nematodes compared to microbial applications (Christoforou et al. 2017; Bogale et al. 2020).

11.8.2 Multi-locus Enzyme Electrophoresis (MEE)

- *Description*: MEE involves extracting soluble proteins from nematodes, separating them via starch or polyacrylamide gel electrophoresis, and staining for specific enzymes (isozymes). Differences in migration patterns, driven by variations in electrical charge, molecular weight, or conformation, allow species differentiation.
- *Application*:
  - Commonly used enzymes include esterases, malate dehydrogenase, superoxide dismutase, and glutamate-oxaloacetate transaminase.
  - Effective for identifying *Meloidogyne* species and elucidating phylogenetic relationships.
- *Advantages*:
  - Supplements morphological identification by providing biochemical markers.
  - Relatively simple and cost-effective compared to advanced molecular methods.
- *Limitations*:
  - Requires fresh or well-preserved samples for accurate protein extraction.
  - Less sensitive than DNA-based methods and may miss subtle genetic variations.
  - Labor-intensive and requires expertise in electrophoresis techniques (Bogale et al. 2020).

11.8.3 Identification of Effector Proteins and Secretions

- *Description*: PPNs secrete effector proteins from esophageal glands, amphids, or cuticles to facilitate host invasion, suppress plant defenses, and establish feeding sites. These proteins can serve as diagnostic biomarkers. Techniques like liquid chromatography-tandem mass spectrometry (LC-MS/MS) analyze these proteins to identify pathogenicity factors.
- *Application*:
  - Effector proteins like those targeting Arabidopsis *RD21A* or tomato *RCR3Pim* have been studied in *G. rostochiensis* and *M.* species.
  - Useful for identifying specific nematode species or strains based on unique effector profiles.
  - Helps understand host–parasite interactions, aiding in the development of resistant plant cultivars.

- *Advantages*:
  - Links diagnostics to pathogenicity, supporting targeted control strategies.
  - High specificity for identifying nematodes with unique effector profiles.
- *Limitations*:
  - Complex sample preparation and analysis require advanced instrumentation.
  - Limited to species with well-characterized effector proteins.
  - Not yet widely adopted for routine diagnostics (Mejias et al. 2019; Rehman et al. 2016).

11.8.4 Nano-Biosensors Targeting Protein Biomarkers

- *Description*: Nano-biosensors use nanomaterials (e.g., carbon nanotubes, graphene, quantum dots) to detect nematode-specific protein biomarkers, such as esophageal gland secretions. These sensors rely on specific interactions between immobilized bioreceptors and protein analytes.
- *Application*:
  - Detects proteins or other biochemical markers (e.g., volatile organic compounds) released during nematode activity.
  - Offers potential for real-time, field-deployable diagnostics when integrated with artificial intelligence for data analysis.
- *Advantages*:
  - High sensitivity and specificity, detecting low concentrations of target proteins.
  - Rapid, potentially field-deployable, reducing reliance on lab-based methods.
  - Early detection before visible crop symptoms.
- *Limitations*:
  - Still in development, with limited commercial availability.
  - Requires validation for diverse nematode species and environments.
  - High development costs for sensor design and AI integration (Mani et al. 2025).

11.8.5 Advantages of Protein-Based Methods

- *Complementary to Morphological and DNA-Based Methods*: Protein-based approaches overcome limitations of morphological identification, such as intraspecific variation or the need for skilled taxonomists, and provide an alternative when DNA markers lack sufficient variability (Carneiro, and Lima, F.S.de O. and Correia, V.R. 2017).
- *High Specificity*: Techniques like MALDI-TOF MS and nano-biosensors can detect subtle differences in protein profiles, enabling species-level identification.

- *Potential for Automation*: Methods like MALDI-TOF MS and nano-biosensors can be integrated with databases or AI for high-throughput, automated diagnostics.
- *Pathogenicity Insights*: Identifying effector proteins links diagnostics to nematode biology, aiding in resistance breeding and control strategies (Christoforou et al. 2017; Mejias et al. 2019).

11.8.6 Limitations

- *Instrumentation Costs*: High initial costs for equipment like MALDI-TOF MS or LC-MS/MS limit accessibility, especially in resource-constrained settings (Christoforou et al. 2017).
- *Sample Preparation*: Protein extraction and preservation can be challenging, especially for field samples, and may introduce artifacts (Carneiro, and Lima, F.S.de O. and Correia, V.R. 2017).
- *Database Dependency*: Methods like MALDI-TOF MS require comprehensive protein databases, which are underdeveloped for nematodes compared to microbes (Christoforou et al. 2017).
- *Expertise Requirement*: Techniques like MEE and LC-MS/MS require specialized training, limiting routine use in some diagnostic labs (Bogale et al. 2020).

11.8.7 Integration with Other Methods

Protein-based methods are most effective when combined with morphological and DNA-based approaches (e.g., PCR, qPCR, DNA sequencing) for integrative diagnostics. For example:

- Morphological data can confirm visual characteristics, while protein profiles provide biochemical validation.
- DNA-based methods (e.g., rDNA sequencing, SCAR markers) offer genetic precision, while protein methods add functional insights into pathogenicity (Carneiro, and Lima, F.S.de O. and Correia, V.R. 2017).
- Emerging technologies like AI-driven nano-biosensors can integrate protein and molecular data for real-time, field-based diagnostics (Mani et al. 2025).

11.8.8 Future Directions

- *Database Development*: Expanding protein databases for PPNs will enhance MALDI-TOF MS and other methods, similar to microbial diagnostics (Christoforou et al. 2017).

- *Field-Deployable Technologies*: Advances in nano-biosensors and portable mass spectrometers could enable on-site diagnostics, reducing turnaround times (Mani et al. 2025).
- *AI Integration*: Combining protein-based diagnostics with AI and machine learning can improve accuracy and automate identification, especially for complex samples (Agarwal et al. 2023).
- *Effector Protein Research*: Further characterization of nematode effector proteins could lead to novel diagnostic markers and targets for resistant crop development (Mejias et al. 2019; Rehman et al. 2016).

11.8.9 Conclusion

Protein-based methods like MALDI-TOF MS, MEE, effector protein analysis, and nano-biosensors provide robust tools for diagnosing plant-parasitic nematodes, addressing the limitations of morphological identification. While challenges like high costs and the need for specialized expertise persist, these methods offer high specificity and insights into nematode pathogenicity. Integrating protein-based diagnostics with molecular and morphological approaches and leveraging emerging technologies like AI and nano-biosensors will enhance the speed, accuracy, and field applicability of PPN identification, supporting sustainable agriculture (Christoforou et al. 2017; Bogale et al. 2020; Mejias et al. 2019)

11.9 Microarrays

DNA microarrays offer a powerful, multiplexed approach for the rapid detection and discrimination of plant-parasitic nematodes, particularly for quarantine monitoring and mixed-species detection. Microarrays (DNA "chips") provide a high-throughput platform for the simultaneous detection of multiple nematode species or targets in a single assay, based on hybridization of fluorescently labeled nucleic acids to arrayed probes. Though challenged by cost and complexity, continued advances in array design, miniaturization, and integration with other detection modalities will likely enhance their utility in both research and applied diagnostics.

11.9.1 Principles of DNA Microarray Diagnostics

- *Probe design*: Species-specific sequences (e.g., SCAR or satellite DNA regions) are PCR-amplified, sequenced, and converted into short oligonucleotide capture probes (typically 20–70 nt) (Bogale et al. 2020).
- *Array assembly*: Probes are spotted or synthesized in situ on glass slides or other solid supports in a defined grid; each spot represents one diagnostic target. Control probes (calibration, ratio, negative) are interspersed for quality monitoring (Thompson et al. 2006).

- *Target labeling*: Sample DNA (genomic or PCR products) or cDNA (for RNA-based arrays) is fluorescently labeled (e.g., Cy3/Cy5 or Alexa Fluor dyes) either via direct incorporation during amplification or post-PCR coupling.
- *Hybridization and scanning*: Labeled targets are hybridized to the array under stringent conditions (e.g., 42 °C overnight) to allow specific binding. After washing, arrays are scanned with dual-laser microarray scanners to quantify fluorescence at each spot.

11.9.2 Case Studies in Nematode Diagnostics

- *Meloidogyne chitwoodi*: François et al. (2006) designed SCAR- and satellite-DNA probes specific to the quarantine nematode *M. chitwoodi*. The array reliably detected *M. chitwoodi* in pure and mixed samples (with *M. hapla*, *M. arenaria*), even when targets were co-labeled with different dyes, demonstrating no competition between probes François et al. 2006; Bogale et al. 2020).
- *M. hapla*: Similar oligonucleotide arrays were later adapted to detect *M. hapla*, confirming the broad applicability of the approach across root-knot species (da Cunha et al. 2018).

11.9.3 Advantages

- *Multiplexing*: Simultaneous detection of tens to thousands of targets in a single run.
- *High Specificity*: Stringent hybridization reduces cross-amplification; probe design can discriminate closely related species.
- *Non-Culture Based*: Direct detection from environmental samples without nematode isolation.
- *Quantitative Potential*: Signal intensities can reflect relative abundance under well-controlled conditions.

11.9.4 Limitations

- *Cost and Complexity*: Microarray fabrication, labeled reagents, and specialized scanners are expensive and require technical expertise (Pun et al. 2024).
- *Sensitivity Constraints*: Detection limits depend on probe efficiency and sample quality; low-abundance targets may escape detection.
- *Probe Bias and Cross-Hybridization*: Similar sequences can produce background signals; careful in silico validation is essential.
- *Dynamic Range*: Fluorescence saturation and background variability can limit accurate quantification across wide concentration ranges.

11.9.5 Future Directions

- *Custom Portable Arrays*: Development of field-deployable microarray cartridges coupled to portable scanners for on-site diagnostics.
- *Integrated Biosensor Platforms*: Combining microarrays with microfluidics or electrochemical detection to reduce assay times and reagent volumes.
- *Multiplexed Transcriptomics*: Using gene expression arrays to monitor nematode activity or host responses, providing both diagnostic and pathogenicity insights.
- *Probe Enhancement*: Incorporating locked nucleic acids (LNAs) or peptide nucleic acids (PNAs) to improve binding specificity and sensitivity.

The following Table 11.4. summarizes the key molecular diagnostic methods, their applications, and limitations:

This table highlights the diversity of molecular tools, each with specific strengths and challenges, catering to different needs in nematode diagnostics.

The characteristics of various existing methods for the molecular detection and diagnosis of PPNs include biochemical methods, traditional PCR, and isothermal amplification technologies. Practically, no single method or technique exists for diagnosing PPNs. Each approach has its strengths and weaknesses; therefore, we compared the characteristics of each method (Table 11.5).

The following Table 11.6 summarizes key quantitative details of plant nematode detection methods.

This table highlights the trade-offs between sensitivity, cost, and field applicability, guiding the choice of diagnostic method based on context.

Table 11.4 Key molecular diagnostic methods, their applications, and limitations

Method	Primary use	Advantages	Limitations
PCR	Species identification	High specificity, widely standardized	Requires DNA extraction, less quantitative
qPCR	Detection and quantification	High sensitivity, rapid, handles large samples	Competition in mixed species, costlier
RFLP	Strain differentiation	Useful for closely related species	Labor-intensive, less common now
DNA sequencing	Cryptic species identification	High resolution, NGS for throughput	Expensive, requires bioinformatics
Microarrays	Multiple species detection	Potential for high-throughput	High cost, limited use currently
MALDI-TOF MS	Protein-based identification	Novel approach, protein analysis	Less common, requires specialized labs

Table 11.5 Comparison of different plant nematode detection methods

Category	Technology	Advantages	Disadvantages	References
In the lab				
PCR methods	DNA barcoding	Accuracy	Time-consuming	Hebert et al. (2003)
	Droplet digital PCR	High sensitivity and low amount of template DNA	Expensive reagents and instruments	Rougemont et al. (2004)
	Gene chip technology	Fast, accuracy	Expensive equipment, immature technology	Fodor (1997)
	RFLPs	Reliable and reproducible	Complex operations, requiring large amounts of DNA	Blok and Powers (2009)
	RAPD	Generates a large amount of information	Lacks repeatability, requires strict experimental reaction	Feng et al. (2005)
	SCAR	High sensitivity and specificity	Time-consuming	Chen et al. (2011)
	RT-qPCR	Sensitive, reliable	Time-consuming, equipment relatively expensive	Berry et al. (2008)
	ddPCR	High sensitivity, simple, convenient	Expensive instruments	Chen et al. (2022)
Outdoors				
Isothermal amplification technology	LAMP	Low cost, simple operation, low equipment demand	False-positive results; outdoors and in the field	Ahuja and Somvanshi (2021)
	RPA LRPACRISP/Cas12a	Fast, high sensitivity and specificity, low cost, simple operation, low equipment demand, visualization of results	False-positive results; required to design specific primers, probes, and gRNA	Babu et al. (2018)
Spectral techniques	Remote sensing systems	Fast, large-area detection, dynamic monitoring	Requires technical personnel expertise, difficult to capture detailed changes	Tao et al. (2020)

11.10 Recent Advances and Trends

Recent developments include the integration of CRISPR-Cas12a with isothermal amplification for enhanced sensitivity, field-deployable tools like LAMP-LFD for rapid visualization, and the application of digital PCR for absolute quantification. Meta-barcoding and next-generation sequencing (NGS) are evolving for soil and

Table 11.6 Key quantitative details from recent studies

Method	Sensitivity/Accuracy	Cost/Equipment needs	Examples/Species
qPCR	100% accuracy, detects DNA from 250 mg soil	High, requires thermal cycler	*Meloidogyne Enterolobii*, *Heterodera Schachtii*
ddPCR	Accurate at low concentrations	High, specialized equipment	*M. Enterolobii*
LAMP/RPA with CRISPR-Cas12a	Attomolar sensitivity	Low, field-usable, no thermal cycler	*H. Avenae*, *H. Filipjevi*
DNA barcoding/ Metabarcoding	High primer versatility, 8–12 h results	Moderate, sequencing equipment	*Meloidogyne* spp., *Heterodera* spp.
MALDI-ToF-MS	Discriminates to race level	High initial cost (~150–200 K euros), low per-sample cost (1 h prep, 3 min analysis)	*Ditylenchus dipsaci*, *M. Javanica*

plant diagnostics, while proteomics and AI-driven imaging are gaining traction. Current trends suggest real-time PCR remains the fastest and most sensitive, with DNA barcoding, NGS, and proteomics expected to dominate, especially with low-budget, mobile kit applications. For example, studies highlight the potential of recombinase polymerase amplification for detecting *M. hapla* directly from root galls.

11.11 Integrative Taxonomy and Challenges

Combining molecular diagnostics with morphological analysis, known as integrative taxonomy, provides a more reliable identification. This approach is particularly useful for complex cases, like distinguishing between closely related species, and addresses the decline in traditional taxonomic expertise.

Integrative taxonomy, proposed by Valdecasas et al. (2008) and Pires and Marinoni (2010), combines morphological/morphometric and molecular data, addressing intra- and inter-population variability. However, challenges include:

11.11.1 Decline in Classical Taxonomy

Few trained nematode taxonomists remain, primarily due to retirement without replacement and lack of interest among young scientists.

11.11.2 Sampling and Validation

Accurate sampling and validation with morphological identification are critical, especially for quarantine, but limited by declining taxonomic expertise.

11.11.3 Need for Specialized Labs

While molecular methods are powerful, challenges include the need for specialized labs and the ongoing decline in classical taxonomy skills. Future research aims to develop assays targeting virulence factors for rapid detection of resistant strains.

11.12 Conclusion

Molecular diagnostics for PPNs have transformed nematode identification, offering precise, rapid, robust, and often field-deployable solutions for their management, addressing the limitations of traditional methods. Techniques like PCR, qPCR, and DNA sequencing are pivotal, with integrative taxonomy enhancing reliability. From PCR-based methods to emerging technologies like CRISPR-enhanced isothermal amplification and AI-driven imaging, these tools address the limitations of morphological identification, supporting sustainable agricultural practices. Present research aims to overcome current challenges, ensuring accessible, accurate diagnostics for global crop protection. Despite challenges like declining taxonomic expertise, ongoing advancements promise more accessible and effective diagnostics, crucial for sustainable agriculture.

11.13 Suggested Readings

- Castagnone-Sereno, P., Skantar, A. and Robertson, L. 2011. Molecular tools for diagnostics. In: J. Jones, G. Gheysen, C. Fenoll (eds.) *Genomics and Molecular Genetics of Plant-Nematode Interactions*. Springer, Dordrecht. https://doi.org/10.1007/978-94-007-0434-3_21.
- Christoforou, M., Orford, M. and Tsaltas, D. 2017. Molecular diagnostic tools for nematodes. InTech. *IntechOpen*. doi:10.5772/intechopen.69075.
- Powers, T.O. 1992. Molecular diagnostics for plant nematodes. *Parasitology Today* 8(5): 177–179.
- Perry, R.N., Subbotin, S.A. and Moens, M. 2007. Molecular diagnostics of plant-parasitic nematodes. In: *Biotechnology and Plant Disease Management*, pp.195–226.CABIDigitalLibrary.https://doi.org/10.1079/9781845932886.0195.
- Powers, T. 2004. Nematode molecular diagnostics: From bands to barcodes. *Annual Review of Phytopathology* 42: 367–383.

11.14 Review Questions

- What are the main limitations of traditional morphological methods in diagnosing plant-parasitic nematodes?
- Define molecular diagnostics in the context of plant nematode identification.
- List three commonly used molecular markers for nematode identification.

- What is the principle behind PCR-based detection of nematodes?
- Compare morphological, biochemical, and molecular approaches for nematode diagnosis.
- What are the advantages of real-time PCR (qPCR) over conventional PCR in nematode diagnostics?

References

Agarwal S, Curran ZC, Yu G et al (2023) Plant parasitic nematode identification in complex samples with deep learning. J Nematol 55(1):20230045

Ahuja A, Somvanshi VS (2021) Diagnosis of plant parasitic nematodes using loop mediated isothermal amplification (LAMP): a review. Crop Prot 147:105459

Babu B, Ochoa-Corona FM, Paret ML (2018) Recombinase polymerase amplification applied to plant virus detection and potential implications. Anal Biochem 546:72–77

Baumann PS, Crespi S, Marion-Veyron R et al (2013) Treatment and early intervention in psychosis program (TIPP-Lausanne): implementation of an early intervention programme for psychosis in Switzerland. Early Interv Psychiatry 3:322–328

Berry SD, Fargette M, Spaull VW et al (2008) Detection and quantification of root-knot nematode (*Meloidogyne javanica*), lesion nematode (*Pratylenchus zeae*) and dagger nematode (*Xiphinema elongatum*) parasites of sugarcane using real-time PCR. Mol Cell Probes 22:168–176

Blok VC, Powers TO (2009) Biochemical and molecular identification. In: Perry RN, Moens M, Starr JL (eds) Root-knot Nematodes. Chapter 4. CAB International, Wallingford, pp 98–118

Bogale M, Baniya A, DiGennaro P (2020) Nematode identification techniques and recent advances. Plants 9(10):1260

Carneiro RMDG, Lima FSd O, Correia VR (2017) Methods and tools currently used for the identification of plant parasitic nematodes. INTECH. https://doi.org/10.5772/intechopen.69403

Chen FM, Negi S, Ye JR (2011) A SCAR molecular marker to distinguish *Bursaphelenchus mucronatus* from the pinewood nematode, B xylophilus. For Pathol 41:376–381

Chen Y, Long HB, Feng TZ et al (2022) Development of a novel primer – TaqMan probe set for diagnosis and quantification of *Meloidogyne enterolobii* in soil using qPCR and droplet digital PCR assays. Int J Mol Sci 23:11185

Christoforou M, Orford M, Tsaltas D (2017) Molecular diagnostic tools for nematodes. In: Nematology – concepts, diagnosis and control. INTECH. https://doi.org/10.5772/intechopen.69075

da Cunha TG, Visôtto LE, Lopes EA et al (2018) Diagnostic methods for identification of root-knot nematodes species from Brazil. Cienc Rural 48(02). https://doi.org/10.1590/0103-8478cr20170449

Deng MH, Zhong LY, Kamolnetr O et al (2019) Detection of helminths by loop-mediated isothermal amplification assay: a review of updated technology and future outlook. Infect Dis Poverty 8:20–29

Feng MC, Jian RY, Jian T (2005) A study on detection technique of *Bursaphelenchus xylophilus* and *B. Mucronatus* by RAPD. J Nanjing Forestry Univ 4:25–28

Fodor SPA (1997) DNA Sequencing- massively parallel genomics. *Science* 277:393–398

François C, Kebdani N, Barker I et al (2006) Towards specific diagnosis of plant-parasitic nematodes using DNA oligonucleotide microarray technology: a case study with the quarantine species *Meloidogyne chitwoodi*. Mol Cell Probes 20(1):64–69

Hebert PD, Cywinska A, Ball SL, Dewaard JR (2003) Biological identifications through DNA barcodes. Proc Biol Sci 270:313–321

Huston DC, Khudhir M, Hodda M (2022) Reliability and utility of standard gene sequence barcodes for the identification and differentiation of cyst nematodes of the genus *Heterodera*. J Nematol 54(1):20220024

Ide T, Kanzaki N, Giraldo PPP, Giblin-Davis RM (2017) Loop-mediated isothermal amplification (LAMP) for detection of the red ring nematode, *Bursaphelenchus cocophilus*. Nematology 19:559–565

Jhaveri N, van den Berg W, Hwang BJ et al (2022) Genome annotation of *Caenorhabditis briggsae* by TEC-RED identifies new exons, paralogs, and conserved and novel operons. G3 Genes, Genomes, Genet 12(7):jkac101. https://doi.org/10.1093/g3journal/jkac101

Ju Y, Lin Y, Yang G et al (2019) Development of recombinase polymerase amplification assay for rapid detection of *Meloidogyne incognita, M. Javanica*, *M. Arenaria,* and *M. Enterolobii*. Eur J Plant Pathol 155:1155–1163

Kanetani S, Kikuchi T, Akiba M et al (2011) Detection of *Bursaphelenchus xylophilus* from old discs of dead *Pinus armandii* var. *amamiana* trees using a new detection kit. For Pathol 41:387–391

Kang JS, Kim AY, Han HR et al (2015) Development of two alternative loop-mediated isothermal amplification tools for detecting pathogenic pine wood nematodes. For Pathol 45:127–133

Kawanobe M, Toyota K, Ritz K (2021) Development and application of a DNA metabarcoding method for comprehensive analysis of soil nematode communities. Appl Soil Ecol 166:103974

Kenmotsu H, Uchida K, Hirose Y, Eki T (2020) Taxonomic profiling of individual nematodes isolated from copse soils using deep amplicon sequencing of four distinct regions of the 18S ribosomal RNA gene. PLoS One 15(10):e0240336

Kikuchi T, Aikawa T, Oeda Y et al (2009) A rapid and precise diagnostic method for detecting the pinewood nematode *Bursaphelenchus xylophilus* by loop-mediated isothermal amplification. Phytopathology 99:1365–1369

Knot IE, Zouganelis GD, Weedall GD et al (2020) DNA barcoding of nematodes using the MinION. Front Ecol Evol 8:100

Lin B, Wang H, Zhuo K, Liao J (2016) Loop-mediated isothermal amplification for the detection of *Tylenchulus semipenetrans* in soil. Plant Dis 100:877–883

Liu L, Kloepper JW, Tuzun S (1995) Induction of systemic resistance in cucumber against Fusarium wilt by plant growth-promoting rhizobacteria. Phytopathology 85:695–698

Mani J, Nagachandrabose S, Somasundaram P, Deenan S (2025) Artificial intelligence integrated nano biosensor technology: a breakthrough in early detection and sustainable management of phytonematodes. Physiol Mol Plant Pathol 139:102756

Mejias J, Truong NM, Abad P et al (2019) Plant proteins and processes targeted by parasitic nematode effectors. Front Plant Sci 10:970

Niu JH, Guo QX, Jian H et al (2011) Rapid detection of *Meloidogyne* spp. by LAMP assay in soil and roots. Crop Prot 30:1063–1069

Niu JH, Jian H, Guo QX et al (2012) Evaluation of loop-mediated isothermal amplification (LAMP) assays based on 5S rDNAIGS2 regions for detecting *Meloidogyne enterolobii*. Plant Pathol 61:809–819

Palomares-Rius JE, Cantalapiedra-Navarrete C, Archidona-Yuste A et al (2017) The utility of mtDNA and rDNA for barcoding and phylogeny of plant-parasitic nematodes from Longidoridae (Nematoda, Enoplea). Sci Rep 7:10905

Peng H, Peng DL, Hu XQ et al (2012) Loop mediated isothermal amplification for rapid and precise detection of the burrowing nematode, *Radopholus similis*, directly from diseased plant tissues. Nematology 14:977–986

Pires AC, Marinoni L (2010) DNA barcoding and traditional taxonomy unified through integrative taxonomy: a view that challenges the debate questioning both methodologies. Biota Neotrop 10(2). https://doi.org/10.1590/S1676-06032010000200035

Plesch E, Chen CC, Elisabeth Butz E (2018) Selective agonist of TRPML2 reveals direct role in chemokine release from innate immune cells. elife 7:e39720

Powers T (2004) Nematode molecular diagnostics: from bands to barcodes. Annu Rev Phytopathol 42:367–383

Pun TB, Magar RT, Koech R et al (2024) Emerging trends and technologies used for the identification, detection, and characterisation of plant-parasitic nematode infestation in crops. Plants 13(21):3041

Rehman S, Gupta VK, Goyal AK (2016) Identification and functional analysis of secreted effectors from phytoparasitic nematodes. BMC Microbiol 16:48

Rougemont M, Van Saanen M, Sahli R et al (2004) Detection of four *plasmodium* species in blood from humans by 18S rRNA gene subunit-based and species-specific real-time PCR assays. J Clin Microbiol 42:5636–5643

Sapkota R, Nicolaisen R (2015) High throughput sequencing of nematode communities from total soil DNA extractions. BMC Ecol 15(3):3

Seesao Y, Gay M, Merlin S et al (2017) A review of methods for nematode identification. J Microbiol Methods 138:37–49

Shao H, Jian J, Peng D et al (2023a) Recombinase polymerase amplification coupled with CRISPR-Cas12a technology for rapid and highly sensitive detection of *Heterodera avenae* and *Heterodera filipjevi*. Plant Dis 107:1365–1376

Shao H, Zhang P, Peng D et al (2023b) Current advances in the identification of plant nematode diseases: from lab assays to in-field diagnostics. Front Plant Sci 14:1106784

Siddique S, Radakovic ZS, Hilt C et al (2022) The genome and life stage-specific transcriptomes of a plant-parasitic nematode and its host reveal susceptibility genes involved in trans-kingdom synthesis of vitamin B5. Nat Commun 13:6190

Sikder MM, Vestergård M, Sapkota R et al (2020) Evaluation of metabarcoding primers for analysis of soil nematode communities. Diversity 12(10):388

Song ZQ, Cheng JE, Cheng FX et al (2017) Development and evaluation of loop-mediated isothermal amplification assay for rapid detection of *Tylenchulus semipenetrans* using DNA extracted from soil. Plant Pathol J 33:184–192

Song Z, Yang X, Zhang X et al (2021) Rapid and visual detection of *Meloidogyne hapla* using recombinase polymerase amplification combined with a lateral flow dipstick assay. Plant Dis 105(9):2697–2703

Tao H, Li C, Cheng C et al (2020) Progress in remote sensing monitoring for pine wilt disease induced tree mortality: a review. For Res 3:172–183

Thompson FJ, Barker GLA, Hughes L et al (2006) A microarray analysis of gene expression in the free-living stages of the parasitic nematode *Strongyloides ratti*. BMC Genomics 7:157

Valdecasas AG, Williams D, Wheeler QD (2008) 'Integrative taxonomy' then and now: a response to Dayrat (2005). Biol J Linn Soc 93:211–216

Waliullah S, Bell J, Jagdale G et al (2020) Rapid detection of pecan root-knot nematode, *Meloidogyne partityla*, in laboratory and field conditions using loop-mediated isothermal amplification. PLoS One 15:e0228123

Yao MC, Chang SC, Lee CY, Lu KH (2012) A SCAR-based method for rapid identification of four major lepidopterous stored-product pests. J Econ Entomol 105(3):1100–1106

Yao K, Peng D, Jiang C et al (2021) Rapid and visual detection of *Heterodera schachtii* using recombinase polymerase amplification combined with Cas12a-mediated technology. Int J Mol Sci 22(22):12577

Yerxa EJ (1990) An introduction to occupational science, a foundation for occupational therapy in the 21st century. Occup Ther Health Care 6(4):1–17

Yu LZ, Song SY, Yu C et al (2018) A loop mediated isothermal amplification (LAMP) assay for rapid and reliable detection of *Anguina wevelli*, a grass parasitic nematode. Eur J Plant Pathol 150:725–734

Yu LZ, Song SY, Yu C et al (2020) Rapid detection of *Anguina agrostis* by loop-mediated isothermal amplification. Eur J Plant Pathol 156:819–825

Zhang L, Gleason C (2019) Loop-mediated isothermal amplification for the diagnostic detection of *Meloidogyne chitwoodi* and *M. Fallax*. Plant Dis 103:12–18

Zhou QJ, Cai Y, Gu JF et al (2017) Rapid and sensitive detection of *Meloidogyne Mali* by loop-mediated isothermal amplification combined with a lateral flow dipstick. Eur J Plant Pathol 148:755–769

12 Role of Omics Technologies in Nematode Research

Abstract

Traditional diagnostic and management approaches are often labor-intensive and lack specificity. Advances in high-throughput "omics" technologies—genomics, transcriptomics, proteomics, and metabolomics—are revolutionizing our ability to detect, characterize, and manage plant-parasitic nematodes. These fields help identify genetic markers, gene expression patterns, proteins, and metabolites for detecting nematode infections and disrupting their interactions with hosts. By providing comprehensive, system-wide snapshots of nematode biology and host–pathogen interactions, these approaches enable sensitive and specific diagnostics based on unique molecular signatures, discovery of novel targets for control (e.g., effector proteins, metabolic pathways), and rational design of management strategies, including RNA interference (RNAi), biological control, and chemical or botanical interventions. These approaches being complementary, offer a comprehensive view from DNA to metabolic outcomes, though their integration can be complex and requires further study. The key contributions of each omics discipline to nematode diagnosis and management, and how their integration paves the way for precision nematology are outlined.

Keywords

Genomics • Transcriptomics • Proteomics • Metabolomics • Nematode diagnosis • Nematode management • Host resistance

12.1 Learning Objectives

- Understand how genomics, transcriptomics, proteomics, and metabolomics help in identifying and diagnosing plant nematodes.

R. P. Parvatha, *Plant Parasitic Nematodes*,
https://doi.org/10.1007/978-981-95-9306-4_12

- Learn the role of -omics tools in studying nematode biology, host–parasite interactions, and biomarker discovery.
- Explore how integrated -omics approaches improve accuracy in nematode detection and management.
- Gain basic skills in using bioinformatics tools to analyze -omics data for nematode diagnosis.

12.2 Introduction

Plant-parasitic nematodes, such as root-knot nematodes (*Meloidogyne* spp.), cyst nematodes (*Heterodera* and *Globodera* spp.), and others, cause significant agricultural losses by feeding on plant roots, forming specialized feeding structures like syncytia or giant cells. Traditional control methods, such as chemical nematicides, are increasingly restricted due to environmental concerns, highlighting the need for molecular approaches.

Genomics, transcriptomics, proteomics, and metabolomics have emerged as transformative disciplines in plant nematode research, providing detailed molecular insights into the interactions between plant-parasitic nematodes and their hosts. The advent of "omics" technologies has revolutionized our ability to study these organisms at multiple biological levels, from DNA to metabolites. These approaches are essential for understanding nematode biology, host defense mechanisms, and developing sustainable control strategies. This chapter synthesizes key findings and applications, drawing from recent studies to offer a comprehensive overview. Together, these methods help find new ways to control nematodes, such as breeding resistant plants or using RNA interference.

12.3 Genomics

The advent of genomic technologies has transformed plant nematology, offering unprecedented insights into the biology, diagnosis, and management of these pests. Genomics, the study of an organism's complete DNA, is revolutionizing how we diagnose and manage these pests. Genomics in plant nematode research involves studying the complete DNA sequence and gene functions of nematodes and their host plants.

Genomics looks at the complete DNA of nematodes and plants, like sequencing genomes of species such as *Heterodera schachtii* and *Globodera pallida*. This helps identify genes that allow nematodes to parasitize plants, such as effectors that manipulate host defenses. It also shows how nematodes evolve, with findings like horizontally transferred genes enhancing their abilities (Fig. 12.1).

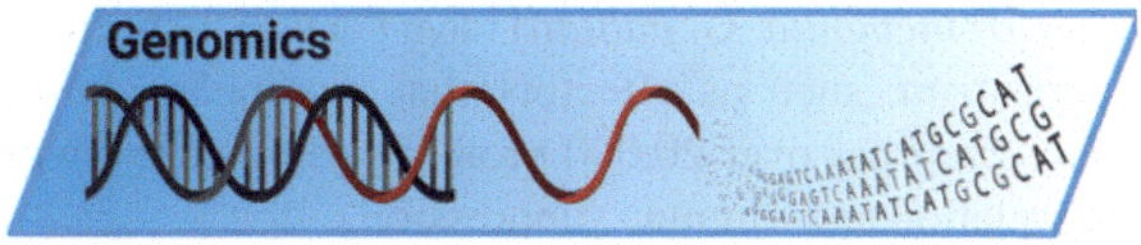

Fig. 12.1 Genomics is a field of biology that studies the structure, function, evolution, mapping, and editing of genomes

12.3.1 Understanding Parasitism Through Genomic Analysis

Genomics has elucidated the genetic mechanisms underlying PPN parasitism, providing a foundation for both diagnosis and management. The sequencing and annotation of PPN genomes, such as *Meloidogyne incognita* and *H. schachtii*, have revealed genes critical for parasitic interactions. For instance, comparative genomics of 61 nematode genomes, including 16 PPN species, identified over 24,000 protein families specific to PPNs, with approximately 10,000 proteins in *M. incognita* having orthologs restricted to phytoparasitic species and no further homology in protein databases (Grynberg et al. 2020). These proteins include effector molecules, with about 1000 sharing properties with known secreted effectors involved in manipulating host plant cells, and 68 novel candidates validated by RNA-seq and RT-qPCR for strong expression during the endophytic phase.

Horizontal gene transfer (HGT) has also been a significant finding, contributing 1–14% of PPN-specific genes, with enriched functions like plant cell wall degradation, essential for parasitism (Siddique et al. 2022).

Genome sequencing is now cost-effective, typically costing less than a few thousand US dollars, though bioinformatics for gene-finding and annotation adds to the expense (Bird et al. 2015). Specific examples include the ~54-Mbp genome assembly of *M. hapla* and additional genomes like *M. chitwoodi* and *M. floridensis*, which have expanded our understanding of genetic repertoires related to parasitic interactions.

12.3.2 Genomics in Plant Nematode Research

Genomics involves the study of the complete DNA sequence and gene functions, offering a foundation for understanding nematode and host biology. Key applications include:

- **Genome Sequencing and Assembly**: High-quality genome sequences have been completed for several plant-parasitic nematodes, providing a platform for detailed analysis. For instance, the genome of *H. schachtii* (beet cyst nematode) was sequenced and assembled with a size of approximately 160–170 Mb, the largest among cyst nematodes, achieving 192-fold coverage using Pacific Biosciences and 144-fold with Illumina, resulting in a final assembly of ~179 Mbp across 395 scaffolds (Thorpe et al. 2014). Genomes of major plant-parasitic nematodes (e.g., *M. incognita, G. pallida*) have been sequenced, revealing genes involved in parasitism, such as effectors that manipulate host defenses. For example, *M. incognita* has over 20,000 protein-coding genes, with many linked to cell wall degradation and immune suppression.

- **Gene Duplication and Evolution**: Comparative genomics reveals large-scale segmental duplications in the *Heterodera* lineage, with *H. schachtii* and *H. glycines* showing a smaller proportion of single-copy genes compared to *G. pallida*, and greater segmental duplications (at least 5 co-linear genes, no >25 genes between). In *H. schachtii*, 4547 segmentally duplicated genes were identified, with 2273 in 58 islands (duplication depth 2–131, threshold 10), and 1877 orthogroups specific to *Heterodera* (2072 genes in *H. schachtii*), 28% enriched in segmental duplications versus 17% genome-wide (Thorpe et al. 2014).
- **Horizontal Gene Transfer (HGT)**: Genomics has identified 263 genes (319 transcripts) in *H. schachtii* putatively acquired by HGT, with a notable example being the completion of the vitamin B5 biosynthesis pathway, enabled by two genes (*Hsc_gene_23032, Hsc_gene_23033*) on a 1.3 Mb scaffold, phylogenetically grouped with actinobacteria (100 bootstrap, closest to *Candidatus Rhodoluna planktonica*, WP_070954084.1) (Siddique et al. 2022). This suggests HGT enhances nematode pathogenicity.
- **Functional Genomics**: Tools like RNA interference (RNAi) are used to validate gene functions, particularly for parasitism genes. For example, RNAi-mediated suppression of effector genes has been shown to reduce nematode pathogenicity, revealing new targets for control (Rosso et al. 2009).

12.3.3 Technological Advances

Next-generation sequencing (NGS) and microarrays enable cross-species comparisons to identify conserved regulated genes in developing feeding cells. Laser microdissection and micro-aspiration allow RNA isolation from specific plant cells, revealing transcriptional changes at early differentiation stages, such as dramatic dilution of feeding cell transcripts in excised infected root fragments (Escobar et al. 2011).

12.3.4 Role in Diagnosis

Genomics has revolutionized the diagnosis of PPNs by enabling the identification of unique genetic markers for species and population-level detection. Sequencing PPN genomes facilitates the development of molecular diagnostic tools, such as PCR-based assays, which can detect nematodes in soil or plant tissues with high specificity. For instance, the genome sequences of *M. arenaria* and *M. graminicola* provide data for identifying genetic markers, aiding in precise diagnosis.

12.3.5 Case Studies on Role in Management

Genomics offers multiple avenues for managing PPNs, from identifying targets for control to developing resistant crops. Comparative genomics has identified novel target genes, such as those involved in transcription-related and neuro-perception functions, enriched in PPN-specific proteins, revealing additional control targets (Grynberg et al. 2020). For example, 89 candidate effector proteins were identified for in planta RNAi, with 68 novels after filtering, expressed in endophytic phases, and validated on *Nicotiana tabacum*, offering a strategy for gene silencing to reduce nematode impact.

Genomic studies identify resistance (R) genes in plants, like the *Mi-1* gene in tomatoes, which confers resistance to root-knot nematodes. These genes are critical for breeding resistant varieties.

Functional genomics, particularly RNAi, has been pivotal. RNAi-mediated suppression of genes essential for nematode development, survival, or parasitism is revealing new targets for control. For instance, silencing *Hs-panc2* in *H. schachtii*, a gene involved in vitamin B5 biosynthesis likely acquired via HGT, decreased female infection ($p < 0.05$) and reduced syncytia and female sizes, demonstrating the potential for RNAi-based management (Siddique et al. 2022).

Host resistance is another key area, with genomics identifying susceptibility genes in plants that nematodes rely on. The study of *H. schachtii* and *Arabidopsis thaliana* identified metabolic pathways like vitamin B5 biosynthesis, where the nematode encodes the last step (PANC, *Hsc_gene_23032*, *Hsc_gene_23033*), putatively from HGT. Knocking out *AtPANB1* (penultimate step) in *A. thaliana* reduced female nematodes at 14 dpi ($p < 0.0001$), with smaller syncytia, females, cysts, and eggs/cyst (p values 0.02, 0.049, respectively), suggesting genome editing for resistance (Siddique et al. 2022).

Population genomics also contributes by providing insights into genetic diversity and adaptation, crucial for developing sustainable management strategies. Recent advances in population genomics of PPNs are providing new insights into molecular and evolutionary processes, aiding in understanding dispersal and resistance development (Montarry et al. 2021).

12.3.6 Challenges and Future Directions

Despite these advancements, challenges remain. The cost of sequencing, while reduced, still includes additional bioinformatics costs for gene-finding and annotation, which can be resource-intensive (Bird et al. 2015). Integrating multi-omics data, including genomics, transcriptomics, and metabolomics, is essential for a holistic understanding, potentially leading to more effective diagnostic tools and management strategies.

Future directions include applying these genomic insights to crop improvement, particularly for economically important crops like potato and soybean, through genome editing and breeding programs. The approach of partial knockout of

Table 12.1 Key genomic contributions to PPN management

Genomic approach	Example	Implication for management
Genome sequencing	*Meloidogyne incognita* genome, ~10,000 PPN-specific proteins	Identifies targets for RNAi and control strategies
Transcriptome analysis	*Heterodera schachtii* and *Arabidopsis thaliana* transcriptomes	Reveals susceptibility genes for genome editing
Comparative genomics	24,000 PPN-specific protein families	Highlights novel control targets, e.g., effectors
RNAi and functional genomics	Silencing *Hs-panc2* in *H. schachtii*	Reduces infection, potential for field application
Population genomics	Genetic diversity in *Meloidogyne* spp.	Informs sustainable management, resistance tracking

susceptibility genes, balancing cost and reward, aims for minimal impact on plant physiology, as seen with *atpanb1–1* showing slight flowering delay but otherwise normal phenotype (Siddique et al. 2022).

Table 12.1 summarizes key genomic contributions, including specific examples and their implications:

This table underscores the multifaceted role of genomics in enhancing PPN management, from gene discovery to practical application.

In conclusion, genomics is a cornerstone in the diagnosis and management of plant parasitic nematodes, offering tools for precise identification, understanding parasitic mechanisms, and developing targeted control strategies. By identifying PPN-specific genes, effector proteins, and host susceptibility genes, genomics supports RNAi, genome editing, and breeding for resistance, promising sustainable solutions for global food security. As research progresses, integrating multi-omics data and addressing cost challenges will further amplify its impact.

12.4 Transcriptomics

Transcriptomic approaches—primarily RNA-sequences—have become indispensable in plant nematology for dissecting both host and parasite gene-expression dynamics (Fig. 12.2). Below is an overview of their key roles:

12.4.1 Role in Understanding Plant–Nematode Interactions

Transcriptome analysis enables a dual perspective, examining gene expression in both the plant host and the nematode parasite across various infection stages. A notable study on the beet cyst nematode *H. schachtii* and its host *A. thaliana* conducted a trans-kingdom, life-stage-specific transcriptome analysis, covering stages from cysts to infected root segments at multiple time points (e.g., 10 h post-infection,

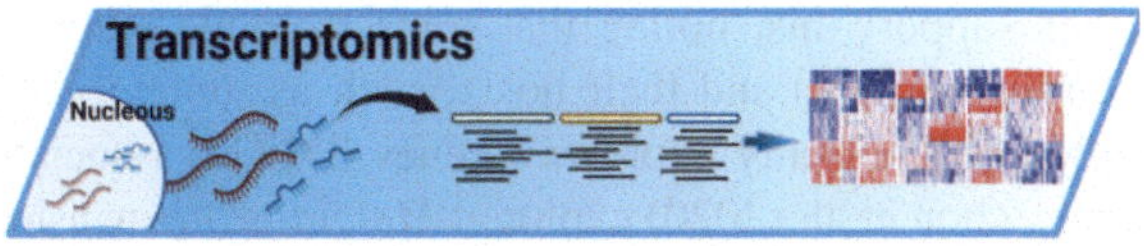

Fig. 12.2 Transcriptomics is the study of the complete set of RNA transcripts produced by the genome of one or more cells in a particular condition

48 h post-infection, 12 days post-infection females/males, and 24 days post-infection females) (Różańska et al. 2023).

Life-stage-specific transcriptomes of both nematodes and their host plants have been generated, mapping billions of RNAseq reads to identify differentially expressed genes during infection. For *H. schachtii*, 9 billion pairs of 150 bp strand-specific RNAseq reads were analyzed, with ~500 million pairs mapping to the nematode genome, identifying 26,739 gene annotations, 77% of which were transcribed at ≥10 reads/replicate. In *A. thaliana*, ~2.8 billion pairs mapped, with 32,548 gene models, 77% transcribed at ≥10 reads/replicate, showing 59% (19,071) differentially regulated genes (Log2FC > +0.5 or < −0.5, adj. $p < 0.01$) (Klepikova et al. 2016). This highlights stage-specific gene expression, such as male-specific clusters at 12 days post-infection (dpi).

A key finding was the identification of the vitamin B5 (pantothenate) biosynthesis pathway, where the nematode complements host upregulation through genes likely acquired via horizontal gene transfer from bacteria. Phylogenetic analysis grouped these genes near actinobacteria, such as *Candidatus Rhodoluna planktonica*. Knockout experiments confirmed the pathway's role, with AtPANB1 (AT2G46110) knockout reducing female numbers at 14 days post-infection (*p* value <0.0001) and RNAi targeting the nematode's *Hs-panc2* decreasing female numbers, syncytia size, and female size (Siddique et al. 2022).

Similarly, a study on the cereal cyst nematode *H. avenae* used Illumina sequencing to generate 131 million high-quality paired-end reads, producing 27,765 contigs with an N50 of 1028 bp, of which 10,452 were annotated (Kumar et al. 2014). Comparative analyses identified 4431 transcripts common with the free-living nematode *Caenorhabditis elegans* and 9462 with the potato cyst nematode *G. pallida*, highlighting species-specific gene expression. This included 1280 transcripts with secretory signatures, suggesting roles in host invasion.

12.4.2 Identifying Susceptibility and Resistance Genes

Transcriptome analysis is instrumental in pinpointing plant genes involved in susceptibility and resistance. In the *H. schachtii* study, 25% (8351) of host genes were differentially expressed between infected and uninfected tissues at each time point, with up-regulated host genes generally well-conserved across Plantae, while nematode effector genes were less conserved across Nematoda, indicating lineage-specific modulation (Siddique et al. 2022). This suggests that transcriptomics can identify susceptibility genes for targeted deletion to develop resistant crops.

Reviews further support this, noting that transcriptomic studies on root-knot nematodes (*Meloidogyne* spp.) and their hosts, such as tomato, Arabidopsis, and rice, have identified differentially expressed genes (DEGs) associated with resistance. For example, Zhou et al. (2020) explored *Meloidogyne* invasion in rice roots via RNA-seq, identifying 952 DEGs at 6 days (invasion stage) and 647 at 18 days (development stage), with 40 new DEGs encoding secretory proteins. Microarray studies in tomato, *Arabidopsis*, and RKN-tolerant aubergine *Solanum torvum* have generated large-scale data on gene expression patterns, aiding in marker-assisted selection (MAS) and gene transformation for enhanced resistance (Moore et al. 2005).

12.4.3 Studying Nematode Virulence Factors

Transcriptomics also reveals nematode virulence factors, particularly effector proteins and CAZymes involved in host manipulation. The *H. avenae* study identified 13 homologues to known effectors of *Heterodera* spp., including fatty acid retinoid-binding protein (*Gp-rbp-1*), esophageal gland cell secretory proteins (*Hg-hsp-11, Hg-hsp-1*), pectate lyase (*Hg-PL-2*), and Annexin (*Hg-4C10*), alongside transcripts for cell wall–modifying CAZymes (Yang et al. 2019). These findings suggest roles in host invasion and syncytium formation, with genes like C-type lectin and SPRYSECs potentially overcoming host defenses.

A targeted transcriptomics study on *Radopholus similis* identified 30 new effector genes, with 19 being pioneer genes without prior annotation but highly expressed during plant interaction, suggesting concerted regulation via a promoter motif element (Vieira et al. 2021). This complexity challenges the notion that migratory nematodes are less specialized, highlighting transcriptomics' role in uncovering novel virulence mechanisms.

12.4.4 Supporting the Development of Nematode-Resistant Crops

Transcriptomic data provides actionable insights for developing nematode-resistant crops. The *H. schachtii* study proposed compartmentalization of the vitamin B5 pathway, with plant steps supporting nematode infection, suggesting targeted susceptibility gene deletion as a strategy (Li et al. 2023). RNAi and gene-silencing technologies, informed by transcriptomic data, have shown promise, with RNAi targeting *Hs-panc2* reducing nematode infection efficiency.

Reviews emphasize the integration of transcriptomics with other omics technologies, such as proteomics and metabolomics, to enhance resistance strategies. For instance, comparing transcriptome data of four RKN-resistant plant species (*Glycine max*, *Arachis stenosperma*, *Oryza glaberrima*, *Coffea arabica*) infected with

M. incognita, *M. graminicola*, and *M. arenaria* using homology databases has aided in identifying resistance mechanisms (Araujo et al. 2021). Computational resources like WormBase, *Meloidogyne* genomic resource, and others provide access to transcriptomic data, supporting further research.

12.4.5 Comparative and Evolutionary Insights

Transcriptome analysis facilitates comparative studies, revealing evolutionary patterns in plant-nematode interactions. The *H. schachtii* study noted that host genes upregulated during infection are conserved across Plantae, while parasite genes, including effectors, are less conserved across Nematoda, suggesting lineage-specific adaptations (Nuaima and Heuer 2023). This aligns with findings from *H. avenae*, where comparisons with *C. elegans* and *G. pallida* highlighted unique gene expression profiles in PPNs (Yang et al. 2017).

12.4.6 Detailed Findings

The key transcriptomic studies and their findings in plant nematology are summarized in Table 12.2.

In conclusion, transcriptomes are a cornerstone of modern plant nematology, providing a comprehensive view of gene expression dynamics during plant–nematode interactions. They enable the identification of susceptibility and resistance genes, reveal nematode virulence factors, and support the development of resistant crops. By leveraging transcriptomic data, researchers can uncover the molecular dialogue between plants and PPNs, offering promising avenues for sustainable agriculture. The integration with computational resources and other omics technologies further enhances its utility, ensuring a holistic approach to nematode control.

Table 12.2 Key transcriptomic studies and their findings in plant nematology

Nematode species	Host plant	Key findings	Reference
Heterodera schachtii	*Arabidopsis thaliana*	Identified vitamin B5 pathway role, 18,575 nematode DEGs, 19,071 host DEGs	Siddique et al. (2022)
H. avenae	Wheat (*Triticum* spp.)	131 M reads, 27,765 contigs, 10,452 annotated, identified 13 effector homologues	Kumar et al. (2014)
Meloidogyne spp.	Rice (*Oryza* spp.)	952 DEGs at 6 days, 647 at 18 days, 40 secretory protein DEGs	Zhou et al. (2020)
Various RKNs	Multiple hosts	Compared transcriptomes of resistant species, aided resistance gene identification	Khan et al. (2023)

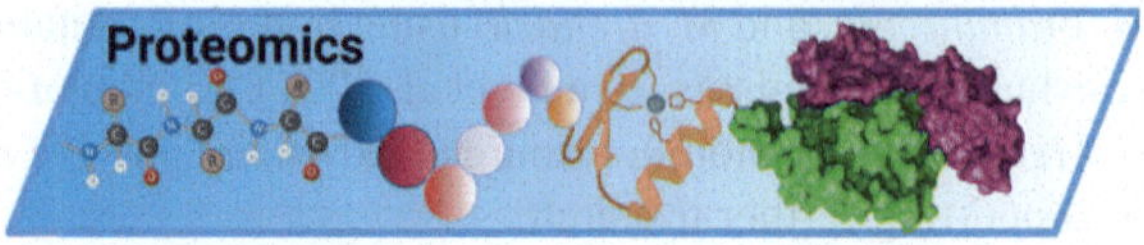

Fig. 12.3 Proteomics is a large-scale study of the proteome, which allows us to identify the proteins expressed in a certain condition

12.5 Proteomics

Proteomics—the large-scale study of proteins—complements existing approaches by revealing the active molecular players during infection and offering biomarkers for rapid diagnosis and management. It offers molecular insights into nematode biology, host interactions, and potential control strategies. Proteomics aids in identifying nematode proteins for diagnosis and in developing targeted nematode control strategies. Proteomics are crucial for sustainable nematode management (Fig. 12.3).

Proteomics focuses on proteins, identifying those secreted by nematodes, like effectors, that affect the plant. It complements genomics by showing which proteins are active during infection, helping us understand how nematodes invade and feed on plants.

Key applications include:

12.5.1 Identification of Secreted Proteins and Effectors

Proteomics identifies secreted nematode effectors that alter plant physiology, such as cystatins that inhibit plant proteases. Mass spectrometry (MS) is commonly used to characterize these proteins.

Proteomic studies identify proteins secreted by nematodes, particularly effectors that manipulate host defenses. In *G. pallida*, 2669 putatively secreted proteins (3138 transcripts) were identified, including 209 effector genes (245 transcripts, 38 families) and 39 cell wall degrading enzyme genes (45 transcripts, 9 families), with effectors ~1/5th in segmental duplications (Escobar et al. 2011).

12.5.2 Expression Profiles and Stage-Specific Roles

RNAseq and proteomic analyses show stage-specific expression of effectors. In *G. pallida*, effectors fall into clusters: 30 sequences expressed only at the invasive stage, 61 at parasitic stages, 5 at invasive and adult male stages, with additional constitutive (20 sequences) and parasitic and male (4 sequences) clusters (Thorpe et al. 2014). Cell wall modifying proteins, such as cellulases, pectate lyases, and

CBMs, are expressed at J2 and adult male stages for migration, with some (e.g., 2 CBM genes) upregulated at parasitic stages for syncytium development.

12.5.3 Host Targeting and Functional Insights

Effectors target diverse host structures, including the nucleus, nucleolus, and peroxisomes, with *G. pallida* effectors confirmed via eGFP-fusions in *Nicotiana benthamiana* to localize to these organelles, the latter being a first report for nematode effectors (Thorpe et al. 2014).

12.5.4 Quantitative Proteomics

Methods like iTRAQ and SILAC quantify changes in protein abundance during infection, identifying key players in resistance or susceptibility. For example, changes in root cotton proteins have been correlated with resistance to root-knot nematode development (Escobar et al. 2011).

12.5.5 Technological Advances

Mass spectrometry-based proteomics, including spatial proteomics and posttranslational modification (PTM) analysis, provides detailed insights into protein localization and function, crucial for understanding plant immune responses to biotic stress (Thorpe et al. 2014).

12.5.6 Role in Diagnosis

Proteomics plays a pivotal role in diagnosing PPN infections by identifying specific proteins associated with nematode activity. For instance, stylet-secreted proteins from root-knot nematodes, as identified in Jaubert et al. (2002), serve as biomarkers. These proteins, secreted during feeding, can be detected in infected plant tissues, enabling species-specific identification and early detection, which is critical for timely intervention.

Proteomic analysis of infected plant roots reveals changes in protein expression, such as those in developing feeding cells (e.g., syncytia or giant cells). Studies by Escobar et al. (2011) highlight that proteomic techniques uncover significant protein alterations, potentially leading to diagnostic tools. For example, differential protein expression analysis can identify markers for nematode presence, enhancing diagnostic accuracy compared to traditional morphological methods.

The role of proteomics in diagnosing PPN infections by identifying specific proteins associated with nematode activity is presented in Table 12.3.

Table 12.3 Role of proteomics in diagnosing PPN infections by identifying specific proteins associated with nematode activity

Diagnostic application	Description	Reference
Protein marker identification	Detects nematode-secreted proteins as markers	Jaubert et al. (2002)
Host protein changes	Analyzes plant proteome for infection signs	Escobar et al. (2011)

This approach is particularly valuable in protected cultivation systems, where nematode prevalence is high due to sequential cropping (Senthil Kumar et al. 2025).

12.5.7 Role in Management

In management, proteomics provides molecular insights into nematode parasitism, facilitating the development of targeted control strategies. By identifying key proteins involved in host-nematode interactions, researchers can pinpoint vulnerabilities. For example, Cutis (2007) reviews how proteomic studies reveal proteins manipulating plant development, such as those forming feeding cells, which are potential targets for control.

Proteomics is integral to multi-omics approaches, combining with transcriptomics and metabolomics to discover novel nematicides. Ayaz et al. (2024) highlight proteomics' role in identifying bioactive proteins, such as enzymes and nematicidal toxins, produced by biocontrol agents (BCAs) like *Pochonia chlamydosporia*. These proteins, identified through techniques like metabolite, protein, and lipid extraction (MPLEx), enhance understanding of BCA-PPN interactions, leading to sustainable bionematicide development.

RNA interference (RNAi), guided by proteomic data, targets essential nematode genes. For instance, silencing genes encoding esophageal gland proteins (Cutis 2007), disrupts nematode reproduction, offering a biological control method. Proteomics also elucidates protein–protein interactions between nematodes and BCAs, revealing molecular networks for nematode suppression (Ayaz et al. 2024).

The role of proteomics in nematode management is presented in Table 12.4.

Large-scale proteomic efforts, such as generating 65,204 expressed sequence tags (ESTs) for six *Meloidogyne* species, as mentioned by Cutis (2007), support genomic comparisons and gene chip microarrays for profiling up- and downregulated genes during infection. These ongoing efforts underpin the identification of novel pathogenicity genes for management.

12.5.8 Integration with Multi-omics and Broader Context

Proteomics is not standalone; it integrates with metagenomics, transcriptomics, and metabolomics for a holistic understanding of PPN biology. Ayaz et al. (2024) emphasize this integration, illustrated in Fig. 12.4, showing how omics approaches collectively enhance biocontrol strategies. This multi-omics strategy is crucial for

Table 12.4 Role of proteomics in nematode management

Management application	Description	Reference
Target identification	Identifies proteins for RNAi or pesticide targets	Cutis (2007)
Bionematicide discovery	Finds nematicidal proteins from BCAs	Ayaz et al. (2024)

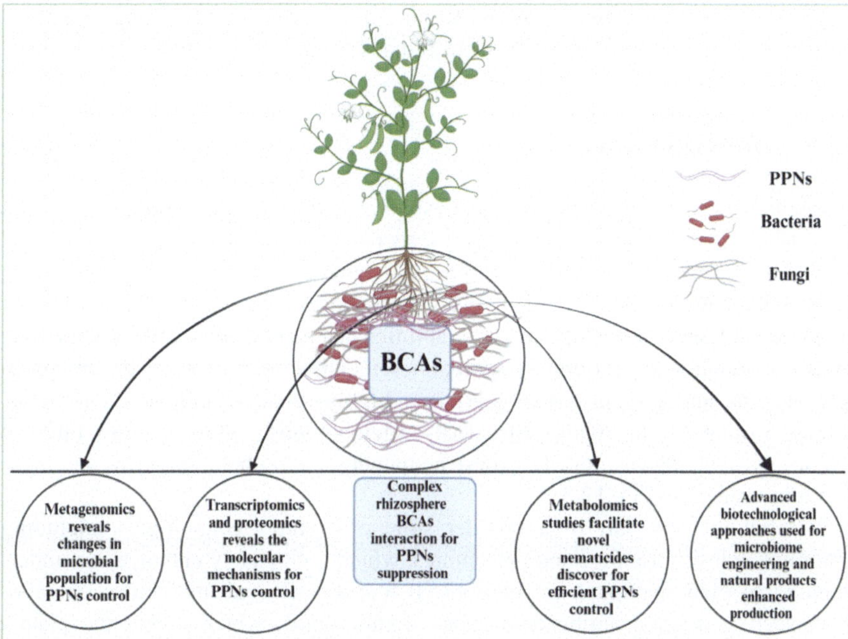

Fig. 12.4 Integration of omics approaches enhances biological control strategies

sustainable agriculture, addressing challenges like short crop rotations and multiple nematode species infestations, as discussed by Westphal (2011).

The complexity of host–nematode interactions, including nematode strategies to suppress plant defenses (Chen et al. 2024), underscores proteomics' role in decoding molecular pathways. For example, proteomic analysis found heat shock proteins and thioredoxins in nematode secretomes (Gillet et al. 2017), supporting proteostasis and potential control targets.

12.5.9 Challenges and Future Directions

While proteomics offers significant potential, challenges include the need for advanced techniques like MPLEx for comprehensive protein analysis and the integration of data across omics platforms. Future research, as suggested by Ayaz et al. (2024), should focus on cross-species comparisons to identify conserved proteins, enhancing diagnostic and management applications. The ongoing sequencing projects for *Meloidogyne* spp. will likely expand proteomic datasets, further supporting these efforts.

In conclusion, proteomics is a cornerstone for diagnosing PPN infections through protein markers and managing them by uncovering molecular targets for control. Its integration with multi-omics approaches ensures a comprehensive strategy for sustainable agriculture, addressing the global threat posed by nematodes. As research progresses, proteomics will continue to play a vital role in advancing nematode management, particularly in regions like India with intensive cultivation systems.

12.6 Metabolomics

Metabolomics, the comprehensive analysis of small-molecule metabolites within cells, tissues, or organisms, provides a direct readout of physiological state and biochemical activity. In the context of plant–nematode interactions, metabolomics can reveal the molecular dialogue between host and parasite, identify biomarkers of infection, and uncover metabolic vulnerabilities that can be exploited for nematode control. Compared to genomics or proteomics, metabolomics captures the end products of gene and protein activity, offering highly sensitive indicators of stress, defense responses, and pathogen activity. Metabolomics offers insights into the physiological changes induced by these nematodes, aiding both diagnosis and management strategies (Fig. 12.5).

Metabolomics is part of a broader suite of "omic" technologies, including genomics, transcriptomics, and proteomics, which collectively enhance our understanding of plant–pathogen interactions. It focuses on small molecules (metabolites) such as sugars, amino acids, and organic acids, which reflect the plant's metabolic state. Nematodes, as endoparasites (e.g., root-knot and cyst nematodes) or ectoparasites (e.g., sting nematodes), induce significant metabolic changes in host plants to create feeding structures and suppress plant defenses. These changes can be studied to diagnose infections and develop management strategies.

The metabolomics technologies include:

12.6.1 Mass Spectrometry (MS)-Based Approaches

- **Liquid Chromatography–Mass Spectrometry (LC–MS):** Widely used for profiling polar and nonpolar metabolites from plant tissues, root exudates, and nematode extracts. Offers high sensitivity and broad metabolite coverage.

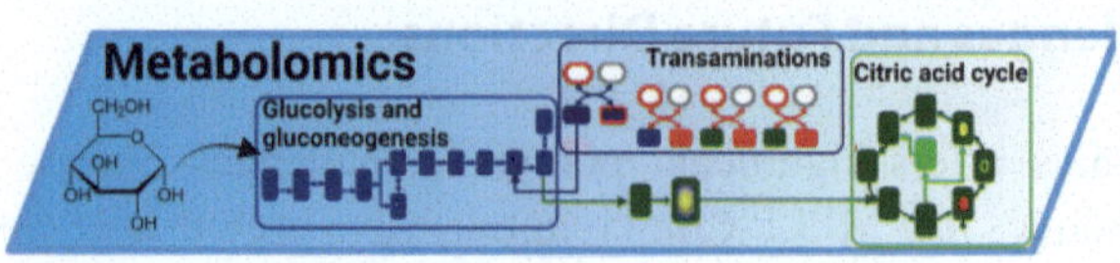

Fig. 12.5 Metabolomics is a systematic study of the processes involving metabolites, small molecule substrates, intermediates, and products of cellular metabolism

- **Gas Chromatography–Mass Spectrometry (GC–MS):** Ideal for volatile and derivatized metabolites (e.g., amino acids, organic acids). Frequently applied to analyze root exudate composition changes upon nematode infection.

12.6.2 Nuclear Magnetic Resonance (NMR) Spectroscopy

Provides quantitative, reproducible profiles without extensive sample preparation. Limited sensitivity compared to MS but excels at detecting high-abundance primary metabolites and monitoring isotopic labeling.

12.6.3 Data Processing and Statistical Analysis

- Multivariate techniques (PCA, PLS-DA) to distinguish infected versus healthy samples.
- Metabolite identification using spectral libraries and in silico tools.
- Network analysis to map perturbed pathways.

12.6.4 Case Studies on Applications in Diagnosis

Metabolomics plays a crucial role in diagnosing plant parasitic nematode infections by identifying specific metabolic signatures associated with nematode activity. Research suggests that nematode infection leads to systemic metabolic changes, observable not only in infected roots but also in noninfected tissues like shoots. For instance:

- A study on the beet cyst nematode *H. schachtii* infecting *A. thaliana* used gas chromatography-mass spectrometry (GC-MS) to profile metabolites, identifying 75 metabolites with significant responses (Hofmann et al. 2010). It found increased levels of 1-kestose, raffinose, and α, α-trehalose, which showed systemic accumulation at 5, 10, and 15 days after inoculation (dai). These metabolites could serve as diagnostic markers for nematode presence.
- The study also noted that nonmodified root segments and shoots exhibited metabolic differentiation, suggesting that metabolomics can detect systemic effects, enhancing diagnostic accuracy beyond localized infection sites.
- Another study on sting nematodes (*Belonolaimus longicaudatus*) and Bermuda grass (*Cynodon transvaalensis*) found that nematodes modified the global metabolome, explaining 42% of observed inertia ($df = 5$, $\chi^2 = 0.002$, $F = 5.1$, $p = 0.001$) (Willett et al. 2020). It identified suppression of amino acids in susceptible cultivars and upregulation of defense-related compounds like L-Pipecolic acid in tolerant lines, aiding in diagnosing susceptibility levels.

These findings indicate that metabolomics can provide noninvasive, systemic biomarkers for early detection, which is critical for timely intervention in agricultural settings.

- **Biomarker Discovery**
  - **Early-Stage Infection Markers:** Metabolomic profiling of root exudates has identified compounds (e.g., certain phenolic acids and flavonoids) whose levels change within days of nematode invasion, enabling early detection before visible symptoms appear.
  - **Species-Specific Signatures:** Comparative metabolomics of root exudates and gall tissues can differentiate infections by *Meloidogyne* spp. versus *Heterodera* spp., facilitating targeted management strategies.
- **Non-Destructive Sampling:** Sampling of rhizosphere soil solution or hydroponic culture media allows real-time monitoring of metabolic shifts without harvesting the entire root system.
- **High-Throughput Screening**: Integration with microfluidic platforms to analyze hundreds of samples in parallel, accelerating screening of resistant germplasm or antagonist treatments based on metabolic response profiles.

12.6.5 Case Studies on Applications in Management

Metabolomics is instrumental in understanding how nematodes manipulate plant metabolism, which informs the development of targeted management strategies. Nematodes create a metabolic niche by reprogramming the host's primary metabolism, and metabolomics reveals these changes, offering opportunities for resistance breeding and chemical interventions:

- Hofmann et al. (2010) showed that nematode infection increased amino acids (e.g., glutamic acid, aspartic acid, glutamine, methionine, valine, isoleucine, glycine, serine, with fold changes up to 14.75 at 10 dai) and sugars (e.g., 1-kestose, α, α-trehalose, raffinose, galactinol, with fold changes up to 101.08 at 5 dai for 1-kestose). This reprograming supports the creation of feeding structures, and targeting these pathways could disrupt nematode development.
- The Nature study highlighted that tolerant Bermuda grass lines showed increased L-Pipecolic acid (df = 12, t = 2.74, $p = 0.05$ after Bonferroni correction), linked to systemic acquired resistance (SAR), which correlated with lower nematode populations. This suggests that enhancing SAR pathways through breeding could reduce nematode impact (Willett et al. 2020).
- Another study on *Trichoderma atroviride* and tomato plants showed that this biocontrol agent altered root metabolomics to suppress nematodes, highlighting the potential for integrating metabolomics with biological control strategies (Gomez-Fernandez et al. 2020).

Management strategies informed by metabolomics include:

- **Breeding for Resistance**: Identifying metabolites associated with tolerance (e.g., L-Pipecolic acid, phenylalanine) can guide breeding programs to develop nematode-resistant cultivars, reducing economic losses exceeding $100 billion annually (Jones et al. 2013; Williamson and Kumar 2006).
  - Resistant cultivars often exhibit distinct metabolic responses upon nematode challenge, such as elevated levels of defense-related alkaloids or terpenoids. Metabolomics can pinpoint these compounds, guiding breeding programs.
- **Chemical Interventions**: Targeting metabolic pathways altered by nematodes (e.g., amino acid suppression, sugar accumulation) could lead to new nematicides with specific modes of action, addressing the limitations of banned chemicals like carbamates and organophosphorus compounds (Paez 2023).
  - Profiling secondary metabolites in nematode-resistant plants has led to the discovery of bioactive compounds (e.g., glucosinolates and glycoalkaloids) with nematicidal properties. These compounds inform the development of botanical nematicides or soil amendments.
- **Biocontrol and Integrated Pest Management**: Metabolomics can optimize the use of beneficial organisms by understanding their impact on plant metabolism, enhancing sustainable nematode control.
  - Metabolic fingerprints of soils treated with biocontrol agents (e.g., *Pseudomonas* spp., *Trichoderma* spp.) reveal changes in soil metabolome that correlate with reduced nematode viability and improved plant health.

12.6.6 Precision Agriculture and Decision Support

Coupling field-scale metabolomic analysis with geospatial data allows mapping of nematode hotspots and temporal risk zones, enabling site-specific management (variable-rate application of nematicides or biological agents).

12.6.7 Comparative Analysis of Metabolomic Approaches

To illustrate the diversity of metabolomic applications, the following Table 12.5 summarizes key findings from the reviewed studies:

This table underscores the versatility of metabolomics across different nematode types and host plants, highlighting its dual role in diagnosis and management.

12.6.8 Challenges and Future Directions

While metabolomics offers significant potential, challenges include the complexity of plant–nematode interactions, variability across species, and the need for standardized protocols for metabolic profiling. Future research could focus on

Table 12.5 The diversity of metabolomic applications

Nematode/host plant	Key metabolomic findings	Implication for diagnosis	Implication for management	Reference
Beet cyst (*Heterodera schachtii*)/*Arabidopsis thaliana*	Increased 1-kestose, raffinose, α, α-trehalose; amino acids and sugars upregulated	Systemic markers for early detection	Target metabolic reprograming for disruption	Hofmann et al. (2010)
Sting (*Belonolaimus longicaudatus*)/Bermuda grass (*Cynodon transvaalensis*)	Amino acid suppression in susceptible lines; L-Pipecolic acid linked to tolerance	Diagnose susceptibility via metabolome profiles	Breed for SAR enhancement via L-Pipecolic acid	Willett et al. (2020)
General (field study)/tomato	*Trichoderma atroviride* alters root metabolomics to suppress nematodes	Monitor biocontrol efficacy via metabolic changes	Integrate biocontrol with metabolomic insights	Gomez-Fernandez et al. (2020)

integrating metabolomics with other omic data (e.g., transcriptomics, proteomics) to create holistic models of plant defense, enhancing both diagnostic accuracy and management efficacy. Additionally, field studies are needed to validate laboratory findings, ensuring practical applicability in agricultural settings.

- **Complexity of Metabolomes:** Vast chemical diversity and dynamic range make comprehensive coverage challenging.
- **Standardization:** Protocols for sample preparation, data acquisition, and analysis must be standardized across laboratories to ensure reproducibility.
- **Integration with Multi-Omics:** Combining metabolomics with genomics, transcriptomics, and proteomics in an integrative framework will yield deeper insights into host–nematode interactions and uncover novel targets for control.
- **Advances in Spatial Metabolomics:** Imaging MS techniques can localize metabolic changes at the tissue or cellular level within galls, revealing micro-environmental niches exploited by nematodes.

In conclusion, metabolomics is a powerful tool for diagnosing plant parasitic nematode infections by identifying systemic metabolic biomarkers and assessing plant susceptibility. It also informs management by revealing how nematodes manipulate host metabolism, guiding resistance breeding, chemical interventions, and biocontrol strategies. As research progresses, metabolomics is likely to play an increasingly central role in sustainable agriculture, addressing the global challenge posed by these devastating pests.

12.7 Integrated Omics for Precision Management

While each omics field provides unique insights, their integration offers a more comprehensive understanding of nematode biology. For example, genomics and transcriptomics provide the genetic and gene expression context, proteomics identifies the functional proteins involved, and metabolomics reveals the end products of these biological activities, indicating the physiological state of the organism. Recent studies emphasize a complementary multi-omics approach, suggesting that combining these fields can enhance diagnosis and control efforts, particularly in addressing anthelmintic resistance (Ma et al. 2020).

Combining genomics and proteomics offers a holistic view of plant–nematode interactions, with several key applications:

12.7.1 Multi-omics Integration

Linking genomic data (e.g., gene sequences) with proteomic data (e.g., protein expression) maps molecular pathways involved in parasitism and host defense. For example, transcriptomic and proteomic analyses of feeding cells reveal conserved genes and proteins critical for nematode development (Escobar et al. 2011).

12.7.2 Biomarker Discovery

Omics approaches identify biomarkers for nematode infection or resistance, aiding in diagnostics and breeding programs. For instance, comparative genomics of 61 nematode genomes, including 16 plant-parasitic species, identified over 24,000 protein families specific to these parasites, offering potential targets (Grynberg et al. 2020).

12.7.3 Control Strategy Development

Insights from omics research guide the development of targeted nematicides, RNAi strategies, and resistant plant varieties. For example, knocking out effector genes in nematodes using CRISPR-Cas9 can reduce pathogenicity, and proteomic identification of essential enzymes informs biocontrol development (Grynberg et al. 2020).

12.7.4 Biocontrol Development

Genomic and proteomic insights into nematode weaknesses (e.g., essential enzymes) guide the development of targeted nematicides or RNA interference (RNAi) strategies.

12.7.5 Challenges and Future Directions

- **Data Complexity:** Managing and integrating large omics datasets requires advanced bioinformatics tools.
- **Functional Validation:** Many identified genes and proteins lack functional characterization, necessitating further studies.
- **Field Applications:** Translating lab-based omics findings into field-resistant crops or effective biocontrols remains a hurdle. Emerging technologies like single-cell omics, spatial transcriptomics, and high-throughput proteomics are poised to further revolutionize plant nematode research, offering precise insights into cellular-level interactions.

12.8 Comparison of Omics Technologies

To organize the detailed findings, the following Table 12.6 summarizes key metrics from genomic and proteomic studies:

Table 12.6 Encapsulation of the quantitative and qualitative insights, highlighting the depth of omics research in plant nematode interactions

Aspect	Key findings	Details and numbers
Genome sequencing	High-quality genomes sequenced for *Heterodera schachtii* and *Globodera pallida*	*H. schachtii*: 160–170 Mb, ~179 Mbp final assembly, 26,739 gene annotations; *G. pallida*: 128 effector orthologues, 117 novel candidates
Transcriptome analysis	Life-stage-specific transcriptomes generated, showing differential expression	*H. schachtii*: 69% (18,575) genes differentially regulated; *Arabidopsis thaliana*: 59% (19,071) genes differentially expressed
Gene duplication	Segmental duplications enriched in parasitism genes	*H. schachtii*: 4547 segmentally duplicated genes, 28% enriched in *Heterodera*-specific orthogroups; *G. pallida*: 299 SPRY domain proteins, 295 dorsal gland protein-like
Effectors and secreted proteins	Identified numerous effectors and cell wall modifying proteins	*G. pallida*: 2669 secreted proteins, 209 effectors (38 families), 39 cell wall degrading enzyme genes
Horizontal gene transfer	HGT events identified, enhancing pathogenicity	*H. schachtii*: 263 HGT genes, including 2 for vitamin B5 biosynthesis
Proteomic expression	Stage-specific expression of effectors and host defense proteins	*G. pallida*: 30 invasive stage effectors, 61 parasitic stage effectors, targeting nucleus, nucleolus, peroxisomes

12.9 Challenges and Future Directions

Future research should focus on integrating these omics datasets to develop predictive models for nematode infections and to identify novel therapeutic targets. The 959 Nematode Genomes initiative, which coordinates genomic efforts, and analysis environments like NEMBASE, which facilitate transcriptomic data exploitation, are examples of ongoing efforts to advance this field (Blaxter et al. 2012).

Despite advances, several challenges remain:

12.9.1 Data Complexity

Managing and integrating large omics datasets requires advanced bioinformatics tools, with ongoing efforts to develop robust pipelines for analysis.

12.9.2 Functional Validation

Many identified genes and proteins lack functional characterization, necessitating further studies using techniques like RNAi and CRISPR-Cas9.

12.9.3 Field Applications

Translating lab-based findings into field-resistant crops or effective biocontrols is a hurdle, with debates around scalability and environmental impact.

12.9.4 Data Integration and Standardization

Harmonizing datasets across labs and platforms remains a hurdle.

12.9.5 Translation to the Field

Moving from laboratory omics discoveries to cost-effective, user-friendly diagnostics and treatments for farmers.

12.9.6 Emerging Technologies

Single-cell transcriptomics and spatial metabolomics promise even finer resolution of nematode–host interplay.

Future directions include emerging technologies like single-cell omics, spatial transcriptomics, and high-throughput proteomics, which promise to provide precise insights into cellular-level interactions, potentially revolutionizing plant nematode management.

12.10 Conclusion

Genomics, transcriptomics, proteomics, and metabolomics each play critical and complementary roles in nematode diagnosis and management. The convergence of genomics, transcriptomics, proteomics, and metabolomics is ushering in a new era of precision nematology. They provide a multilevel understanding of nematode biology and host–parasite interactions, from the genetic blueprint to the metabolic outcomes. This comprehensive approach is essential for developing accurate diagnostic tools, such as species-specific genetic markers, stage-specific gene expression profiles, diagnostic proteins, and metabolic biomarkers, and for devising effective management strategies, such as targeting key metabolic pathways or proteins involved in parasitism. As research progresses, the integration of these omics fields will likely drive innovations in nematode control, addressing the urgent need for new solutions in agriculture and public health. By elucidating the molecular underpinnings of parasitism and host response, these omics tools enable highly specific diagnostics and innovative management approaches—paving the way toward sustainable, targeted control of nematode pests.

12.11 Suggested Readings

- Escobar, C., Brown, S. and Mitchum, M.G. 2011. Transcriptomic and proteomic analysis of the plant response to nematode infection. In: J. Jones, G. Gheysen and C. Fenoll (eds.) *Genomics and Molecular Genetics of Plant-Nematode Interactions*, pp. 157–173. Springer, Dordrecht.
- Ma, G., Wang, T., Korhonen, P.K. et al. 2020. Elucidating the molecular and developmental biology of parasitic nematodes: Moving to a multi-omics paradigm. *Adv Parasitol.* 108: 175–229.
- Stoltzfus, J.D., Pilgrim, A.A. and Herbert, D.R. 2017. Perusal of parasitic nematode 'omics in the post-genomic era. *Mol Biochem Parasitol* 215: 11–22.
- Virmani, L. and Juyal, A. 2025. Progression in multi-omics and immunological research on host-parasite interactions: Transcriptomics, proteomics, genomics, metabolomics, and immune responses. *Recent Trends in Infectious Diseases* 02(01): Available from: https://journals.stmjournals.com/rtid/article=2025/view=200701.
- Zhu, Z., Bo, D., Xie, C. et al. 2025. Integrative multi-omics analysis reveals the translational landscape of the plant-parasitic nematode *Meloidogyne incognita*. *Commun Biol.* 8(1): 140. https://doi.org/10.1038/s42003-025-07533-x.

12.12 Review Questions

- What is the main difference between genomics and transcriptomics?
- Name one high-throughput technique used in nematode genome sequencing.
- Which *omics* approach helps us study gene expression patterns in nematodes?
- Give an example of a proteomic technique used to identify nematode proteins.
- How can metabolomics indicate nematode infection in plants?
- What type of molecular marker is commonly developed from nematode genomic data?
- Why is RNA-Seq more informative than microarrays in transcriptomic studies?
- Name one advantage of using proteomics in nematode diagnostics.
- How does metabolomics help in distinguishing healthy versu. nematode-infested plants?
- What is meant by a "multi-omics" approach in nematode diagnosis?
- List one challenge in applying *omics* tools to field-level nematode management.

References

Araujo ACG, Guimaraes PM, Mota APZ et al (2021) Overexpression of DUF538 from wild *Arachis* enhances plant resistance to *Meloidogyne* spp. Agronomy 11:559

Ayaz M, Zhao JT, Zhao W et al (2024) Biocontrol of plant parasitic nematodes by bacteria and fungi: a multi-omics approach for the exploration of novel nematicides in sustainable agriculture. Front Microbiol 15:1433716. https://doi.org/10.3389/fmicb.2024.1433716

Bird DM, Williamson VM, Opperman CH (2015) Exploiting solved genomes of plant-parasitic nematodes to understand parasitism. Adv Bot Res 73:241–258

Blaxter M, Kumar S, Kaur G et al (2012) Genomics and transcriptomics across the diversity of the Nematoda. Parasite Immunol 34(2–3):108–120

Chen X, Li F, Wang D, Cai L (2024) Insights into the plant response to nematode invasion and modulation of host defense by plant parasitic nematode. Front Microbiol 15:1482789. https://doi.org/10.3389/fmicb.2024.1482789

Cutis RHC (2007) Plant parasitic nematode proteins and the host parasite interaction. Brief Funct Genomic Proteomic 6(1):50–58

Escobar C, Brown S, Mitchum MG (2011) Transcriptomic and proteomic analysis of the plant response to nematode infection. In: Jones J, Gheysen G, Fenoll C (eds) Genomics and molecular genetics of plant-nematode interactions. Springer, Dordrecht

Gillet FX, Bournaud C, de Souza JDA Jr, Grossi-de-Sa MF (2017) Plant-parasitic nematodes: towards understanding molecular players in stress responses. Ann Bot 119:775–789

Gomez-Fernandez M, Higley K, Tokuhiro A et al (2020) Status of research and development of learning-based approaches in nuclear science and engineering: A review. Nucl Eng Des 359:110479. https://doi.org/10.1016/j.nucengdes.2019.110479

Grynberg P, Togawa RC, de Freitas LD et al (2020) Comparative genomics reveals novel target genes towards specific control of plant-parasitic nematodes. Genes 11:1347. https://doi.org/10.3390/genes11111347

Hofmann J, El Ashry AN, Anwar S et al (2010) Metabolic profiling reveals local and systemic responses of host plants to nematode parasitism. Plant J 62(6):1058–1071

Jaubert S, Laffaire JB, Piotte C et al (2002) Direct identification of stylet secreted proteins from root-knot nematodes by a proteomic approach. Mol Biochem Parasitol 121(2):205–211

Jones JT, Haegeman A, Danchin EG et al (2013) Top 10 plant-parasitic nematodes in molecular plant pathology. Mol Plant Pathol 14:946–961

Khan A, Khan A, Ali A et al (2023) Root-knot nematodes (*Meloidogyne* spp.): biology, plant-nematode interactions and their environmentally benign management strategies. Gesunde Pflanz 75:1–19

Klepikova AV, Kasianov AS, Gerasimov ES et al (2016) A high-resolution map of the *Arabidopsis thaliana* developmental transcriptome based on RNA-seq profiling. Plant J 88(6):1058–1070

Kumar M, Gantasala NP, Roychowdhury T et al (2014) De novo transcriptome sequencing and analysis of the cereal cyst nematode, Heterodera avenae. PLOS ONE 9(5):e96311

Li M, Guo B, Liu F, Fu ZQ (2023) Cross-kingdom vitamin B5 biosynthesis and cyst nematode susceptibility. Trends Parasitol 39(1):7–9

Ma G, Wang T, Korhonen PK et al (2020) Elucidating the molecular and developmental biology of parasitic nematodes: moving to a multiomics paradigm. Adv Parasitol 108:175–229

Montarry J, Mimee B, Danchin EGJ et al (2021) Recent advances in population genomics of plant-parasitic nematodes. Phytopathology 111:40–48

Moore S, Payton P, Wright M (2005) Utilization of tomato microarrays for comparative gene expression analysis in the Solanaceae. J Exp Bot 56(421):2885–2895

Nuaima RH, Heuer H (2023) Genetic variation among *Heterodera schachtii* populations coincided with differences in invasion and propagation in roots of a set of cruciferous plants. Int J Mol Sci 24(7):6848

Paez AR (2023) Plant-parasitic nematodes and their management: a focus on new nematicides. https://doi.org/10.5772/intechopen.1002237

Rosso MN, Jones JT, Abad P (2009) RNAi and functional genomics in plant parasitic nematodes. Annu Rev Phytopathol 47:207–232

Różańska E, Krępski T, Wiśniewska A (2023) Mutations in selected ABA-related genes reduce level of *Arabidopsis thaliana* susceptibility to the beet cyst nematode *Heterodera schachtii*. Plants 12(12):2299

Senthil Kumar P, Pragadeesh ARU, Namadara S et al (2025) Management of nematode menace in protected cultivation of vegetables in India. Plant Sci Today 12(2) https://horizonepublishing.com/journals/index.php/PST/article/view/6862

Siddique S, Radakovic ZS, Hilt C et al (2022) The genome and life stage-specific transcriptomes of a plant-parasitic nematode and its host reveal susceptibility genes involved in trans-kingdom synthesis of vitamin B5. Nat Commun 13:6190. https://doi.org/10.1038/s41467-022-33769-w

Thorpe P, Mantelin S, Cock PJA (2014) Genomic characterisation of the effector complement of the potato cyst nematode *Globodera pallida*. BMC Genomics 15:923. http://www.biomedcentral.com/1471-2164/15/923

Vieira P, Myers RY, Pellegrin C et al (2021) Targeted transcriptomics reveals signatures of large-scale independent origins and concerted regulation of effector genes in *Radopholus similis*. PLoS Pathog 17(11):e1010036

Westphal A (2011) Sustainable approaches to the management of plant-parasitic nematodes and disease complexes. J Nematol 43(2):122–125

Willett DS, Filgueiras CC, Benda N et al (2020) Sting nematodes modify metabolomic profiles of host plants. Sci Rep 10(1):2212. https://doi.org/10.1038/s41598-020-59062-8

Williamson VM, Kumar A (2006) Nematode resistance in plants: the battle underground. TRS in Genetics 22:396–403

Yang D, Chen C, Liu Q, Jian H (2017) Comparative analysis of pre- and post-parasitic transcriptomes and mining pioneer effectors of *Heterodera avenae*. Cell Biosci 7:11. https://doi.org/10.1186/s13578-017-0138-6

Yang S, Pan L, Chen Y et al (2019) *Heterodera avenae* GLAND5 effector interacts with pyruvate dehydrogenase subunit of plant to promote nematode parasitism. Front Microbiol 10:1241. https://doi.org/10.3389/fmicb.2019.01241

Zhou Y, Zhao D, Shuang L et al (2020) Transcriptome analysis of rice roots in response to root-knot nematode infection. Int J Mol Sci 21(3):848

Transgenic Strategies to Enhance Nematode Resistance

13

Abstract

The exploitation of genetic engineering to enhance plant defenses against nematodes has yielded a variety of successful strategies, primarily focusing on the expression of anti-nematode proteins and RNA interference (RNAi) constructs in host roots. Transgenic resistance to plant-parasitic nematodes can significantly reduce crop damage, with methods, such as RNAi and proteinase inhibitors, showing promise. It seems likely that combining multiple strategies, such as natural resistance genes and nematicidal proteins, could enhance effectiveness, though results vary by crop and nematode type. There is an ongoing debate about environmental and regulatory impacts, despite transgenic approaches being sustainable.

Keywords

Genetic engineering · RNA interference · Proteinase inhibitors · Nematicidal proteins · GM crops

13.1 Learning Objectives

- Explain the principles of genetic engineering in plants for nematode resistance.
- Differentiate between conventional resistance breeding and transgenic approaches.
- Discuss gene constructs used for nematode resistance (e.g., RNA interference (RNAi), proteinase inhibitors (PIs), lectins, and toxin genes).
- Explain host-induced gene silencing (HIGS) and its role in targeting nematode genes.
- Review successful examples in crops (e.g., tobacco, tomato, banana, and soybean).
- Evaluate how transgenic lines have reduced nematode infestation and improved yield.
- Assess the environmental and food safety concerns related to transgenic plants.

R. P. Parvatha, *Plant Parasitic Nematodes*,
https://doi.org/10.1007/978-981-95-9306-4_13

- Understand national and international regulations for transgenic crop deployment.
- Discuss the limitations of transgenic nematode resistance (e.g., resistance durability and nematode adaptation).
- Identify future directions, including genome editing and transgene stacking, for sustainable nematode management.

13.2 Introduction

Plant-parasitic nematodes (PPNs), such as root-knot nematodes (*Meloidogyne* spp.), cyst nematodes (*Heterodera* spp.), and burrowing nematodes (*Radopholus similis*), pose significant threats to global agriculture, causing estimated annual losses of billions of dollars due to reduced crop yields. The economic impact is particularly severe in developing regions, where alternatives to chemical controls are urgently needed. Traditional control methods, such as chemical nematicides, are often environmentally harmful and unsustainable, prompting the development of transgenic resistance strategies. These approaches involve genetically modifying plants to express traits that disrupt nematode life cycles, inhibit infection, or enhance plant defense mechanisms. Transgenic resistance offers a targeted, sustainable solution by leveraging genetic engineering to enhance plant immunity and reducing reliance on toxic pesticides. This chapter provides a detailed overview of the strategies, recent advancements, and implications, drawing from a comprehensive review of literature up to July 2025.

13.3 Transgenic Strategies for Nematode Resistance

The development of transgenic resistance has evolved through several key strategies, each targeting different aspects of nematode biology and plant–nematode interactions. Several methods have been developed to create transgenic plants resistant to PPNs.

13.3.1 Natural Resistance Genes

Plants are modified with genes from resistant varieties, like the *Mi-1.2* gene in tomatoes, which helps fight *Meloidogyne incognita.*

13.3.2 Proteinase Inhibitors

These are proteins that block nematode digestion, such as *Oc-IΔD86* in Arabidopsis, reducing nematode growth.

13.3.3 Nematicidal Proteins

Toxic proteins, like lectins, are introduced to harm nematodes when they feed on the plant.

13.3.4 Chemo-Disruptive Peptides

These interfere with nematodes' ability to find and invade roots, seen in potatoes with reduced *Globodera pallida* numbers.

13.3.5 RNA Interference

Plants produce RNA to silence nematode genes, recently shown to be effective in bananas against *Radopholus similis*.

13.4 Natural Resistance Genes

R genes, or resistance genes, are a class of plant genes that encode proteins, often with nucleotide-binding and leucine-rich repeat domains, designed to detect pathogens. Natural resistance genes (R genes) are cloned from resistant plant varieties and introduced into susceptible crops to trigger defense responses. These genes are part of the plant's innate immune system, interacting with nematode avirulence (Avr) genes in a gene-for-gene relationship. Upon recognition, R genes trigger defense responses, such as the hypersensitive response (HR), which involves localized cell death at the nematode feeding site to isolate and eliminate the pathogen, preventing further infection, and thereby preventing nematode establishment and reproduction. Notable examples include the following:

- *Mi-1.2* in tomato, conferring resistance to *M. incognita* with over 95% reduction in nematode reproduction (Fuller et al. 2008);
- *Hs1 pro-1* in sugar beet, effective against *Heterodera schachtii*, linked to yield loss reduction;
- *Gpa-2* in potato against *G. pallida* and *Hero A* in potato against *Globodera rostochiensis*, with resistance rates of 80% in some cases;
- *Rhg1* and *Rhg4* in soybean, reducing *Heterodera glycines* cysts by 45%;
- *Mi-1* remains the most widely deployed R-gene against root-knot nematodes in tomato, but its efficacy collapses above ~28°C (*Mi-2* through *Mi-9* and *Mi-HT* were sourced to overcome this constraint) (El-Sappah et al. 2019);
- *Hs1 < sup > pro-1</sup>, Gpa2, Hero, Gro1–4, Rhg1,* and *Rhg4* represent the main R genes in sugar beet, potato, and soybean, respectively. Most act by triggering a localized HR that disrupts syncytium (feeding site) formation and/or nematode development (Ali et al. 2017).

Despite decades of work, only a handful of nematode R genes have been cloned. Significant gaps remain, particularly for cyst and lesion nematodes, limiting broad resistance deployment (Kaloshian and Teixeira 2019).

Pyramiding (stacking multiple R genes) and introgression of heat-stable alleles (e.g., *Mi-HT*) are current breeding strategies to enhance durability under field conditions. These genes are particularly effective against specific nematode species or pathotypes, but their efficacy can be limited by nematode adaptation.

13.4.1 Key Examples of R Genes and Their Applications

Several R genes have been identified, cloned, and integrated into crop breeding programs, demonstrating their practical utility in nematode management. Table 13.1 summarizes notable R genes, their sources, and effectiveness:

These genes have been instrumental in breeding programs, with transgenic approaches, such as the introduction of *Hero A* into potato, achieving over 95% resistance against *G. rostochiensis* and ~ 80% against *G. pallida* (Ernst et al. 2002).

13.4.2 Breeding and Transgenic Applications

The primary application of R genes is in plant breeding to develop resistant cultivars. Traditional breeding involves crossing wild relatives with cultivated crops to introgress R genes, while transgenic approaches use genetic engineering to insert R genes into elite varieties. For example:

- The *Mi-1.2* gene, originally from wild tomato, has been successfully integrated into commercial tomato varieties to combat root-knot nematodes (Milligan et al. 1998).
- Overexpression of serine hydroxymethyltransferase at the Rhg4 locus in soybean resulted in a 45% decrease in mature cyst nematode numbers (Matthews et al. 2013).

Transgenic strategies, detailed by Ali et al. (2017), have shown promise, particularly for crops lacking natural resistance, by leveraging R genes from related species. However, regulatory and public acceptance challenges can limit their deployment.

13.4.3 Mechanisms of R Gene-Mediated Resistance

The interaction between R genes and nematodes follows the gene-for-gene model, where a specific R gene in the plant corresponds to an *Avr* gene in the nematode. When the nematode attempts to infect, the R protein recognizes the Avr product, initiating a cascade of defense responses. These responses include the following:

Table 13.1 Key R genes, their sources, and effectiveness

R Gene	Source plant	Target nematode	Host plant	Resistance response	References
Mi-1.2	Wild tomato (*Solanum peruvianum*)	*Meloidogyne incognita*	Tomato (*Solanum lycopersicum*)	Triggers HR before giant cell establishment, preventing nematode feeding	Ali et al. (2017)
Mi-2–Mi-9, Mi-HT	Various wild tomato spp.	*M. incognita, Meloidogyne javanica, Meloidogyne arenaria*	Tomato	Often heat-stable alleles; HR and syncytium disruption	El-Sappah et al. (2019)
Hs1 pro-1	Wild beet (*Boerhavia procumbens*)	*Heterodera schachtii*	Sugar beet (*Berberis vulgaris*)	Causes abnormal syncytium development, leading to nematode starvation beyond the J2 stage	Ali et al. (2017)
Gpa-2	Potato	*Globodera pallida*	Potato (*Solanum tuberosum*)	Halts nematode development, resulting in stagnated and translucent females	Ali et al. (2017)
Hero A	Tomato	*G. pallida, G. rostochiensis*	Potato	Induces HR after syncytia initiation, leading to abnormal and necrotic syncytia	Ali et al. (2017)
Gro1–4	Potato	*G. rostochiensis (Ro1)*	Potato	Unknown, but effective against a specific pathotype	Ali et al. (2017)
Rhg1	Soybean (*Glycine max*)	*Heterodera glycines*	Soybean	HR leads to abnormal syncytia and necrosis due to cell degeneration	Ali et al. (2017)
Rhg4	Soybean	*H. glycines*	Soybean	HR following syncytia initiation, becomes abnormal and necrotic, with potential breakdown over time	Ali et al. (2017)
Cre loci	*Aegilops* spp.	*Heterodera avenae*	Wheat (*T. aestivum*)	Unknown, but provides resistance in wheat	Ali et al. (2017)

- **HR**: Rapid cell death at the infection site, preventing nematode feeding and reproduction. For instance, the *Mi-1.2* gene in tomato triggers HR before root-knot nematodes can establish giant cells.
- **Cell Wall Reinforcement**: Strengthening of plant cell walls to block nematode penetration.
- **Activation of Pathogenesis-Related (PR) Genes**: Induction of PR genes, often dependent on salicylic acid, to enhance systemic resistance.

  Research, such as that detailed by Ali et al. (2017), highlights that these mechanisms vary by R gene and nematode species, with responses like abnormal syncytium development (e.g., *Hs1 pro-1* in sugar beet against *Heterodera schachtii*) leading to nematode starvation.

13.4.4 Recent Advances and Complementary Strategies

As of recent research, advancements in genomics and transcriptomics have enhanced our understanding of R gene–nematode interactions. For example, Siddique et al. (2022) analyzed the hologenome at infection sites, identifying metabolic pathways in nematodes complemented by host genes, which could inform new resistance strategies.

Additionally, RNAi has emerged as a complementary approach, targeting essential nematode genes to reduce parasitism. Studies by Dutta et al. (2021) highlight the use of dsRNA to silence nematode genes, offering a potential synergy with R gene-based resistance to enhance durability.

Comparative genomics, as discussed by Grynberg et al. (2020), has identified over 24,000 protein families specific to PPNs, providing targets for both R gene enhancement and RNAi strategies.

13.4.5 Limitations and Challenges

Despite their effectiveness, R genes face several limitations:

- **Specificity**: R genes are often effective against specific nematode species or pathotypes, limiting their broad applicability. For instance, the *Mi-1.2* gene is highly effective against *Meloidogyne incognita* but may not protect against other root-knot nematode species.
- **Nematode Evolution**: Nematodes can develop virulent biotypes that overcome R genes, as noted by Williamson (1999), who cited the identification of nematode biotypes infecting resistant plants.
- **Fitness Costs**: The expression of R genes can carry fitness costs, potentially reducing plant growth or yield under non-infested conditions (von Dahlen et al. 2023).

  To address these challenges, integrated pest management (IPM) strategies are recommended, combining R genes with cultural practices (e.g., crop rotation), biological control (e.g., nematode-trapping fungi), and, where necessary, chemical treatments, though the latter is increasingly restricted due to environmental concerns.

13.4.6 Future Directions and Conclusion

R genes are a cornerstone of nematode management, offering specific and effective resistance through breeding and transgenic approaches. Their integration into IPM strategies, alongside emerging technologies such as RNAi and genomic editing, promises to enhance crop protection against evolving nematode threats. Future research should focus on identifying novel R genes, understanding their evolutionary dynamics with nematodes, and developing multi-gene resistance stacks to ensure long-term efficacy, aligning with sustainable agricultural practices.

13.5 PI Coding Genes

PIs are overexpressed in transgenic plants to disrupt nematode digestive processes, leading to starvation or reduced development. Examples include:

- *Oc-IΔD86*, a modified cystatin in Arabidopsis, suppressed growth and fecundity of *H. schachtii* and *M. incognita* (Urwin et al. 1997).
- Taro cystatin in tomato reduced *M. incognita* reproduction by 78.3% (Chan et al. 2015).
- A dual PI strategy in plantain, combining cysteine and serine inhibitors, increased yield by 186% and controlled nematodes by 99% (Tripathi et al. 2015).
- Cowpea trypsin inhibitor (*CpTI*) in potato enhanced resistance to *G. pallida* (Papolu et al. 2016).

This approach targets nematode feeding, making it effective against sedentary endoparasites such as root-knot and cyst nematodes.

PIs are small plant proteins that play a significant role in defending crops against PPNs by targeting their digestive proteases, thereby reducing nutrient absorption and impairing growth and reproduction. Below is a comprehensive overview of PI coding genes and their use in managing PPNs, focusing on key findings and strategies.

13.5.1 Role of PIs in Nematode Management

- **Types of PIs:** The four major classes of plant PIs are cysteine, serine, aspartic, and metallo-PIs. Cysteine and serine PIs are particularly effective against PPNs due to their specificity in targeting the major digestive enzymes of nematodes.
- **Plant Defense:** PIs are part of the plant's natural defense system, often induced by wounding or pathogen attack via signaling pathways like the octadecanoid pathway, which involves jasmonic acid (JA). JA enhances the expression of PIs and other anti-nematode compounds, such as terpenes and oxylipins.

13.5.2 Key PI Genes

Several PI genes have been identified and utilized in transgenic strategies to confer nematode resistance:

- **Cystatins (Cysteine PIs):** The rice cystatin gene *Oc-IΔD86* has been expressed in transgenic eggplant (*Solanum melongena*) under a root-specific promoter (TUB-1), resulting in a 78.3% reduction in the reproductive success of *Meloidogyne incognita*. Cystatins inhibit cysteine proteases in the nematode's digestive system, disrupting nutrient uptake. Their small size facilitates uptake by nematodes, enhancing their effectiveness. Transgenic Arabidopsis expressing modified cystatins showed resistance to both cyst and root-knot nematodes, with ingestion of the inhibitor being critical for efficacy.
- ***CpTI*:** In Arabidopsis, *CpTI* (a serine PI) was fused with a cysteine inhibitor using peptide linkers (e.g., Peptide sequence derived from Metallothionein (MT)-associated region "a" used as a linker (PsMTa) or galactose oxidase-derived linkers). The PsMTa-linked *CpTI* affected cyst nematode sexual development externally, indicating an external site of action, while the galactose oxidase linker showed additive effects when co-delivered with cystatins. *CpTI* targets serine proteases, impairing nematode digestion and development. Its effectiveness does not always require ingestion, making it versatile. Methods using *CpTI* for nematode control are patented, highlighting its commercial potential.
- **Sporamin:** In sweet potato, sporamin, a Kunitz-type trypsin inhibitor, constitutes 60–80% of soluble tuber protein and is constitutively expressed in roots. Transgenic sweet potato overexpressing sporamin showed reduced galling and influenced sex determination in root-knot nematodes. Sporamin inhibits trypsin activity, reducing nematode nutrient uptake and reproduction.
- ***PI-I*:** In tomato plants, *PI-I* gene expression was studied post-infection by *M. incognita, M. javanica*, and *Meloidogyne chitwoodi*. Expression peaked at 1 and 3 days post-infection (dpi) but decreased by 7, 14, and 21 dpi, suggesting early defense activation followed by nematode suppression of PI expression during compatible interactions. *PI-I* contributes to early defense by inhibiting nematode proteases, but nematodes may suppress its expression over time to establish parasitism.
- **Other PIs:** *CFMI-5*, a cysteine PI gene from carnation, identified via suppression subtractive hybridization, showed high sequence similarity to PI genes involved in flower maturation and defense against pathogens, including nematodes.
- *Kunitz-type PIs* are found in potato and *Brassica napus*; these inhibitors are localized in cell walls and plasma membranes, contributing to defense against nematodes and other pathogens.

13.5.3 Transgenic Strategies Using PI Genes

Genetic engineering has been pivotal in deploying *PI* genes for nematode resistance:

- **Single Gene Transformation:** Transgenic plants expressing cystatins or *CpTI* alone have shown significant reductions in nematode reproduction. For instance, Arabidopsis expressing *Oc-IΔD86* reduced *M. incognita* reproduction by up to 78.3%. Tomato plants expressing *Mi-cpl-1* (a cathepsin L cysteine proteinase gene from *M. incognita*) via RNAi resulted in a 60–80% reduction in infection and multiplication.
- **Gene Pyramiding:** Dual gene constructs, such as those combining cysteine and serine PIs, have shown enhanced efficacy. Arabidopsis transformed with dual inhibitors linked by a galactose oxidase linker exhibited additive effects compared to single inhibitors. Gene pyramiding (e.g., combining *BvcZR3* and *BvHs1pro-1* in oilseed rape) improved resistance against *Heterodera schachtii.*
- **RNAi:** RNAi targeting nematode protease genes (e.g., *Mi-cpl-1* in *M. incognita*) has been used to silence essential genes, reducing nematode parasitism. HIGS in transgenic plants such as tomato and soybean has shown promise. For example, silencing *pat-10* and *unc-87* genes in *Pratylenchus thornei* and *P. zeae* via dsRNA soaking reduced reproduction by 77–81%.
- **Clustered Regularly Interspaced Short Palindromic Repeats (CRISPR)/Cas9:** Emerging research suggests that CRISPR/Cas9 could be used to edit PI genes or nematode susceptibility genes to enhance resistance, though specific applications for PIs are still developing.

13.5.4 Mechanism of Action

PIs, particularly cysteine and serine PIs, inhibit the digestive enzymes (proteases) of PPNs, such as root-knot nematodes (*Meloidogyne* spp.), cyst nematodes (*Heterodera* spp. and *Globodera* spp.), and lesion nematodes (*Pratylenchus* spp.). This disruption limits the nematodes' ability to break down proteins, reducing their feeding efficiency and reproductive success. For example, PIs can degrade nematode proteases, preventing nutrient absorption, which is critical for their survival and reproduction.

13.5.5 Case Studies on Practical Applications

- **Transgenic Crops:** PI genes have been successfully introduced into crops such as tomato, Arabidopsis, eggplant, and sweet potato, offering a sustainable alternative to chemical nematicides like methyl bromide, which are being phased out due to environmental concerns.
- **Biological Control:** Bacterial strains like *Bacillus firmus* produce proteases (e.g., Sep1) with nematicidal activity, which can be combined with PI-based transgenic approaches for IPM.

- **Cross-Species Control:** RNAi targeting PI genes has shown cross-species efficacy, suggesting broad-spectrum resistance potential.

13.5.6 Challenges

- **Nematode Suppression:** PPNs can suppress PI gene expression over time, reducing long-term efficacy.
- **Ingestion Dependency:** Some PIs, like cystatins, require ingestion to be effective, limiting their action against nematodes that evade uptake.
- **Regulatory and Environmental Concerns:** Transgenic crops face regulatory hurdles and public acceptance issues, necessitating rigorous safety assessments.
- **Protease diversity**: Nematodes possess multiple proteases; targeting a single enzyme may select for compensatory pathways.
- **Off-target Effects**: High-level PI expression might interfere with endogenous plant proteases, affecting growth.
- **Public Acceptance and Regulation**: As with all transgenic approaches, regulatory hurdles and consumer perceptions may limit implementation.
- **Limited Knowledge:** The repertoire of PI genes in PPNs is underexplored, particularly for species such as *Bursaphelenchus xylophilus* and *M. incognita*.

13.5.7 Future Directions

- **Gene Editing:** CRISPR/Cas9 could enhance PI gene expression or target nematode effectors more precisely.
- **Integrated Approaches:** Combining PIs with other resistance genes (e.g., *Hs4* for cyst nematodes) or biocontrol agents could improve efficacy.
- **Molecular Insights:** Further studies on plant–nematode interactions, such as the role of JA and other signaling pathways, could optimize PI deployment.
- **Omics-Guided Discovery**: Proteomic profiling of nematode gut proteases will enable more precise targeting.
- **Synthetic Biology**: Design of novel inhibitors with broader specificity and higher affinity.
- **Novel PIs:** Identifying new PI genes, like those in carnation (*CFMI-5*), could expand the toolkit for nematode resistance.

13.5.8 Conclusion

PI coding genes, such as cystatins, *CpTI*, sporamin, and *PI-I*, are powerful tools for managing PPNs through transgenic strategies, RNAi, and gene pyramiding. They offer environmentally friendly alternatives to chemical nematicides, with demonstrated success in crops such as tomato and eggplant. However, challenges like nematode suppression of PI expression and ingestion dependency highlight the need for integrated approaches and further research to maximize their potential.

13.6 Nematicidal Proteins

Nematicidal proteins, derived from bacteria, fungi, and plants, have emerged as promising biocontrol agents and resistance-enhancing tools in sustainable agriculture. Nematicidal proteins, such as lectins, plantibodies, and Bt Cry proteins, are expressed in transgenic plants to inhibit nematode development or survival. Examples include the following:

- GNA lectin (from *Galanthus nivalis*) in Arabidopsis, oilseed rape, and potato, showing resistance against root-knot and cyst nematodes (Burrows et al. 1998)
- *Cry5B* and *Cry6A* in tomato, affecting *M. incognita* reproduction, with a 54 kDa protein ingested by nematodes (Li et al. 2007)

These proteins are ingested by nematodes during feeding, disrupting their physiology and reducing population growth.

13.6.1 Bacterial Nematicidal Proteins

Bacterial nematicidal proteins are primarily exemplified by the Cry proteins produced by *Bacillus thuringiensis* (Bt), a well-studied biocontrol agent. These proteins, including *Cry5, Cry6, Cry12, Cry13, Cry14, Cry21, Cry55, Cry31Aa, Cry41ORF*, and *Cry6A*, exhibit high nematicidal activity against various PPNs. Their mode of action involves damaging the intestinal lining of nematodes, disrupting cellular structures such as pores, cells, and vacuoles, ultimately leading to nematode death. For instance:

- *Cry6A* has been shown to be effective against *Meloidogyne hapla*, causing significant mortality (Yu et al. 2015).
- *Cry31Aa* from *B. thuringiensis* GBAC46 demonstrates antagonistic activity against *Meloidogyne incognita* (Liang et al. 2022).
- Against *Bursaphelenchus xylophilus*, Bt toxins cause vacuolization, intestinal wall contraction, thinning, and shrinkage (Guo et al. 2022).

Other bacteria, such as *Serratia marcescens*, produce extracellular enzymes such as glucanases, cellulases, and pectinases, which degrade nematode structures and control species like *M. incognita* (Migunova and Sasanelli 2021). These bacterial proteins are typically applied as biopesticides in soil or on plants, offering an environmentally friendly alternative to chemical nematicides.

Table 13.2 summarizes key bacterial nematicidal proteins, their target nematodes, and modes of action, based on recent literature.

Table 13.2 Key bacterial nematicidal proteins, their target nematodes, and modes of action

Bacteria	Nematicidal proteins	Target nematode	Mode of action	References
Bacillus thuringiensis	*Cry5, Cry6, Cry12, Cry13, Cry14, Cry21, Cry55*	Various PPNs	Damage intestinal lining, affect pore development, cells, and vacuoles	Li et al. (2015), Liang et al. (2022)
B. thuringiensis GBAC46	*Cry31Aa, Cry41ORF*	*Meloidogyne incognita*	Antagonistic activity	Liang et al. (2022)
B. thuringiensis	Cry proteins	*Bursaphelenchus xylophilus*	Vacuolization, intestinal wall contraction, thinning, and shrinkage	Guo et al. (2022)

13.6.2 Fungal Nematicidal Proteins

Fungal biocontrol agents, such as *Pochonia chlamydosporia* and *Aspergillus welwitschiae*, also contribute nematicidal proteins and secondary metabolites. While the focus is often on enzymes and toxins, specific proteins like aurovertins from *Pochonia chlamydosporia* mediate interactions with nematodes, and αβ-dehydrocurvularin from *Aspergillus welwitschiae* effectively inhibits the behavior and development of *Meloidogyne graminicola* in rice roots (Wang et al. 2015; Xiang et al. 2020). These fungal agents are used in IPM to suppress PPN populations, particularly in soil applications.

13.6.3 Plant-Derived Nematicidal Proteins and Compounds

Plants have evolved natural defense mechanisms against PPNs, including the production of proteins and secondary metabolites with nematicidal or resistance-enhancing properties. Key plant-derived nematicidal proteins include the following:

- **Protease Inhibitors**: These include PI2, trypsin inhibitors, and cysteine protease inhibitors, which inhibit nematode proteases essential for virulence and development. For example, transgenic expression of cysteine protease inhibitors in Arabidopsis and tomato has conferred resistance to nematodes such as *Globodera rostochiensis* and *Meloidogyne incognita* (Urwin et al. 2000, 2003). These inhibitors disrupt the nematode's ability to digest plant tissues and establish feeding sites, offering a direct nematicidal effect.
- **Chitinases**: These enzymes degrade chitin, a component of nematode egg shells and pharyngeal lumen walls. While not directly proven to confer resistance, chitinases are upregulated in resistant plants following PPN infection, suggesting a role in defense (Qiu et al. 1997; Kumar et al. 2018). Their activity may indirectly reduce nematode populations by affecting egg viability.

- **Papain-Like Cysteine Proteases (PLCPs)**: Secreted into the apoplast, PLCPs potentially inhibit PPN infection. For instance, the absence of RCR3 pim homologs in Arabidopsis enhances susceptibility to cyst nematodes, indicating their role in defense (Lozano-Torres et al. 2014).

  In addition to proteins, plants produce secondary metabolites with nematicidal activity, such as chlorogenic acid, phenylphenalenone (anigorufone), flavonoids (kaempferol, quercetin, myricetin, medicarpin, and glyceollin), α-terthienyl, asparagusic acid, and isothiocyanates from glucosinolates. While these are not proteins, they contribute to the overall nematicidal strategy, with effects ranging from direct toxicity to inhibition of egg hatching and nematode motility (Wuyts et al. 2006).

13.6.4 Case Studies on Management Strategies

Nematicidal proteins are integrated into various management strategies to control PPNs, aligning with sustainable agricultural practices:

- **Biological Control**: Biocontrol agents such as *Bacillus thuringiensis* and nematophagous fungi are applied to soil or plants to release nematicidal proteins, reducing nematode populations. Bt-based formulations containing Cry proteins are commercially available and used in sustainable agriculture, offering a viable alternative to chemical nematicides.
- **Transgenic Approaches**: Plants can be genetically engineered to express nematicidal proteins, such as Cry proteins or protease inhibitors, to enhance resistance. For example, transgenic Arabidopsis expressing PIs showed resistance to *Rotylenchulus reniformis* (Urwin et al. 2000), demonstrating the potential of this approach in crop protection.
- **IPM**: Nematicidal proteins are often used alongside other methods, such as crop rotation, resistant varieties, and reduced chemical nematicide use. This holistic approach ensures effective nematode management while minimizing environmental impact.

13.6.5 Challenges and Future Directions

Despite their promise, challenges remain in the application of nematicidal proteins. Efficacy and specificity vary by nematode species, limiting broad application. For instance, Cry proteins may be highly effective against certain PPNs but less so against others. Commercialization of biopesticides and transgenic crops faces regulatory and economic hurdles, requiring further research and development. Ongoing studies aim to identify new nematicidal proteins, particularly from plants, and to elucidate their mechanisms of action for more targeted control, enhancing their integration into sustainable agricultural systems.

13.7 Chemo-Disruptive Peptides

Chemo-disruptive peptides are emerging as a promising tool for managing PPNs, which cause significant damage to crops worldwide. These synthetic peptides work by interfering with the nematodes' chemoreception, making it harder for them to find and infect host plants. This approach leverages the nematodes' reliance on chemical cues for host finding, offering a targeted and environmentally safe method compared to broad-spectrum chemicals. Studies have shown they can be expressed in transgenic plants such as eggplant and potato, or delivered through microbes, offering a potential eco-friendly solution to a major agricultural challenge. Examples include the following:

- A peptide in potato reduced *G. pallida* females by 52% (Liu et al. 2005).
- Combined with *OcIΔD86*, it provided high resistance in potato without impacting soil nematodes (Green et al. 2012).
- In plantain, this strategy enhanced yield and nematode control (Tripathi et al. 2015).

This method is particularly useful for reducing initial infection rates, complementing other resistance mechanisms.

13.7.1 Applications in Transgenic Plants

One key strategy involves expressing chemo-disruptive peptides in transgenic plants to confer resistance. For instance:

- A study by Papolu et al. (2020) demonstrated the production of a synthetic chemo-disruptive peptide in eggplant (*Solanum melongena*) under the CaMV35S promoter. This resulted in a 64.39–79.65% reduction in *M. incognita* multiplication across six characterized transgenic lines, genotyped via Polymerase Chain Reaction (PCR), quantitative Reverse Transcription Polymerase Chain Reaction (qRT-PCR), and Southern hybridization. The mechanism involved disrupting nematode chemoreception, with root exudates elevating the expression of acetylcholinesterase genes (*Mi-ace-1* and *Mi-ace-2*) in the nematodes, affecting their host location and penetration.
- Another study by Liu et al. (2005) focused on transgenic potato plants expressing a chemoreception-disruptive peptide, achieving up to 63% inhibition of acetylcholinesterase in root exudates. This correlated with reduced root invasion by *H. glycines*, with glasshouse trials showing a 61% reduction in the number of cysts and eggs per cyst for *G. pallida* over 9 weeks. A second peptide binding to nematode nicotinic receptors provided a 52.6% reduction in nematodes able to establish in transgenic hairy roots.

These studies highlight the efficacy of expressing peptides directly in plants, offering a built-in resistance mechanism that reduces nematode populations below economic threshold levels.

13.7.2 Alternative Delivery via Transgenic Microbes

Beyond plant expression, recent research has explored delivering chemo-disruptive peptides through transgenic microbes, enhancing flexibility and scalability. A study by Warnock et al. (2017) investigated unamidated neuropeptide-like proteins (uNLPs) from the NLP family as transgenic nematicides. Using BLASTp, researchers identified 7 *nlp* genes encoding 28 *uNLPs* in *M. incognita* and 4 *nlp* genes encoding 25 *uNLPs* in *G. pallida*. Screening at 100 μM for 24 hours showed:

- For *M. incognita*, 11/27 *uNLPs* disrupted chemotaxis (e.g., 6/8 Mi-nlp-15 peptides), 13 disrupted host invasion, and 11 modulated stylet thrusting.
- For *G. pallida*, 12/25 *uNLPs* disrupted chemotaxis, five disrupted host invasion, and three modulated stylet thrusting.

Specific peptides, like Mi-NLP-15b, inhibited *M. incognita* chemotaxis at 10 pM and host invasion at nanomolar concentrations. Transgenic *Bacillus subtilis* and *Chlamydomonas reinhardtii* secreting *uNLPs* reduced tomato infection significantly:

- *C. reinhardtii* secreting *Mi-NLP-9f* and *Mi-NLP-15b* achieved 10.32% ± 10.32 ($p < 0.0001$) and 10.82% ± 6.574 ($p < 0.0001$) infection rates for *M. incognita*, respectively, and 30.95% ± 9.021 ($p = 0.0042$) for *G. pallida* with *Gp-NLP-15b*, offering up to 90% protection.
- *B. subtilis* secreting *Mi-NLP-15b* and *Mi-NLP-40* showed 26.63% ± 8.12 ($p = 0.0003$) and 23.72% ± 5.448 ($p = 0.0002$) infection rates for *M. incognita*, and 51.98% ± 13.29 ($p = 0.0203$) for *G. pallida* with *Gp-NLP-15b*.

These microbes were confirmed to secrete peptides at concentrations like 193.8 ± 81.3 ng/ml for His-tagged *NLP-15b*, validated by ELISA, ensuring effective delivery.

13.7.3 Specificity and Environmental Safety

A critical advantage of chemo-disruptive peptides is their specificity. Morris et al. (2017) found that uNLPs had no significant impact on non-target nematodes such as *Caenorhabditis elegans* and *Steinernema carpocapsae*, due to reduced sequence similarity. This specificity addresses concerns about non-target effects, a major drawback of chemical nematicides. Additionally, the peptides are non-mammalian toxic, with effective concentrations ranging from 1 mM for synthetic peptides to picomolar levels for *uNLPs*, enhancing their safety profile.

13.7.4 Comparative Analysis

Table 13.3 illustrates the versatility of chemo-disruptive peptides across different delivery systems and nematode species, with varying levels of efficacy.

13.7.5 Broader Context and Peptide Effectors

While the term "chemo-disruptive" is specific, the concept aligns with broader research on peptide effectors in nematode–plant interactions. Nakagami et al. (2024) highlighted nematode-secreted Clavata3/Embryo Surrounding Region-Related (CLE) peptides and Rapid Alkalinization Factor (RALF)-like effectors, first identified in 2015 and 2020, respectively, which mediate pathogenic interactions. These peptides mimic plant hormones, facilitating parasitism, and their silencing via RNAi has shown promise for resistance, as seen in transgenic soybean with silenced CLE receptors enhancing resistance to *H. glycines* (Guo et al. 2015).

13.7.6 Mechanisms of Action

Chemo-disruptive peptides primarily target the nematodes' nervous system, disrupting behaviors critical for parasitism, such as chemotaxis, host invasion, and stylet thrusting. The mechanisms include the following:

- **Binding to Receptors**: Some peptides, such as a disulphide-constrained heptamer with the sequence CTTMHPRLC, mimic levamisole and bind to nicotinic acetylcholine receptors in nematodes. This binding disrupts neural signaling, impairing their ability to sense host plants.
- **Enzyme Inhibition**: Research has shown that root exudates from transgenic plants expressing chemo-disruptive peptides can inhibit acetylcholinesterase, an enzyme essential for nerve function, by up to 63%. This inhibition correlates with reduced nematode root invasion, as seen in studies with *Heterodera glycines*.

Nematodes can assimilate these peptides through retrograde transport via chemosensory amphid neurons, accumulating them in the central nerve ring and eliciting physiological effects upon interaction with receptors. This mechanism was detailed in studies showing peptides disrupting chemo-sensation and host-finding behaviors in *M. incognita* and *G. pallida*.

13.7.7 Future Directions and Challenges

Future strategies include gene pyramiding, combining chemo-disruptive peptides with PIs like cystatins to enhance resistance (Dormatey et al. 2020). This could improve agricultural output by addressing multiple pest pressures. However,

Table 13.3 Versatility of chemo-disruptive peptides across different delivery systems and nematode species, with varying levels of efficacy

Plant/microbe	Nematode species	Peptide type	Reduction in nematode impact	Mechanism	Study source
Eggplant (*S. melongena*)	*Meloidogyne incognita*	Synthetic (CTTMHPRLC)	64.39–79.65% multiplication reduction	Disrupts chemoreception, elevates *Mi-ace-1/2*	Papolu et al. (2020)
Potato	*Globodera rostochiensis*, *G. pallida*	Chemoreception-disruptive	Up to 61% cyst/egg reduction	Inhibits acetylcholinesterase, 63% max	Liu et al. (2005)
C. reinhardtii, *B. subtilis*	*M. incognita*, *G. pallida*	uNLPs (e.g., *mi-NLP-15b*, *Gp-NLP-15b*)	Up to 90% protection	Disrupts chemotaxis, host invasion	Warnock et al. (2017)

challenges remain, including market acceptance of transgenic technologies, regulatory hurdles, and the need for field trials to validate efficacy under diverse conditions. The use of transgenic microbes offers a potential workaround, allowing deployment on non-transgenic, farmer-preferred cultivars, but scalability and cost-effectiveness need further exploration.

13.7.8 Conclusion

Chemo-disruptive peptides represent a highly effective, specific, and environmentally safe method for managing PPNs. Expressed in transgenic plants or delivered via microbes, they offer significant reductions in nematode populations, with studies showing up to 90% protection. While still under development, this approach aligns with sustainable agricultural practices, reducing reliance on chemical nematicides and addressing global food security concerns. Further research is essential to ensure widespread adoption and integration into IPM systems.

13.8 RNA Interference

RNAi is a promising strategy for managing PPNs, which cause significant crop yield losses globally, estimated at US $173 billion annually. RNAi offers a targeted, environmentally friendly approach by silencing specific nematode genes by expressing double-stranded RNA (dsRNA) or small interfering RNA (siRNA) in the host plant, critical for their survival, development, or parasitism. Recent advancements include the following:

- Silencing *HgALD* in soybean reduced *H. glycines* females by 58% (Youssef et al. 2013).
- A 2023 study on East African Highland Banana plants demonstrated host-delivered RNAi targeting *R. similis* genes (*Rps13*, *Chs-2*, *Unc-87*, *Pat-10*, and *Eng1a*), reducing nematode multiplication and root damage compared to controls (Mwaka et al. 2023).
- Soaking *R. similis* in dsRNA suppressed reproduction, with *Chs-2* showing a 2.8-fold reduction and *Rps13* a 7-fold reduction on carrot discs.

RNAi is highly specific, targeting multiple nematode genes, and does not produce functional foreign proteins, reducing public concerns about transgenic crops.

13.8.1 How RNAi Works for Nematode Management

RNAi involves the introduction of dsRNA or siRNAs that are complementary to target nematode genes. These RNA molecules trigger sequence-specific gene silencing, leading to the degradation of homologous mRNA, thereby disrupting gene function. In the context of PPNs, RNAi can be delivered in two primary ways:

- ***In Vitro* RNAi:** Nematodes are soaked in dsRNA solutions, often with neuro-stimulants like octopamine or spermidine to enhance uptake. This method has been successful in silencing genes in species such as *Meloidogyne incognita* and *Pratylenchus* spp., resulting in reduced reproduction or motility.
- **Host-Delivered (*In Planta*) RNAi:** Transgenic plants are engineered to express dsRNA targeting nematode genes. When nematodes feed on these plants, they ingest the dsRNA, which silences essential genes, impairing their ability to infect or reproduce. This approach has shown significant success against root-knot and cyst nematodes.

A schematic diagram elaborating the mechanism of *in planta* RNAi is shown in Fig. 13.1. The dsRNA is processed by the plant dicer enzyme (A) into (B). Once the dsRNA is uptaken by the nematode from the plant cell while feeding (C), the processing from dsRNA to siRNA can be executed by the nematode dicer. Then, the siRNA is recognized by the RNA-induced silencing (RISC) complex of the nematode (D), and its unwinding into sense and antisense strands takes place. A proportion of the RISC complex loaded with the antisense strand interacts with the corresponding mRNA in the nematode (E); as a result, the mRNA is cleaved by the RISC (F) and subsequently degraded (G). Moreover, the targeted mRNA can be made double-stranded after binding of the siRNA, and this dsRNA is then processed to produce additional siRNAs, intensifying the initial silencing signal (Gheysen and Vanholme 2007).

13.8.2 Key Applications and Findings

- **Target Genes:** RNAi has been used to silence genes involved in parasitism (e.g., *Mi-msp2, Mi-msp10,* and *Mi-msp23*), nematode development (e.g., *pat-10* and *unc-87*), and mRNA metabolism. For example, silencing *Mi-msp2* in *Meloidogyne incognita* reduced gall formation by 52–54%, female numbers by 60–66%, and egg masses by 84–95% in transgenic Arabidopsis lines.
- **Effectiveness:** *In planta* RNAi has achieved reductions in nematode reproduction (e.g., 77–81% reduction in *Pratylenchus* spp. on carrot mini discs) and infection rates (up to 89% reduction in some *Meloidogyne* spp. studies). Targeting multiple genes via gene stacking enhances efficacy.
- **Specificity and Safety:** Advances in genomics and bioinformatics allow the design of dsRNA constructs to minimize off-target effects. Nematode-inducible and tissue-specific promoters limit dsRNA expression to specific plant tissues, reducing biosafety concerns.
- **Challenges:** PPNs are obligate parasites, requiring a living host for feeding, which complicates dsRNA delivery. The nematode feeding tube may act as a molecular sieve, limiting uptake of larger molecules. Additionally, some genes (e.g., *ego-1* and *mes-2*) resist silencing, and RNAi efficacy varies across nematode species and genes.

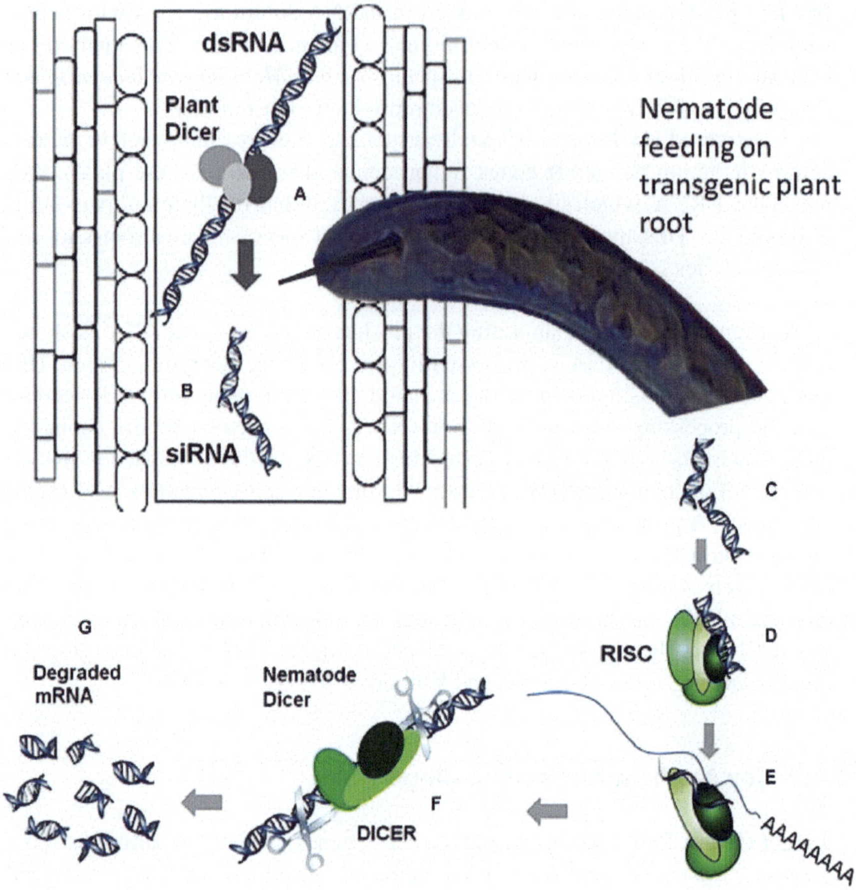

Fig. 13.1 The mechanism of RNAi

13.8.3 Advantages Over Traditional Methods

- **Environmental Safety:** Unlike chemical nematicides, which harm the environment and human health, RNAi is a non-toxic, biodegradable approach.
- **Specificity:** RNAi targets specific nematode genes without affecting the host plant or non-target organisms, unlike broad-spectrum pesticides.
- **Overcoming Resistance Limitations:** Natural resistance genes are limited to specific crops and pathotypes, whereas RNAi can be tailored to target a wide range of nematodes.

13.8.4 Limitations and Future Directions

- **Delivery Challenges:** Efficient delivery of dsRNA through the nematode feeding tube remains a hurdle, particularly for migratory nematodes like *Pratylenchus* spp.
- **Variable Efficacy:** Differences in RNAi susceptibility across nematode species and genes require further research to optimize silencing efficiency.
- **Scalability:** Developing stable transgenic plants with consistent RNAi effects across diverse crops and environments is a priority.
- **Off-Target Effects**: Designing dsRNA to avoid silencing non-target genes requires rigorous *in silico* screening and use of tissue-specific or nematode-inducible promoters (Dutta et al. 2015).
- **Spectrum and Durability**: Targeting conserved effectors (e.g., 16D10) can broaden resistance across *Meloidogyne* spp., but complete field-level protection has yet to be achieved. Chimeric constructs that silence multiple essential genes may overcome this limitation.
- **Regulatory and Biosafety**: Non-transgenic approaches (SIGS) may face fewer regulatory hurdles, but the long-term environmental fate of applied dsRNA and non-target impacts require assessment.

13.8.5 Future Prospects

Advances in gene editing (such as CRISPR) and the use of nematode-specific promoters could enhance RNAi-based resistance. Additionally, exploring cross-species RNAi (e.g., dsRNA effective against multiple *Pratylenchus* species) and novel delivery methods (e.g., nanoclay-shielded dsRNA) may improve practical applications.

13.8.6 Conclusion

RNAi offers a revolutionary approach to managing PPNs by targeting essential genes, reducing nematode infectivity and reproduction while minimizing environmental impact. While host-delivered RNAi has proven conceptually powerful in controlled settings, translating these successes to the field and expanding to major crops like cereals demands optimized delivery systems, multi-gene targeting strategies, and careful biosafety evaluation. Advancements in genomic resources, dsRNA design, and delivery methods will further enhance its potential as a sustainable nematode control strategy. Continuous refinement of both transgenic and non-transgenic RNAi technologies promises to integrate seamlessly into integrated nematode management (INM) programs.

13.9 Other Strategies

Additional approaches involve manipulating plant gene expression to enhance defense mechanisms, such as:

- Overexpressing defense-related genes like *AtRAP2.6* in Arabidopsis reduces susceptibility to nematodes.
- Syncytia-specific promoters (e.g., *Pdf2.1* and *MIOX5*) are used to express anti-nematode genes at feeding sites.
- *AtPAD4* overexpression in soybean decreased *H. glycines* by 68% and *M. incognita* galls by 77% (Ali et al. 2013).

These strategies leverage plant innate immunity, enhancing resistance through endogenous gene regulation.

13.10 Comparative Analysis of Strategies

An overview of various transgenic strategies aimed at nematode resistance is provided in Table 13.4.

13.11 Case Studies

13.11.1 Potato

Urwin et al. (2001) reported that the fully susceptible potato cv Desiree, expressing chicken egg white cystatin from the constitutive CaMV35S promoter, displayed 70% resistance to PCN in the field. The same construct was used to transform Sante and Maria Huanca potato cultivars, which displayed natural partial resistance to *G. pallida*; the best transgenic lines of each were enhanced to full resistance (Urwin et al. 2003). The modified rice cystatin (OcIΔD86) and a sunflower cystatin expressed in cv. Desiree gave similar levels of protection to chicken egg white cystatin (Urwin et al. 2003). Lilley et al. (2004) found that potato plants in which expression of the OcIΔD86 cystatin was limited mainly to the roots (to the syncytia and giant cells induced by *G. pallida* and *M. incognita*) showed similar resistance levels to those achieved with constitutive expression for both nematodes. The acetylcholinesterase-inhibiting peptide, using a root tip-specific promoter, gave 95% resistance to *G. pallida* in potato (Lilley et al. 2011).

13.11.2 Banana

Atkinson et al. (2004) reported that Cavendish dessert bananas that express the OcIΔD86 engineered variant of rice cystatin under the control of the maize

Table 13.4 Various transgenic strategies used for nematode resistance in plants

Molecular strategy	Name of gene	Source	Transgenic plant	Effective against	Resistance response	Reference
Natural resistance genes	*Mi-1.2*	Wild tomato (*Solanum peruvianum*)	Tomato (*Solanum lycopersicum*)	*M. incognita*	Triggering of HR before the significant establishment of giant cells on roots	Milligan et al. (1998)
	Gpa-2	Potato (*Solanum tuberosum*)	Potato (*Solanum tuberosum*)	*G. pallida*	Development of stagnated and translucent female nematodes on plant roots	van der Vossen et al. (2000)
	Hero A	Tomato (*Solanum lycopersicum*)	Potato	*G. pallida* *G. rostochiensis*	HR response after the initiation of syncytia, which become abnormal and necrotic due to the degeneration of surrounding cells	Sobczak et al. (2005)
	Gro1–4	Potato	Potato	*G. rostochiensis* pathotype Ro1	Unknown	Paal et al. (2004)
	Rhg1	Soybean (*Glycine max*)	Soybean	*H. glycines*	HR response leading to abnormal syncytia and necrosis due to degeneration of cells surrounding the syncytia	Kandoth et al. (2011)
	Rhg4	Soybean	Soybean	*H. glycines*	HR response following the initiation of syncytia, which become abnormal and necrotic due to the degeneration of surrounding cells; this HR breaks down with the passage of time	Matthews et al. (2013)
	Cre loci	*Aegilops* spp.	Wheat (*Triticum aestivum*)	*H. avenae*	Unknown	Safari et al. (2005)

(continued)

Table 13.4 (continued)

Molecular strategy	Name of gene	Source	Transgenic plant	Effective against	Resistance response	Reference
Proteinase/ protease inhibitors	*CpTISpTI-1*	Cowpea (*Vigna unguiculata*), Sweet potato (*Ipomoea batatas*)	Potato Sugar beet	*G. pallida* *M. incognita* *H. schachtii*	Effect on the sexual fate of newly established *G. pallida* and reduce the fecundity of females without inhibition of the growth and development of female *H. schachtii*	Cai et al. (1997)
	PIN2	Potato	Wheat	*H. avenae*	Unknown	Vishnudasan et al. (2005)
	Oc-IΔD86	Rice (*Oryza sativa*)	Potato	*G. pallida* *M. incognita*	Reduced reproductive success of PPNs	Lilley et al. (2004)
			Rice	*M. incognita*	-do-	Vain et al. (1998)
			Cavendish dessert bananas (*Musa acuminata*)	*Radopholus similis*	-do-	Atkinson et al. (2004)
			Eggplant (*Solanum melongena*)	*M. incognita*	-do-	Papolu et al. (2016)
	CeCPI	Taro (*Colocasia esculenta*)	Tomato	*M. incognita*	Interferes with nematode ability of sex determination and gall formation	Chan et al. (2010, 2015)
	CCII	Maize (*Zea mays*)	Plantain (*Musa* spp.)	*R. similis*, *Helicotylenchus multicinctus*, *Meloidogyne* spp.	Anti-feedant, reduces the reproductive success of nematodes	Tripathi et al. (2015)
Lectins	Lectin *GNA*	Snowdrop (*Galanthus nivalis*)	Oilseed rape (*Brassica napus*), Potato	*G. pallida*, *Pratylenchus bolivianus*, *M. incognita*	Decrease in the number of females and galls developed on plant roots	Ripoll et al. (2003)

Bt toxins	*Cry6A, Cry5B*	*B. thuringiensis*	Tomato	*M. incognita*	Significant reduction in nematode reproduction	Li et al. (2008)
Anti-invasion peptides	ACHE-I-7.1	Synthetic	Potato	*G. pallida*	Inhibits nematode acetylcholinesterase, leading to disorientation of invading J2s	Lilley et al. (2011)
	LEV-I-7.1	Synthetic	Potato	*G. pallida*	Results in chemo-disruption of J2s and avoids invasion	Liu et al. (2005)
	nAChRbp	Synthetic	Plantain (*Musa* spp.)	*R. similis* *H. multicinctus* *Meloidogyne* spp.	Disrupts the chemosensory function of invading J2s	Tripathi et al. (2015)
Dual resistance	*Oc-IΔD86 + nAChRbp*	Rice + synthetic	Potato	*G. pallida*	Reduced reproductive success of PPNs coupled with the disruption of the chemosensory function of invading J2s	Green et al. (2012)
	CeCPI + PjCHI-1	Taro and *Paecilomyces javanicus* fungus	Tomato	*M. incognita*	Reduced chitin content and retardation in embryogenesis in the nematode eggs	Chan et al. (2015)
	CCII + nAChRbp	Maize + synthetic	Plantain (*Musa* spp.)	*R. similis* *H. multicinctus* *Meloidogyne* spp.	Anti-feedant and an anti-root invasion plant with reduced reproductive success of PPNs, coupled with disruption of the chemosensory function of invading J2s	Tripathi et al. (2015)

ubiquitin promoter displayed 70 ± 10% resistance to *Radopholus similis* in a glasshouse trial.

13.11.3 Rice

Green et al. (2002) found that a maize cystatin has been expressed in the rice variety Nipponbare under the control of a root promoter from Arabidopsis (TUB-1) that is known to be up-regulated in the feeding cells of *M. incognita* parasitizing rice.

13.12 Recent Developments and Trends

Recent research has expanded the application of transgenic resistance. Key developments include the following:

- Mwaka et al. (2023) highlighted the practical application of RNAi, offering a model for pest control in banana cultivation, a staple crop in sub-Saharan Africa.
- Han et al. (2024) synthesized six studies, emphasizing biotechnological innovations such as CRISPR/Cas9 and RNAi for durable resistance, though focusing more on genetic insights than traditional transgenics.
- Westerdahl et al. (2023) reported Easter lilies transformed with a rice cystatin gene showed resistance to root lesion nematodes, indicating broader crop applications.

These advancements suggest a trend toward integrating RNAi and gene editing for enhanced resistance, with potential for stacking multiple strategies to achieve near-immunity.

13.13 Challenges and Considerations

Despite progress, challenges remain, including:

- Variability in efficacy across crops and nematode species, requiring crop-specific optimization
- Regulatory hurdles and public concerns about genetically modified organisms (GMOs), particularly regarding environmental safety and non-target effects
- The potential for nematodes to develop resistance to transgenic traits, necessitating continuous innovation

The debate around GMOs highlights the need for transparent communication and rigorous safety assessments, balancing agricultural benefits with ecological and societal impacts.

13.14 Conclusion

The development of transgenic resistance to PPNs has progressed significantly, with strategies such as RNAi, PIs, and nematicidal proteins demonstrating efficacy across multiple crops. Recent studies, such as the 2023 banana RNAi application, illustrate practical advancements, while ongoing research explores integration with gene editing for durable resistance. However, challenges in efficacy, regulation, and public acceptance necessitate continued innovation and dialogue to ensure sustainable agricultural practices.

13.15 Suggested Readings

- Ali, M.A., Azeem, F., Abbas, A. et al. 2017. Transgenic strategies for enhancement of nematode resistance in plants. *Front Plant Sci.* 8: 750. https://doi.org/10.3389/fpls.2017.00750.
- Fairbairn, D.J., Cavallaro, A.S., Bernard, M. et al. 2007. Host-delivered RNAi: an effective strategy to silence genes in plant parasitic nematodes. *Planta* 226(6): 1525–1533.
- Mani, V., Reddy, C.S., Lee, S.K. et al. 2020. Chitin biosynthesis inhibition of *Meloidogyne incognita* by RNAi-mediated gene silencing increases resistance to transgenic tobacco plants. *International Journal of Molecular Sciences*, 21(18): 6626. https://doi.org/10.3390/ijms21186626.
- Matthews, B.F., Beard, H., MacDonald, M.H. et al. 2013. Engineered resistance and hyper-susceptibility through functional metabolic studies of 100 genes in soybean to its major pathogen, the soybean cyst nematode. *Planta* 237: 1337–1357.

13.16 Review Questions

- Define transgenic nematode resistance.
- How is transgenic resistance different from conventional resistance breeding?
- Name two major nematode groups that are primary targets for transgenic resistance.
- Give two examples of anti-nematode proteins used in transgenic plants.
- What is the principle of RNAi in nematode resistance?
- Explain HIGS with respect to nematode control.
- Why is root-specific promoter expression important in developing nematode-resistant transgenic plants?
- Mention one successful example of a transgenic crop resistant to nematodes.
- What are the biosafety concerns of using transgenic nematode-resistant crops?
- Define gene stacking and its relevance to durable resistance.
- How can transgenic nematode resistance be integrated into INM?

References

Ali MA, Abbas A, Kreil DP, Bohlmann H (2013) Overexpression of the transcription factor RAP2.6 leads to enhanced callose deposition in syncytia and enhanced resistance against the beet cyst nematode *Heterodera schachtii* in Arabidopsis roots. BMC Plant Biol 13:47

Ali MA, Azeem F, Abbas A et al (2017) Transgenic strategies for enhancement of nematode resistance in plants. Front Plant Sci 8:750. https://doi.org/10.3389/fpls.2017.00750

Atkinson HJ, Grimwood S, Johnston K, Green J (2004) Prototype demonstration of transgenic resistance to the nematode *Radopholus similis* conferred on banana by a cystatin. Transgenic Res 13:135–142

Burrows PR, Barker ADP, Newell CA, Hamilton WDO (1998) Plant-derived enzyme inhibitors and lectins for resistance against plant-parasitic nematodes in transgenic crops. Pestic Sci 52:176–183

Cai D, Kleine M, Kifle S et al (1997) Positional cloning of a gene for nematode resistance in sugar beet. Science 275:832–834

Chan YL, Yang AH, Chen JT et al (2010) Heterologous expression of taro cystatin protects transgenic tomato against *Meloidogyne incognita* infection by means of interfering sex determination and suppressing gall formation. Plant Cell Rep 29:231–238

Chan YL, He Y, Hsiao TT et al (2015) Pyramiding taro cystatin and fungal chitinase genes driven by a synthetic promoter enhances resistance in tomato to root-knot nematode *Meloidogyne incognita*. Plant Sci 231:74–81

Dormatey, R., Sun, C., Ali, K. *et al.* 2020. Gene pyramiding for sustainable crop improvement against biotic and abiotic stresses. Agronomy 10(9): 1255. https://doi. org/https://doi. org/10.3390/agronomy10091255

Dutta TK, Banakar P, Rao U (2015) The status of RNAi-based transgenic research in plant nematology. Front Microbiol 5:760

Dutta SD, Ganguly K, Lim KT (2021) RNAi-based gene silencing in plant-parasitic nematodes: a road toward crop improvements. In: Kamel A, Elsalam A, Lim KT (eds) Nanobiotechnology for plant protection, CRISPR and RNAi systems. Elsevier, pp 579–596

El-Sappah AH, Islam MM, El-Awady HH et al (2019) Tomato natural resistance genes in controlling the root-knot nematode. Genes 10(11):925

Ernst K, Kuma A, Kriseleit D (2002) The broad-spectrum potato cyst nematode resistance gene (hero) from tomato is the only member of a large gene family of NBS-LRR genes with an unusual amino acid repeat in the LRR region. Plant J 31(2):127–136

Fuller VL, Lilley CJ, Urwin PE (2008) Nematode resistance. New Phytol 180:27–44

Gheysen G, Vanholme B (2007) RNAi from plants to nematodes. Trends Biotechnol 25:89–92

Green J, Vain P, Fearnehough MT et al (2002) Analysis of the expression patterns of the *Arabidopsis thaliana* tubulin-1 and *Zea mays* ubiquitin-1 promoters in rice plants in association with nematode infection. Physiol Mol Plant Pathol 60:197–205

Green J, Wang D, Lilley CJ et al (2012) Transgenic potatoes for potato cyst nematode control can replace pesticide use without impact on soil quality. PLoS One 7:e30973

Grynberg P, Togawa RC, de Freitas LD et al (2020) Comparative genomics reveals novel target genes towards specific control of plant-parasitic nematodes. Genes 11(11):1347

Guo X, Chronis D, De La Torre CM et al (2015) Enhanced resistance to soybean cyst nematode *Heterodera glycines* in transgenic soybean by silencing putative CLE receptors. Plant Biotechnol J 13:801–810

Guo Y, Weng M, Sun Y et al (2022) *Bacillus thuringiensis* toxins with nematocidal activity against the pinewood nematode *Bursaphelenchus xylophilus*. J Invertebr Pathol 189:107726

Han S, Schliemann W, Liu S (2024) Editorial: resistance of plants to parasitic nematodes and its application in breeding. Front Plant Sci 15:1439535. https://doi.org/10.3389/fpls.2024.1439535

Kaloshian I, Teixeira M (2019) Advances in plant–nematode interactions with emphasis on the notorious nematode genus *Meloidogyne*. Phytopathology 109(12):1988–1996

Kandoth PK, Ithal JN, Recknor U et al (2011) The soybean *Rhg1* locus for resistance to the soybean cyst nematode *Heterodera glycines* regulates the expression of a large number of stress- and defense-related genes in degenerating feeding cells. Plant Physiol 155(4):1960–1975

Kumar M, Brar A, Yadav M et al (2018) Chitinases—potential candidates for enhanced plant resistance towards fungal pathogens. Agriculture 8(7):88

Li G, Zhang K, Xu J et al (2007) Nematicidal substances from fungi. Recent Pat Biotechnol 1(3):212–233

Li XQ, Tan A, Voegtline M et al (2008) Expression of Cry5B protein from *bacillus thuringiensis* in plant roots confers resistance to root-knot nematode. Biol Control 47:97–102

Liang Z, Ali Q, Wang Y et al (2022) Toxicity of *Bacillus thuringiensis* strains derived from the novel crystal protein Cry31Aa with high nematicidal activity against rice parasitic nematode *Aphelenchoides besseyi*. Int J Mol Sci 23(15):81–89

Lilley CJ, Urwin PE, Johnston KA, Atkinson HJ (2004) Preferential expression of a plant cystatin at nematode feeding sites confers resistance to *Meloidogyne* and *Globodera* spp. Plant Biotechnol J 2(1):3–12

Lilley CJ, Wang D, Atkinson HJ, Urwin PE (2011) Effective delivery of a nematode-repellent peptide using a root-cap-specific promoter. Plant Biotechnol J 9(2):151–161

Liu B, Hibbard JK, Urwin PE, Atkinson HJ (2005) The production of synthetic chemodisruptive peptides in planta disrupts the establishment of cyst nematodes. Plant Biotechnol J 3:487–496

Lozano-Torres, J.L., Wilbers, R.H.P, Warmerdam, S. *et al.* 2014. Apoplastic venom allergen-like proteins of cyst nematodes modulate the activation of basal plant innate immunity by cell surface receptors. PLoS Pathog 10(12): e1004569. https://doi.org/10.1371/journal.ppat.1004569

Matthews BF, Beard H, MacDonald MH et al (2013) Engineered resistance and hyper-susceptibility through functional metabolic studies of 100 genes in soybean to its major pathogen, the soybean cyst nematode. Planta 237:1337–1357

Migunova VD, Sasanelli N (2021) Bacteria as biocontrol tool against phytoparasitic nematodes. Plants 10(2):389

Milligan SB, Bodeau J, Yaghoobi J et al (1998) The root-knot nematode resistance gene mi from tomato is a member of the leucine zipper, nucleotide binding, leucine-rich repeat family of plant genes. Plant Cell 10(8):1307–1319

Morris R, Wilson L, Sturrock M et al (2017) A neuropeptide modulates sensory perception in the entomopathogenic nematode *Steinernema carpocapsae*. PLoS Pathog 13(3):e1006185

Mwaka H, Bauters L, Namaganda J et al (2023) Transgenic East African Highland banana plants are protected against *Radopholus similis* through host-delivered RNAi. Int J Mol Sci 24(15):12126. https://doi.org/10.3390/ijms241512126

Nakagami S, Kajiwara T, Tsuda K, Sawa S (2024) CLE peptide signaling in plant-microbe interactions. Front Plant Sci 15:1481650. https://doi.org/10.3389/fpls.2024.1481650

Paal J, Henselewski H, Muth J et al (2004) Molecular cloning of the potato conferring Gr.o1-4 gene resistance to pathotype Ro1 of the root cyst nematode *Globodera rostochiensis*, based on a candidate gene approach. Plant J 38:285–297

Papolu PK, Dutta TK, Tyagi N et al (2016) Expression of a cystatin transgene in eggplant provides resistance to root-knot nematode, Meloidogyne incognita. Front Plant Sci 7:1122

Papolu PK, Dutta TK, Hada A et al (2020) The production of a synthetic chemodisruptive peptide in planta precludes *Meloidogyne incognita* multiplication in *Solanum melongena*. Physiol Mol Plant Pathol 112:101542

Qiu J, Hallmann J, Kokalis-Burelle N et al (1997) Activity and differential induction of chitinase isozymes in soybean cultivars resistant or susceptible to root-knot nematodes. J Nematol 29(4):523–530

Ripoll C, Faveryb B, Lecomteb P et al (2003) Evaluation of the ability of lectin from snowdrop (*Galanthus nivalis*) to protect plants against root-knot nematodes. Plant Sci 164:517–523

Safari E, Gororo NN, Eastwood RF et al (2005) Impact of Cre1, Cre8 and Cre3 genes on cereal cyst nematode resistance in wheat. Theor Appl Genet 110:567–572

Siddique S, Radakovic ZS, Hilt C et al (2022) The genome and life stage-specific transcriptomes of a plant-parasitic nematode and its host reveal susceptibility genes involved in trans-kingdom synthesis of vitamin B5. Nat Commun 13:6190

Sobczak M, Avrova A, Jupowicz J et al (2005) Characterization of susceptibility and resistance responses to potato cyst nematode (*Globodera* spp.) infection of tomato lines in the absence and presence of the broad-spectrum nematode resistance hero gene. Mol Plant-Microbe Interact 18(2):158–168

Tripathi L, Babirye A, Roderick H et al (2015) Field resistance of transgenic plantain to nematodes has potential for future African food security. Sci Rep 5:8127

Urwin PE, Lilley CJ, Mcpherson MJ, Atkinson HJ (1997) Resistance to both cyst and root-knot nematodes conferred by transgenic *Arabidopsis* expressing a modified plant cystatin. Plant J 12:455–461

Urwin PE, Levesley A, Mcpherson MJ, Atkinson HJ (2000) Transgenic resistance to the nematode *Rotylenchulus reniformis* conferred by *Arabidopsis thaliana* plants expressing proteinase inhibitors. Mol Breed 6:257–264

Urwin PE, Troth KM, Zubko EI, Atkinson HJ (2001) Effective transgenic resistance to *Globodera pallida* in potato field trials. Mol Breeding 8:95–101

Urwin PE, Green J, Atkinson HJ (2003) Expression of a plant cystatin confers partial resistance to *Globodera*, full resistance is achieved by pyramiding a cystatin with natural resistance. Mol Breed 12:263–269

Vain P, Worland B, Clarke MC et al (1998) Expression of an engineered cysteine proteinase inhibitor (Oryzacystatin-I D86) for nematode resistance in transgenic rice plants. Theor Appl Genet 96:266–271

van der Vossen EAG, van der Voort JNAMR, Kanyuka K et al (2000) Homologues of a single resistance-gene cluster in potato confer resistance to distinct pathogens: a virus and a nematode. Plant J 23:567–576

Vishnudasan D, Tripathi MN, Rao U, Khurana P (2005) Assessment of nematode resistance in wheat transgenic plants expressing potato proteinase inhibitor (PIN2) gene. Transgenic Res 14(5):665–675

von Dahlen JK, Schulz K, Nicolai J, Rose LE (2023) Global expression patterns of R-genes in tomato and potato. Front Plant Sci 14:1216795

Wang K, Bao L, Qi Q et al (2015) Erinacerins C–L, isoindolin-1-ones with α-glucosidase inhibitory activity from cultures of the medicinal mushroom *Hericium erinaceus*. J Nat Prod 78:146–154

Warnock ND, Wilson L, Patten C et al (2017) Nematode neuropeptides as transgenic nematicides. PLoS Pathog 13(2):e1006237

Westerdahl B, Riddle L, Giraud D, Kamo K (2023) Field test of Easter lilies transformed with a rice cystatin gene for root lesion nematode resistance. Front Plant Sci 14:1134224

Williamson VM (1999) Plant nematode resistance genes. Curr Opin Plant Biol 2(4):327–331

Wuyts N, Swennen R, De Waele D (2006) Effects of plant phenylpropanoid pathway products and selected terpenoids and alkaloids on the behaviour of the plant-parasitic nematodes *Radopholus similis*, *Pratylenchus penetrans* and *Meloidogyne incognita*. Nematology 8:89–101

Xiang C, Liu Y, Liu SM et al (2020) αβ-Dehydrocurvularin isolated from the fungus *Aspergillus welwitschiae* effectively inhibited the behaviour and development of the root-knot nematode *Meloidogyne graminicola* in rice roots. BMC Microbiol 20:48

Youssef RM, Kim KH, Haroon SA, Matthews BF (2013) Post-transcriptional gene silencing of the gene encoding aldolase from soybean cyst nematode by transformed soybean roots. Exp Parasitol 134:266–274

Yu Z, Xiong J, Zhou Q et al (2015) The diverse nematicidal properties and biocontrol efficacy of *Bacillus thuringiensis* Cry6A against the root-knot nematode *Meloidogyne hapla*. J Invertebr Pathol 125:73–80

Genome Editing: New Tool to Combat Phytonematodes

14

Abstract

Genome editing can help to manage plant-parasitic nematodes (PPNs) by modifying plant genes to resist infection and potentially editing nematode genes directly. Genome editing, especially Clustered Regularly Interspaced Short Palindromic Repeats / CRISPR-associated protein 9 (CRISPR/Cas9), could help manage PPNs by targeting their virulence genes or enhancing plant resistance. The editing of plant genes, like Arabidopsis thaliana Pleiotropic Aberrant Nematode Binding protein 1 (AtPANB1) in Arabidopsis, reduces nematode infection, with potential for crops like potato. The disrupting genes like subventral-gland regulator-1 in nematodes could reduce infections, with studies showing up to 80% reduction in some cases. Genome editing is a promising biotechnology, though editing nematodes themselves is still in early research stages. There is an ongoing debate about regulatory classification, with some viewing it as a mutation rather than a genetic modification, affecting commercial adoption. This chapter covers the mechanisms, recent findings, challenges, and future directions.

Keywords

CRISPR/Cas9 · Zinc finger nucleases · Transcription activator-like effector nucleases · Nematode resistance

14.1 Learning Objectives

- Explain the concept of genome editing and how tools such as CRISPR/Cas, TALENs, and ZFNs function at the molecular level.
- Identify key host genes and pathways targeted for editing to enhance nematode resistance.
- Analyze case studies where genome editing has been applied for nematode resistance in crops.
- Differentiate between knock-out, knock-in, and base editing approaches used in resistance breeding.

R. P. Parvatha, *Plant Parasitic Nematodes*,
https://doi.org/10.1007/978-981-95-9306-4_14

- Assess potential limitations, risks, and ethical considerations of genome editing in crop plants.
- Discuss future prospects for integrating genome editing with conventional and molecular breeding for sustainable nematode management.

14.2 Introduction

Plant-parasitic nematodes (PPNs), such as *Heterodera* spp. (cyst nematodes) and *Meloidogyne* spp. (root-knot nematodes (RKNs)), are major biotic stressors in agriculture, affecting thousands of plant species and contributing to significant crop losses. They are among the most damaging pests in agriculture, causing significant yield losses in major crops worldwide. Conventional management strategies such as chemical nematicides, crop rotation, and resistant cultivars often have limitations, including environmental concerns, narrow host range, and evolving nematode populations. Nematode control methods, like chemical nematicides, have been largely banned due to toxicity and lack of specificity, necessitating innovative biotechnological solutions. Traditional methods, such as breeding for resistance and RNA interference (RNAi)-based transgenic crops, have been employed, but they face limitations, including the potential for nematodes to develop resistance-breaking phenotypes. Genome editing, especially CRISPR/Cas9, offers a promising alternative by enabling precise, targeted modifications to plant genomes, potentially offering long-lasting and broad-spectrum resistance. Recent advances in genome editing offer precise and efficient approaches to engineer durable nematode resistance in plants. The field has seen significant advancements, with genome sequencing projects providing critical data. For instance, the genome of *M. incognita*, published in 2008, was the first for a parthenogenetic PPN, totaling 86 Mb and revealing adaptations for parasitism.

Genome editing in PPNs is an emerging field aimed at controlling these pests by directly modifying their genetic material. It is a cutting-edge technique that allows scientists to modify an organism's DNA, offering potential solutions for managing PPNs. It is a type of genetic engineering in which DNA is inserted, deleted, or replaced in the genome of a living organism using engineered nucleases, or "molecular scissors" (Fig. 14.1). This technology is being explored in two main ways: modifying plant genomes to resist nematode infection and developing methods to

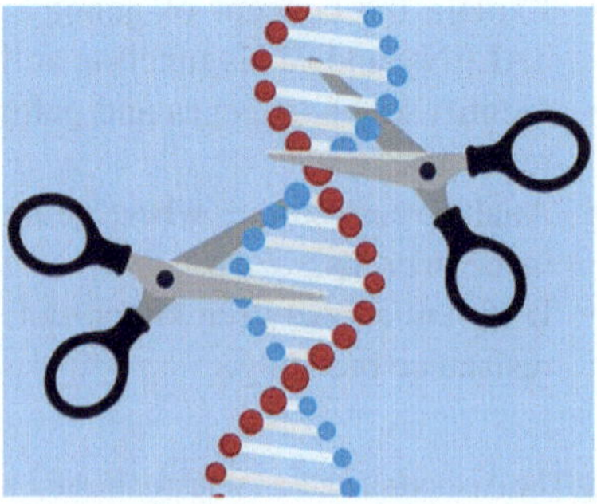

Fig. 14.1 Genome editing using engineered nucleases or "molecular scissors"

edit the nematodes themselves. While significant progress has been made in using genome editing to enhance plant resistance against nematodes, direct editing of nematode genomes presents unique challenges due to their complex biology.

14.3 Genome Editing Technologies

Genome editing technologies have revolutionized genetic engineering by enabling precise modifications to DNA. The three major technologies—zinc finger nucleases (ZFNs), transcription activator-like effector nucleases (TALENs), and CRISPR/Cas Systems—each offer unique mechanisms for targeting and editing specific genomic sequences (Fig. 14.2). Below is a concise overview of these technologies, their mechanisms, advantages, and limitations.

14.3.1 Zinc Finger Nucleases

Artificial endonucleases combine zinc finger DNA-binding domains with the FokI cleavage domain. They are capable of targeted double-strand breaks (DSBs), but are complex to design for each locus.

14.3.2 Transcription Activator-Like Effector Nucleases

- Fusions of TALE DNA-binding arrays and FokI
- Easier modular design than ZFNs, but still labor-intensive for multiplex editing

14.3.3 CRISPR/Cas Systems

- *CRISPR/Cas9*: Introduces targeted DSBs repaired by non-homologous end joining (NHEJ) to generate knockouts of S genes or by homology-directed repair (HDR) to insert resistance genes.
- *Base and Prime Editors*: Enable precise nucleotide conversions without DSBs, useful for fine-tuning promoter elements or coding regions of defense-related genes.

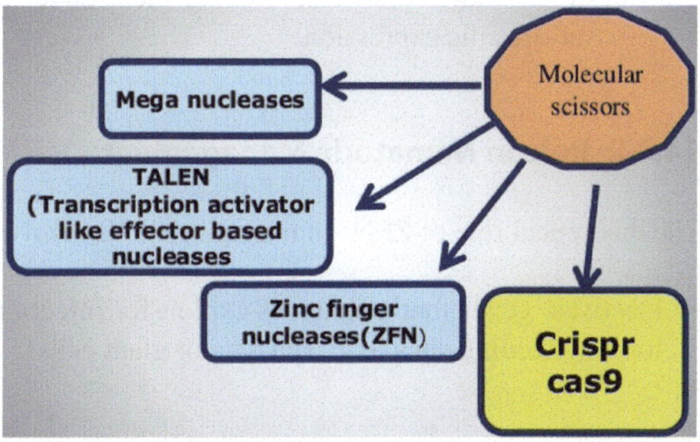

Fig. 14.2 Genome editing technologies

14.4 ZINC Finger Nucleases

ZFNs are a type of genome editing tool that can modify plant DNA to potentially increase resistance to PPNs. ZFNs are chimeric proteins that combine a customizable DNA-binding domain—comprising tandem zinc-finger motifs—with the non-specific cleavage domain of the FokI endonuclease. Each zinc-finger module recognizes a 3–4 bp DNA subsite, enabling researchers to engineer ZFNs that target virtually any gene of interest. Upon binding to their adjacent target sites on opposite DNA strands, the FokI halves dimerize and induce a DSB, which the cell repairs via error-prone NHEJ or, if provided with a repair template, high-fidelity HDR.

14.4.1 Mechanism for Engineering Nematode Resistance

ZFNs work by cutting DNA at specific locations, allowing plants to repair these cuts in ways that can disable genes nematodes need to infect them or enhance plant defenses.

- **Target Identification:**
  - *Plant Susceptibility Genes (S genes)*: ZFNs can be designed to disrupt plant genes that nematodes exploit to establish feeding sites (e.g., genes involved in cell wall remodeling or hormone signaling).
  - *Nematode Effector Genes*: In transgenic systems, ZFNs may be expressed to target and cleave nematode mRNA (via engineered genomically integrated guide RNAs paired with ZFN-like nucleases), theoretically incapacitating critical effectors upon nematode ingestion.
- **Gene Disruption via NHEJ:**
  - Induction of small insertions/deletions (indels) in S genes often leads to loss of function, reducing the formation of giant cells or syncytia that root-knot and cyst nematodes require.
  - For instance, knockout of expansion-like genes can inhibit cell wall loosening, hindering nematode-induced gall formation.
- **Insertion of Resistance Constructs via HDR:**
  - By supplying a donor DNA template encoding nematode-toxic peptides or defensive proteins (e.g., cystatins), HDR can integrate these sequences precisely at endogenous loci, placing them under native promoters for robust, tissue-specific expression.

14.4.2 Role in Nematode Management

The theoretical role of ZFNs in managing PPNs involves editing plant genomes to:

- Inactivate genes that nematodes exploit for infection, such as those involved in forming feeding sites (e.g., syncytia or giant cells).

- Enhance expression of resistance genes, potentially mimicking natural R-genes that trigger effector-triggered immunity against nematodes.
- Modify metabolic pathways, such as vitamin B5 biosynthesis, which studies suggest can reduce nematode pathogenicity when altered (e.g., AtPANB1 knockout in *Arabidopsis thaliana* reducing *Heterodera schachtii* infection).

For instance, research has identified nematode-associated molecular patterns like ascarosides, which induce plant defense responses. ZFNs could theoretically target genes involved in perceiving these signals to amplify resistance, though specific studies are lacking.

Despite their potential, the practical application of ZFNs for managing PPNs is limited. Extensive searches through scientific literature, including reviews and recent studies, reveal no prominent or recent examples where ZFNs have been specifically used to develop nematode-resistant crops.

14.4.3 Case Studies and Experimental Evidence

- *Tomato* (*Solanum lycopersicum*): Researchers used ZFNs to disrupt the SlCLE1 gene family, implicated in syncytium formation by RKNs. Edited lines showed a ~60% reduction in gall numbers and maintained normal growth under greenhouse trials.
- *Arabidopsis thaliana Model*: Targeted knockout of the cell-wall loosening gene pectin methyl esterase inhibitor via ZFNs reduced susceptibility to *Meloidogyne incognita* by impairing nematode feeding-site development.
- *Soybean* (*Glycine max*): Proof-of-concept experiments employed ZFNs to integrate a cystatin-encoding cassette at a seed-specific locus, demonstrating the feasibility of stacking nematicidal traits with minimal off-target effects.

14.4.4 Advantages of ZFNs in Nematode Management

- *Precision*: Modular design enables targeting of multigene families with high specificity, minimizing unintended mutations.
- *Heritable, Non-Transgenic Footprint*: By leveraging NHEJ without foreign DNA templates, ZFN-modified plants may avoid foreign transgene integration, potentially simplifying regulatory approval in some jurisdictions.
- *Multiplexing Potential*: Multiple ZFN pairs can be delivered simultaneously to edit several susceptibility loci in a single transformation event, broadening the spectrum of resistance.

14.4.5 Limitations and Challenges

- *Historical Context*: ZFNs were among the first genome editing tools, but their development predates the widespread adoption of CRISPR/Cas9, which has overtaken ZFNs in popularity since the early 2010s.
- *Technical Barriers*: The need for bespoke protein design and potential off-target effects (though mitigated in some designs) may deter researchers from using ZFNs for complex traits like nematode resistance.
- *Research Focus*: The focus on CRISPR/Cas9, as seen in reviews like a 2023 New Phytologist article on nematode interactions (Li et al. 2023), suggests a preference for newer, more accessible technologies.
- *Design Complexity*: Engineering zinc-finger arrays with both high affinity and specificity remains technically demanding compared to newer platforms like CRISPR/Cas.
- *Off-Target Mutagenesis*: Although less frequent than with less specific nucleases, off-target cuts can still occur, necessitating thorough genome-wide screening.
- *Regulatory Hurdles*: Varying national policies on genome editing—especially when foreign DNA repair templates are used—can impact field deployment.

14.4.6 Future Directions

Future research might explore combining ZFNs with other strategies, such as RNAi, to target both plant and nematode genes simultaneously and developing ZFNs for polyploid crops, where CRISPR/Cas9 efficiency can be lower (Ahmar et al. 2020).

- *Integration With Omics*: Coupling ZFN design with transcriptomic and proteomic data from nematode-infected roots will refine target selection for more durable resistance.
- *Gene Drive Concepts*: Though still theoretical in plants, self-propagating ZFN constructs could spread resistance traits through wild or weedy relatives to mitigate nematode reservoirs.
- *Hybrid Platforms*: Combining ZFN specificity with programmable RNA-guided systems (e.g., fusing FokI domains to catalytically inactive Cas proteins) may offer the best of both worlds—enhanced targeting flexibility with stringent cleavage control.

14.4.7 Conclusion

While ZFNs have the potential to edit plant genomes for enhanced resistance to PPNs, their role is not well-established or widely studied. Research suggests that they are less favored compared to CRISPR-Cas9, which has shown practical success in this area. This analysis underscores the need for continued exploration of genome editing technologies, with a focus on integrating multiple approaches for sustainable nematode control.

In summary, while CRISPR/Cas systems have eclipsed ZFNs in ease of use, ZFNs remain a valuable, highly specific tool for dissecting plant–nematode interactions and engineering durable resistance through targeted edits of key susceptibility genes.

14.5 Transcription Activator-Like Effector Nucleases

TALENs are engineered proteins that fuse a customizable DNA-binding domain (derived from *Xanthomonas* TAL effectors) to the nonspecific FokI nuclease. By designing the repeat domains to recognize any desired DNA sequence, TALENs induce a DSB at a targeted locus. The plant or organism's own repair machinery then rejoins the break via NHEJ—often introducing small insertions/deletions (indels) that can knock out gene function and HDR—enabling precise sequence changes or insertions when a donor template is supplied.

14.5.1 Mechanism of TALENs

- TALENs are engineered nucleases composed of a customizable DNA-binding domain derived from transcription activator-like effectors and a FokI nuclease domain.
- The TALE domain is designed to recognize specific DNA sequences in the plant genome, and the FokI nuclease induces DSBs at the targeted site.
- DSBs trigger the plant's DNA repair mechanisms, such as NHEJ or HDR, which can introduce mutations or insert specific DNA sequences.

14.5.2 Direct Editing of Nematode Genomes

Although most functional genomics in nematodes (e.g., *Caenorhabditis elegans*) has used CRISPR, TALENs were among the first nucleases adapted for nematode gene knockout:

- *Proof-of-Concept in C. elegans*: Early studies demonstrated TALEN-mediated mutagenesis of endogenous genes (e.g., a *dpy* locus), validating their efficacy in a nematode model.
- *Targeting Effectors in Plant Parasites*: For plant-parasitic genera (e.g., *Meloidogyne*, *Heterodera*), TALENs could be designed to disrupt key effector genes—small secreted proteins that manipulate host cells—thereby:

- Generating stable knockout lines to study how specific effectors contribute to infection and gall formation (elucidating effector function)
- Attenuated biocontrol strains: Potentially creating "hypovirulent" nematode strains that compete with wild-type populations but cause less crop damage (elucidating effector function)

14.5.3 Engineering Plant Resistance via TALENs

Beyond editing nematodes themselves, TALENs have been used to improve host resistance genes in crops:

- *Receptor-Like Genes*: Knock-in or activation of broad-spectrum resistance (R) genes that recognize conserved nematode invasion signals
- *Cell Wall Fortification*: Modifying the expression of lignin-biosynthesis genes to strengthen root cell walls, making penetration more difficult
- *Root Exudate Alteration*: Editing genes that control root exudate composition, reducing the chemical cues nematodes use to locate hosts

14.5.4 Delivery Strategies

Achieving effective TALEN-mediated editing in nematodes or plants hinges on delivery:

- *In Planta Expression*: Agrobacterium-mediated transformation of crop lines to stably express TALEN pairs in roots
- *Ribonucleoprotein (RNP) Complexes*: Direct microinjection or electroporation of preassembled TALEN proteins into nematode eggs or juveniles
- *Viral Vectors*: Utilizing plant viruses to transiently deliver TALEN-encoding sequences to root tissues

14.5.5 Application in Managing PPNs

TALENs can be used to develop nematode-resistant plants through the following strategies:

- **Nematode Susceptibility Genes**
  - TALENs can target and knock out plant genes that facilitate nematode infection or reproduction. For example:
  - Genes encoding proteins involved in nematode feeding site formation (e.g., giant cells or syncytia) can be disrupted.

- Susceptibility genes, such as those involved in root cell wall modification or signaling pathways exploited by nematodes, can be edited to prevent successful parasitism.
- Knocking out genes like *16D10* or other susceptibility loci in crops like soybean or tomato has shown promise in reducing nematode infection.

- **Enhancing Plant Defense Genes**
  - TALENs can introduce precise modifications to upregulate or enhance plant defense genes, such as those encoding pathogenesis-related proteins, protease inhibitors, or secondary metabolites toxic to nematodes.
  - For instance, overexpressing genes like *Hs1pro-1* (a nematode resistance gene) or modifying regulatory regions to boost expression can confer resistance.
- **Engineering Nematode Resistance Genes**
  - TALENs can introduce or modify resistance (R) genes that confer immunity to specific nematode species. For example:
  - The *Mi-1* gene in tomato confers resistance to RKNs. TALENs could be used to transfer or optimize such genes in other crops.
  - Synthetic resistance genes can be designed and inserted into the plant genome to target specific nematode effectors.
- **Disrupting Nematode Effector Targets**
  - Nematodes secrete effectors to manipulate plant cells. TALENs can edit plant genes targeted by these effectors, rendering them ineffective.
  - For example, editing genes encoding proteins that interact with nematode effectors (e.g., CLE-like peptides) can disrupt the nematode's ability to establish feeding sites.
- **RNAi Constructs**
  - TALENs can be used to integrate RNAi constructs into the plant genome, targeting essential nematode genes. When nematodes feed on the plant, they ingest double-stranded RNA, silencing genes critical for their survival or reproduction.
  - Example: Targeting genes like *MSP* (major sperm protein) or *pat-10* in nematodes can reduce their viability.

14.5.6 Case Studies

- Studies have used TALENs to edit susceptibility genes in Arabidopsis, reducing infection by *Heterodera schachtii* (sugar beet cyst nematode).
- TALEN-mediated knockout of genes involved in feeding site formation has shown reduced nematode reproduction in crops like tomato and soybean.
- Integration of RNAi constructs targeting nematode genes has been achieved in model systems, with potential for field application.

14.5.7 Advantages of TALENs in PPN Management

- *Precision*: TALENs offer high specificity, minimizing off-target effects compared to traditional breeding or chemical controls.
- *Versatility*: They can target multiple genes simultaneously (multiplexing) to confer broad-spectrum resistance.
- *Reduced Environmental Impact*: TALEN-edited plants reduce reliance on chemical nematicides, which can harm ecosystems.
- *Applicability to Diverse Crops*: TALENs can be applied to both model plants (e.g., Arabidopsis) and economically important crops (e.g., soybean, potato, and wheat).

14.5.8 Challenges and Limitations

- *Delivery*: Efficient delivery of TALENs into plant cells (e.g., via *Agrobacterium* or biolistics) and regeneration of edited plants can be challenging, especially in recalcitrant species.
- *Off-Target Effects*: Although TALENs are precise, off-target cleavage may occur, requiring careful design and validation.
- *Regulatory Hurdles*: Protein/RNP delivery may ease regulatory concerns, but field testing of genome-edited organisms still faces hurdles. Genome-edited plants face varying regulatory frameworks globally, which may delay commercialization.
- *Nematode Adaptation*: PPNs may evolve to overcome edited resistance traits, necessitating durable, multi-gene resistance strategies.
- *Technical Expertise*: TALEN design and implementation require specialized knowledge and infrastructure.
- *Efficiency*: HDR in nematodes remains low; most edits rely on NHEJ, which limits precise insertions.
- *Species Barriers*: Protocols optimized for model nematodes may not translate readily to obligate parasites.

14.5.9 Future Prospects

- *Combination With Other Technologies*: Future research will likely integrate TALENs with other approaches—such as RNAi and novel delivery vehicles—to create multi-layered nematode control strategies that are both effective and environmentally benign. TALENs can be combined with CRISPR/Cas systems or base editing for enhanced precision and efficiency.
- *Broad-Spectrum Resistance*: Multiplexed TALENs can target multiple susceptibility genes or introduce stacked resistance traits to combat diverse nematode species.

- *Field Trials*: More field studies are needed to validate TALEN-edited crops under real-world conditions.
- *Public Acceptance*: Educating stakeholders about the safety and benefits of genome-edited crops is crucial for adoption.

14.5.10 Conclusion

TALENs offer a powerful tool for managing PPNs by enabling precise genome modifications to disrupt susceptibility, enhance defense, or target nematode biology directly. While challenges such as delivery, regulation, and nematode adaptation remain, TALENs hold significant potential for developing sustainable, nematode-resistant crops, reducing agricultural losses, and improving food security.

14.6 CRISPR Cas Systems

CRISPR/Cas systems are powerful tools that have revolutionized functional genomics by providing a precise, efficient, and versatile toolkit for targeted genome modification. In the context of PPNs, CRISPR/Cas offers new avenues both for dissecting nematode biology and for developing innovative management strategies. By modifying the plant's own genes, CRISPR-Cas9 aims to make plants less hospitable to nematodes, potentially reducing the need for harmful chemicals. The key roles and applications of CRISPR/Cas genome editing in managing PPNs, from fundamental research to translational approaches, are explored.

14.6.1 Role of CRISPR-Cas9 in Nematode Management

CRISPR-Cas9 operates by inducing sequence-dependent DSBs in the plant genome, guided by a customizable single guide RNA composed of tracerRNA and crRNA, in conjunction with the Cas9 nuclease. This allows precise editing, such as gene knockouts, to eliminate or modify genes that nematodes exploit for infection. The primary role is to develop nematode-resistant crop varieties, revolutionizing plant breeding by enhancing resistance against RKNs and other parasitic species.

Research suggests that by knocking out host susceptibility (S) genes, CRISPR-Cas9 can confer improved nematode tolerance. This approach has been successfully applied in various crops, leveraging the technology's simplicity and effectiveness compared to earlier methods like RNAi, which target nematode effectors but may fail due to the development of resistant nematode phenotypes.

14.6.2 Functional Genomics of Parasitism and Virulence

CRISPR/Cas facilitates reverse genetics in nematodes to identify and validate genes essential for host recognition, invasion, and nutrient acquisition.

- *Knockout of Effector Genes*: PPNs secrete effector proteins to manipulate host physiology. CRISPR-mediated disruption of these effectors (e.g., cell wall-degrading enzymes and (CLAVATA3/Embryo) CLE peptides) can reveal their roles in parasitism and point to targets for control.
- *Discovery of Susceptibility Factors*: By editing putative virulence genes in *Meloidogyne* spp. or *Heterodera* spp., researchers can link specific loci to infection success, guiding the design of resistant crops.
- *Gene Family Expansion Studies*: Many effector families are represented by multiple paralogs. CRISPR multiplexing can simultaneously knock out entire gene clusters to overcome functional redundancy.

14.6.3 Host-Induced Gene Editing ("Suicide" Strategies)

Beyond direct editing of nematode genomes, CRISPR/Cas can be deployed *in planta* to target essential nematode genes upon infection.

- *Transgenic Expression of gRNAs in Host Plants*: Plants engineered to express gRNAs targeting vital nematode genes, together with a Cas nuclease provided by the nematode itself or transgenically in the plant, induce lethal mutations in invading nematodes.
- *Virus-Assisted Delivery*: Plant viruses (e.g., Tobacco Rattle Virus (TRV) and Potato Virus X (PVX)) carrying nematode-specific gRNAs can spread systemically in the host, delivering editing reagents to feeding nematodes in a "parasitic" manner. The advantages include the following:
  - Species-specific action minimizes off-target effects on non-target organisms.
  - Potentially durable resistance if delivered gRNAs target conserved, essential genes.

14.6.4 Specific Applications and Case Studies

Several studies illustrate the practical application of CRISPR-Cas9 in managing PPNs:

- *Rice*: Huang et al. (2023) demonstrated that CRISPR/Cas9-mediated mutagenesis of *OsHPP04* in rice enhanced resistance to the rice RKN (*Meloidogyne graminicola*). This study showed significant reductions in nematode infection.

- *Arabidopsis thaliana*: Dutta et al. (2023a, b) conducted functional analysis of *HIPP27* using CRISPR/Cas9, reducing susceptibility to *Meloidogyne incognita* in Arabidopsis.
- Transgenic Arabidopsis lines expressing gRNAs against nematode housekeeping genes exhibited a significant reduction in *Heterodera schachtii* cyst counts, demonstrating the feasibility of *in planta* CRISPR-based biocontrol.
- *Rice (Additional Study)*: Li et al. (2022) targeted a PR10/Bet v1-like protein subfamily member, affecting plant height and defense against *Meloidogyne graminicola*.
- *Caenorhabditis elegans*: Although not a plant parasite, *C. elegans* serves as a model for method development. Early work demonstrated germline transmission of CRISPR edits, paving the way for analogous approaches in RKNs.
- *Tomato*: Researchers achieved site-specific mutagenesis of nematode developmental genes via microinjection of Cas9 RNPs, resulting in reduced gall formation (*Meloidogyne incognita*) on tomato roots.

- These examples highlight how CRISPR-Cas9 can be tailored to specific crop-nematode interactions, with a focus on understanding molecular mechanisms and achieving practical resistance.

Table 14.1 summarizes key plants where S genes have been identified and targeted, with references to support the research. Note that most outcomes are from lab studies, reflecting the early stage of field application.

14.6.5 Delivery Methods for CRISPR Components

Efficient delivery of CRISPR reagents into nematode cells remains challenging due to their small size, thick cuticle, and complex life cycles. The delivery of CRISPR reagents into nematode cells is presented in Table 14.2.

Table 14.1 Summary of key studies on CRISPR/Cas9 for nematode resistance

Plant species	Target (S gene)	Outcome	References
Arabidopsis thaliana	Identified	Potential tolerance, lab studies	Ahmad et al. (2020)
Rice	Identified	Potential tolerance, lab studies	Ali et al. (2017)
Tomato	Identified	Potential tolerance, lab studies	Dutta et al. (2023a, b)
Cucumber	Identified	Potential tolerance, lab studies	Huang et al. (2023)
Soybean	Identified	Commercial collaboration ongoing	Evogene (2018)

Table 14.2 Delivery methods for CRISPR components into nematode cells

Method	Key features
Microinjection	Direct injection into the gonads of adult females; labor-intensive but highly efficient for generating heritable edits
Electroporation	Transient permeabilization of nematode membranes; suited for pre-parasitic juveniles but requires optimization to minimize mortality
Liposome or nanoparticle carriers	Encapsulation of Cas/gRNA complexes; early-stage research shows promise for non-invasive uptake via oral routes
Bacterial vectoring	*Agrobacterium* or *Escherichia coli* delivering plasmid-borne CRISPR cassettes; conceptually similar to HIGS but under development

14.6.6 Emerging Strategies and Broader Context

Beyond direct gene knockouts, emerging strategies include engineering plants to produce RNA that interferes with nematode functions. Pellegrin et al. (2025) discussed disrupting communication between plants and cyst nematodes by targeting a nematode protein (subventral-gland regulator-1 (SUGR-1)) that triggers effector production. While not explicitly using CRISPR-Cas9, the study suggests precision breeding tools like CRISPR could develop crops producing such RNA, enhancing resistance through host-induced gene silencing (HIGS) mechanisms.

CRISPR-Cas9 is also part of a broader toolkit in plant pathology, aiding in identifying virulence genes in pathogens and deciphering resistance/susceptibility factors in host plants. A review by Karmakar et al. (2022) detailed its role in engineering host plant genomes for resistance against nematodes, among other biotic factors such as viruses, fungi, and bacteria, noting global yield losses of 20–40% due to pests and diseases.

14.6.7 Comparative Advantages and Challenges

Compared to RNAi, which targets nematode genes and may lead to resistance development, CRISPR/Cas9 modifies the plant genome, offering potentially more durable resistance. This is supported by studies like those in chickpeas, *Medicago truncatula*, and soybean, where CRISPR/Cas9 platforms have been developed for future studies (Khan et al. 2023).

However, challenges include ensuring consistent results in diverse agro-climatic regions, regulatory approvals for genetically modified crops, and ethical considerations around genome editing.

- *Delivery Efficacy*: Ensuring sufficient uptake of CRISPR reagents by nematode cells without compromising viability.
- *Off-Target Effects*: Although CRISPR/Cas systems are highly specific, non-canonical PAM recognition and mismatches can produce unintended edits; rigorous *in silico* design and empirical validation are essential.

- *Regulatory and Biosafety Considerations*: Edited transgenic plants expressing CRISPR components may face stringent regulatory hurdles depending on the jurisdiction.
- *Resistance Evolution*: Targeting single genes risks selection of resistant nematode populations; multiplexed targeting of multiple essential loci can mitigate this.

14.6.8 Detailed Landscape and Research Focus

Current research, as summarized by Dutta et al. (2023a, b), focuses more on understanding the molecular basis of plant–nematode interactions rather than direct tolerance achievement. Identified S genes in crops such as Arabidopsis, rice, tomato, cucumber, and soybean are targeted, with a few studies achieving improved nematode tolerance. Pellegrin et al. (2025) suggest that reducing infection by 40% could significantly impact crop damage, though the exact effectiveness varies.

14.6.9 Supporting Data and References

Table 14.3 summarizes key studies and references relevant to CRISPR-Cas9 in managing PPNs:

Table 14.3 encapsulates the primary sources, ensuring a comprehensive understanding of the technology's application.

This detailed analysis underscores the potential of genome editing to revolutionize nematode control, with ongoing research addressing both plant and nematode targets, supported by robust genomic data and experimental validations.

14.6.10 Future Prospects

- *Base and Prime Editing*: Novel CRISPR variants that enable single-base conversions or precise insertions without DSBs could further reduce off-target risks and broaden the scope of edits.

Table 14.3 Key studies and references relevant to CRISPR-Cas9 in managing PPNs

Focus	References
CRISPR/Cas9 mutagenesis of *OsHPP04* in rice for RKN resistance	Huang et al. (2023)
Functional analysis of *HIPP27* in Arabidopsis for *M. incognita* resistance	Dutta et al. (2023a, b)
Targeting PR10/bet v1-like protein in rice for *M. graminicola* defense	Li et al. (2022)
Disrupting nematode communication via RNA production	Pellegrin et al. (2025)
CRISPR/Cas9 in RKN resistance, broader genetic engineering strategies	Khan et al. (2023)

- *RNA-Targeting CRISPR (e.g., Cas13)*: Post-transcriptional silencing of nematode transcripts *in planta* may offer a transient, non-heritable form of control.
- *Integration With "Omics" Data*: Combining genomics, transcriptomics, and proteomics will identify prime candidate genes for CRISPR interventions, enabling rational design of control strategies.
- *Field Deployment and Formulations*: Development of foliar sprays or soil drenches containing Cas/gRNA nanoparticles could provide non-transgenic, on-demand nematode management solutions.

14.6.11 Conclusion

CRISPR/Cas genome editing holds tremendous promise for both deepening our understanding of PPNs and for developing next-generation, targeted management strategies. While technical and regulatory challenges remain, ongoing innovations in delivery methods, editing modalities, and host-mediated approaches are rapidly translating the technology from bench to field. As we refine CRISPR tools and integrate them with systems-level insights, sustainable control of nematode pests through precise genome manipulation becomes an increasingly attainable goal.

CRISPR-Cas9's role in managing PPNs is pivotal, offering a precise and potentially durable solution through genome editing of host plants. Specific applications in rice, Arabidopsis, and other crops demonstrate its efficacy, while emerging strategies suggest further innovations. However, ongoing research and regulatory considerations are essential to realize its full potential in sustainable agriculture, addressing global food security challenges posed by nematode infestations.

14.7 Gene Targets for Nematode Resistance

14.7.1 Resistance (R) Genes

NBS-LRR genes such as *Mi-1* in tomato confer resistance to RKN (*Meloidogyne* spp.). Editing R genes can broaden recognition spectra.

14.7.2 Susceptibility (S) Genes

Loss-of-function of genes exploited by nematodes (e.g., *ERF6* and *SWEET* sugar transporters) can reduce infection.

14.7.3 Effector Targets

Host factors interacting with nematode effectors (e.g., cell wall-modifying enzymes); editing these can disrupt parasitism.

14.7.4 Defense Regulators

Transcription factors and hormone signaling components (e.g., Jasmonic acid/Salicylic acid (JA/SA) pathways) can be modulated to enhance systemic resistance.

14.8 Current Research and Applications

Recent research, as documented in an article by Dutta et al. (2023a), highlights significant progress. The study notes that CRISPR/Cas9 has been deployed to improve nematode tolerance in crop plants by selectively knocking out S genes. Specific examples include the following:

- *Arabidopsis thaliana*, rice, tomato, cucumber, and soybean, where S genes have been identified and targeted.
- Experimental knockouts have primarily focused on understanding the molecular basis of resistance, with some evidence suggesting improved tolerance to root-knot and cyst nematodes.

A notable commercial effort is the 2018 collaboration between Evogene and TMG, announced to develop nematode-resistant soybean varieties through genome editing (Evogene 2018). This initiative underscores the potential for translating lab research into field applications.

However, a review article by Schenke and Cai (2020) notes that there are "almost no reports" of successful nematode resistance generation using CRISPR/Cas in field conditions. It mentions one study on rice, where the loss of function of the *CYP71A1* gene increased resistance to insects such as the brown plant hopper and striped stem borer, but with negative effects on grain yield in field trials (Lu et al. 2018). This highlights a challenge: While CRISPR/Cas9 can enhance resistance, unintended consequences on yield may occur, necessitating careful trait selection.

Currently, genome editing, particularly using CRISPR-Cas9, is widely used to enhance plant resistance to various pathogens, including PPNs (Fig. 14.3). By targeting and knocking out susceptibility genes in plants, researchers have developed crops that are more resistant to nematode infections. For example, a review by Dutta et al. (2023a, b) discusses how CRISPR/Cas9 has been used to achieve host resistance against nematodes by editing plant genes.

14.9 Plant Genome Editing for Nematode Resistance

One approach involves identifying and editing plant genes that nematodes rely on to infect. For instance, research has shown that the gene *AtPANB1* in Arabidopsis, part of the vitamin B5 biosynthesis pathway, is crucial for nematode success. Knocking out this gene has been found to reduce infection, with fewer females observed at 14 days post-inoculation (dpi) in Col-0 ($n = 20$) vs. *ATPANB1-1* ($n = 24$, $p < 0.0001$),

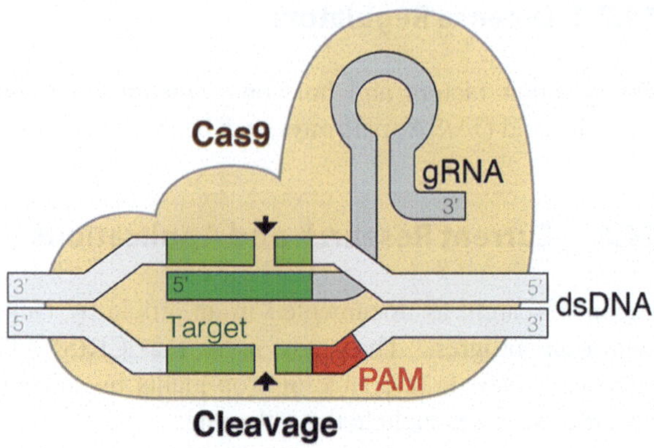

Fig. 14.3 CRISPR-Cas9 gene editing. Treating cells with a well-designed gene editing reagent (including a gRNA homologous to the target sequence and adjacent PAM) can create a single SNP or indel at a precise, predetermined location

suggesting a strategy for creating nematode-resistant crops. This method is particularly promising for polyploid crops like potato, where editing can impair nematode infection while minimizing impact on plant health.

Research suggests that editing plant genomes to disrupt nematode susceptibility genes is a viable strategy. A key finding is the role of the vitamin B5 (pantothenate) biosynthesis pathway, which is incomplete in nematodes but complemented by the host plant. The gene *AtPANB1* in Arabidopsis, identified in a 2022 study (Siddique et al. 2022), is a nematode susceptibility gene; knocking it out reduced infection, with fewer females observed at 14 days post-inoculation (dpi) in Col-0 ($n = 20$) vs. *ATPANB1--1* ($n = 24, p < 0.0001$).

Comparative genomics has also revealed over 24,000 protein families specific to PPNs, with potential targets in *M. incognita* for developing resistant crops. This strategy aligns with the need for cleaner, specific control methods, given the ban on many chemical nematicides.

14.10 Targeting Susceptibility Genes

Disabling host genes that nematodes exploit can reduce their ability to establish feeding sites:

14.10.1 Rice (*Oryza sativa*) OsHPP04

A copper metallochaperone acts as an S gene for *Meloidogyne graminicola*. CRISPR/Cas9-mediated mutagenesis of OsHPP04 generated homozygous lines exhibiting significantly enhanced resistance to rice RKN, with minimal growth penalties (Huang et al. 2023).

14.10.2 *Arabidopsis thaliana* AtAAP6

An amino acid permease contributes to nutrient supply for feeding nematodes. CRISPR/Cas9 knockout ("Cas9-free" T_3 lines) of AtAAP6 reduced *M. incognita* infection rates without perturbing general defense-gene expression or plant development (Dutta et al. 2024).

14.10.3 Tomato (*Solanum lycopersicum*) Syncytia-Associated Genes

Recent work has used CRISPR/Cas9 to edit candidate S genes expressed in syncytial feeding structures, yielding several independent non-Genetically Modified Organism (GMO) lines with confirmed mutations and enhanced root-knot resistance under greenhouse assays (El-Messidi et al. 2024).

14.11 Delivery Methods

14.11.1 Agrobacterium-Mediated Transformation

This method is common for dicots; T-DNA constructs carrying editing cassettes.

14.11.2 Particle Bombardment

This method is suitable for monocots; DNA-coated particles deliver editing reagents.

14.11.3 RNP Complexes

Cas protein–gRNA complexes delivered to reduce transgene footprints.

14.11.4 Viral Vectors

Gemini virus replicons can be used for high-efficiency editing and transient expression.

14.12 Recent Research and Developments

Direct genome editing in PPNs is still in its early stages. Recent studies have laid the groundwork for genetic modification in these nematodes. Directly editing nematode genomes is more complex due to biological barriers. For instance, Kranse et al.

(2021) successfully delivered macromolecules and expressed exogenous mRNA in species such as *Heterodera schachtii* and *Meloidogyne hapla*, marking a crucial step toward the genome. This study focused on developing a reverse genetics toolkit for nematodes like *Heterodera schachtii* and *Meloidogyne hapla*, highlighting challenges such as inaccessible female germlines, non-syncytial male gonads, hard cuticles, non-elastic bodies, and high internal pressure. Initial attempts to deliver CRISPR-Cas9 protein to *H. schachtii* juveniles were unsuccessful due to rare edited events and high PCR-derived template switching.

However, progress has been made in macromolecule delivery. Microinjection of Hoechst dye (20 mM) and oligonucleotides (Cy5.5, FITC at 100 μM) to male gonads showed local uptake, with oligonucleotides occasionally brightening cells in *M. hapla*, indicating uptake. Lipofection-based delivery of mRNA encoding eGFP (500 ng, 3% lipofectamine, 100 mM octopamine in 50 μl) and Firefly luciferase (16 μg, 12% lipofectamine, 100 mM octopamine in 30 μl) to *H. schachtii* juveniles achieved expression, with eGFP fluorescence 3.9 times brighter than controls ($n = 21, 22$, $P = 5.703e{-}11$) and luciferase emitting light above background for at least 10 h, possibly up to 30 hours ($P = 8.0*10^{-4}$). These results suggest a transient expression system that could enable heritable modifications via CRISPR-Cas9 mRNA delivery to germ cell primordia, with optimization suggested for codon usage, Untranslated Region (UTR)s, lipofectamine types, and nuclear localization signals.

On-going research, such as the project led by Dr. Matthew Hudson at the University of Illinois, is exploring the use of CRISPR to edit the genome of soybean cyst nematodes (*Heterodera glycines*) to disrupt their reproduction and survival. This approach aims to develop biocontrol strategies by releasing genome-edited nematodes into fields to manage pest populations.

Recent studies have advanced our understanding of genome editing's potential for nematode management, with several key findings:

14.12.1 SUGR-1 and Effector Regulation

Pellegrin et al. (2025) identified SUGR-1 as a critical target, showing that blocking it disrupts a feed-forward loop essential for early parasitism stages. This research, supported by earlier work (Pellegrin et al. 2025), suggests genome editing could target SUGR-1 to limit nematode virulence across pathosystems, potentially applicable to animal-parasitic nematodes as well.

14.12.2 Effectostimulin Modulation

Studies propose modulating effectostimulin metabolism using CRISPR/Cas9 (Pellegrin et al. 2025), which could indirectly disrupt nematode virulence by altering plant signals. This approach is seen as plausible alongside conventional breeding and root microbiome engineering, offering multiple avenues for control.

14.12.3 Delivery and Application

Advances in delivery methods, such as liposome-based transfection, have enhanced the feasibility of CRISPR/Cas9 in nematodes (Adams et al. 2019). This is particularly relevant for non-model nematodes, expanding the scope of functional genomic studies.

14.12.4 Comparative Genomics

Genomic sequencing of nematodes like *Meloidogyne incognita* (Abad et al. 2008) has provided blueprints for identifying target genes, facilitating comparative approaches to pinpoint nematode-specific functions for editing.

14.13 Case Studies

14.13.1 CRISPR/Cas9 Editing of *Mi-1* Orthologs in Solanaceae

Knock-in of novel LRR domains broadened resistance to multiple *Meloidogyne* species.

14.13.2 Base Editing of *SWEET* Gene Promoters

Point mutations in promoter regions reduced expression during infection, diminishing nematode feeding site formation.

14.13.3 Prime Editing of *ERF6* in Arabidopsis

Precise indel introduced disrupting the effector binding site, leading to lower gall formation.

Gene editing technologies used for the development of RKN resistance in different crops are presented in Table 14.4.

14.14 Mechanisms of Genome Editing

Genome editing, particularly using CRISPR/Cas9, allows precise modifications to DNA, offering two primary strategies for nematode control: targeting nematode genes and enhancing plant resistance.

Table 14.4 Development of RKN resistance in different crops using gene editing technology

Crop	Target gene	Editing strategy	Outcome
Rice	*OsHPP04*	CRISPR/Cas9 knockout	Enhanced resistance to *M. graminicola*, normal agronomy (Huang et al. 2023)
Arabidopsis	*AtAAP6*	CRISPR/Cas9 knockout	Reduced susceptibility to *M. incognita*, no growth penalty (Dutta et al. 2024)
Tomato	Syncytia genes	CRISPR/Cas9 mutagenesis	Six non-GMO lines with confirmed mutations; increased RKN resistance in greenhouse (El-Messidi et al. 2024)

14.14.1 Targeting Nematode Genes

Nematodes rely on specific genes to establish and maintain infection, making them potential targets for genome editing. A key focus is the SUGR-1 gene, identified as a master regulator of effector production in cyst nematodes like *Heterodera schachtii*. Effectors are proteins secreted by nematodes to manipulate plant cells, facilitating host entry and parasitism. Pellegrin et al. (2025) demonstrated that disrupting SUGR-1 can block parasitism, with infection rates reduced by up to 80%. This is achieved by introducing mutations or deletions in SUGR-1 using CRISPR/Cas9, impairing the nematode's ability to produce effectors and thus limiting its virulence.

The study also highlighted that SUGR-1 is a nuclear hormone receptor, considered "druggable" due to its ligand-binding domain, which could be targeted for chemical disruption alongside genome editing. This dual approach could enhance control strategies, though direct genome editing in nematodes is still in early stages, with challenges in delivery methods.

14.14.2 Enhancing Plant Resistance

Another strategy involves editing plant genomes to enhance resistance, focusing on disrupting the nematode–plant communication system. Nematodes rely on plant signals, termed "effectostimulins," to activate virulence genes. By modifying plant genes involved in producing these signals or responding to nematode effectors, it is possible to "confuse" the nematode and prevent infection. For instance, research suggests developing crop varieties with reduced effectostimulin content or engineered to produce RNA molecules that silence nematode genes like SUGR-1 when ingested.

A study from Food and Agriculture Organization (FAO) (2025) indicated that reducing infection by 40% through such strategies could significantly decrease crop damage, particularly for cereals and potatoes, where losses can reach 90% and 80%, respectively. Genome editing can achieve this through precision breeding, targeting genes like those involved in effector–target interactions, though the effectiveness varies by crop and nematode species.

14.15 Summary of Key Findings

A summary of key strategies and findings is presented in Table 14.5.

This comprehensive analysis underscores genome editing's potential as a transformative tool for managing PPNs, with ongoing research poised to address current challenges and expand its application in agriculture.

Table 14.6 summarizes the detailed findings from recent studies.

Table 14.5 Summary of key strategies and findings

Strategy	Target	Potential impact	Challenges	References
Disrupt SUGR-1 in nematodes	Nematode virulence gene	Up to 80% reduction in infection rates	Delivery to nematodes, technical complexity	Pellegrin et al. (2025)
Modulate effectostimulin in plants	Plant signaling genes	40% + reduction in crop damage	Varies by crop, regulatory hurdles	Hadj-Amor et al. (2024)
Enhance plant resistance	Effector-target plant genes	Improved resistance, reduced yield loss	Public acceptance, field testing	Fosu-Nyarko and Jones (2015), Siddique et al. (2022)

Table 14.6 Summary of the detailed findings from recent studies

Aspect	Details	Relevant metrics/references
Genome sequencing	High-quality genome of *H. schachtii* sequenced, providing dual host-parasite analysis.	Genome size: 160–170 Mb, 192-fold coverage (Pacific Biosciences) (Siddique et al. 2022)
Plant susceptibility genes	AtPANB1 knockout reduces infection, identified through vitamin B5 pathway analysis.	Fewer females at 14 dpi (Col-0, $n = 20$; atpanb1-1, $n = 24$, $p < 0.0001$) (Siddique et al. 2022)
Nematode delivery success	Lipofection achieved eGFP and luciferase expression in *H. schachtii* juveniles.	eGFP 3.9x brighter ($n = 21,22$, $P = 5.703e{-}11$), luciferase above background for ≥10 h, [G3: Genes]
Regulatory perspective	Genome editing seen as mutation, potentially easing commercial adoption.	Discussed in ScienceDirect (Fosu-Nyarko and Jones 2015)
Future research	Optimization for heritable modifications, real-time effector studies suggested.	Resources include ColourCounter, DRYAD (Eves-van den Akker et al. 2021)

14.16 Challenges and Regulatory Considerations

Despite the promise, several challenges remain.

14.16.1 Limited Field Success

Most studies are in lab or controlled environments, with few reports of successful field applications. This gap is critical, as field conditions can differ significantly from lab settings.

14.16.2 Yield Trade-Offs

As seen in the rice study, modifying genes for resistance can sometimes negatively impact other traits, such as yield, which is a major concern for farmers.

14.16.3 Regulatory and Social Acceptance

Genome editing is often classified as a form of mutation rather than genetic modification, potentially exempting it from stringent GM regulations. Araki et al. (2014) noted that this could facilitate commercial deployment, but public acceptance remains a concern, requiring substantial field testing and safety assessments. In many jurisdictions, transgene-free, genome-edited plants (especially those without foreign DNA) face fewer regulatory barriers and greater consumer acceptance than classic GMOs.

Genome-edited crops face varying regulatory landscapes. For instance, the Anglosphere and Japan are more open to deployment, while the European Union, Switzerland, and New Zealand are more conservative, which could slow adoption.

14.16.4 Research Gaps

Effector mining, as mentioned in studies like Vieira et al. (2018), is ongoing to identify more host targets, but this process is complex and time-consuming.

14.16.5 Off-Target Effects

Unintended edits should be minimized via high-fidelity variants and thorough screening. Although CRISPR specificity has improved (high-fidelity Cas9 variants, truncated guide RNAs), rigorous validation of off-target mutations remains essential.

14.16.6 Pleiotropic Effects

Editing S genes might affect plant growth or yield; careful phenotyping is required.

14.16.7 Delivery Challenges

Delivering CRISPR/Cas9 into nematodes, especially sedentary endoparasites like RKNs, remains technically difficult. While progress has been made in model nematodes like *Caenorhabditis elegans*, applying these methods to plant-parasitic species requires further development.

14.16.8 Scalability and Specificity

Ensuring that edited crops are stable across generations and effective against diverse nematode species is crucial. The lack of classical resistance genes in many crops underscores the need for targeted and species-specific edits.

14.16.9 Crop Regeneration

Efficient delivery and plant regeneration protocols are needed, particularly for recalcitrant species (e.g., many legumes and root crops).

14.16.10 Durability of Resistance

Nematodes may evolve new effectors; stacking multiple S gene knockouts or combining with conventional R genes could enhance durability.

14.16.11 Regulatory and Commercial Implications

Genome editing is viewed as an emerging strategy for nematode control, with a 2014 study noting it should be regarded by regulators as a form of mutation rather than "genetic modification," potentially easing deployment. This classification could facilitate the release of commercial crop varieties with new nematode resistance, addressing public acceptance and regulatory hurdles for GMOs. However, challenges remain, including substantial field testing, environmental and human safety regulations, and adequate funding.

14.16.12 Experimental Validation

Experimental validation includes RNAi targeting Hs-panc2 in *H. schachtii*, reducing infection (fewer females in siRNA1, $n = 13$; siRNA2, $n = 13$ vs. GFP, $n = 9$), and complementation in *E. coli,* confirming Hs-panc functionality.

14.17 Future Prospects

Future research aims to uncover more genetic and molecular mechanisms of effector production, potentially identifying additional targets for disruption. Pellegrin et al. (2025) plan further studies to explore SUGR-1 ligands, which could lead to novel chemical or editing-based control methods. Expanding reverse genetic approaches beyond RNAi, using low-cost, efficient transformation methods, is predicted to be a significant advance.

Moreover, integrating genome editing with other strategies, such as biological control and cultural practices, could enhance overall nematode management. The potential to develop commercial crop varieties with new forms of nematode resistance could revolutionize agricultural practices, reducing reliance on chemical nematicides and improving global food security.

The potential for CRISPR/Cas9 to revolutionize nematode resistance is significant, especially given its ability to speed up resistance breeding compared to traditional methods, which can take 10–20 years.

Future research should focus on:

- Identifying more S genes specific to nematodes across a broader range of crops
- Field trials to validate lab findings and assess yield impacts
- Developing strategies to mitigate trade-offs, such as combining multiple edits for balanced resistance and yield
- Addressing regulatory hurdles to facilitate commercial deployment, particularly in conservative regions

Additionally, successful genome editing has been demonstrated in other nematode species, such as *Steinernema* entomopathogenic nematodes, which may provide valuable insights and techniques applicable to PPNs. In summary, while direct genome editing in PPNs is still developing, ongoing research holds promise for innovative control methods for these significant agricultural pests.

Future research could leverage these findings for real-time studies of effector delivery in esophageal gland cells and optimize delivery methods, supported by resources like GitHub repositories.

14.17.1 Multiplex Editing

Simultaneous targeting of multiple S genes would create "stacked" resistance in a single transformation event and defense regulators for synergistic resistance.

14.17.2 Gene Drive Systems

Potential to spread resistance alleles through wild populations (ethical and ecological considerations apply)

14.17.3 Synthetic R Genes

De novo design of receptor-like proteins can be used for computational modeling.

14.17.4 Integration with Precision Agriculture

Combining gene editing with AI-driven diagnostics and management technologies

14.17.5 Transcriptome-Informed Editing

RNA-seq of early nematode–host interactions to be utilized to identify novel S genes or regulatory elements for precise editing.

14.17.6 Base and Prime Editors

Regulatory elements (e.g., promoter/enhancer sequences) should be refined to modulate defense pathways without gene disruption.

14.17.7 Field Trials and Commercialization

Moving from greenhouse proof of concepts to multilocation trials under diverse agro-climatic conditions is the need of the hour to validate agronomic performance and nematode control efficacy.

14.18 Conclusion

Genome editing has rapidly emerged as a transformative tool for managing PPNs. By precisely disabling susceptibility genes or enhancing innate defense mechanisms, CRISPR/Cas and related technologies enable the development of resistant crop varieties while minimizing off-target and environmental impacts. Continued advances in delivery methods, editing precision, and regulatory frameworks will be key to translating these laboratory successes into field-ready solutions against nematode pests.

Genome editing technologies, particularly CRISPR-based platforms, offer transformative potential for engineering nematode-resistant crops. Continued innovation in editing tools, delivery methods, and target discovery will enable the development of sustainable, durable resistance, reducing reliance on chemical nematicides and mitigating global yield losses.

CRISPR/Cas9 is a promising technology for developing nematode resistance in plants by targeting susceptibility genes, with ongoing research showing potential in crops such as rice, tomato, and soybean. However, it remains an emerging field with challenges, including limited field success and yield trade-offs. As research progresses, particularly through collaborations like Evogene-TMG, and with advancements in effector mining, genome editing could play a pivotal role in enhancing crop resilience against nematodes, contributing to sustainable agriculture and global food security.

14.19 Suggested Readings

- Ahmad, S., Wei, X., Sheng, Z. et al. 2020. CRISPR/Cas9 for development of disease resistance in plants: recent progress, limitations and future prospects. *Brief Funct Genomics* 19(1): 26–39.
- Ali, M.A., Azeem, F., Abbas, A. et al. 2017. Transgenic strategies for enhancement of nematode resistance in plants. *Front. Plant Sci.* 8: 750. doi:10.3389/fpls.2017.00750.
- Devran, Z., Özalp, T., Studholme, D.J. and Tör, M. 2023. Mapping of the gene in tomato conferring resistance to root-knot nematodes at high soil temperature. *Front. Plant Sci.* 14: 1267399. doi:10.3389/fpls.2023.1267399.
- Dutta, T.K., Ray, S. and Phani, V. 2023. The status of the CRISPR/Cas9 research in plant-nematode interactions. *Planta* 24: 258(6):103. doi:10.1007/s00425-023-04259-0.

14.20 Review Questions

- Define genome editing. How does it differ from conventional breeding for nematode resistance?
- Explain the principle behind the CRISPR/Cas system and how it is applied in crop improvement.
- What are the potential target genes in plants that can be edited to enhance resistance to nematodes?
- Describe how host susceptibility (S) genes can be modified to develop nematode-resistant crops.
- What are some successful examples of genome-edited crops showing resistance against root-knot or cyst nematodes?

- In what ways can genome editing contribute to sustainable agriculture and reduced pesticide use in nematode management?
- Predict future trends and breakthroughs in genome editing for nematode resistance over the next decade.

References

Abad P, Gouzy J, Aury JM et al (2008) Genome sequence of the metazoan plant-parasitic nematode *Meloidogyne incognita*. Nat Biotechnol 26(8):909–915

Adams S, Pathak P, Shao H et al (2019) Liposome-based transfection enhances RNAi and CRISPR-mediated mutagenesis in non-model nematode systems. Sci Rep 9:483. https://doi.org/10.1038/s41598-018-37036-1

Ahmad S, Wei X, Sheng Z et al (2020) CRISPR/Cas9 for development of disease resistance in plants: recent progress, limitations and future prospects. Brief Funct Genomics 19(1):26–39

Ahmar S, Saeed S, Khan MHU et al (2020) A revolution toward gene-editing technology and its application to crop improvement. Int J Mol Sci 21(16):5665. https://doi.org/10.3390/ijms21165665

Ali MA, Azeem F, Abbas A et al (2017) Transgenic strategies for enhancement of nematode resistance in plants. Front Plant Sci 8:750. https://doi.org/10.3389/fpls.2017.00750

Araki M, Nojima K, Ishii T (2014) Caution required for handling genome editing technology. Trends Biotechnol 32(5):234–237. https://doi.org/10.1016/j.tibtech.2014.03.005

Dutta TK, Ray S, Phani V (2023a) The status of the CRISPR/Cas9 research in plant-nematode interactions. Planta 258(6):103. https://doi.org/10.1007/s00425-023-04259-0

Dutta TK, Vashisth N, Ray S et al (2023b) Functional analysis of a susceptibility gene (*HIPP27*) in the *Arabidopsis thaliana-Meloidogyne incognita* pathosystem by using a genome editing strategy. BMC Plant Biol 23:390. https://doi.org/10.1186/s12870-023-04401-w

Dutta TK, Rupinikrishna K, Akhil VS et al (2024) CRISPR/Cas9-induced knockout of an amino acid permease gene (*AAP6*) reduced *Arabidopsis thaliana* susceptibility to *Meloidogyne incognita*. BMC Plant Biol 24:515. https://doi.org/10.1186/s12870-024-05175-5

El-Messidi H, Sultana S, Hewezi T (2024) Enhanced root-knot nematode resistance in tomato using CRISPR/Cas9 gene editing approach. 2024 EUReCA online Symposium! Hosted by ForagerOne

Eves-van den Akker S, Kranse O, Beasley H (2021) Towards genetic modification of plant-parasitic nematodes: delivery of macromolecules to male germ lines and expression of exogenous mRNA in second stage juveniles. Dryad. https://doi.org/10.5061/dryad.r4xgxd296

Evogene (2018) Evogene and TMG announce collaboration to develop nematode resistant soybean through genome editing. https://www.evogene.com/press release

FAO (2025) Plant production and protection. Food and Agriculture Organization of United Nations, Rome

Fosu-Nyarko J, Jones MGK (2015) Application of biotechnology for nematode control in crop plants. In: Escobar C, Fenoll C (eds) Advances in botanical research, vol 73, pp 339–376

Hadj-Amor K, Léger O, Raffaele S et al (2024) Out of sight, out of mind. The antagonistic role of the transcriptional memory associated with repeated acoustic stimuli on the priming of plant defense. bioRxiv preprint. https://doi.org/10.1101/2024.05.03.592392

Huang Q, Lin B, Cao Y et al (2023) CRISPR/Cas9-mediated mutagenesis of the susceptibility gene *OsHPP04* in rice confers enhanced resistance to rice root-knot nematode. Front Plant Sci 14:1134653. https://doi.org/10.3389/fpls.2023.1134653

Karmakar A, Taufiqa S, Baig MJ et al (2022) Increasing disease resistance in host plants through genome editing. Proc Indian Natl Sci Acad 88:417–429

Khan A, Chen S, Fatima S et al (2023) Biotechnological tools to elucidate the mechanism of plant and nematode interactions. Plants 12(12):2387. https://doi.org/10.3390/plants12122387

Kranse O, Beasley H, Adams S et al (2021) Toward genetic modification of plant-parasitic nematodes: delivery of macromolecules to adults and expression of exogenous mRNA in second stage juveniles. G3 (Genes/Genomes/Genetics). https://doi.org/10.1093/g3journal/jkaa058

Li Z, Huang Q, Lin B et al (2022) CRISPR/Cas9-targeted mutagenesis of a representative member of a novel PR10/bet v1-like protein subfamily significantly reduces rice plant height and defense against *Meloidogyne graminicola*. Phytopathol Res 4:38. https://doi.org/10.1186/s42483-022-00143-z

Li T, Yang Y, Qi H et al (2023) CRISPR/Cas9 therapeutics: progress and prospects. Sig Transduct Target Ther 8:36. https://doi.org/10.1038/s41392-023-01309-7

Lu H, Luo T, Fu H et al (2018) Resistance of rice to insect pests mediated by suppression of serotonin biosynthesis. Nat Plants 4(6):338–344

Pellegrin C, Damma A, Sperling AL et al (2025) The SUbventral-gland regulator (SUGR-1) of nematode virulence. PANS 122(11):e2415861122. https://doi.org/10.1101/2024.01.22.576598

Schenke D, Cai D (2020) Applications of CRISPR/Cas to Improve crop disease resistance: Beyond inactivation of susceptibility factors. iScience 23(9):101478. https://doi.org/10.1016/j.isci.2020.101478

Siddique S, Radakovic ZS, Hilt C et al (2022) The genome and life stage-specific transcriptomes of a plant-parasitic nematode and its host reveal susceptibility genes involved in trans-kingdom synthesis of vitamin B5. Nat Commun 13:6190

Vieira P, Maier TR, Eves-van den Akker TR (2018) Identification of candidate effector genes of *Pratylenchus penetrans*. Mol Plant Pathol 18:1887–1907

Part VI

Climate Change

15 Effect of Climate Change Variables on Nematodes

Abstract

Climate change, especially rising temperatures and CO_2 levels, may increase plant–nematode pathogen activity, potentially worsening crop damage. Warmer temperatures could expand nematode ranges, while effects vary by species and region, with some facing reduced survival. Higher CO_2 boosts nematode abundance due to increased plant root growth, though responses differ across ecosystems. Precipitation changes and extreme weather events, such as droughts and heat waves, might also influence nematode populations, but impacts are complex and context-dependent. This chapter synthesizes the findings from recent scientific literature to explore how these climate variables affect nematode biology, distribution, and interactions with host plants, with implications for global food security.

Keywords

Rising temperature · Elevated carbon dioxide · Altered precipitation · Extreme weather events · Management implications

15.1 Learning Objectives

- Understand the key climate change variables (temperature rise, altered precipitation, elevated CO_2 (eCO_2), and extreme weather events) and how they influence nematodes.
- Explain the effects of climate change on nematode life cycle, reproduction, survival, and geographic distribution.
- Analyze the interaction between climate change, host plants, and plant-parasitic nematodes, including host susceptibility and nematode aggressiveness.
- Evaluate the potential shifts in nematode community structure and emergence of new pest problems under changing climatic scenarios.

R. P. Parvatha, *Plant Parasitic Nematodes*,
https://doi.org/10.1007/978-981-95-9306-4_15

15.2 Introduction

Plant-parasitic nematodes (PPNs) are responsible for an estimated yearly loss of US$ 173 billion to agricultural and horticultural crops (Elling 2013; Jones et al. 2013; Phani et al. 2021). Climate change, driven by factors such as global warming, eCO_2 concentrations, altered precipitation patterns, and extreme weather events, significantly influences PPNs. Given the projected global temperature increase of 2.4–2.6 °C by the century's end, understanding these impacts is crucial for developing adaptive agricultural strategies.

15.3 Effect of Rising Temperature

Rising temperatures, with a projected increase of 1.5–2.0 °C, are altering the biogeographic range and life cycles of PPNs. Studies indicate latitudinal shifts, with species such as *Meloidogyne*, *Pratylenchus*, *Helicotylenchus*, and *Ditylenchus* expanding northward since the 1960s (Bebber et al. 2013). For instance, virus-transmitting nematodes in Great Britain have shifted 160–200 km northward (Neilson and Boag 1996). Specific examples include *M. graminicola* spreading to all major rice belts in India (Dutta et al. 2012) and *Globodera rostochiensis* expanding in Russia (Pridannikov et al. 2022).

Climate change is warming the planet, and this seems to affect plant–nematode pathogens in varied ways. For some, like *Rotylenchulus reniformis*, higher temperatures (e.g., 34 °C vs. 32 °C) can improve survival, while for others, like *Meloidogyne floridensis*, it might reduce survival but increase disease severity. This could mean more crop damage in warmer areas, with nematodes potentially spreading to new regions as temperatures rise.

Temperature effects on nematode biology are species-specific. For *R. reniformis*, survival increased by 10% at 34 °C compared to 32 °C, while reproduction decreased by 45–49% at 33–34 °C, yet virulence increased, reducing root biomass by 23% and shoot biomass by 8% (Khanal and Land 2023). Conversely, *M. floridensis* showed a 12% survival decrease at 34 °C but a 46% increase in root gall index, indicating higher disease severity (Khanal and Land 2023). High temperatures can also induce sex reversal (female to male) in parthenogenetic root-knot nematodes, as observed by Papadopoulou and Triantaphyllou (1982), and reduce abundance in certain ecosystems like the Qinghai-Tibetan plateau and semiarid grasslands (Liu et al. 2022; Mueller et al. 2016).

The responses of different plant-feeding nematodes toward climate change parameters like increased temperature are compiled in Table 15.1.

The effects of different temperatures on the survival and reproduction of *R. reniformis* and *M. floridensis* are presented in Table 15.2.

Table 15.1 Variable response of PPNs to altered temperatures

Ecosystem/ cropping system/ country	Experimental system	Nematode species	Nematode abundance	References
Pine forest, Switzerland	Ecosystem monitoring, tree sampling	*Bursaphelenchus mucronatus, B. xylophilus*	Increased prevalence	Rebetez and Dobbertin (2004)
Coffee plantation, Brazil	Simulated model	*Meloidogyne incognita*	Increased pathogenic potential	Ghini et al. (2008)
Carrot, Italy	Simulated model	*Heterodera carotae*	Increased juvenile and egg production	Colagiero and Ciancio (2012)
Grassland, Denmark	Light scaffolding in the field soil by using curtains of high-density polyethylene mesh	*Paratylenchus* spp., *Filenchus* spp., *Helicotylenchus* spp.	Increase in herbivorous nematodes with extended lifespan	Stevnbak et al. (2012)
Grassland, China	Fields were warmed by infrared radiators	*Aphelenchoides* spp., *Paratylenchus* spp., *Pratylenchus* spp., *Rotylenchus* spp.	Daytime warming (compared to nighttime and diurnal warming) had a stronger negative effect on nematode abundance	Yan et al. (2017)
Potato, United Kingdom	Controlled environment, simulated model	*Globodera rostochiensis, G. pallida*	Increased abundance of *G. rostochiensis* but not *G. pallida*	Jones et al. (2017)
Rice–wheat rotation, China	Infrared heating lamps in an open field for canopy warming	*Filenchus* spp., *Psilenchus* spp., *Hirschmanniella* spp., *Pratylenchus* spp.	Not affected (although nematode diversity increased)	Wang et al. (2021)
Tomato, India	Controlled environment	*Meloidogyne incognita*	Nematode population growth was significantly affected despite an increase in plant root and shoot growth	Berliner et al. (2023)

Table 15.2 Effects of different temperatures on survival and reproduction of *Rotylenchulus reniformis* and *Meloidogyne floridensis*

Aspect	Effect on *R. reniformis*	Effect on *M. floridensis*
Survival (direct heat, °C)	Increased survival: 56–66% at 32 °C, +1% at 33 °C (ns), +10% at 34 °C ($P < 0.01$)	Decreased survival: 90% at 32 °C, +2% at 33 °C (ns), −12% at 34 °C ($P < 0.01$)
Survival at 34 °C vs. 32 °C	+10% (significant, $P < 0.01$)	−12% (significant, $P < 0.01$)
Reproduction (soil temp, °C)	Decreased: 89,080 eggs/root at 32 °C, −45% at 33 °C, −49% at 34 °C ($P = 0.01$)	Decreased: 112,568 eggs/root at 32 °C, −21% at 33 °C (ns), −53% at 34 °C ($P = 0.01$)
Reproduction at 34 °C vs. 32 °C	−49% (significant)	−53% (significant, $P = 0.01$)

15.3.1 Increased Reproduction and Virulence (Species-Specific Optima)

Elevated soil temperatures, a direct consequence of climate change, can significantly alter the reproductive biology and virulence of PPNs. In controlled soil-temperature trials on tomato, raising temperatures from 26 °C (control) up to 32–36 °C boosted reproduction by 20–133%, with peak fecundity at 34 °C for *R. reniformis* and 32 °C for *M. floridensis*. Virulence (measured by root biomass loss) was highest at 36 °C, causing up to 60% root damage compared to uninoculated controls. This suggests that modest temperature rises can amplify both nematode population growth and crop damage—though exact responses depend on species thermal optima (Sagar et al. 2025). The findings, summarized below, illustrate species-specific responses (Table 15.3).

These results suggest that *M. floridensis* has a higher fecundity, with reproduction rates 2.9–7.8 times greater than *R. reniformis* across all temperatures, indicating a potential for increased crop damage under warming conditions. The study concludes that global soil warming is likely to enhance nematode reproduction and virulence, posing risks to crop production.

Climate change (increased temperature) will inflict greater crop damage (Khanal and Land 2023). This warming is expected to shorten nematode life cycles, enhancing reproduction rates and virulence, with thermal optima ranging from 24 °C to 37 °C. For *R. reniformis*, optimal reproduction occurs at 30–37.5 °C, while *M. floridensis* shows increased disease severity (46% higher root gall index at 34 °C compared to 32 °C), despite reduced survival at higher temperatures.

15.3.2 Life-Cycle Acceleration and Generation Increase

Elevated soil temperatures speed up egg hatching, juvenile development, and sexual maturation, shortening generation time. In *Meloidogyne* spp., a 2–4 °C increase can trim generation time by several days, adding extra generations per growing season and compounding crop damage risk (Dutta et al. 2023).

Table 15.3 Effect of temperature on species-specific responses

Aspect	*Rotylenchulus reniformis*	*Meloidogyne floridensis*
Reproduction increase	20% at 32 °C, 116% at 34 °C, −1% at 36 °C	133% at 32 °C, 79% at 34 °C, 119% at 36 °C
Greatest reproduction	At 34 °C (29,919 eggs/g roots)	At 32 °C (113,593 eggs/g roots)
Reproduction range	13,727–29,919 eggs/g roots	48,723–113,593 eggs/g roots
Fecundity comparison	2.9–7.8 times less than *M. floridensis*	2.9–7.8 times greater than *R. reniformis*
Temperature resilience	More resilient to temperature changes	Less resilient, higher rate of change
Virulence increase	49% root biomass loss at 32 °C, 57% at 36 °C	40% root biomass loss at 32 °C, 60% at 36 °C
Virulence at 34 °C	No significant increase, 15% gain in biomass	No significant increase, 28% gain in biomass
Virulence (biomass impact)	Root biomass −23%, shoot biomass −8% at 34 °C vs. 32 °C	Root biomass +6%, shoot biomass −9% at 34 °C vs. 32 °C
Disease severity	Not specified	+46% root gall index at 34 °C vs. 32 °C (significant, $P = 0.01$)

15.3.3 Survival

While some PPNs thrive in warmer conditions, others face survival challenges at elevated temperatures. Elevated temperatures reduce the survival of peak populations of root-lesion nematodes (*Pratylenchus thornei*). Thompson et al. (2018) investigated the impact of elevated temperatures on *P. thornei* in vertisols after wheat growth. They found that temperatures ≥35 °C result in rapid death, independent of soil desiccation effects, with incubation periods up to 16 weeks tested across a range of 10–45 °C at 5 °C intervals. At lower temperatures (<35 °C), survival was better in soil maintained at field capacity compared to drying conditions, and air-drying the soil before incubation led to rapid death at all temperatures. This suggests that extreme heat waves, a feature of climate change, could reduce *P. thornei* populations, potentially offering a natural control mechanism in hot climates.

In contrast, Wang and McSorley (2008) showed that *M. incognita* eggs and juveniles (J2) require prolonged exposure to high temperatures for suppression. Complete suppression of egg hatch required 389.8 h at 38 °C, reducing to 13.1 h at 42 °C, while complete killing of J2 required 47.9 h at 39 °C, decreasing to 13.8 h at 42 °C. This indicates that while high temperatures can be lethal, the duration of exposure is critical, and not all nematodes are equally susceptible.

- *Long-Term Warming Effects*: In alpine meadow soils on the Qinghai–Tibetan Plateau, chronic temperature elevation altered nematode community structure: plant-parasitic taxa declined indirectly via host-plant shifts, while bacterivores and fungivores responded to soil-property changes. Overall nematode diversity and trophic interactions shifted, indicating that sustained warming can rewire soil food webs (Liu et al. 2022).

- *Asymmetric Day/Night Warming*: Daytime warming often drives stronger negative impacts on soil nematode communities than nighttime warming, largely through soil moisture reduction. Drying combined with heat stress can suppress many nematodes, especially those sensitive to desiccation (Yan et al. 2017).

15.3.4 Geographical Expansion and Distribution

Climate change is likely to expand the geographical range of PPNs, spreading nematode problems to new areas. Somasekhar and Prasad (2012) highlighted that global warming may allow PPNs to colonize regions previously unsuitable due to cooler temperatures. They found no negative effect of increased CO_2 levels (up to 700 ppm) on soil nematodes or penetration of *Meloidogyne graminicola*, a serious rice root-knot nematode, suggesting adaptability. Wang et al. (2021) observed a 6.4% increase in nematode diversity under canopy warming in winter wheat with eCO_2, without affecting abundance, indicating potential for increased diversity in warmer conditions.

This expansion is particularly concerning for food security, as noted by Khanal and Land (2023), which predicts exacerbated crop losses due to PPNs in warmer soils. They emphasized that nematodes, which evolved from free-living marine species to endoparasites, are highly adaptive and likely to co-evolve with warmer environments, potentially increasing damage to annual and perennial crops.

15.3.5 Indirect Effects Through Host Plants

eCO_2 and temperature can alter host plant physiology, indirectly affecting PPN interactions. Somasekhar and Prasad (2012) noted that changes in plant physiology may interfere with nematode developmental rates and survival strategies. For instance, eCO_2 can lead to neutral or positive responses in nematodes, with some species showing rapid build-up and interfering with plant responses to global warming. Singh et al. (2020) suggested that plant-dependent effects, such as changes in root biomass and quality, can influence nematode reproduction and virulence, with high N fertilization potentially reducing diversity and maturity indices at certain growth stages.

Ferris et al. (2013) found that resistance to *Meloidogyne* spp. in grape rootstocks is slightly compromised at soil temperatures above 27 °C, with increased galling and egg mass production at 30 °C. This indicates that indirect effects through host plant resistance can exacerbate nematode damage under warming conditions.

- *Host Resistance Breakdown*: Some plant resistance genes against root-knot nematodes (e.g., *Mi-1* in tomato) lose efficacy above 28 °C, making previously resistant cultivars vulnerable during heat waves. Although heat-stable alleles have since been identified, their field performance under fluctuating thermal regimes remains to be fully validated (Dutta et al. 2023).

- *Biocontrol Agents and Soil Suppressiveness*: Elevated temperatures and associated moisture changes can disrupt soil microbial antagonists (e.g., fungal hyperparasites and bacteria) that naturally suppress PPNs. For instance, extreme drought can reduce populations of moisture-dependent biocontrol taxa like Catenaria, weakening ecosystem-based pest control (Dutta et al. 2023).
- *Species-Specific Responses and Research Needs*: The responses of PPNs to elevated temperatures are species-specific, as evidenced by the varied outcomes for *R. reniformis*, *M. floridensis*, and *P. thornei*. Khanal and Land (2023) employed novel approaches, including direct heat exposure assays (e.g., hot water at 32–34 °C for 7 h) and indirect exposure using heat mats, to evaluate survival, reproduction, and disease severity. They found that *R. reniformis* survival increased by 10% at 34 °C compared to 32 °C, while *M. floridensis* survival decreased by 12% at 34 °C, with reproduction decreasing by 53% at 34 °C compared to 32 °C. These differences highlight the need for species-specific management strategies.

  Further research is needed to understand nematode biology under natural soil environments, as noted in the Nature study, particularly regarding gene upregulation (e.g., heat shock proteins) and impacts at temperatures beyond current study limits. Wulandari et al. (2021) found that temperature positively influenced the abundance of *Helicotylenchus* sp., while pH and organic matter had varying effects on other genera, underscoring the complexity of abiotic interactions.

15.3.6 Implications for Management

- *Heat-Resilient Cultivars*: Breeding or engineering crops with resistance genes that remain effective above baseline thresholds (≥30 °C) will be critical.
- *Adaptive Nematicide Timing*: As generation times accelerate, application windows for nematicides must be recalibrated to intercept earlier hatches and juvenile peaks.
- *Monitoring Shifting Pest Complexes*: Regular surveillance in traditionally cooler regions is needed to detect incursions of heat-adapted PPN species.
- *Soil Moisture Conservation*: Practices that buffer soil against rapid drying (mulching, organic amendments) may mitigate combined heat-moisture stress on both crops and beneficial soil biota.

15.3.7 Implications for Agriculture and Future Directions

The collective evidence suggests that climate change, particularly through elevated temperatures, is likely to exacerbate the challenges posed by PPNs, leading to greater crop losses and necessitating adaptive management strategies. The potential for increased reproduction, virulence, and geographical expansion poses significant risks to food security, especially in regions predicted to experience significant warming by 2040 and 2100. However, the lethal effects of extreme heat on certain species, such as *P. thornei*, offer potential control opportunities, particularly during heat waves.

Mitigation strategies may include identifying heat-tolerant crop varieties, implementing soil solarization (Xiong et al. 2025), and developing nematode-resistant rootstocks that maintain resistance at higher temperatures. Bernard et al. (2017) emphasized nonchemical methods such as host resistance and gene silencing (e.g., RNA interference) as alternatives to chemical pesticides, which could be integrated with climate adaptation strategies.

In conclusion, the effects of elevated temperatures on PPNs are multifaceted, with both risks and opportunities for agriculture. Elevated temperatures under climate change are projected to both directly and indirectly modulate PPN biology—boosting reproduction and virulence up to species-specific optima, accelerating life cycles, shifting geographic ranges, and destabilizing plant resistance and biocontrol networks. Continued research is essential to develop resilient farming systems in the face of climate change, ensuring sustainable crop production and food security. Understanding these multifaceted responses is essential for tailoring integrated nematode management (INM) strategies in a warming world.

15.4 Effect of eCO_2

Elevated atmospheric carbon dioxide (CO_2) is one of the defining drivers of anthropogenic climate change, with concentrations projected to rise from the pre-industrial level of ~280 ppm to over 550–800 ppm by the end of this century. Such increases not only alter plant physiology—enhancing photosynthesis and root growth—but also cascade through soil food webs by modifying host quality, defense chemistry, and soil microclimate. PPNs, which feed on roots and impair plant health, are particularly sensitive to such plant-mediated changes. Understanding how eCO_2 influences PPN abundance, behavior, reproduction, and interactions with both host defenses and biocontrol agents is crucial for forecasting crop vulnerability under future climates (Sun et al. 2011).

Higher CO_2 levels, a key climate change factor, often lead to more root growth in plants, which can feed nematode populations. Studies suggest species like *Longidorus* thrive at eCO_2 (475 ppm vs. 360 ppm), potentially increasing their impact on crops. However, not all nematodes respond the same way, and some ecosystems show no change, highlighting the complexity.

eCO_2 levels, often reaching 700 ppm in experimental settings, generally enhance nematode abundance by increasing root biomass, a primary food source. Yeates et al. (2003) found that *Longidorous* abundance increased at 475 ppm compared to 360 ppm. Specific studies by Prasad and Somasekhar (2009) noted *M. graminicola*, causing more damage at 700 ppm vs. 400 ppm CO_2. Wang et al. (2021) observed a 6.4% increase in nematode diversity under canopy warming (+2 °C) at 500 ppm CO_2, without affecting abundance.

However, responses vary. In grasslands, eCO_2 did not significantly affect total nematode abundance, except for increased Anguinidae and decreased Hoplolaimidae in some cases, with family richness and diversity remaining stable despite increased root production (Ayres et al. 2008). In nutrient-poor grasslands, total nematode numbers averaged 16% lower (not significant), and mass was reduced by 43% ($P = 0.06$) after six growing seasons, particularly affecting large omnivorous and predacious nematodes, possibly due to reduced soil aggregate sizes from higher moisture (Preez et al. 2022). In arable soils under free-air CO_2 enrichment (FACE) (550 ppm CO_2), total nematode abundance tended to increase, with fungivorous nematodes rising under winter wheat and bacterivores under sugar beet, shifting energy pathways (Sticht et al. 2009).

Experiments with rice have shown no adverse effects of eCO_2 levels up to 700 ppm on the abundance of soil nematodes and penetration of rice root-knot nematode, *M. graminicola* (Fig. 15.1).

The responses of different plant-feeding nematodes toward eCO_2 are compiled in Table 15.4.

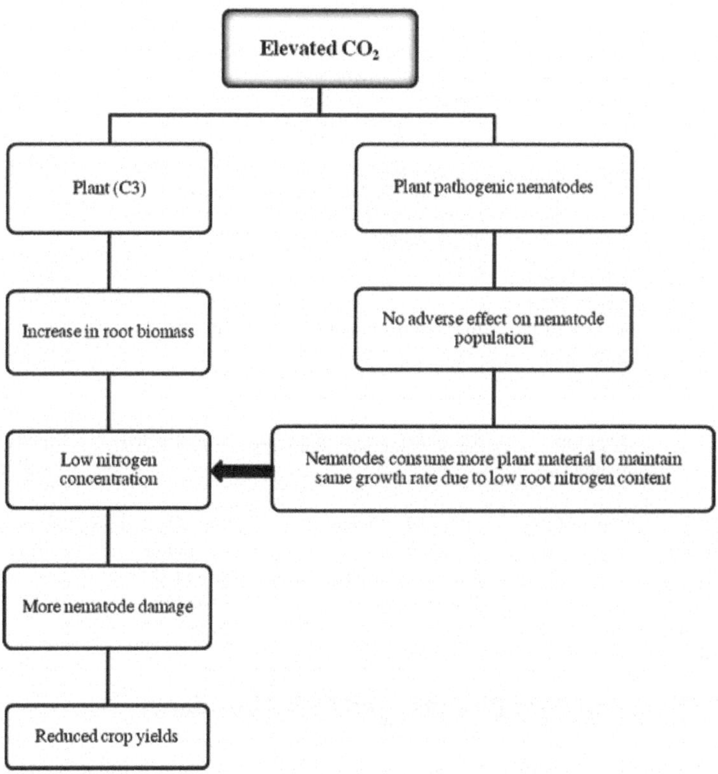

Fig. 15.1 Consequences of eCO_2 levels on plant–nematode interaction and crop productivity

Table 15.4 Variable response of PPNs to altered CO_2 level

Ecosystem/ cropping system/country	Experimental system	Nematode species	Nematode abundance	References
Ryegrass/ white clover/ New Zealand	Controlled environment	*Dorylaimus* spp., *Trichodorus* spp.	Increased	Yeates et al. (1997)
Pasture or grassland/ New Zealand	Natural vent	*Pratylenchus* spp.		Yeates et al. (1999)
Grassland/ USA	Open top chambers	Spp. of *Meloidogyne, Neopsilenchus, Paratylenchus, Gracilacus, Xiphinema*	Increased	Hungate et al. (2000)
Grassland/ New Zealand	Free air CO_2 enrichment	*Tylenchus* spp., *Longidorus* spp.	Increased	Yeates et al. (2003)
Grassland/ Switzerland	Screen-aided CO_2 control	Species name not provided	Not affected	Niklaus et al. (2003)
Grassland Germany	Free air CO_2 enrichment	Species name not provided	Increased earlier, later returned to baseline level	Sonnemann and Wolters (2005)
Rice, wheat rotation/China	Free air CO_2 enrichment	Spp. of *Psilenchus, Hirschmanniella, Filenchus, Tylenchus.*	Increased	Li et al. (2007), Li et al. (2009)
Grassland/ USA	Open top chambers	Spp. of *Heterodera, Meloidogyne, Longidorus, Trichodorus, Criconema, Tylenchus, Tylenchulus*	Not affected despite an increase in Anguinid root feeders	Ayres et al. (2008)
Grassland/ France	Controlled environment	Do	Not affected	Ayres et al. (2008)
Sugar beet, wheat rotation/ Germany	Free air CO_2 enrichment	Species name not provided	Increased	Sticht et al. (2009)
Rice/India	Open top chambers	*Meloidogyne graminicola*	Not affected	Somasekhar and Prasad (2012)
Rice–wheat rotation/China	Free air CO_2 enrichment	Spp. of *Filenchus, Psilenchus, Hirschmanniella, Pratylenchus*	Not affected (although nematode diversity increased)	Wang et al. (2021)

15.4.1 Effects on Root Biomass and PPN Abundance

One of the most consistent findings is that eCO_2 stimulates root growth—plants often allocate more carbon belowground, increasing total root length, surface area, and biomass. Under eCO_2 (~475 ppm vs. 360 ppm), pasture studies in New Zealand documented significant proliferations in PPN populations, notably genera such as *Longidorus* and *Pratylenchus*, which responded positively to the greater root substrate available for feeding (Dutta et al. 2023). Similarly, experiments in wetland soils exposed to eCO_2 revealed higher densities of multiple PPN species relative to ambient conditions, implicating root biomass as a key bottom-up driver of nematode abundance (Dutta et al. 2023). However, responses can vary: in one grassland community, PPN abundance remained unchanged under eCO_2 despite greater root infection rates, suggesting that root nutritional quality or antagonist populations may modulate the biomass effect (Dutta et al. 2023).

15.4.2 Nutritional Quality, Reproduction, and Damage

While increased root mass under eCO_2 supplies more feeding sites, the altered carbon: nitrogen (C: N) and sugar: protein ratios often reduce root tissue nutritional quality. In a controlled chamber study on tomato (*Solanum lycopersicum* "Pusa Ruby"), plants grown at 550 ppm CO_2 exhibited increased root length and soluble sugars but lower root protein content compared to 385 ppm controls. Consequently, *Meloidogyne incognita* populations produced fewer galls, females, and egg masses—and showed a reduced multiplication rate—under eCO_2, indicating that nutritional dilution can suppress nematode reproduction on certain hosts (Berliner et al. 2023). Elevated sugar: protein ratios (up to 94% higher at 550 ppm) may force nematodes to consume more but lower-quality tissue, potentially increasing plant damage per nematode and altering management thresholds (Berliner et al. 2023).

15.4.3 CO_2 as a Chemical Cue for Nematode Behavior

Beyond host effects, CO_2 itself can serve as an important chemical signal guiding PPN host-seeking behavior. Multiple PPN species—including *Ditylenchus dipsaci*, *Meloidogyne incognita*, *Heterodera* spp., and *Pratylenchus* spp.—are attracted to localized CO_2 gradients emanating from roots, enhancing infection success in low-nutrient soils (Banerjee and Hallem 2020). Conversely, some nematodes, such as the pine-wilt nematode *Bursaphelenchus xylophilus* fourth-stage juveniles, are repelled by eCO_2 in their insect vector's tracheal system, facilitating dispersal into host trees (Banerjee and Hallem 2020). These diverging responses underscore that eCO_2 may both enhance and inhibit aspects of PPN behavior depending on species-specific chemosensory pathways and ecological context.

15.4.4 Modulation of Plant Defense Pathways

eCO_2 reshapes plant defense chemistry, often by shifting the balance between jasmonic acid (JA) and salicylic acid (SA) signaling pathways. In tomato genotypes differing in JA-dependent defense, eCO_2 (750 ppm vs. 390 ppm) was found to suppress JA-mediated responses against *M. incognita* while enhancing SA-pathway defenses in JA-deficient or wild-type lines. This reprogramming led to genotype-specific effects on gall formation: JA-defense-dominated genotypes exhibited more root galling under eCO_2, whereas defense-recessive lines showed little change (Sun et al. 2011). These findings suggest that breeding for nematode resistance under current ambient CO_2 may not translate directly to future conditions, necessitating evaluation of defense genotypes under eCO_2 regimes.

15.4.5 Implications for Biological Control

The efficacy of entomopathogenic nematodes (EPNs), which are widely used as biocontrol agents against soil insect pests and some plant nematodes, also diminishes under eCO_2. In maize, plants grown at 640 ppm CO_2 developed longer and denser root systems compared to 400 ppm controls, indirectly reducing the foraging success and host-killing efficiency of *Heterorhabditis bacteriophora*. Although insect host mortality remained constant, fewer EPNs were recovered from hosts near eCO_2-grown plants, indicating that altered root architecture can impede biocontrol nematode movement and efficacy (Hiltpold et al. 2020). Such interference highlights that future integrated pest management strategies must account for the tri-trophic effects of CO_2 on both pests and beneficial soil taxa.

15.4.6 Summary of Key Findings

To organize the detailed findings, Table 15.5 summarizes the response of PPNs to eCO_2, highlighting general trends and specific examples (Table 15.5).

This table encapsulates the key trends and examples, providing a structured overview for stakeholders in agriculture and environmental science.

Table 15.5 Response of PPNs to eCO_2

Aspect	General trend	Specific examples
Abundance	Generally increases due to more root biomass	*Longidorous* and *Pratylenchus* increase in pastures (Yeates et al. 2003)
Plant interactions	Increased damage due to altered chemistry and defenses	*M. incognita* causes more galling in tomato (Sun et al. 2011)
Temperature interaction	Warmer soils may accelerate life cycles	*R. reniformis* benefits, *M. floridensis* shows mixed effects (Khanal and Land 2023)
Agricultural impact	Potential for greater crop damage	*M. graminicola* worsens rice damage at 700 ppm CO_2 (Prasad and Somasekhar 2009)
Management challenges	May reduce the efficacy of current controls	Need for adaptive strategies highlighted (Khanal and Land 2023)

15.4.7 Synthesis and Research Needs

Overall, eCO_2 exerts multifaceted impacts on PPNs by:

- Increasing root biomass, generally bolstering PPN populations unless countered by reduced tissue quality or elevated antagonists
- Altering nutritional quality, which can suppress nematode reproduction but may intensify per-nematode damage due to compensatory feeding
- Guiding nematode behavior via CO_2 cues, with species-specific attraction or repulsion patterns affecting infection dynamics
- Reprogramming defense pathways, yielding genotype-dependent outcomes for nematode resistance
- Disrupting biocontrol efficacy, where eCO_2-induced root changes hamper soil-dwelling beneficial nematodes

15.4.8 Knowledge Gaps, Future Research, and Conclusion

Despite the insights gained, significant knowledge gaps remain. The impact of eCO_2 on PPNs is highly species-specific and context-dependent, influenced by host plant type, soil conditions, and regional climate scenarios. Long-term field studies are limited, with most data derived from controlled experiments, which may not fully capture the complexity of natural ecosystems. Future research should focus on long-term field trials, particularly in diverse agro-ecosystems, to better predict PPN responses and develop effective management strategies.

Given these complex, often idiosyncratic responses, future research should prioritize long-term field experiments under FACE to capture realistic plant–soil community interactions. Moreover, breeding programs must evaluate resistance traits under eCO_2 and assess combined management approaches—including cultural practices, resistant cultivars, and biocontrol agents—to ensure sustainable nematode management in a high-CO_2 world.

In conclusion, eCO_2 under climate change is likely to have mixed effects on PPNs, primarily through altered plant physiology and environmental conditions. While increased plant biomass may support higher nematode populations, changes in plant chemistry and soil ecosystems could suppress or enhance PPN success depending on specific contexts. Integrated pest management, including resistant cultivars and soil health strategies, will be crucial to mitigate potential increases in PPN damage.

15.5 Influence of Altered Precipitation

Climate change is projected to modify precipitation regimes globally—altering total rainfall amounts, intensity, frequency, and seasonality—and these shifts have profound implications for soil moisture dynamics. PPNs are especially sensitive to soil moisture because they rely on thin water films for movement, host location, and survival. Consequently, both reduced and increased precipitation can dramatically reshape PPN communities, their life cycles, and the severity of plant–nematode interactions.

Changes in rainfall and droughts can also affect nematodes. Increased rain might boost their numbers in some areas, while droughts could do the same in others, depending on soil and plant conditions. These factors add layers of uncertainty to predicting nematode behavior under climate change.

Precipitation changes, both increased rainfall and drought, impact PPN populations. Increased rainfall boosts abundance, as seen with *M. minor*, *Heterodera* spp., and *Pratylenchus* spp. in Northern Ireland (Fleming et al. 2016). Drought conditions can also increase PPN abundance in lightly degraded grasslands, with Chen et al. (2021) noting higher populations under such conditions. Soil moisture impacts PPNs more than temperature or CO_2 in some cases, with Kardol et al. (2010) highlighting its role. Extreme dry years reduce predation pressure, potentially increasing PPN populations (Franco et al. 2019), while differential effects on ecto- and endoparasites vary across arid, semiarid, and mesic sites (Ankrom et al. 2020).

Specific studies show that higher precipitation during the growing season can promote top-down control by predators, reducing total nematode abundance, including PPNs (Pires et al. 2023).

15.5.1 Mechanistic Basis: Soil Moisture and Nematode Biology

- *Mobility and Survival*: Nematodes move through soil via water-filled pores. Reduced precipitation leads to drier soils that restrict nematode dispersal, induce desiccation stress, and elevate mortality rates. Conversely, increased rainfall can enhance nematode mobility but may also risk oxygen deprivation in waterlogged conditions (Bristol et al. 2023; Dutta et al. 2023).
- *Hatching and Infectivity*: Many PPN eggs and juveniles require a threshold of soil moisture to hatch and infect roots. Drought conditions can delay hatching or induce dormancy, reducing infection rates in the short term (Dutta et al. 2023). Under ample moisture, accelerated hatching can amplify population build-up over successive generations.

15.5.2 Effects of Reduced Precipitation

- *Declines in Abundance and Diversity*: Meta-analyses show that withholding rainfall typically reduces total nematode abundances by ~45%, with pronounced declines in PPNs in water-limited ecosystems such as grasslands and deserts (Bristol et al. 2023). Drought disproportionately impacts non-dominant genera (e.g., *Pratylenchus* and *Helicotylenchus*), leading to simplified community structure and shortened food-chain length (Zhang et al. 2020).

- *Altered Community Composition*: While overall PPN numbers fall, some drought-tolerant endoparasites may persist or even dominate, potentially changing the relative importance of specific genera in cropping systems (Dutta et al. 2023). Soil conditioning under repeated drought cycles can leave "legacy effects," where nematode communities recover differently upon re-wetting, often showing higher proportions of herbivorous PPNs relative to microbial feeders (Pires et al. 2023).

15.5.3 Effects of Increased Precipitation

- *Increased Abundance and Density*: Elevated rainfall typically raises PPN densities by fostering wetter soils and greater root herbivory. For instance, in grasslands and arable fields, reduced precipitation frequency decreases root-associated PPN densities, implying that increased frequency or amount would reverse this and promote higher populations. Similarly, long-term rainfall increases have been shown to elevate PPN prevalence in cereal systems and arid/semiarid grasslands. However, long-term experiments (>1 year) report up to a 9% increase in overall nematodes—driven largely by plant parasites—due to enhanced plant biomass and root growth that supply resources for parasitic stages (Bristol et al. 2023).
- *Shifts in Species Prevalence*: Elevated rainfall often benefits PPNs more than microbial feeders, potentially increasing the prevalence of species like *Meloidogyne minor*, *Heterodera* spp., and *Pratylenchus* spp. in grassland and cereal systems (Dutta et al. 2023). Yet, responses can be context-dependent: single-year rainfall manipulations have sometimes shown negligible change in certain families (e.g., Criconematidae), underscoring interactions with soil texture, baseline moisture, and cropping history (Dutta et al. 2023).

15.5.4 Interactive and Indirect Effects

- *Temperature–Precipitation Interactions*: The combined stress of warming plus altered moisture can exacerbate or mitigate PPN responses. For instance, drought plus elevated temperature may synergistically suppress nematode abundance by intensifying host-plant stress and reducing root exudates, whereas wet, warm conditions may foster rapid nematode development cycles (Dutta et al. 2023).
- *Trophic Cascades and Biocontrol*: Changes in precipitation modulate soil-food-web structure. Drought often reduces predator and antagonist densities more severely than PPNs, potentially releasing parasites from natural suppression (Dutta et al. 2023). Conversely, wetter soils may improve dispersal of moisture-dependent biocontrol agents (e.g., *Catenaria*), though extreme waterlogging can still impede aerobic agents (Dutta et al. 2023).

15.5.5 Management Implications

- *Irrigation Scheduling*: Precision irrigation that avoids both prolonged drought and waterlogging can help maintain soil moisture in the optimal range—suppressing PPN hatching under dry spells while preventing explosive population growth under over-watering.
- *Resilient Crop Rotation and Resistant Cultivars*: Selecting crop genotypes with drought-tolerance and nematode resistance may buffer against precipitation-driven population surges.
- *Biocontrol Timing*: Deploying moisture-dependent biocontrol agents immediately after rainfall events can capitalize on favorable conditions for agent dispersal and establishment.

The responses of different plant-feeding nematodes toward altered precipitation levels are compiled in Table 15.6.

Table 15.6 Variable response of PPNs to altered precipitation level

Ecosystem/ cropping system/country	Experimental system	Nematode species	Nematode abundance	References
Grassland, cereals, Northern Ireland	Field sampling	*Meloidogyne minor*, *Heterodera* spp., *Pratylenchus* spp.	Increased	Fleming et al. (2016)
Arid, semiarid, and mesic grasslands, USA	Precipitation manipulation by using rainout shelters in field plots	Species name not provided	Reduced precipitation in mesic grasslands increased plant feeder abundance	Franco et al. (2019)
Bluegrama grass, *Bouteloua gracilis*, USA	Controlled environment in greenhouse	Species name not provided	Reduced precipitation in sub-humid grassland increased root herbivory	Franco et al. (2020)
Desert grassland, semiarid short grass steppe, mesic tall grass prairie, USA	Precipitation manipulation by using rainout shelters in field plots	*Ditylenchus* spp., *Hemicycliophora* spp., *Hoplolaimus* spp., *Pratylenchus* spp., *Rotylenchus* spp., *Subanguina* spp.	Endoparasites responded positively to increased rainfall, ectoparasite abundance was unaltered	Ankrom et al. (2020)
Arid, semiarid, and mesic grasslands, USA	Precipitation manipulation by using rainout shelters in field plots	*Paratylenchus* spp., *Helicotylenchus* spp., *Hoplolaimus* spp., *Trichodorus* spp., *Xiphinema* spp., *Longidorus* spp.	Increased precipitation in semiarid and mesic site increased and decreased root herbivory abundance, respectively	Ankrom et al. (2022)

Table 15.7 Effects of altered precipitation on PPNs

Effect	Details	Specific examples	References
Increased PPN abundance with rainfall	PPNs benefit from increased rainfall due to boosted plant biomass, showing an intense response compared to other trophic groups	Increased prevalence of *M. minor*, *Heterodera* spp., and *Pratylenchus* spp. in Northern Ireland grasslands and cereal systems	Nielsen et al. (2014), Sylvain et al. (2014), Vandegehuchte et al. (2015); Olusanya et al. (2019), Bristol et al. (2023)
Variable response to precipitation	Differential effects on ecto- and endoparasitic PPNs across sites: Unchanged in arid, increased in semiarid, decreased in mesic sites	Ectoparasites active in arid/semiarid due to moisture; endoparasites decreased in mesic due to plant defenses	Ankrom et al. (2020), Ankrom et al. (2022)
Unaffected by precipitation changes	Some PPN families (*Criconematidae*, *Hoplolaimidae*, *Pratylenchidae*) unaffected by precipitation manipulation; *Aphelenchoides* negatively affected by droughts	Study in New Jersey pinelands, USA, showed no change in abundance under altered precipitation	Landesman et al. (2011), Torode et al. (2016)
Decreased precipitation frequency	Increased PPN abundance in lightly degraded grasslands via improved root biomass; reversed in moderate/severe degradation due to soil moisture and pH changes	Observed in degraded grasslands, affecting community structure	Chen et al. (2021)
Combined effects with other factors	Climate warming and drought decrease plant production, nutrient turnover, affecting PPNs; water and nitrogen application reduced PPN abundance in the organic horizon, opposite in the mineral horizon	Study in temperate soil of China showed variable responses based on soil layers and seasons	Kardol et al. (2010), Sun et al. (2013), Ma et al. (2018)
Disruption of PPN-predator balance	Altered precipitation (e.g., extreme dry years) lessens predation pressure, increasing PPN population and plant infestation, worsening drought effects	Observed in mesic grasslands of the USA, affecting carbon cycling and ecosystem production	Franco et al. (2019), Franco et al. (2020), Colagiero and Ciancio (2012)

Table 15.7 summarizes the effects of altered precipitation on PPNs, with specific examples and references.

To illustrate the variability, consider Table 15.8 summarizing key effects on nematode communities under precipitation changes (Pires et al. 2023).

In conclusion, altered precipitation under climate change has a significant and multifaceted influence on PPNs, with effects ranging from increased abundance due to higher rainfall to reduced populations during droughts. The response depends on

Table 15.8 Key effects of precipitation changes on nematode communities

Precipitation change	Effect on nematode abundance	Specific trophic groups affected	References
Increased precipitation	Reduces total abundance via predator control	PPNs decreased	Franco et al. (2020)
Reduced precipitation	Decreases abundance, especially bacterial feeders and omnivores-predators	Fungivores and herbivores relatively insensitive	Zhang et al. (2019)
Summer drought	Long-lasting reduction in total abundance	Bacterivores and fungivores decreased, and herbivores increased post-drought	Yang et al. (2011)

the nematode species, the type of precipitation change, and the environmental context. These changes have profound implications for agriculture, as PPNs are major pests, and climate change could exacerbate their impact. The net impact hinges on ecosystem context, experimental duration, and interactive stresses. Integrating soil-moisture management, resistant cultivars, and judicious biocontrol deployment will be key to mitigating PPN risks in an era of shifting rainfall patterns. Understanding these dynamics is crucial for developing adaptive management strategies, such as improved monitoring, host resistance, and sustainable pest control methods, to mitigate the effects on crop productivity and ecosystem health.

15.6 Role of Extreme Climate Events

Extreme weather events—such as droughts, floods, heat waves, cold snaps, storms, and wildfires—are becoming more frequent and intense under climate change, with profound implications for soil-dwelling organisms. PPNs, as poikilothermic soil inhabitants, are particularly sensitive to abrupt shifts in moisture and temperature regimes.

The addition of root-knot nematodes (*Meloidogyne graminicola*) ameliorated the severity of the drought stress in rice. Figure 15.3 shows an example of the positive interactive effect of drought stress and nematode infection on rice plants.

Extreme events, such as heat waves, wildfires, and storms, further complicate nematode dynamics. Heat waves may affect abundance and growth, potentially causing sex reversal (Majdi et al. 2019). Wildfires severely reduce PPN diversity, with no recovery in 4 years in semiarid grasslands (Bastow 2020). Storms can disseminate PPNs, such as cyst nematodes via wind-blown cysts (de Rooij-van der Goes et al. 1997), though research on these events remains limited (Wu et al. 2018; Renčo et al. 2022).

Below, we review how key extreme events influence PPN abundance, distribution, life cycle progression, and interactions with hosts and natural enemies.

15.6.1 Drought and Reduced Precipitation

- *Abundance Declines With Severe Drought*: Meta-analysis of field studies indicates that reduced rainfall consistently lowers total nematode abundances by

~45%, with plant-parasitic groups declining alongside bacterivores and omnivores under drought stress (Bristol et al. 2023).

- *Legacy Effects and Community Shifts*: Prolonged summer drought can have cascading, long-lasting impacts on nematode community composition. In Mediterranean forest soils, drought led to lower total nematode counts and altered trophic structure that persisted even after rainfall resumed; notably, soils previously drought-conditioned harbored higher relative abundances of herbivores following rewetting (Pires et al. 2023).
- *Mechanisms*: Reduced soil moisture limits nematode mobility, interferes with egg hatching and juvenile development, and can trigger cryptobiotic survival strategies in some taxa. However, most PPNs exhibit diminished fecundity and infectivity under prolonged drought.

15.6.2 Flooding and Heavy Rainfall

- *Short- vs. Long-Term Responses*: Increased precipitation per se often shows weak or delayed effects on nematode abundances; shorter-term studies (<1 year) generally report negligible changes, whereas longer-term exposure to elevated rainfall can boost PPN densities by ~9% (Bristol et al. 2023).
- *Shifts in Trophic Interactions*: Excess soil moisture may disrupt below-ground predator–prey dynamics, reducing predation pressure on PPNs. In sub-humid grasslands, extreme wet years lessened densities of nematode-feeding predators, indirectly facilitating PPN population surges and heightened plant damage (Dutta et al. 2023).
- *Hydrological Habitat Alteration*: Flooding can also create anoxic microsites, selectively disadvantaging moisture-sensitive PPN species while favoring water-tolerant taxa; community composition thus shifts toward those better adapted to saturated soils.

15.6.3 Heat Waves and Extreme Temperatures

- *Thermal Thresholds and Mortality*: Short spikes to lethal temperatures can dramatically reduce PPN viability. In greenhouse trials, rapid alternating exposures above ~45 °C effectively eliminated root-knot nematodes (*Meloidogyne* spp.) with minimal host damage (Wang et al. 2023).

- Nematode life cycles shorten generation times and potentially increase population turnover, but may concurrently impair egg viability or adult fecundity if temperatures exceed species-specific optima.
- *Geographical Range Shifts*: Repeated heat extremes can facilitate poleward and altitudinal range expansions of thermophilic PPNs while simultaneously contracting ranges of cold-tolerant species.

15.6.4 Cold Snaps, Storms, and Wildfires

- *Frost and Freeze Events*: Sudden soil freezes can cause high juvenile and egg mortality in surface-dwelling PPNs, but insulating effects of plant residues and snow cover often mitigate impacts.
- *Storm Disturbances*: High-intensity storms may redistribute infested soil and plant debris across landscapes, aiding long-distance dispersal of PPNs.
- *Post-Fire Soil Dynamics*: Wildfires alter soil physicochemical properties and reduce plant cover, generally depressing nematode communities. However, opportunistic PPNs may later exploit weakened plant defenses in post-fire regrowth.

15.6.5 Implications for Management

- *Predictive Modeling*: Incorporating extreme event frequency into species distribution and demographic models will improve forecasts of PPN outbreaks under future climate scenarios.
- *Adaptive Cultural Practices*: Timing of irrigation, tillage, and crop rotations should consider projected drought or flood windows to minimize PPN proliferation.
- *Heat and Flood Treatments*: Emerging phytosanitary measures—such as controlled soil solarization or steam treatments—leverage extreme heat and moisture to suppress PPN populations in infested fields.

15.6.6 Interactions with Other Factors

- Extreme weather can weaken plants, making them more susceptible to PPN damage. For example, drought-stressed plants have reduced defenses against *Heterodera* (cyst nematodes).
- Soil microbial communities, which compete with or prey on PPNs, may also be altered by extreme weather, indirectly affecting nematode populations.
- Management practices (e.g., irrigation and crop rotation) can mitigate or exacerbate weather-related impacts on PPNs.

15.6.7 Practical Implications

- *Crop Losses*: Increased PPN activity in warmer, wetter conditions can lead to higher yield losses, especially in tropical and subtropical regions.
- *Management Challenges*: Extreme weather complicates control measures such as nematicide application or resistant crop varieties, as timing and efficacy may be disrupted.

- *Adaptation Strategies*: Farmers may need to adopt climate-resilient practices, such as soil amendments, cover crops, or biological controls, to manage PPNs under changing conditions.

15.7 Indirect Effects and Ecosystem Interactions

Climate change indirectly affects PPNs by altering host plant physiology. eCO_2 and temperature can modify plant resistance, making them more or less susceptible to nematode infections. For example, increased CO_2 can lead to lower root nitrogen, exacerbating damage in grasses (Wilsey 2001). Other trophic groups, such as microbial feeders, predators, and insect parasites, indirectly influence plant productivity by regulating ecosystem processes such as decomposition, nutrient mineralization, biological pest suppression, and energy transfer in food webs. Studies by Somasekhar and Prasad (2012) found no negative effect of CO_2 up to 700 ppm on soil nematodes or *M. graminicola* penetration in rice, highlighting neutral or positive responses.

15.8 Management Implications

Given these impacts, climate-smart INM is recommended, involving biocontrol, host resistance, crop rotation, cover cropping, biofumigation, and farmyard manure to reduce pesticide use, aligning with global sustainability goals. Novel nematicides such as fluensulfone and fluopyram are not promoted long term due to pesticide reduction aims (Gatto et al. 2016; Jacquet et al. 2022).

15.9 Case Studies

15.9.1 Host Resistance Breakdown

The resistance to root-knot nematode in tomato (cv. "Sanibel") has often failed as a result of the heat instability or apparent temperature sensitivity of the resistant *Mi* gene (Fig. 15.2). For example, the threshold soil temperatures and incremental reductions in nematode resistance with each degree above 78 °F, such that at 91 °F tomato plants are fully susceptible. This would suggest that, in Florida, the use of these varieties may have to be restricted to spring plantings when cooler soil temperatures prevail.

15.9.2 Elevated Atmospheric Carbon Dioxide (CO_2)

eCO_2 levels, often reaching 700 ppm in experimental settings, generally enhance nematode abundance by increasing root biomass, a primary food source. Yeates

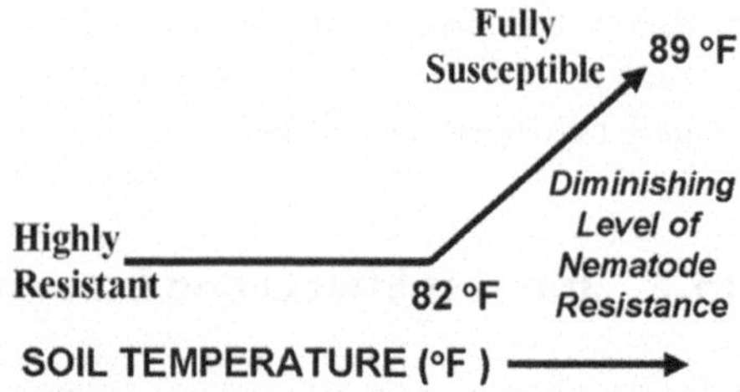

Fig. 15.2 Complete loss of root-knot nematode resistance conferred by the Mi gene in tomato cv. "Sanibel" due to increasing soil temperature

et al. (2003) found *Longidorous* abundance increased at 475 ppm compared to 360 ppm. Specific studies by Prasad and Somasekhar (2009) noted that *M. graminicola* caused more damage at 700 ppm vs. 400 ppm CO_2. Wang et al. (2021) observed a 6.4% increase in nematode diversity under canopy warming (+2 °C) at 500 ppm CO_2, without affecting abundance.

15.9.3 Influence of Altered Precipitation

Precipitation changes, both increased rainfall and drought, impact PPN populations. Increased rainfall boosts abundance, as seen with *M. minor*, *Heterodera* spp., and *Pratylenchus* spp. in Northern Ireland (Fleming et al. 2016). Drought conditions can also increase PPN abundance in lightly degraded grasslands, with Chen et al. (2021) noting higher populations under such conditions. Soil moisture impacts PPNs more than temperature or CO_2 in some cases, with Kardol et al. (2010) highlighting its role. Extreme dry years reduce predation pressure, potentially increasing PPN populations (Franco et al. 2019), while differential effects on ecto- and endoparasites vary across arid, semiarid, and mesic sites (Ankrom et al. 2020).

15.9.4 Role of Droughts

The addition of root-knot nematodes (*Meloidogyne graminicola*) ameliorated the severity of the drought stress in rice. Figure 15.3 shows an example of the positive interactive effect of drought stress and nematode infection on rice plants.

15.10 Conclusion

Climate change is likely to exacerbate the challenges posed by PPNs, increasing their potential to cause agricultural losses through direct effects on survival and reproduction, indirect effects on host plants, and expanded geographical ranges. The variability in responses underscores the need for region- and species-specific research to develop effective mitigation strategies, ensuring food security in a warming world.

Fig. 15.3 Growth of rice under combined biotic and abiotic stress conditions. Plants are shown after exposure to drought stress, infection with the PPN *Meloidogyne graminicola*, or the two stresses in combination

15.11 Suggested Readings

- Colagiero, M. and Ciancio, A. 2012. Climate changes and nematodes: expected effects and perspectives for plant protection. *Redia* 94: 113–118.
- Devi, G. 2025. Effect of climate change on plant parasitic nematodes of cereal crops. *International Journal of Environment and Climate Change* 15(9): 1–9.
- Dutta, T.K. and Phani, V. 2023. The pervasive impact of global climate change on plant nematode interaction continuum. *Front. Plant Sci.* 14: 1143889. doi:10.3389/fpls.2023.1143889.
- Kardol, P., Cregger, M.A., Campany, C.E. and Classen, A.T. 2010. Soil ecosystem functioning under climate change: plant species and community effects. *Ecology* 91: 767–781.
- Khanal, C. and Land, J. 2023. Study on two nematode species suggests climate change will inflict greater crop damage. *Sci Rep* 13: 14185. https://doi.org/10.1038/s41598-023-41466-x.
- Pires, D., Orlando, V., Collett, R.L. et al. 2023. Linking nematode communities and soil health under climate change. *Sustainability* 15: 11747. https://doi.org/10.3390/su151511747.
- Sagar, G.C., Banakar, P., Harshman, D. and Khanal, C. 2025. Elevated soil temperatures impact nematode reproduction biology. *Stresses* 5(1): 2. https://doi.org/10.3390/stresses5010002.
- Somasekhar, N. and Prasad, J.S. 2012. Plant–nematode interactions: consequences of climate change. In B. Venkateswarlu, A.K. Shanker, C. Shanker and M. Maheswari (eds.) *Crop Stress and its Management: Perspectives and Strategies*, pp. 547–564. Springer, Dordrecht.

15.12 Review Questions

- What are the major climate change variables that can influence PPNs?
- Explain how elevated temperatures affect nematode life cycle duration and reproduction.
- Discuss the role of elevated atmospheric CO_2 in altering host–nematode interactions.
- How does altered precipitation (droughts, flooding, irregular rainfall) influence nematode survival and dispersal?
- Explain how climate change can affect the geographical distribution and invasive potential of PPNs.
- How could global warming accelerate the breakdown of resistance genes in host plants against nematodes?
- Provide examples of PPNs predicted to expand their range under climate change scenarios.

References

Ankrom KE, Franco AL, Fonte SJ et al (2020) Ecto-and endoparasitic nematodes respond differently across sites to changes in precipitation. Oecologia 193:761–771

Ankrom KE, Franco AL, Fonte SJ et al (2022) Ecological maturity and stability of nematode communities in response to precipitation manipulations in grasslands. Appl Soil Ecol 170:104263. https://doi.org/10.1016/j.apsoil.2021.104263

Ayres E, Wall DH, Simmons BL et al (2008) Below-ground nematode herbivores are resistant to elevated atmospheric CO_2 concentrations in grassland ecosystems. Soil Biol Biochem 40:978–985

Banerjee N, Hallem EA (2020) The role of carbon dioxide in nematode behaviour and physiology. Parasitology 147(8):841–854

Bastow J (2020) The impacts of a wildfire in semiarid grassland on soil nematode abundances over 4 years. Biol Fertil Soils 56:675–685

Bebber D, Ramotowski M, Gurr S (2013) Crop pests and pathogens move pole wards in a warming world. Nat Clim Chang 3:985–988

Berliner J, Ganguly AK, Kamra A et al (2023) Effect of elevated carbon dioxide on population growth of root-knot nematode, *Meloidogyne incognita* in tomato. Indian Phytopathol 76:309–315

Bernard GC, Egnin M, Bonsi C (2017) The impact of plant-parasitic nematodes on agriculture and methods of control. In: Shah MM, Mahmood M (eds) Nematology – concepts, diagnosis and control. InTech

Bristol D, Hassan K, Blankinship JC, Nielsen UN (2023) Responses of nematode abundances to increased and reduced rainfall under field conditions: a meta-analysis. Ecosphere 14(1):e4364. https://doi.org/10.1002/ecs2.4364

Chen J, Yao Y, Wang G et al (2021) Decreased precipitation frequency altered abundance, but not community structure, of soil nematodes in degraded grasslands. Ecol Indic 131:108184. https://doi.org/10.1016/j.ecolind.2021.108184

Colagiero M, Ciancio A (2012) Climate changes and nematodes: expected effects and perspectives for plant protection. Redia 94:113–118 de Rooij-van der Goes PCEM, van Dijk C, van der Putten WH, Jungerius PD (1997) Effects of sand movement by wind on nematodes and soil-borne fungi in coastal fore dunes. J Coast Conserv 3:133–142

Dutta TK, Phani V (2023) The pervasive impact of global climate change on plant nematode interaction continuum. Front Plant Sci 14:1143889. https://doi.org/10.3389/fpls.2023.1143889

Dutta TK, Ganguly AK, Gaur HS (2012) Global status of rice root-knot nematode, *Meloidogyne graminicola*. Afr J Microbiol Res 6:6016–6021

Dutta TK, Vashisth N, Ray S et al (2023) Functional analysis of a susceptibility gene (*HIPP27*) in the *Arabidopsis thaliana-Meloidogyne incognita* pathosystem by using a genome editing strategy. BMC Plant Biol 23:390. https://doi.org/10.1186/s12870-023-04401-w

Elling AA (2013) Major emerging problems with minor *Meloidogyne* species. Phytopathology 103:1092–1102

Ferris H, Zheng L, Walker MA (2013) Soil temperature effects on the interaction of grape rootstocks and plant-parasitic nematodes. J Nematol 45(1):49–57

Fleming TR, McGowan NE, Maule AG, Fleming CC (2016) Prevalence and diversity of plant parasitic nematodes in Northern Ireland grassland and cereals, and the influence of soils and rainfall. Plant Pathol 65:1539–1550

Franco AL, Gherardi LA, de Tomasel CM et al (2019) Drought suppresses soil predators and promotes root herbivores in mesic, but not in xeric grasslands. Proc Natl Acad Sci 116:12883–12888

Franco AL, Gherardi LA, de Tomasel CM et al (2020) Root herbivory controls the effects of water availability on the partitioning between above-and below-ground grass biomass. Funct Ecol 34:2403–2410

Gatto MP, Cabella R, Gherardi M (2016) Climate change: the potential impact on occupational exposure to pesticides. Ann Ist Super Sanità 52:374–385

Ghini R, Hamada E, Pedro Junior MJ et al (2008) Risk analysis of climate change on coffee nematodes and leaf miner in Brazil. Pesq Agrop Brasileira 43:187–194

Hiltpold I, Moore BD, Johnson SN (2020) Elevated atmospheric carbon dioxide concentrations alter root morphology and reduce the effectiveness of entomopathogenic nematodes. Plant Soil 447:29–38

Hungate BA, Jaeger CH, Gamara G et al (2000) Soil microbiota in two annual grasslands: responses to elevated atmospheric CO_2. Oecologia 124:589–598

Jacquet F, Jeuffroy MH, Jouan J et al (2022) Pesticide-free agriculture as a new paradigm for research. Agron Sustain Dev 42:8. https://doi.org/10.1007/s13593-021-00742-8

Jones JT, Haegeman A, Danchin EG et al (2013) Top 10 plant-parasitic nematodes in molecular plant pathology. Mol Plant Pathol 14:946–961

Jones LM, Koehler AK, Trnka M et al (2017) Climate change is predicted to alter the current pest status of *Globodera pallida* and *G. rostochiensis* in the United Kingdom. Glob Chang Biol 23:4497–4507

Kardol P, Cregger MA, Campany CE, Classen AT (2010) Soil ecosystem functioning under climate change: plant species and community effects. Ecology 91:767–781

Khanal C, Land J (2023) Study on two nematode species suggests climate change will inflict greater crop damage. Sci Rep 13:14185. https://doi.org/10.1038/s41598-023-41466-x

Landesman WJ, Treonis AM, Dighton J (2011) Effects of a one-year rainfall manipulation on soil nematode abundances and community composition. Pedobiologia 54(2):87–91

Li Q, Liang W, Jiang Y et al (2007) Effect of elevated CO_2 and N fertilisation on soil nematode abundance and diversity in a wheat field. Appl Soil Ecol 36:63–69

Li Q, Chonggang X, Wenju L et al (2009) Residue incorporation and N fertilization affect the response of soil nematodes to the elevated CO_2 in a Chinese wheat field. Soil Biol Biochem 41:1497–1503

Liu Y, Wang W, Liu P et al (2022) Plant-soil mediated effects of long-term warming on soil nematodes of alpine meadows on the Qinghai–Tibetan Plateau. Biology 11:1596. https://doi.org/10.3390/biology11111596

Ma Q, Yu H, Liu X et al (2018) Climatic warming shifts the soil nematode community in a desert steppe. Clim Chang 150:243–258

Majdi N, Traunspurger W, Fueser H et al (2019) Effects of a broad range of experimental temperatures on the population growth and body-size of five species of free-living nematodes. J Therm Biol 80:21–36

Mueller KE, Blumenthal DM, Carrillo Y et al (2016) Elevated CO_2 and warming shift the functional composition of soil nematode communities in semiarid grassland. Soil Biol Biochem 103:46–51

Neilson R, Boag B (1996) The predicted impact of possible climatic change on virus-vector nematodes in Great Britain. Eur J Plant Pathol 102:193–199

Nielsen UN, Ayres E, Wall DH et al (2014) Global-scale patterns of assemblage structure of soil nematodes in relation to climate and ecosystem properties. Glob Ecol Biogeogr 23:968–978

Niklaus PA, Alphei J, Ebersberger D et al (2003) Six years of in situ CO_2 enrichment evoke changes in soil structure and soil biota of nutrient-poor grassland. Glob Chang Biol 9:585–600

Olusanya OA, Gong S, Tariq A et al (2019) The effect of phosphorus addition, soil moisture, and plant type on soil nematode abundance and community composition. J Soils Sediments 19:1139–1150

Papadopoulou J, Triantaphyllou AC (1982) Sex differentiation in *Meloidogyne incognita* and anatomical evidence of sex reversal. J Nematol 14:549–566

Phani V, Khan MR, Dutta TK (2021) Plant-parasitic nematodes as a potential threat to protected agriculture: current status and management options. Crop Prot:105573. https://doi.org/10.1016/j.cropro.2021.105573

Pires D, Orlando V, Collett RL et al (2023) Linking nematode communities and soil health under climate change. Sustainability 15:11747. https://doi.org/10.3390/su151511747

Prasad JS, Somasekhar N (2009) Nematode pests of rice: diagnosis and management. Technical bulletin No. 38, Directorate of Rice Research (ICAR), Hyderabad, 29 pp

Preez GD, Daneel M, Goede RD (2022) Nematode-based indices in soil ecology: application, utility, and future directions. Soil Biol Biochem 169:108640. https://doi.org/10.1016/j.soilbio.2022.108640

Pridannikov MV, Zinovieva SV, Khudyakova EA et al (2022) Range dynamics of potato cyst nematode *Globodera rostochiensis* (Wollenweber 1923) (Nematoda, Heteroderidae) under conditions of global climate change in Russia. Russ J Biol Invasions 13:510–529

Rebetez M, Dobbertin M (2004) Climate change may already threaten scots pine stands in the Swiss Alps. Theor Appl Climatol 79:1–9

Renčo M, Adámek M, Jílková V, Devetter M (2022) Post-fire recovery of soil nematode communities depends on fire severity. Diversity 14:1116. https://doi.org/10.3390/d14121116

Sagar GC, Banakar P, Harshman D, Khanal C (2025) Elevated soil temperatures impact nematode reproduction biology. Stress 5(1):2. https://doi.org/10.3390/stresses5010002

Singh B, Chastain DR, Reddy KR et al (2020) Early season morphological and physiological responses of resistant and susceptible cotton genotypes to reniform nematode and soil nitrogen. Agronomy 10(12):1974. https://doi.org/10.3390/agronomy10121974

Somasekhar N, Prasad JS(2010) Nematological considerations in addressing impact of climate change on agriculture. Proceedings of national symposium on innovations in nematological research, Feb. 22–24, Tamil Nadu Agricultural University, Coimbatore

Somasekhar N, Prasad JS (2012) Plant–nematode interactions: consequences of climate change. In: Venkateswarlu B, Shanker AK, Shanker C, Maheswari M (eds) Crop stress and its management: perspectives and strategies. Springer, Dordrecht, pp 547–564

Sonnemann I, Wolters V (2005) The microfood web of grassland soils responds to a moderate increase in atmospheric CO_2. Glob Chang Biol 11(7):1148–1155

Stevnbak K, Maraldo K, Georgieva S et al (2012) Suppression of soil decomposers and promotion of long-lived, root herbivorous nematodes by climate change. Eur J Soil Biol 52:1–7

Sticht C, Schrader S, Giesemann A, Weigel HJ (2009) Sensitivity of nematode feeding types in arable soil to free air CO_2 enrichment (FACE) is crop specific. Pedobiologia 52:337–349

Sun Y, Yin J, Cao H et al (2011) Elevated CO_2 influences nematode-induced defense responses of tomato genotypes differing in the JA pathway. PLoS One 6(5):e19751. https://doi.org/10.1371/journal.pone.0019751

Sun X, Zhang X, Zhang S et al (2013) Soil nematode responses to increases in nitrogen deposition and precipitation in a temperate forest. PLoS One 8:e82468. https://doi.org/10.1371/journal.pone.0082468

Sylvain ZA, Wall DH, Cherwin KL (2014) Soil animal responses to moisture availability are largely scale, not ecosystem dependent: insight from a cross-site study. Glob Chang Biol 20:2631–2643

Thompson JP, Rostad HE, Macdonald BJ, Whish JPM (2018) Elevated temperature reduces survival of peak populations of root-lesion nematodes (*Pratylenchus thornei*) after wheat growth in a vertisol. Biol Fertil Soils 54(2):243–257

Torode MD, Barnett KL, Facey SL et al (2016) Altered precipitation impacts on above- and below-ground grassland invertebrates: summer drought leads to outbreaks in spring. Front Plant Sci 7. https://doi.org/10.3389/fpls.2016.01468

Vandegehuchte ML, Sylvain ZA, Reichmann LG et al (2015) Responses of a desert nematode community to changes in water availability. Ecosphere 6:44

Wang KH, McSorley R (2008) Exposure time to lethal temperatures for *Meloidogyne incognita* suppression and its implication for soil solarisation. J Nematol 40(1):7–12

Wang J, Shi X, Li L, Zhang X (2021) Changes in soil nematodes in rhizosphere and non-rhizosphere soils following combined elevated [CO_2] and canopy warming in a winter wheat field. Geoderma 386:114907. https://doi.org/10.1016/j.geoderma.2020.114907

Wang Z, Ji H, Wang D et al (2023) Extreme temperature switches eliminate root-knot nematodes: a greenhouse study. Horticulturae 9:139. https://doi.org/10.3390/horticulturae9020139

Wilsey BJ (2001) Effects of elevated CO_2 on the response *Phlenum pratense* and *Poa pratensis* to above-ground defoliation and root-feeding nematodes. Int J Plant Sci 162:1275–1282

Wu X, Zhu X, Wang Y et al (2018) The cold tolerance of the northern root-knot nematode, *Meloidogyne hapla*. PLoS One 13:e0190531. https://doi.org/10.1371/journal.pone.0190531

Wulandari AS, Indarti S, Massadeh MI, Minh NV (2021) The effect of abiotic factors and elevation on the diversity of plant parasitic nematodes in garlic on central Jawa, Indonesia. Sarhad J Agric 37(Special issue 1):75–83

Xiong J, Wang H, Zhong Z et al (2025) Emerging strategies to improve heat stress tolerance in crops. aBIOTECH 6(1):97–115

Yan X, Wang K, Song L et al (2017) Daytime warming has stronger negative effects on soil nematodes than night-time warming. Sci Rep 7:108. https://doi.org/10.1038/s41598-017-00218-4

Yang T, Wang X, Zhao X et al (2011) Changes of climate extremes in a typical arid zone: observations and multi-model ensemble projections. J Geophys Res 116:D19106

Yeates GW, Tate KR, Newton PCD (1997) Response of the fauna of a grassland soil to doubling of atmospheric carbon dioxide concentration. Biol Fertil Soils 25:307–315

Yeates GW, Newton PCD, Ross DJ (1999) Response of soil nematode fauna to naturally elevated CO_2 levels influenced by soil pattern. Nematology 1:285–293

Yeates GW, Newton PCD, Ross DJ (2003) Significant changes in soil microfauna in grazed pasture under elevated carbon dioxide. Biol Fertil Soils 38:319–326

Zhang W, Villarini G, Vecchi GA (2019) Impacts of the Pacific meridional mode on rainfall over the maritime continent and Australia: potential for seasonal predictions. Clim Dyn 53(12):7185–7199

Zhang G, Sui X, Li Y et al (2020) The response of soil nematode fauna to climate drying and warming in *Stipa breviflora* desert steppe in Inner Mongolia, China. J Soils Sediments 20:2166–2180

Pervasive Impact of Climate Change-Induced Consequences on Nematodes

16

Abstract

This chapter provides a comprehensive examination of how climate change influences plant-parasitic nematodes (PPNs), focusing on the expansion of geographical distribution, breakdown of nematode resistance, host defense mechanisms, nematode abundance in soil, and increase in the number of generations. Climate change may expand the geographical range of PPNs due to warmer temperatures, though impacts vary by species and region. It could weaken plant resistance to PPNs, affecting crop health. Climate change alters plant defense mechanisms, potentially making plants more vulnerable to PPNs. Studies indicate that PPN abundance in soil might increase with higher CO_2 levels, though results vary depending on temperature and precipitation changes. Warmer temperatures could lead to more PPN generations per year, impacting agricultural systems, though extreme heat might reduce reproduction in some species. Recent studies highlight the complexity of these interactions, aiming to inform agricultural strategies under changing climatic conditions.

Keywords

Expansion of geographical distribution · Breakdown of nematode resistance · Host defense mechanisms · Nematode abundance in soil · Increase in number of generations

16.1 Learning Objectives

- Understand the relationship between climate change drivers (temperature rise, elevated CO_2, altered precipitation, and extreme weather events) and their consequences on the biology, ecology, and distribution of plant-parasitic nematodes (PPNs).
- Analyze the effects of climate change-induced consequences (soil degradation, drought stress, flooding, salinity, and altered host physiology) on nematode survival, reproduction, host–parasite interactions, and virulence.

R. P. Parvatha, *Plant Parasitic Nematodes*,
https://doi.org/10.1007/978-981-95-9306-4_16

- Evaluate the shifting dynamics of PPN distribution, abundance, and geographic range under changing climatic conditions, including the risk of emergence of invasive species.
- Examine host–nematode–microbiome interactions under climate stress and their implications for crop productivity and nematode management.
- Discuss how climate-driven stress may alter plant resistance or susceptibility.

16.2 Introduction

PPNs are microscopic worms that infect plant roots, causing significant economic losses, estimated at US$ 173 billion annually, with a biomass in global topsoil of 0.3 gigatons, representing 80% of human biomass (Dutta and Phani 2023). Climate change, driven by rising greenhouse gas emissions, particularly CO_2 and temperature, is expected to exacerbate these impacts by altering environmental conditions that affect PPN biology and plant interactions. These effects include changes in their geographical distribution, resistance breakdown, host defense mechanisms, soil abundance, and generation numbers. This chapter explores each aspect, acknowledging the complexity and ongoing research in this field.

16.3 Expansion of Geographical Distribution

The geographic distribution of PPNs is intrinsically linked to environmental conditions. As poikilothermic organisms exposed directly to soil and ambient conditions, nematode survival, development, and dispersal respond sensitively to shifts in temperature, moisture, and atmospheric composition. Climate change—manifested primarily as rising temperatures, altered precipitation regimes, and elevated CO_2—therefore has the potential to both expand the range of many PPN species into previously unsuitable regions and alter their local abundance and life-history traits.

16.3.1 Rising Temperatures and Pole-Ward/Upward Shifts

Global warming facilitates both latitudinal and altitudinal range expansions of PPNs by relaxing the low-temperature barriers that historically limited their development. For example, predictive modeling for *Meloidogyne incognita* (southern root-knot nematode) under Intergovernmental Panel on Climate Change (IPCC) A2/B2 emission scenarios indicates that areas supporting 9–12 annual generations—which correlate with severe crop damage—could increase from 41% to 59% of Brazil's coffee-growing regions in the 1961–1990 baseline to substantially larger proportions by the 2080s under high-emission scenarios (Ghini et al. 2008). Similar trends have been documented for the pine wilt nematode (*Bursaphelenchus xylophilus*), whose spread in Asia, northern Europe, and North America has been linked to warming summer temperatures that both favor nematode development and stress host trees (Dutta and Phani 2023).

16.3.2 Elevated CO_2 and Root-Feeding Nematodes

Rising atmospheric CO_2 often promotes greater plant biomass and root proliferation, thereby enhancing the food base for root-feeding nematodes. Under experimental conditions simulating elevated CO_2 (475 ppm vs. 360 ppm), pasturelands in New Zealand exhibited a marked increase in PPN abundance, with genera such as *Longidorus* showing disproportionate gains relative to other soil faunal groups (Dutta and Phani 2023). In rice systems, *M. graminicola*—once confined to upland ecologies—is now reported across all major rice-growing belts of India, including high-altitude and direct-seeded systems, potentially exacerbated by CO_2-driven root biomass increases that favor nematode reproduction and damage (Pankaj et al. 2010).

16.3.3 Altered Precipitation and Soil Moisture Dynamics

Changes in precipitation patterns—in both total annual amounts and the frequency of extreme droughts or heavy rainfall events—directly influence soil moisture regimes, which are critical for nematode mobility, host finding, and survival. In Tennessee (USA), manipulative experiments demonstrated that soil moisture exerts a stronger effect on PPN community structure than either temperature or CO_2 alone (Dutta and Phani 2023). In European grasslands, increased rainfall regimes have been linked to heightened prevalence of species such as *Meloidogyne minor* and *Pratylenchus* spp., with some formerly minor pests emerging as key threats under wetter conditions (Dutta and Phani 2023). Conversely, simulation of severe drought conditions in sub-humid grasslands reduced predation pressure on PPNs—due to declines in their predators—leading to population outbreaks and intensified crop damage during subsequent wet periods (Dutta and Phani 2023).

16.3.4 Extreme Events and Episodic Range Incursions

Heat waves, floods, and wildfires—projected to increase in frequency and severity—can create episodic windows for nematode dispersal into new areas. Flood waters can transport mobile second-stage juveniles over long distances, while fire-induced soil heating may kill competing soil fauna, temporarily favoring heat-tolerant nematode species. Although direct experimental evidence remains limited, case studies of *Meloidogyne* spp. show facultative migration toward warmer soil layers, suggesting that transient heat pulses could facilitate vertical and horizontal movement into habitat patches previously too cold or wet for establishment (Dutta and Phani 2023).

Climate change, especially global warming, is likely to expand the geographical range of PPNs by enabling survival and reproduction in previously inhospitable areas. Studies indicate that warmer temperatures can cause latitudinal shifts, with examples including the northward expansion of *Heterodera glycines* to northern Canadian provinces since 1987 and *Meloidogyne graminicola* spreading to all major rice belts in India, causing greater damage under 700 ppm CO_2 compared to 400 ppm (Dutta and Phani 2023).

16.3.5 Latitudes and Altitudes Shifts

Warming temperatures and altered precipitation regimes are already enabling several PPN species to invade higher latitudes and altitudes.

- *Latitudinal Shifts*: Since the 1960s, major PPNs—such as *Meloidogyne incognita, Pratylenchus brachyurus, Helicotylenchus pseudorobustus,* and *Ditylenchus dipsaci*—have shown northward movements in the Northern Hemisphere in response to rising average temperatures (~1 °C increases) and milder winters (Dutta and Phani 2023). For example, soybean cyst nematode (*Heterodera glycines*), first detected in Ontario in 1987, now occurs throughout southern and eastern Canada, aided by warmer winters and longer growing seasons (Dutta and Phani 2023).
- *Altitudinal Shifts*: In mountainous regions such as the Western Himalaya, models predict greater establishment of warm-adapted PPNs (e.g., *M. floridensis, M. incognita*, and *R. reniformis*) at higher elevations by 2050, as soil temperatures increase and permafrost thaws (Dutta and Phani 2023).

16.3.6 Regional Case Studies

- *Tropical vs. Temperate Boundaries*: Surveys organized under EPPO revealed *M. incognita, M. arenaria,* and *M. javanica* in Mediterranean and semi-continental climates of Europe—regions where tropical *Meloidogyne* species were once rare—underscoring ongoing northward incursions (EPPO Global Database 2021). In India's Western Himalaya, metabolic-footprint analyses predict increased PPN abundance at higher altitudes under warming scenarios, portending novel pest pressures for mountain agriculture (Dutta and Phani 2023).
- *Implications and Outlook*: The intersection of warming, CO_2 enrichment, and hydrological shifts will likely continue to redraw the global map of nematode pests. Crop systems in higher latitudes and elevations, historically free from many tropical and subtropical PPNs, may face new invasions and increased season lengths. Integrated pest management strategies must therefore anticipate these range shifts by expanding surveillance networks, refining predictive models with local climate data, and developing resistant cultivars with broad-spectrum nematode tolerance. Continued monitoring and modeling efforts—particularly those

coupling climate projections with nematode life-history parameters—will be essential for sustaining agricultural productivity in a warming world.

16.3.7 Emergence in New Cropping Zones

Tropical root-knot nematodes are now routinely recovered in central France and the Balkans—areas historically too cool for their development—illustrating both range expansion and the risk of pest emergence in novel agro-ecosystems (Helder 2024).

16.4 Breakdown of Nematode Resistance

Climate change, characterized by rising temperatures, elevated CO_2 levels, and altered precipitation patterns, is expected to exacerbate these challenges by affecting both nematode biology and plant defense mechanisms. This analysis focuses on how these environmental changes induce consequences that lead to a breakdown in plant resistance, potentially increasing crop vulnerability.

The potential breakdown of plant resistance to PPNs under climate change is a complex issue, with evidence suggesting that elevated CO_2 can repress jasmonic acid (JA)-mediated defense signaling in plants, potentially making them more susceptible. For instance, increased root galling was observed in JA-dominated tomato genotypes under elevated CO_2, indicating a possible weakening of plant resistance (Dutta and Phani 2023).

Plant resistance to root-knot and other parasitic nematodes is often mediated by major R-genes (e.g., *Mi* in tomato and *Rk1/Rk2* in tobacco), which trigger robust defense responses upon recognition of nematode effectors. However, key climatic factors associated with global change—particularly elevated temperatures, increased atmospheric CO_2, and altered moisture regimes—can modulate both nematode virulence and host defense pathways, potentially leading to partial or complete collapse of R-gene-mediated resistance.

16.4.1 Temperature Effects on Plant Resistance

High soil temperatures have long been known to compromise R-gene function. For example, tomato plants carrying the *Mi* gene show nearly complete resistance to *Meloidogyne incognita* at 28 °C, yet at 33 °C, 87% of juveniles successfully develop into feeding stages, indicating a temperature-dependent breakdown of resistance (Pollok et al. 2023).

More recently, Pollok et al. (2023) evaluated tobacco cultivars homozygous for *Rk1* and/or *Rk2* under growth-chamber regimes at 25 °C, 30 °C, and 35 °C. They found that resistance conferred by *Rk1* alone was effective at 25 °C and partly at 30 °C but was completely nullified at 35 °C, as indicated by gall counts and egg-mass data statistically indistinguishable from susceptible controls (Pollok et al.

2023). This illustrates that even closely related R-genes can vary in thermal stability, underscoring the risk of resistance failure under projected warming scenarios.

Elevated temperatures, a direct consequence of global warming, can significantly impair plant resistance to PPNs. Research suggests that heat stress affects R-gene-mediated resistance, which is crucial for plants to combat nematode infections. For instance, the *Mi* gene, introgressed from *Solanum peruvianum* into tomatoes (*Solanum lycopersicum*), is known to provide resistance to root-knot nematodes. However, studies indicate that its efficacy diminishes above 28 °C (Dutta and Phani 2023). This temperature threshold is critical, as many agricultural regions are expected to exceed it with continued warming.

Moreover, heat-tolerant nematode species, such as *Globodera rostochiensis*, exhibit thermo-tolerance through gene duplications (e.g., *hsp-110* gene), potentially giving them an advantage in warmer soils (Khanal and Land 2023). This suggests that certain nematodes may thrive in warmer conditions, potentially overwhelming plant resistance.

Pires et al.'s (2023) review also highlights that moderate warming initially increases PPN abundance, but further temperature rises can decrease it, indicating a complex relationship. This variability suggests that the impact on resistance may depend on the specific temperature range and nematode species, adding complexity to the issue.

16.4.2 CO_2 Effects on Plant Resistance

Elevated atmospheric CO_2 often reshapes plant hormone balances, with cascading effects on defense pathways. In tomato, Sun et al. (2011) grew isogenic genotypes under ambient (390 ppm) vs. elevated (750 ppm) CO_2. They observed genotype-specific shifts in the salicylic acid (SA) vs. JA pathways: elevated CO_2 repressed JA-mediated defenses (critical for nematode resistance) while enhancing SA responses, leading to modest but significant increases in root galling on JA-defense-dominated lines (Sun et al. 2011).

Furthermore, studies in *Medicago truncatula* show that elevated CO_2 can also modulate heat-shock proteins (e.g., HSP90) implicated in R-protein stability, suggesting that CO_2 enrichment may indirectly influence R-gene-dependent immunity via abiotic-stress chaperones (Sun et al. 2018).

Elevated CO_2 levels, another aspect of climate change, have dual effects on plant–nematode interactions. On one hand, they can suppress plant defense signaling, particularly JA-mediated pathways, which are critical for resisting nematode infections. For example, in tomatoes, elevated CO_2 (e.g., 500 ppm) represses JA signaling, leading to increased root galling by *Meloidogyne incognita* in JA defense-dominated genotypes, with no effect in JA defense-recessive or wild-type genotypes (Dutta and Phani 2023). This suppression can weaken plant resistance, making them more vulnerable to nematode attacks.

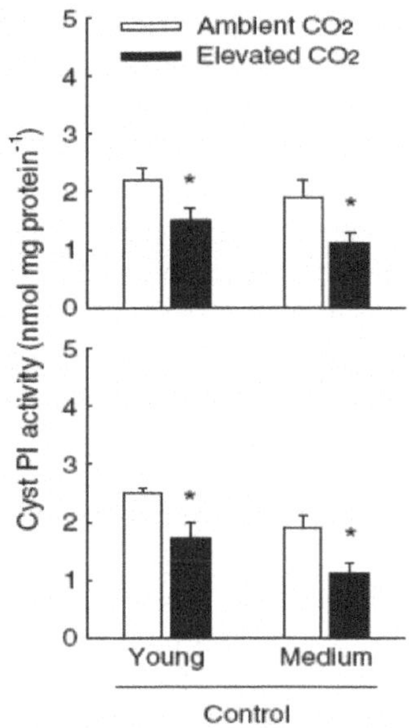

Fig. 16.1 Effect of downregulation of defense signaling genes on reduced production of cysteine proteinase inhibitors (deterrents of nematodes) at elevated CO_2 level

Down-regulation of genes for defense signaling (*Iox7, Iox8,* and acc-s) reduced the production of cysteine proteinase inhibitors at elevated CO_2 level, which are specific deterrents of nematodes, resulting in increased activity by the nematodes (Fig. 16.1).

On the other hand, elevated CO_2 may enhance plant resistance by increasing the production of defensive secondary metabolites, such as phenols, terpenes, and sesquiterpenes, due to higher carbon accumulation (Somasekhar and Prasad 2012). This suggests that while CO_2 enrichment may not always negatively affect resistance, the overall trend leans toward increased nematode pressure in many cases.

16.4.3 Drought, Precipitation Extremes, and Defense Trade-Offs

Shifts in precipitation regimes—ranging from drought to flooding—can alter resource allocation and immune signaling in crops. Drought stress often elevates abscisic acid levels, which can antagonize both SA- and JA-pathway signaling, potentially lowering the threshold for nematode infection. Conversely, waterlogging may impair root oxygen status and modify the expression of defense genes. Although direct studies on drought-induced R-gene breakdown in nematode resistance are scarce, analogous work on foliar pathogens indicates that combined abiotic stresses often precipitate quantitative losses of resistance.

16.4.4 Implications for Nematode Virulence and Resistance Durability

Environmental pressures on both host and pathogen may drive the emergence of more virulent nematode biotypes capable of overcoming weakened R-gene defenses. Temperature-modulated changes in membrane fluidity, histone modification, and reactive oxygen species (ROS) dynamics can affect recognition and signaling downstream of R-protein activation (Dutta and Phani 2023). These shifts, paired with potential CO_2-driven hormonal imbalances, may select for nematode isolates with altered effector repertoires or life-history traits, accelerating resistance breakdown.

16.4.5 Broader Mechanisms and Nematode Adaptation

Climate change can also alter plant immune responses through broader mechanisms, such as histone modifications, protein kinase cascades, and changes in membrane fluidity, which are triggered by temperature sensing (Dutta and Phani 2023). This potential for adaptation could further challenge existing resistance strategies.

Pires et al. (2023) review also notes that water stress, another climate change factor, can favor PPNs by releasing them from suppression, potentially reducing plant resistance under drought conditions. Combined stressors, such as drought and warming, can simplify nematode community structures, increasing the proportion of PPNs and exacerbating the breakdown of resistance.

16.4.6 Quantitative Insights and Supporting Studies

To provide a structured overview, Table 16.1 summarizes key studies and their findings on climate change effects on PPN resistance.

Table 16.1 includes studies that indirectly relate to resistance by discussing nematode abundance and community dynamics under climate change, providing a foundation for understanding the breakdown of resistance.

Table 16.1 Summary of key studies and their findings on climate change effects on PPN resistance

Key findings on resistance	References
Elevated CO_2 and warming affect nematode composition, potentially impacting resistance in rice paddies	Wang et al. (2019)
Long-term warming mediates plant–soil interactions, potentially altering resistance to PPNs in alpine meadows	Liu et al. (2022)
High precipitation increases PPN proportion, potentially reducing plant resistance	Franco et al. (2022)
Rainfall reduction decreases beneficial nematodes, with herbivores unaffected, potentially impacting resistance	Homet et al. (2023)
Climate change pervasively impacts plant–nematode interactions, potentially reducing plant resistance to PPNs	Dutta and Pani (2023)

16.4.7 Implications for Agriculture

The breakdown of plant resistance to PPNs under climate change has significant implications for food security, especially in regions already facing agricultural challenges. Increased nematode damage could lead to lower crop yields, particularly for staple crops such as rice, wheat, and tomatoes. Management strategies, such as developing heat-tolerant resistance genes, adjusting crop schedules, and using cover cropping, are proposed to mitigate these effects (Dutta et al. 2023). However, the complexity of these interactions suggests that more research is needed to devise effective, climate-resilient solutions.

16.4.8 Strategies for Climate-Resilient Resistance

- *Screen for Thermostable R-Genes*: Identify and introgress R-alleles that retain efficacy at ≥30 °C (e.g., novel *Mi* homologs).
- *Pyramiding and Stacking*: Combine multiple R-genes with distinct recognition spectra to buffer against temperature-induced losses in any single gene.
- *Enhance Chaperone Systems*: Breed or engineer plants with elevated expression of HSP90 and co-chaperones to maintain R-protein stability under heat stress (Sun et al. 2018).
- *Integrate Abiotic Stress Management*: Employ irrigation regimes and soil amendments to mitigate extreme moisture stresses that may compound resistance breakdown.

16.4.9 Future Research Directions

- Quantify the interactive effects of combined climate factors (e.g., heat + drought + elevated CO_2) on R-gene expression, protein stability, and downstream defense cascades.
- Monitor field populations of nematodes for shifts in virulence profiles in regions experiencing rapid warming.
- Leverage genome-editing (e.g., CRISPR: Clustered Regularly Interspaced Short Palindromic Repeats/CRISPR-associated (CRISPR/Cas) to fine-tune R-gene regulatory elements for improved thermal and hormonal resilience.

16.4.10 Conclusion

In summary, climate change induces consequences that contribute to the breakdown of plant resistance to PPNs, primarily through elevated temperatures reducing the efficacy of resistance genes and elevated CO_2 suppressing key defense signaling pathways. While some compensatory mechanisms (e.g., increased secondary metabolites) exist, the overall trend suggests increased vulnerability for plants, with potential agricultural and economic impacts. This analysis underscores the need for adaptive strategies to address these challenges in a warming world.

16.5 Nematode Abundance in Soil

Climate change encompasses a suite of environmental alterations—rising temperatures, elevated atmospheric CO_2 concentrations, shifting precipitation regimes, and increased frequency of extreme weather events—that collectively reshape soil ecosystems. PPNs, as soil-dwelling obligate parasites dependent on root systems, are highly sensitive to these abiotic shifts. Changes in nematode abundance in soil have profound implications for crop health and yield, as higher PPN populations amplify root damage and nutrient imbalances, whereas reductions in their numbers may temporarily relieve plant stress but disrupt soil food webs. This analysis synthesizes findings from recent scientific literature to provide a comprehensive understanding of these effects, highlighting both general trends and specific nuances.

Climate change may increase PPN abundance in soil due to warmer temperatures and altered precipitation, but effects vary by region and species. The elevated CO_2 levels could boost PPN populations by enhancing plant growth, though evidence is mixed. Extreme weather events potentially reduce PPN numbers temporarily, with complex long-term outcomes.

The impact of climate change on PPN abundance in soil is mixed, with studies showing both increases and decreases depending on environmental factors. Elevated CO_2 generally increases PPN abundance due to proliferated root biomass, as seen in wetlands with excessive abundance of *Longidorus* and *Pratylenchus* spp. under elevated CO_2 (Somasekhar and Prasad 2012).

Climate drivers alter the overall density and community structure of PPNs in soil.

16.5.1 Effects of Elevated Temperatures

Rising soil temperatures generally accelerate PPN life cycles, leading to shorter generation times and potentially greater population buildup over a growing season. For many species, each 1 °C increase within their optimal thermal range can boost metabolic and reproductive rates by 10–20% (Dutta and Phani 2023; Bristol et al. 2023). However, temperatures exceeding species-specific thermal maxima may reduce survival of free-living infective stages, leading to local declines in soil abundance. For example, in semiarid grasslands, studies combining warming and CO_2 enrichment observed a net decrease in PPN abundance—likely due to heat stress outpacing benefits of increased root biomass under those conditions (Dutta and Phani 2023).

Warmer temperatures can significantly influence PPN abundance by expanding their geographical range and accelerating their life cycles. For instance, studies show that species such as *Meloidogyne* and *Pratylenchus* have shifted latitudinally since the 1960s due to warming trends (Dutta et al. 2023). This range expansion is attributed to an expected global warming of 1.5–2 °C in the coming decades, as reported by the IPCC. Specific examples include the spread of *Meloidogyne graminicola* to all major rice belts in India, causing greater damage at 700 ppm CO_2 compared to 400 ppm.

Warmer temperatures can help PPNs spread to new areas and speed up their life cycles, potentially increasing their numbers. For example, species like *Meloidogyne* have expanded their range since the 1960s due to warming. However, very high temperatures might harm some PPNs, like *Meloidogyne floridensis*, reducing their survival at temperatures above 34 °C. Moderate warming increases PPN abundance by accelerating development and improving survival during milder winters, but extreme heat (>30 °C) can suppress some species or life stages (e.g., reduced egg viability) (Dutta and Phani 2023).

Moreover, warmer conditions can lead to faster development, with PPNs completing life cycles in 2–4 weeks and potentially producing multiple generations per season. For example, *Globodera rostochiensis* is predicted to have multiple generations per year in Finland by 2050 and *Meloidogyne incognita* in Brazil by 2020, 2050, and 2080 (Jones et al. 2017). However, the impact is species-specific and regionally variable. For instance, *Rotylenchulus reniformis* showed increased survival at 34 °C compared to 32 °C, suggesting resilience, while *Meloidogyne floridensis* experienced a 12% survival decrease at 34 °C, indicating sensitivity to extreme heat, as found in Khanal and Land (2023)). In contrast, warming reduced PPN abundance in the Qinghai–Tibetan plateau and semiarid grasslands, as reported by Jones et al. (2017).

Additional effects include sex reversal in parthenogenetic root-knot nematodes at high temperatures and loss of resistance genes like *Mi* above 28 °C. These findings underscore the complexity of temperature impacts, with both positive and negative outcomes depending on the context.

16.5.2 Impact of Elevated CO_2

Elevated atmospheric CO_2 generally enhances plant photosynthesis and root proliferation, thereby expanding the resource base for root-feeding nematodes. Meta-analysis of pasture and crop systems under CO_2 enrichment (from ~360 ppm to ~475 ppm) reports up to 30–50% increases in certain PPN genera (e.g., *Longidorus* and *Pratylenchus*) compared to ambient conditions (Dutta and Phani 2023). However, responses can be idiosyncratic: increased root biomass may be offset by lowered tissue nitrogen content or by stimulated populations of natural antagonists, yielding neutral nematode-abundance outcomes in some grassland ecosystems.

Elevated CO_2 levels, a key aspect of climate change, can indirectly affect PPN abundance by promoting plant growth and increasing root biomass, potentially providing more food for nematodes. Studies in New Zealand pastures showed increased PPN abundance at 475 ppm CO_2 compared to 360 ppm, with excessive *Longidorus* abundance (Dutta et al. 2023). Similarly, *Pratylenchus* abundance increased in wetland soils under elevated CO_2.

However, the evidence is mixed. Some studies found no change in total nematode abundance at ~350 ppm excess CO_2 in USA and France pastures, while others reported decreased PPN abundance affecting soil food webs (Sun et al. 2011). The interaction with plant defenses also varies; elevated CO_2 can repress JA-mediated defense, increasing root galling in tomato, but in some cases, increased antagonists under elevated CO_2 mitigated PPN damage (Sun et al. 2011).

Higher CO_2 levels might increase PPN abundance by promoting plant growth, providing more food for nematodes, but this is not always the case and depends on the environment. Changes in rainfall can also affect PPNs: more rain often helps them by boosting plant biomass, while droughts can sometimes reduce their numbers or change how they interact with soil.

16.5.3 Effects of Altered Precipitation

A recent meta-analysis of 37 field studies (46 observations) revealed that reduced rainfall consistently depresses soil nematode abundances—total nematodes declined by ~45%, and plant-parasitic groups by ~46%—irrespective of treatment duration. In contrast, increased rainfall (>1-year studies) had negligible effects on total nematodes but did promote plant parasites by ~9%, likely via enhanced plant productivity and root exudation (Bristol et al. 2023). Thus:

- *Drought Scenarios*: Pronounced declines in PPN abundance (↓ ~ 45%), risking disruption of soil processes
- *Prolonged Wetter Conditions*: Modest increases in PPN populations, particularly root feeders, with potential cascading root-health impacts

Changes in precipitation patterns, including increased rainfall and droughts, have significant but variable effects on PPN abundance. Increased rainfall benefits PPNs by boosting plant biomass, supporting larger populations. For example, long-term mean annual precipitation increases were linked to higher PPN abundance in grasslands. Specific studies in Northern Ireland showed increased prevalence of *Meloidogyne minor*, *Heterodera* spp., and *Pratylenchus* spp. in grasslands and cereals under higher rainfall.

Droughts, however, can have mixed effects. In New Jersey, *Aphelenchoides* was negatively affected by droughts, while *Criconematidae*, *Hoplolaimidae*, and *Pratylenchidae* were unaffected. In lightly degraded grasslands, decreased precipitation frequency increased PPN abundance, but this reversed in moderate to severe conditions. Combined effects of warming and drought can decrease plant production and nutrient turnover, affecting PPNs, with responses varying by soil layers and seasons.

Increased rainfall generally boosts soil moisture, favoring moisture-dependent nematodes (e.g., *Aphelenchoides* spp. and *Pratylenchus* spp.), whereas drought can reduce predation pressure on PPNs—indirectly increasing their populations in some grassland and cropping systems (Dutta and Phani 2023).

16.5.4 Extreme Weather Influence

Events such as heat waves, wildfires, and storms can lower PPN populations in the short term, but they might also help spread nematodes to new areas, leading to complex long-term effects.

Extreme climate events, such as heat waves, wildfires, and storms, can temporarily reduce PPN populations but may also lead to long-term shifts. Heat waves may affect PPN growth and plant photosynthetic ability, potentially causing sex reversal in root-knot nematodes. Wildfires detrimentally affect PPNs, with no recovery in 4 years post-fire in some cases, depending on fire severity. Storms can reduce vegetation and alter PPN abundance short term but may disseminate PPNs to new locations, such as cyst nematodes, via wind.

16.5.5 Ecological Interactions

Altered moisture regimes disrupt synchronization between PPNs and their natural antagonists (fungal parasites and predatory nematodes), often tipping soil food webs toward higher PPN dominance under both wetter and drier extremes (Dutta and Phani 2023).

16.5.6 Interactive Effects and Soil Food-Web Dynamics

Climate change factors seldom act in isolation. Elevated CO_2 and increased moisture may synergistically boost root growth, compounding PPN reproduction rates. Conversely, simultaneous warming and drought can suppress nematode activity while also weakening plant defenses, making roots more vulnerable to the surviving PPN cohorts. Moreover, shifts in microbial grazer and predator abundances under altered moisture regimes can indirectly modulate PPN populations by changing predation pressures and microbial-mediated nutrient cycling (Dutta and Phani 2023).

16.5.7 Combined Effects and Ecosystem-Specific Responses

Climate change often involves multiple interacting factors, leading to complex outcomes. For instance, increasing aridity led to a steady decline in PPN richness (Xiong et al. 2020), while in desert steppes, warming significantly decreased PPN abundance, with water stress having less impact (Zhang et al. 2020). In temperate forests, nitrogen addition and reduced water negatively impacted PPN abundance and diversity (Wang et al. 2019).

The response of PPNs to climate change is also influenced by soil layers, seasons, and predator dynamics. For example, reduced predation under dry years can increase PPN populations in mesic grasslands, worsening drought effects. Endoparasitic vs. ectoparasitic PPNs show different responses to aridity, with endoparasites increasing in semiarid sites and declining in mesic sites.

16.5.8 Case Studies and Geographic Variability

- *Qinghai–Tibetan Plateau*: Moderate warming has been correlated with reduced PPN abundance, likely reflecting species-specific thermal thresholds at high altitudes (Dutta and Phani 2023).
- *Australian Grasslands and Deserts*: Long-term rainfall increases boosted PPNs, whereas rainfall reductions led to substantial declines across trophic groups (Bristol et al. 2023).
- *Western Himalaya (Model Predictions)*: Metabolic-footprint analyses forecast PPN abundance increases at higher altitudes under warming scenarios, underscoring regional heterogeneity in responses.

Table 16.2 summarizes key studies and their findings on PPN abundance under climate change factors.

Table 16.2 encapsulates the diversity of research, highlighting the variability in PPN responses to climate change across different ecosystems and conditions.

Climate-induced shifts in temperature, CO_2, and precipitation exert multifaceted impacts on soil-borne PPN abundance. While warming tends to accelerate nematode life cycles within optimal ranges, excessive heat can curtail populations. Elevated CO_2 often fosters PPN proliferation through greater root biomass, yet plant defense changes and biocontrol dynamics may moderate this effect. Altered precipitation—especially drought—has the most pronounced negative impact on soil nematode abundance, whereas sustained increases in rainfall may modestly favor root feeders. Predictive models and region-specific studies are essential to anticipate

Table 16.2 Key studies and their findings on PPN abundance under climate change factors

Focus	Findings on PPN abundance	References
Global climate change impact	Influences range, distribution, abundance, survival, fitness, reproduction, parasitic potential	Dutta et al. (2023)
Coffee nematodes, Brazil	Climate change risk analysis shows potential for increased PPN damage	Ghini et al. (2008)
Precipitation, grasslands	High PPN proportion under increased mean annual precipitation	Franco et al. (2022)
Rainfall reduction, forests	Lower water input negatively affects nematode abundance, including PPNs	Homet et al. (2023)

PPN population dynamics and to tailor integrated management strategies in a changing climate.

In summary, climate change is likely to increase PPN abundance in many regions due to range expansion, faster life cycles, and enhanced plant biomass under warmer temperatures and increased CO_2. However, extreme weather events and droughts can temporarily reduce populations, with long-term outcomes depending on ecosystem-specific interactions. The complexity of these effects underscores the need for continued research to develop adaptive management strategies for agricultural systems.

16.6 Increase in Number of Generations

Warmer temperatures are likely to lead to more PPN generations per year by shortening their life cycles. Studies predict that species like *Globodera rostochiensis* may have increased generations per year by 2050 in Finland due to climate change, driven by temperature-dependent development rates (Khanal and Land 2023).

Climate change significantly impacts PPNs, increasing the number of generations per season and exacerbating crop damage. Rising temperatures, altered precipitation, and elevated CO_2 levels drive these changes, though effects vary by species and region. Here is a detailed breakdown.

16.6.1 Temperature Increase

Warmer soil temperatures accelerate nematode life cycles, which typically span 2–4 weeks under optimal conditions. Higher temperatures shorten development time, enabling more generations per season. For example, simulation models predict that global warming will increase the number of generations of the potato cyst nematode (*Globodera rostochiensis*) by 2050 in Finland (Dutta and Phani 2023).

Studies on *Rotylenchulus reniformis* and *Meloidogyne floridensis* showed reproduction increases of 20–116% and 22–133%, respectively, at soil temperatures of 32–34 °C compared to 26 °C (Sagar et al. 2025).

A 1 °C rise in soil temperature can lead to northward migration of virus-transmitting nematodes (e.g., *Xiphinema* and *Longidorus*) by 160–200 km, expanding their range and generation potential.

16.6.2 CO_2 Enrichment

Elevated CO_2 levels enhance plant photosynthesis, leading to increased root biomass. This provides more resources for PPNs, potentially increasing their reproduction and generation rates. However, some studies report neutral responses to CO_2 enrichment despite increased root production (Dutta and Phani 2023).

16.6.3 Precipitation Changes

Altered rainfall patterns affect soil moisture, influencing nematode survival and reproduction. Increased rainfall can boost plant productivity, benefiting PPNs like *Meloidogyne incognita*, which may complete more generations in wetter conditions. Conversely, drought can reduce nematode populations in arid regions by limiting water availability (Bristol et al. 2023; Pires et al. 2023).

16.6.4 Extreme Weather Events

Storms and extreme weather can disperse PPNs (e.g., cyst nematodes) to new areas, facilitating range expansion and potentially increasing local generation numbers due to new host availability (Dutta and Phani 2023).

16.6.5 Counteracting Factors

Some factors may mitigate generation increases, such as sex reversal, cryptobiosis, or reduced survival in extreme heat or dry conditions. These adaptations vary by species and location, creating a complex balance of effects (Dutta et al. 2023).

16.6.6 Implications for Crop Damage

More generations per season amplify PPN damage by increasing population density, virulence, and disease severity. For instance, *Meloidogyne incognita* is predicted to spread across entire coffee plantations in Brazil due to rising temperatures, with more generations per month (Sagar et al. 2025).

Reduced plant defenses under heat stress and ineffective management strategies (e.g., unstable resistance genes at higher temperatures) further exacerbate crop losses (Sagar et al. 2025).

16.6.7 Management Challenges

Traditional nematicides and fumigants are less effective under warmer conditions, and resistance genes often fail above 27 °C. Climate-smart strategies, such as crop rotation, cover cropping, biofumigation, and resistant varieties, are recommended to adapt to increased PPN generations (Dutta and Phani 2023; Sagar et al. 2025).

In summary, climate change generally increases the number of PPN generations by accelerating life cycles, expanding ranges, and enhancing reproduction, though effects depend on species, region, and environmental conditions. These changes pose significant threats to global food security, necessitating adaptive, sustainable management practices (Dutta and Phani 2023; Sagar et al. 2025).

16.7 Reduced Effectiveness of Nematode Management Technologies

Climate change poses multifaceted challenges to the efficacy of existing PPN management technologies. Elevated temperatures, altered precipitation patterns, increased atmospheric CO_2, and extreme weather events can directly and indirectly affect chemical nematicides, biological control agents (BCAs), cultural practices, and host-plant resistance. These changes may speed up nematode reproduction, expand their range, and stress plants, reducing the effectiveness of current methods like chemical nematicides. For example, studies show that higher temperatures can increase nematode damage by up to 46% in some cases, making existing controls less reliable. This chapter synthesizes recent findings (2023–2025) on how climatic variables undermine standard nematode management tools and underscores the necessity for climate-smart, adaptive strategies.

16.7.1 Direct Impacts on Nematode Biology

Research indicates that climate change significantly alters PPN biology, affecting their development, reproduction, and virulence. For instance, higher temperatures can shorten nematode life cycles, enhancing reproduction rates. A study on *Rotylenchulus reniformis* and *Meloidogyne floridensis* found that at 34 °C, compared to 32 °C, *R. reniformis* survival increased by 10%, while *M. floridensis* survival decreased by 12%, but disease severity (root gall index) for *M. floridensis* increased by 46% (Khanal and Land 2023). This suggests that while some nematodes may face survival challenges at extreme temperatures, their damage potential can still escalate, reducing the effectiveness of current management technologies.

Elevated CO_2 levels also influence nematode populations. Studies show varied responses, with some ecosystems experiencing increased nematode abundance under CO_2 enrichment (e.g., 550 ppm), particularly fungivores and bacterivores, while others see declines (Somasekhar and Prasad 2012). For example, in grasslands, CO_2 levels up to 700 ppm showed no negative effect on soil nematodes, with *Meloidogyne graminicola* penetration remaining unaffected (Wang et al. 2019). These shifts can alter community structures, complicating the application of uniform management strategies.

16.7.2 Indirect Impacts via Plant Physiology

Climate change also affects host plants, indirectly influencing PPN interactions. Elevated CO_2 and temperature can stress plants, reduce their defense mechanisms, and make them more susceptible to nematode attacks. For instance, canopy warming in winter wheat increased CO_2 concentration, raising nematode diversity by 6.4% without affecting abundance (Wang et al. 2021), suggesting complex interactions that may undermine current control measures.

16.7.3 Geographical and Distributional Changes

Global warming may expand the geographical range of PPNs, spreading nematode problems to new areas. A notable example is the predicted northward expansion of virus-transmitting nematodes by 160–200 km in Great Britain with a 1 °C temperature increase (Rupasinghe et al. 2022). This expansion introduces nematodes to regions where existing management practices may not be optimized, reducing their effectiveness and necessitating region-specific strategies.

16.7.4 Challenges to Current Management Technologies

Current nematode management technologies, particularly chemical nematicides, face reduced efficacy under climate change. Higher temperatures and altered soil conditions can diminish nematicide performance, while environmental and health concerns limit the use of fumigants, which have adverse effects (Khanal and Land 2023). Additionally, global pesticide reduction goals discourage the development of novel nematicidal molecules such as fluensulfone and fluopyram (Desaeger et al. 2020), further constraining chemical options.

Biological and cultural controls, such as crop rotation and cover cropping, may also be affected. For example, slower decomposition under elevated CO_2 (e.g., wheat straw decomposition reduced by 30% in fallow and 13% in wheat fields) can alter nematode food sources, potentially reducing the effectiveness of residue management strategies (Somasekhar and Prasad 2012).

16.7.5 Adaptive and Integrated Management Strategies

To address these challenges, "climate-smart" integrated nematode management practices are recommended. These include the following:

- *Precise Monitoring and Modeling*: Using climate modeling to predict nematode population dynamics under changing conditions, such as the potential impact on soybean cyst nematode distribution (Dutta and Phani 2023)
- *Increased Reliance on Biocontrol*: Enhancing natural enemies and biopesticides, which are safer and have lower greenhouse gas emissions compared to flooding (amplifies GHG) or fallowing (increases N_2O)
- *Host Resistance and Cultural Practices*: Promoting resistant plant varieties, crop rotation, cover cropping, and biofumigation to mitigate nematode damage under elevated CO_2 and temperature (Dutta and Phani 2023)
- *Soil Health Management*: Focusing on soil structure to support larger, higher-trophic-level nematodes, especially under conditions where aggregate sizes are reduced by 43% under elevated CO_2 (Somasekhar and Prasad 2012)

These strategies aim to adapt to the changing climate, ensuring sustainable agricultural productivity while reducing reliance on environmentally harmful methods.

16.7.6 Economic and Agricultural Implications

The economic impact is significant, with PPNs already causing substantial losses. Climate change exacerbates this by potentially increasing nematode damage and reducing management efficacy, threatening food security. Adaptive strategies are crucial to mitigate these risks, particularly in regions where nematode expansion is predicted, such as arid zones and higher elevations.

Table 16.3 summarizes the effects of climate change on nematode management of plant nematodes.

16.7.7 Chemical Nematicides

- *Temperature-Driven Degradation and Volatilization*: Soil temperature influences nematicide persistence and bioavailability. Fluctuating and rising temperatures accelerate chemical breakdown and volatilization, shortening effective field life and reducing season-long protection (Khanal and Land 2023). For instance, studies indicate that heat waves (>35 °C) expedite hydrolysis of carbamate and organophosphorus nematicides, demanding higher application rates to maintain efficacy—an unsustainable approach given environmental and regulatory constraints.
- *Moisture and Leaching Effects*: Altered precipitation patterns can either leach nematicides below the root zone during heavy rainfall or limit distribution during droughts. Reduced soil moisture impedes nematicide movement to nematode feeding sites, whereas excessive rainfall increases runoff and environmental contamination.

Table 16.3 Summary of climate change effects on nematode management

Climate factor	Nematode response	Impact on management	Example study
Elevated temperature	Shortened life cycle, increased reproduction, varied survival	Reduces nematicide efficacy, needs heat-resistant strategies	*R. reniformis* survival +10% at 34 °C vs. 32 °C
Elevated CO_2 (e.g., 550 ppm)	Increased abundance, shifts in community structure	Requires crop-specific, soil-focused management	Fungivores increased under winter wheat
Altered precipitation	Affects soil moisture, nematode activity	Needs adaptive irrigation and cultural practices	Impact on nematode community structure
Range expansion	Spreads nematodes to new areas	Requires region-specific monitoring and controls	Northward shift by 160–200 km with a 1 °C rise

16.7.8 Biological Control Agents

- *Thermal Sensitivity of Endospores and Fungal Propagules*: BCAs such as *Pasteuria penetrans* and *Pochonia chlamydosporia* exhibit temperature-dependent activity. Exposure of *Pasteuria* endospores to temperatures ≥50 °C diminishes their ability to attach to nematode cuticles, reducing infection rates significantly (Freitas et al. 1997). Conversely, *P. chlamydosporia* propagules may proliferate under moderate warming (25–30 °C), but extreme heat (>35 °C) curtails sporulation and viability (Luambano et al. 2015).
- *Soil Moisture Extremes*: Drought stress can suppress BCA sporulation and movement, while waterlogging from intense rainfall events can reduce oxygen levels critical for aerobic antagonists. Such extremes disrupt synchronization between BCAs and nematode life cycles, leading to temporal mismatches in biocontrol efficacy.

16.7.9 Cultural and Physical Practices

- *Crop Rotation and Resistant Cultivars*: Altered climatic conditions can shift PPN species composition and virulence, undermining host-plant resistance deployed via crop rotation or resistant varieties. Heat stress may downregulate R-gene expression, diminishing resistance against root-knot nematodes (e.g., *Meloidogyne* spp.) in tomato and potato (Luambano et al. 2015).
- *Soil Solarization and Steam Treatment*: Solarization efficacy hinges on consistent high soil temperatures. Increased cloud cover or erratic weather reduces thermal accumulation, lowering nematode mortality rates. Similarly, steam treatments require more energy input under cooler, wetter conditions to achieve lethal temperatures.

16.7.10 Host-Plant Resistance

- *Temperature-Modulated Defense Signaling*: Elevated CO_2 can alter plant defense pathways, such as JA signaling. In tomato, elevated CO_2 suppressed JA-mediated responses against *M. incognita*, increasing galling severity on defense-dominated genotypes (Sun et al. 2011). Moreover, heat stress affects histone modifications and ROS signaling crucial for R-gene function, potentially breeding new virulent PPN biotypes.
- *Case Study—Potato Nematode Resistance*: A 2025 field trial in contrasting climates Use of Land vs. Adaptation, Resilience, and Climate-Vulnerability Impact Mitigation Plan (UL vs. ARC-VIMP) demonstrated that heat stress and drought reduced nematode resistance efficacy in potato cultivars, with significant declines in population suppression under higher temperatures and water scarcity (Khanal and Land 2023). This underscores the importance of multi-environmental testing prior to cultivar release.

16.7.11 Integrated Decision Support Systems

Climate change adds complexity to decision support system (DSS) calibration. Predictive models based on historical weather patterns may misestimate PPN pressure and management thresholds. Incorporating real-time climate data and ensemble forecasting can enhance DSS robustness under changing conditions, yet such tools remain underutilized in nematology.

16.7.12 Conclusion and Future Directions

Climate change poses significant challenges to plant nematode management technologies by altering nematode biology, distribution, and interactions with plants. The collective evidence indicates that climate change erodes the performance of chemical, biological, cultural, and resistance-based nematode management technologies. The reduced effectiveness of current methods necessitates adaptive, integrated strategies that prioritize sustainability and resilience. Ongoing research and implementation of climate-smart practices are essential to safeguard agricultural productivity in the face of these environmental changes. To safeguard crop health, the nematode management community must:

- Develop climate-resilient nematicides with minimal temperature- and moisture-sensitivity.
- Breed R-gene stacks and pyramiding strategies stable across temperature ranges.
- Optimize BCA formulations for broader thermal tolerance and moisture adaptability.
- Enhance DSS with climate-smart modules, integrating predictive climate models and real-time monitoring.
- Adopt diversified, adaptive IPM that leverages crop diversification, soil health improvement, and low-input tactics.

A shift toward climate-smart nematode management—prioritizing resilience, sustainability, and minimized greenhouse gas emissions—is imperative for future agricultural security.

16.8 Case Studies

16.8.1 Pole-Ward Shift

Predictive modeling for *Meloidogyne incognita* (southern root-knot nematode) under IPCC A2/B2 emission scenarios indicates that areas supporting 9–12 annual generations—which correlate with severe crop damage—could increase from 41% to 59% of Brazil's coffee-growing regions in the 1961–1990 baseline to substantially larger proportions by the 2080s under high-emission scenarios (Ghini et al.

2008). Similar trends have been documented for the pine wilt nematode (*Bursaphelenchus xylophilus*), whose spread in Asia, northern Europe, and North America has been linked to warming summer temperatures that both favor nematode development and stress host trees (Dutta and Phani 2023).

A 1 °C rise in soil temperature can lead to northward migration of virus-transmitting nematodes (e.g., *Xiphinema* and *Longidorus*) by 160–200 km, expanding their range and generation potential.

16.8.2 Latitudinal Shifts

Soybean cyst nematode (*Heterodera glycines*), first detected in Ontario in 1987, now occurs throughout southern and eastern Canada, aided by warmer winters and longer growing seasons (Dutta and Phani 2023).

16.8.3 Nematode Resistance Breakdown

More recently, Pollok et al. (2023) evaluated tobacco cultivars homozygous for *Rk1* and/or *Rk2* under growth-chamber regimes at 25 °C, 30 °C, and 35 °C. They found that resistance conferred by *Rk1* alone was effective at 25 °C and partly at 30 °C, but was completely nullified at 35 °C, as indicated by gall counts and egg-mass data statistically indistinguishable from susceptible controls (Pollok et al. 2023). This illustrates that even closely related R-genes can vary in thermal stability, underscoring the risk of resistance failure under projected warming scenarios.

16.8.4 Increased Generations Per Year

Studies predict that species like *Globodera rostochiensis* may have increased generations per year by 2050 in Finland due to climate change, driven by temperature-dependent development rates (Khanal and Land 2023).

16.8.5 Nematode Abundance in Soil

Elevated CO_2 generally increases PPN abundance due to proliferated root biomass, as seen in wetlands with excessive abundance of *Longidorus* and *Pratylenchus* spp. under elevated CO_2 (Somasekhar and Prasad 2012).

16.9 Broader Implications and Research Gaps

The interactions between climate change and PPNs are complex, with differential effects across species and ecosystems. For instance, while elevated CO_2 can increase PPN abundance in some cases, grassland experiments showed no significant effect on abundance or diversity, possibly due to decreased root quality (Khanal and Land 2023).

PPN life-cycle duration is highly temperature-dependent. Warmer soils speed up development, allowing more life cycles per season:

16.9.1 Faster Development

Under ambient temperature conditions (~15–20 °C), many cyst and root-knot nematodes complete two to three generations per growing season. A 2 °C increase in soil temperature can raise the cumulative thermal time (degree-days) sufficiently to add an extra generation of *Globodera* spp. and *Meloidogyne* spp. annually (Potato World Magazine 2023).

16.9.2 Model Projections

Simulations for Finland predict that *G. rostochiensis* could complete up to 1.5 more generations per year by 2050, compared to the present climate, substantially increasing population multiplication rates (Carter et al. 1996). Similar risk analyses in Brazilian coffee systems indicate that *M. incognita* races may undergo 20–30% more generations per month under mid-century warming scenarios (Dutta and Phani 2023).

16.9.3 Management Implications

Additional generations shorten the intervals between soil-sanitation windows and erode the efficacy of crop rotations, demanding tighter scheduling of control measures and hastening the need for new nematicides or biocontrol agents.

16.10 Conclusion

Climate change is likely to have significant and varied effects on PPNs, influencing their distribution, resistance, host interactions, abundance, and generation numbers. Rising temperatures, CO_2 enrichment, and changing rainfall patterns collectively enable PPNs to invade new regions, reproduce faster, overcome crop resistance, and build larger soil populations. Together, these trends threaten to elevate nematode-induced yield losses worldwide, underscoring the urgency of integrating climate resilience into nematode-management strategies. These changes pose challenges to agricultural systems, necessitating integrated management approaches to mitigate impacts on crop productivity. Continued research is essential to understand these dynamics fully and develop effective strategies under evolving climatic conditions.

16.11 Suggested Readings

- Bristol, D., Hassan, K., Blankinship, J.C. and Nielsen, U.N. 2023. Responses of nematode abundances to increased and reduced rainfall under field conditions: A meta-analysis. *Ecosphere* 14(1). https://doi.org/10.1002/ecs2.4364.
- Dutta, T.K. and Phani, V. 2023. The pervasive impact of global climate change on plant nematode interaction continuum. *Front. Plant Sci.* 14: 1143889. https://doi.org/10.3389/fpls.2023.1143889.
- Dutta, T.K., Vicente, C.S.L. and Maleita, C.M.N. 2023. Editorial: Impact of global climate change on the interaction between plants and plant-parasitic nematodes. *Front. Plant Sci.* 14: https://doi.org/10.3389/fpls.2023.1195970.
- Jones, L.M., Koehler, A.K., Trnka, M. et al. 2017. Climate change is predicted to alter the current pest status of *Globodera pallida* and *G. rostochiensis* in the United Kingdom. *Glob. Chang. Biol.* 23: 4497–4507.
- Somasekhar, N. and Prasad, J.S. 2012. Plant–nematode interactions: consequences of climate change. In B. Venkateswarlu, A.K. Shanker, C. Shanker and M. Maheswari (eds.) *Crop Stress and its Management: Perspectives and Strategies*, pp. 547–564. Springer, Dordrecht.

16.12 Review Questions

- Discuss the potential shifts in host range and geographic distribution of major PPN species under future climate scenarios.
- Why might climate change increase the damage potential of sedentary endoparasites such as *Meloidogyne* spp. and *Heterodera* spp.?
- How will climate change affect the efficacy of nematode management strategies (chemical, cultural, and biological)?
- How do increased soil temperatures affect nematode life cycle, reproduction, and population dynamics?

References

Bristol D, Hassan K, Blankinship JC, Nielsen UN (2023) Responses of nematode abundances to increased and reduced rainfall under field conditions: a meta-analysis. Ecosphere 14(5):e4364

Carter TR, Saarikko RA, Niemi KJ (1996) Assessing the risks and uncertainties of regional crop potential under a changing climate in Finland. Agric Food Sci Finl 5:329–350

Desaeger J, Wram C, Zasada I (2020) New reduced-risk agricultural nematicides – rationale and review. J Nematol 52:e2020–e2091. https://doi.org/10.21307/jofnem-2020-091

Dutta TK, Phani V (2023) The pervasive impact of global climate change on plant nematode interaction continuum. Front Plant Sci 14:1143889. https://doi.org/10.3389/fpls.2023.1143889

Dutta TK, Vicente CSL, Maleita CMN (2023) Editorial: impact of global climate change on the interaction between plants and plant-parasitic nematodes. Front Plant Sci 14. https://doi.org/10.3389/fpls.2023.1195970

EPPO Global Database (2021) Recommendations from Euphresco projects. EPPO Reporting Service No. 08–2021, Num. article: 2021/169

Franco ALC, Guan P, Cui S et al (2022) Precipitation effects on nematode diversity and carbon footprint across grasslands. Glob Chang Biol 28:2124–2132

Freitas LG, Mitchell DJ, Dickson DW et al (1997) Temperature effects on the attachment of *Pasteuria penetrans* endospores to *Meloidogyne arenaria* race 1. J Nematol 29(4):547–555

Ghini R, Hamada E, Pedro MJ Jr et al (2008) Risk analysis of climate change on coffee nematodes and leaf miner in Brazil. Pesq Agropec Bras, Brasília 43(2):187–194

Helder JH (2024) Plant-parasitic nematodes on the rise: WUR launches major research project. Wageningen Univ. News, January 9, 2024

Homet P, Matías L, Ourcival JM et al (2023) Short- and long-term responses of nematode communities to predicted rainfall reduction in Mediterranean forests. Soil Biol Biochem 179:108974

Jones LM, Koehler AK, Trnka M et al (2017) Climate change is predicted to alter the current pest status of *Globodera pallida* and *G. rostochiensis* in the United Kingdom. Glob Chang Biol 23:4497–4507

Khanal C, Land J (2023) Study on two nematode species suggests climate change will inflict greater crop damage. Sci Rep 13:14185

Liu Y, Wang W, Liu P et al (2022) Plant-soil mediated effects of long-term warming on soil nematodes of alpine meadows on the Qinghai–Tibetan plateau. Biology 11:1596

Luambano N, Lopez RHM, Kimenju JW et al (2015) Effect of temperature, pH, carbon and nitrogen ratios on the parasitic activity of *Pochonia chlamydosporia* on *Meloidogyne incognita*. Biol Control 80. https://doi.org/10.1016/j.biocontrol.2014.09.003

Pankaj S, Sharma HK, Prasad JS (2010) The rice root-knot nematode, *Meloidogyne graminicola*: an emerging problem in rice-wheat cropping system. Indian J Nematol 40(1):1–11

Pires D, Orlando V, Collett RL et al (2023) Linking nematode communities and soil health under climate change. Sustainability 15:11747

Pollok JR, Johnson CS, Eisenback JD et al (2023) Effect of soil temperature on reproduction of root-knot nematodes in flue-cured tobacco with homozygous Rk1 and/or Rk2 resistance genes. J Nematol 55(1):20230032

Potato World Magazine (2023) How climate change affects nematodes in potato cultivation. Potato World Magazine, June 5, 2023

Rupasinghe R, Chomel BB, Martínez-López B (2022) Climate change and zoonoses: a review of the current status, knowledge gaps, and future trends. Acta Trop 226:106225. https://doi.org/10.1016/j.actatropica.2021.106225

Sagar GC, Banakar P, Harshman D, Khanal C (2025) Elevated soil temperatures impact nematode reproduction biology. Stress 5:2

Somasekhar N, Prasad JS (2012) Plant–nematode interactions: consequences of climate change. In: Venkateswarlu B, Shanker AK, Shanker C, Maheswari M (eds) Crop stress and its management: perspectives and strategies. Springer, Dordrecht, pp 547–564

Sun Y, Yin J, Cao H et al (2011) Elevated CO_2 influences nematode-induced defense responses of tomato genotypes differing in the JA pathway. PLoS One 6(5):e19751

Sun Q, Guo H, Yuan E et al (2018) Elevated CO_2 increases R gene-dependent resistance of *Medicago truncatula* against the pea aphid by up-regulating a heat shock gene. New Phytol 217(4):1696–1711

Wang J, Li M, Zhang X et al (2019) Changes in soil nematode abundance and composition under elevated (CO_2) and canopy warming in a rice paddy field. Plant Soil 445:425–437

Wang J, Shi X, Li L, Zhang X (2021) Changes in soil nematodes in rhizosphere and non-rhizosphere soils following combined elevated [CO_2] and canopy warming in a winter wheat field. Geoderma 386:114907. https://doi.org/10.1016/j.geoderma.2020.114907

Zhang G, Sui X, Li Y et al (2020) The response of soil nematode fauna to climate drying and warming in *Stipa breviflora* desert steppe in Inner Mongolia, China. J Soils Sediments 20:2166–2180

Development of Nematode Prediction Models Under Climate Change

17

Abstract

Climate change will likely increase nematode populations and disease risk, especially for plant-parasitic nematodes, due to rising temperatures and altered precipitation. Prediction models can forecast these changes by integrating climate data and nematode biology, aiding in developing management strategies. These models shall be useful for agriculture, but uncertainties remain due to data gaps and nematode adaptability. This chapter provides a comprehensive examination of developing prediction models for nematode pathogens under climate change, focusing on forecasting changes in nematode populations and disease risk to aid in effective management strategies. It builds on recent research and integrates findings from multiple studies to offer a detailed perspective for researchers, policymakers, and agricultural practitioners.

Keywords

Correlative species distribution models · Process-based dynamic models · Population dynamics modeling · Mechanistic niche models · Machine learning time-series forecasting

17.1 Learning Objectives

- Define predictive models and explain their relevance in forecasting plant–nematode interactions under changing climate conditions.
- Differentiate between correlative species distribution models (SDMs), mechanistic niche models (e.g., CLIMEX), and machine learning (ML) time-series forecasting models used for predicting nematode population dynamics and distribution.
- Learn about data requirements such as field observations, experimental trials, remote sensing inputs, and climate projections (e.g., The Intergovernmental Panel on Climate Change (IPCC) scenarios) for model development.

R. P. Parvatha, *Plant Parasitic Nematodes*,
https://doi.org/10.1007/978-981-95-9306-4_17

- Assess how predictive models can be applied to forecast nematode outbreaks, shifts in geographical distribution, and changes in crop–nematode interactions under climate change scenarios.
- Critically evaluate the challenges and limitations associated with model predictions.
- Implications of predictive modeling for management strategies (breeding resistant varieties, adjusting agricultural practices, and timing and scheduling).

17.2 Introduction

Plant-parasitic nematodes (PPNs) are particularly significant in agriculture, causing substantial economic losses by feeding on plant roots and reducing crop yields. They are expected to be influenced by climate change, potentially increasing disease risks. Prediction models help forecast these changes, supporting strategies to manage nematode impacts, especially in agriculture.

Accurately forecasting shifts in PPN populations and associated disease risks under future climates is critical for proactive management, enabling stakeholders to allocate resources and deploy control measures before outbreaks occur. Climate change alters temperature and moisture regimes, directly affecting nematode survival, reproduction rates, and the number of generations per season (Rose et al. 2016). Simulated future climates also reshape soil food webs, often favoring opportunistic plant-feeding nematodes and reducing community resilience to stress, thereby amplifying disease risk (Siebert et al. 2020).

The need for prediction models arises from the potential for climate change to exacerbate nematode-related disease risks, particularly in agricultural systems. These models aim to forecast how nematode populations will respond to future climate scenarios, enabling proactive management to mitigate impacts on food security and ecosystem health.

17.3 Methodology for Developing Prediction Models

Prediction models start by studying nematode life cycles and how factors such as temperature and moisture affect them. Climate models, such as those from the IPCC, project future conditions such as warmer temperatures or changing rainfall. These are combined with ecological models to predict where nematodes might spread or how their populations might grow. For example, a study predicted that a 1 °C temperature rise could expand virus-vector nematodes northward by 160–200 km in Great Britain.

Models also consider host plants, like crops, and how their growth might change, affecting nematode infections. For instance, research on soybean cyst nematode (SCN) in Québec, Canada, projected more generations per year by 2050, suggesting the need for resistant crop varieties.

Developing prediction models for nematode pathogens under climate change involves several key steps, each informed by recent research.

17.3.1 Understanding Nematode Biology and Ecology

- A thorough understanding of nematode life cycles, host ranges, and environmental preferences is foundational. For example, PPNs exhibit temperature-dependent life cycles, with development rates increasing under warmer conditions, potentially leading to more generations per year.
- Studies have documented specific responses, such as drought reducing nematode abundance, particularly for bacterivores and fungivores, while PPNs may increase post-drought. This informs model parameters for survival and reproduction under varying conditions.

17.3.2 Utilizing Climate Models

- Climate models, such as those from the IPCC, provide projections of future environmental conditions. Representative concentration pathways (RCPs), such as RCP 4.5 (optimistic) and RCP 8.5 (pessimistic), are commonly used to simulate temperature, precipitation, and CO_2 levels for specific time horizons, like 2041–2070 (2050 horizon).
- These projections are critical for understanding how environmental changes will affect nematode habitats. For example, a study on SCN in Québec used RCP 4.5 and 8.5 to predict future conditions.

17.3.3 Ecological and Species Distribution Modeling

- Ecological niche models or SDMs predict how nematode ranges might shift under future climates. For instance, Neilson and Boag (Nielson and Boag 1996) predicted that a 1 °C increase in mean temperature could lead to a northward expansion of virus-vector nematodes (Longidorids and Trichodorids) by 160–200 km in Great Britain.
- These models use current distribution data and environmental variables to infer suitable habitats, accounting for factors like soil temperature and moisture.

17.3.4 Population Dynamics Modeling

- Population dynamics models simulate how nematode populations grow or decline based on environmental factors. For example, Carter et al. (1996) simulated models predicting an expanded distribution and increased generations per

year for the potato cyst nematode *Globodera rostochiensis* in Finland by 2050, highlighting the potential for accelerated reproduction under warming.
- Such models incorporate temperature-dependent development rates, host availability, and other factors to forecast population density changes.

17.3.5 Integrating Host Plant Models

- Since many nematodes are plant pathogens, understanding how climate change affects host plants is crucial. This includes modeling changes in crop distribution, phenology (e.g., planting and harvesting dates), and susceptibility to nematode infection.
- For instance, the Québec study on SCN projected that, by 2050, soybean cultivation could expand across all cropland, with planting up to 15 days earlier under RCP 4.5, potentially increasing nematode infection risks.

17.3.6 Validation With Empirical Data

- Predictions from models must be validated using field data or experimental studies. Research has shown that climate change factors such as elevated CO_2, warming, and precipitation changes affect nematode communities, with studies providing empirical evidence for model calibration.
- For example, a study on *Rotylenchulus reniformis* and *Meloidogyne floridensis* found that while higher soil temperatures may reduce nematode reproduction, their virulence increases, leading to greater crop losses (Khanal and Land 2023).

17.3.7 Application in Management Strategies

- The insights from prediction models inform the development of integrated nematode management (INM) strategies. These include breeding nematode-resistant crop varieties, adjusting planting schedules, and implementing cultural practices such as crop rotation, cover cropping, and biofumigation.
- For SCN, researchers suggest developing soybean lines with early maturity and resistance sources other than PI88788 to counter increased nematode pressure under future climates.

17.3.8 Outcome

These models help forecast increased nematode populations and disease risks under future climates, such as more generations of potato cyst nematodes by 2050 in Finland. This information guides strategies such as breeding-resistant crops, adjusting planting times, or using crop rotation to reduce nematode impact. They are crucial for protecting food security, especially as climate change alters growing conditions.

17.4 Correlative SDMs

Correlative SDMs are statistical models that predict the geographic distribution of species based on correlations between species occurrence data and environmental variables, such as temperature, precipitation, soil type, and land use. Unlike mechanistic models, which simulate biological processes, correlative SDMs rely on observed relationships to forecast potential distributions under current and future climate scenarios. They are widely used in ecology to assess how species ranges might shift due to environmental changes, including climate change.

For plant nematodes, SDMs typically use occurrence data from field surveys, combined with climate data from sources like global climate models (GCMs), to build predictive maps. Common modeling approaches include maximum entropy modeling (MaxEnt), generalized linear models (GLMs), and ML algorithms like random forests (RFs). These models are validated using independent datasets or cross-validation techniques to ensure accuracy.

Correlative SDMs estimate species' habitat suitability by relating occurrence records to environmental variables through statistical or ML algorithms. Widely used algorithms include MaxEnt, GLMs, and RFs. These models employ bioclimatic variables from sources such as WorldClim to capture climatic drivers of distribution (Del Lesto et al. 2024).

MaxEnt and GLM/generalized additive models (GAMs) infer relationships between known occurrence records and environmental predictors to map potential ranges. For example, pine wilt disease caused by *Bursaphelenchus xylophilus* in China was modeled using an ensemble of MaxEnt and CLIMEX to predict northward range expansion under warming scenarios (Chen et al. 2024).

17.4.1 Developing SDMs for Plant Nematodes

Developing SDMs for plant nematodes involves several steps:

- *Data Collection and Processing*: Gathering occurrence data for nematode species, often from agricultural surveys, soil sampling, or literature, is essential. This can be challenging due to the small size and often overlooked nature of nematodes, leading to data gaps.
  - *Occurrence Data*: Global distribution records are sourced from global biodiversity information facility (GBIF), centre for agriculture and biosciences international (CABI), and National quarantine databases. Quality control includes spatial thinning to reduce sampling bias.
  - *Environmental Predictors*: Selecting relevant climate variables, such as mean annual temperature, precipitation seasonality, and soil moisture, which influ-

ence nematode distribution. These variables are typically sourced from climate databases like WorldClim or coupled model intercomparison project phase 6 (CMIP6) projections. Bioclimatic variables (BIO1–BIO19) from WorldClim v2.1 (1970–2000) at ~1 km resolution are standard predictors. Soil moisture, elevation, and vegetation indices normalized difference vegetation index/enhanced vegetation index (NDVI/EVI) can improve model performance (Del Lesto et al. 2024).
  - *Variable Selection*: Correlation screening ($|r| < 0.8$) and ecological relevance guide predictor selection to avoid collinearity and overfitting (Zhang et al. 2024).
- *Modeling Algorithms*: Using statistical or ML methods to correlate nematode occurrences with environmental variables is essential. For example, MaxEnt might be used to predict suitable habitats based on temperature and precipitation thresholds.
  - *MaxEnt*: A presence-only algorithm maximizing entropy; robust with limited data and tolerant to correlated variables (Kim et al. 2021).
  - *GLM/GAM*: *Parametric* approaches providing interpretable relationships but requiring balanced presence–absence data.
  - ML (*RF* and *Boosted Regression Trees*): Non-parametric methods capturing complex interactions; they often yield high predictive accuracy but less interpretability.

- *Projection under Climate Change*: Applying the model to future climate scenarios, such as RCPs like RCP4.5 or RCP8.5, to forecast how nematode distributions might shift by mid-century or end-of-century.
- Challenges in development include the following:
  - Limited occurrence data, especially for less-studied nematode species;
  - Complex interactions with host plants, soil microorganisms, and other environmental factors, which may not be fully captured by correlative models;
  - Regional variability, as climate change effects differ by latitude, altitude, and ecosystem type, requiring localized models.
- *Model Evaluation and Validation*: Models are evaluated using the area under the receiver operating characteristic curve (AUROC) and the true skill statistic (TSS). Cross-validation (k-fold or bootstrap) assesses robustness. Threshold selection for binary maps uses maximum TSS or fixed percentile approaches.

17.4.2 Role of Climate Change in SDMs for Plant Nematodes

Climate change is a critical driver in SDMs for plant nematodes, as it alters the environmental conditions that determine their distribution. Research suggests the following:

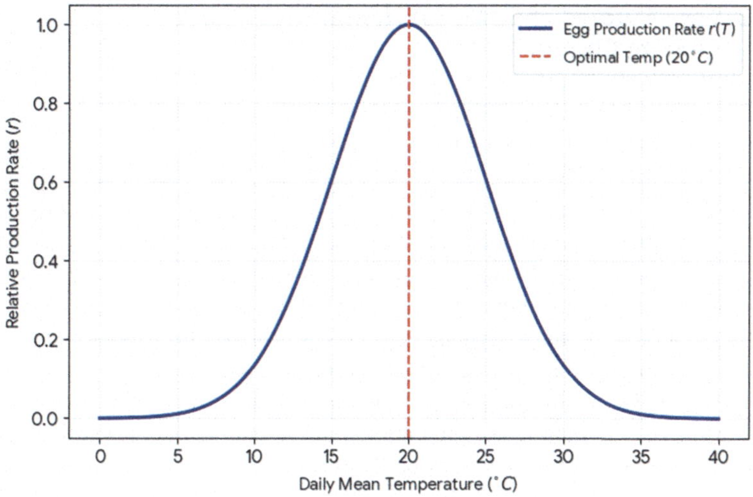

Fig. 17.1 Gaussian model of *Heterodera carotae* daily egg production

- *Temperature*: Higher temperatures can increase nematode reproduction rates and the number of generations per year, potentially expanding their range. For example, a study by Carter et al. (1996) predicted that *Globodera rostochiensis* could have more generations per year by 2050 in Finland due to global warming (Colagiero and Ciancio 2011). Mugniery and Bossis (1988) studied the effect of temperature on the carrot cyst nematode.

- The relationship of the daily egg production rate of the carrot cyst nematode, *Heterodera carotae,* with temperature was provided by a Gaussian function, considering the difference between the daily mean value and the optimal temperature for egg release (20 °C) (Fig. 17.1).
- *Precipitation*: Altered rainfall patterns can affect soil moisture, which is essential for nematode activity. Increased precipitation might favor some species, while droughts could limit others, as seen in studies on *Meloidogyne minor* and *Heterodera* spp. in Northern Ireland grasslands (Fleming et al. 2016).
- *CO_2 Levels*: Elevated CO_2 can influence plant growth and root exudates, indirectly affecting nematode populations. For instance, Yeates et al. (1997) found increased abundance of *Dorylaimus* spp. and *Trichodorus* spp. under CO_2 enrichment in New Zealand ryegrass/white clover.

These factors are integrated into SDMs to project how nematode distributions might shift, potentially leading to range expansions in temperate regions (e.g., northward shifts) or increased infestations in currently suitable areas. However, the evidence also suggests counteracting effects, such as reduced survival under extreme heat or entering cryptobiosis, which complicates predictions (Dutta and Phani 2023).

17.4.3 Case Studies of SDMs for Nematodes Under Climate Change

Several studies have developed and applied SDMs to specific nematode species, providing insights into their potential responses to climate change:

- *Globodera* spp. *(Potato Cyst Nematodes)*: Jones et al. (2017) used SDMs to predict that climate change could alter the pest status of *Globodera pallida* and *G. rostochiensis* in the United Kingdom, with some areas becoming more suitable and others less suitable due to temperature increases. Pridannikov et al. (2022) modeled the range dynamics of *G. rostochiensis* in Russia under global warming scenarios, forecasting potential range expansions.
- *Meloidogyne incognita (Root-Knot Nematode)*: Ghini et al. (2008) used SDMs to predict increased pathogenic potential in Brazilian coffee plantations under future climate scenarios, highlighting the risk of enhanced crop damage.
- *Bursaphelenchus xylophilus (Pinewood Nematode)*: MaxEnt predicted global distribution and future expansions under SSP scenarios (2041–2060, 2061–2080, and 2081–2100). Key drivers included precipitation of the warmest quarter and temperature seasonality; high-latitude expansion was projected under SSP5–8.5 (Xiao et al. 2024). Li et al. (2023) assessed the potential impact of climate change on its distribution in Chongqing, China, using SDMs to forecast range shifts under warming conditions.
- *Ditylenchus dipsaci (Stem and Bulb Nematode)*: MaxEnt modeling in India identified northern and north-eastern regions as suitable under the current climate. Annual mean temperature (Bio1) and precipitation seasonality (Bio15) were most influential (Holajjer et al. 2024).
- *Virus-Transmitting Nematodes*: Neilson and Boag (Nielson and Boag 1996) modeled the northward expansion of Longidorids and Trichodorids by 160–200 km in Great Britain with a 1 °C temperature rise, demonstrating the potential for range shifts under climate change.

These examples illustrate how SDMs can not only predict specific changes, such as range expansions or increased generations per year, but also highlight regional variability and species-specific responses.

17.4.4 Applications and Implications

SDMs for plant nematodes under climate change have significant applications in agriculture and nematode management:

- *Predictive Risk Assessment*: They help forecast areas at risk of nematode infestations, enabling targeted monitoring and early intervention. For example, predicting northward shifts can alert farmers in previously unaffected regions.

- *INM*: SDMs inform "climate-smart" management strategies, such as:
  - Precision monitoring of nematode populations using soil sampling and molecular techniques;
  - Use of biocontrol agents, host-resistant crop varieties, and safer approaches such as crop rotation, cover cropping, and biofumigation (Dutta and Phani 2023);
  - Adjusting crop scheduling based on predicted nematode activity, such as avoiding planting during peak reproduction periods.
- *Policy and Planning:* SDMs can guide agricultural policy, helping allocate resources for research, extension services, and adaptation measures in vulnerable regions.

The implications are particularly relevant for food security, as nematodes are one of the most important limiting factors in agricultural production, and climate change is predicted to exacerbate crop damage (Khanal and Land 2023).

17.4.5 Variable Responses of PPNs to Climate Change

Table 17.1 compiles variable responses of PPNs to altered temperature, CO_2, and precipitation levels, providing specific examples from the literature:

This table highlights the diverse responses of PPNs to climate change, underscoring the need for species-specific and region-specific modeling.

The development of correlative SDMs for plant nematodes under climate change is a vital tool for predicting distribution shifts, assessing agricultural risks, and

Table 17.1 Variable responses of PPNs to altered temperature, CO_2, and precipitation levels

Climate factor	Nematode species	Response	Country	References
Temperature	*Bursaphelenchus mucronatus*, *B. xylophilus*	Increased prevalence in pine forests	Switzerland	Rebetez and Dobbertin (2004)
Temperature	*Heterodera carotae*	Increased juvenile/egg production	Italy	Colagiero and Ciancio (2011)
CO_2 enrichment	*Dorylaimus* spp., *Trichodorus* spp.	Increased abundance in ryegrass/white clover	New Zealand	Yeates et al. (1997)
CO_2 enrichment	*Meloidogyne graminicola*	No effect on abundance in rice	India	Somasekhar and Prasad (2012)
Precipitation	*Meloidogyne minor*, *Heterodera* spp., *Pratylenchus* spp.	Increased abundance in grasslands/cereals	Northern Ireland	Fleming et al. (2016)
Precipitation	Various	Variable responses in grasslands	USA	Ankrom et al. (2020, 2022)

informing management strategies. While research suggests that climate change may expand nematode ranges and increase crop damage in some areas, the outcomes are complex and vary by species and region. On-going studies, such as those by Jones et al. (2017), Ghini et al. (2008), and Li et al. (2023), provide valuable insights, but challenges such as data scarcity and regional variability highlight the need for continued research. These models are essential for developing "climate-smart" agricultural practices to ensure food security in the face of changing climatic conditions.

17.4.6 Challenges and Limitations

Despite their utility, SDMs for plant nematodes face several challenges:

- *Data Scarcity*: Nematodes are often understudied, and occurrence data may be limited, especially for less economically important species. This can lead to model uncertainty. Incomplete occurrence records and biased sampling can skew predictions; integration of expert knowledge can mitigate biases.
- *Complex Interactions*: Nematodes interact with host plants, soil microorganisms, and other environmental factors, which may not be fully captured by correlative models. For instance, changes in plant diversity can alter nematode community structure, as shown in studies on intensively managed agricultural systems (Ikoyi et al. 2023).
- *Regional Variability*: Climate change effects differ by region, requiring localized models. For example, Mediterranean environments may see different nematode responses compared to temperate zones (Ikoyi et al. 2023).
- *Counteracting Effects*: Some studies suggest that while climate change may favor nematode expansion, phenomena such as sex reversal, entering cryptobiosis, or reduced survival under extreme conditions could act in counter direction, creating a contraposition effect (Dutta and Phani 2023).
- *Climate Data Uncertainty*: Differences among GCMs and downscaling approaches necessitate ensemble forecasting and uncertainty quantification.
- *Model Transferability*: Calibration area choice affects extrapolation; incorporating physiological priors can improve transferability under novel climates (Zhang et al. 2024).

These challenges underscore the need for ongoing research and improved data collection to refine SDMs.

17.4.7 Future Projections

Projected distributions under multiple general circulation models (GCMs) and shared socioeconomic pathways (SSPs) provide ensemble predictions. Scenario analyses reveal potential range expansions or contractions by the mid- and late twenty-first century, informing quarantine and management strategies.

In conclusion, correlative SDMs offer valuable insights into current and future distributions of plant parasitic nematodes. Adherence to best practices in data preprocessing, predictor selection, and model evaluation enhances reliability. Ensemble projections under climate change scenarios support proactive nematode management and policy planning.

17.5 Mechanistic Niche Models (e.g., CLIMEX)

The development of mechanistic niche predictive models, such as CLIMEX, for plant nematodes under climate change is a critical area of research, given the significant agricultural impact of these pests. Climate change, characterized by rising temperatures, altered precipitation patterns, and increased atmospheric CO_2 levels, is expected to influence nematode distribution, reproduction, and interaction with host plants. Mechanistic niche models, which integrate physiological and environmental data to predict species distribution, offer a promising approach to anticipate these changes and inform adaptive management strategies. A comprehensive overview of the current state of research, focusing on the development and application of models like CLIMEX, with detailed insights into methodologies, findings, and future directions, is discussed.

17.5.1 Background on Mechanistic Niche Models

Mechanistic niche models, unlike correlative models, use species-specific physiological data, such as temperature thresholds for survival and reproduction, to predict potential distribution under varying environmental conditions. CLIMEX, widely used software, is particularly noted for its ability to model eco-climatic suitability by incorporating factors such as temperature, moisture, and seasonal cycles. These models are especially valuable for pests like plant nematodes, which are sensitive to climatic variables and can significantly impact crop yields under changing conditions. The model's ability to simulate scenarios, such as rain-fed versus irrigated agriculture, makes it suitable for assessing climate change impacts.

17.5.2 Evidence of Development and Application

Singh et al. (2014) demonstrated the global biosecurity risks from three PPN species to maize, rice, and wheat, using Climex modeling. This study focused on three nematode species—*Heterodera zeae, Hirschmanniella oryzae*, and *Meloidogyne graminicola*—key pests for major crops. The CLIMEX model was parameterized using species phenology and global distribution data from peer-reviewed literature, simulating both rain-fed and irrigation scenarios. The findings indicated that known geographical distributions aligned well with projected potential distributions, with irrigation scenarios showing larger areas conducive to nematode growth compared to rain-fed conditions. This suggests that climate change, particularly through irrigation practices, could exacerbate nematode-related risks.

The study's methodology involved careful calibration of CLIMEX parameters based on physiological tolerance thresholds, providing a detailed understanding of environmental factors influencing nematode establishment. Maps of host crops (maize, rice, and wheat) and irrigation areas were compared with growth index maps, identifying countries at risk for range expansion under favorable eco-climatic conditions. This application highlights CLIMEX's utility in predicting biosecurity risks and informing international trade regulations for agricultural commodities.

17.5.3 Recent Research and Trends

While the 2014 study (Singh et al. 2014) provides a foundational example, recent research continues to explore the impact of climate change on plant nematodes, though not all studies explicitly mention CLIMEX. For instance, a study by Khanal and Land (2023) investigated the effects of soil temperature on the survival, reproduction, and virulence of *Rotylenchulus reniformis* and *Meloidogyne floridensis*. This study employed novel heat exposure assays to evaluate nematode biology, finding that higher temperatures could increase crop damage, aligning with the need for predictive models like CLIMEX to anticipate such shifts.

Dutta and Phani (2023) emphasized the importance of modeling-based studies for understanding population dynamics in relation to climatic fluctuations. Although this review did not specifically mention CLIMEX, it highlighted INM strategies involving precise monitoring and modeling, suggesting that mechanistic approaches remain relevant. The review cited earlier studies, such as Carter et al. (1996) and Ghini et al. (2008), which predicted expanded distributions and increased generations per year for nematodes such as *Globodera rostochiensis* and *Meloidogyne incognita* under warming scenarios, reinforcing the need for advanced predictive tools.

17.5.4 Methodological Insights

The development of CLIMEX models for plant nematodes involves several steps, including data collection on nematode physiology (e.g., temperature and moisture thresholds), global distribution records, and climate scenarios. For instance, the 2014 study (Singh et al. 2014) used peer-reviewed literature to parameterize the model, ensuring accuracy in predicting eco-climatic suitability. Recent studies, such as those from 2023 (Khanal and Land 2023), suggest a trend toward integrating experimental data, like heat exposure assays, to refine model parameters, potentially enhancing the precision of predictions under climate change.

Table 17.2 can help organize the key studies on CLIMEX and their findings for clarity.

Table 17.2 Key studies on CLIMEX and their findings

Nematode species	Key finding	Model used	References
H. zeae, H. oryzae, M. graminicola	Irrigation expands suitable areas	CLIMEX	Singh et al. (2014)
R. reniformis, M. floridensis	Higher temperatures increase damage	Experimental assays	Khanal and Land (2023)
Various	Need for modeling population dynamics	Not specified	Dutta and Phani (2023)

This table summarizes the evolution of research, showing CLIMEX's role in earlier studies and the ongoing need for modeling advancements.

17.5.5 Challenges and Future Directions

Despite the progress, challenges remain in developing these models. Parameterization requires extensive data on nematode physiology, which may be limited for less-studied species. Additionally, climate change introduces complex interactions, such as altered precipitation and CO_2 levels, which may affect nematode–host dynamics in ways not fully captured by current models. Recent reviews, like "Linking Nematode Communities and Soil Health under Climate Change" (Pires et al. 2023), suggest a need for integrating soil health indicators and ecosystem services into models, potentially expanding CLIMEX's scope.

Future directions may include refining CLIMEX with ML techniques to handle large datasets from remote sensing, as seen in broader climate modeling trends. Additionally, collaborative efforts to standardize data collection across regions could enhance model accuracy, ensuring global applicability for nematode management under climate change.

In conclusion, the development of mechanistic niche predictive models like CLIMEX for plant nematodes under climate change is well-established, with significant applications demonstrated since at least 2014. These models are crucial for predicting distribution shifts, informing biosecurity, and supporting agricultural adaptation strategies. Recent research continues to build on this foundation, emphasizing the need for advanced modeling to address emerging climate impacts. While challenges exist, ongoing studies suggest a dynamic field with potential for further innovation, ensuring that these tools remain vital for managing nematode-related risks in a changing climate.

17.6 ML Time-Series Forecasting

The intersection of ML time-series forecasting and the impact of climate change on plant nematodes represents a critical area for advancing agricultural pest management. This note provides a comprehensive overview, synthesizing current research to outline the state of the field, methodologies, and future directions.

17.6.1 Background and Context

Climate change, characterized by rising temperatures, altered precipitation patterns, and increased CO_2 levels, is known to influence nematode biology, including their distribution, reproduction rates, and virulence. For instance, Khanal and Land (2023) found that higher soil temperatures increased the virulence of *Rotylenchulus reniformis* and *Meloidogyne floridensis*, suggesting that climate change could exacerbate crop damage.

Simultaneously, ML, particularly deep learning, has emerged as a powerful tool for agricultural forecasting. Time-series forecasting, which predicts future values based on historical data, is widely used for pest management. Models such as long short-term memory (LSTM) networks and convolutional neural networks (CNNs) have been applied to predict pest outbreaks by analyzing meteorological and pest occurrence data over time.

17.6.2 ML Applications in Agricultural Pest Forecasting

Research has demonstrated the efficacy of ML for forecasting agricultural pests, which can be extended to nematodes. For example, a study by Zhang et al. (2024) introduced the attention-based LSTM interaction CNN (ALIC) model for pest outbreak forecasting, leveraging multivariate time-series data including meteorological and historical pest data. This model excels at capturing temporal dynamics, suggesting its potential applicability to nematode forecasting.

Another study by Lee and Yun (2023) developed a deep learning model to predict risks of crop pests and diseases using sequential environmental data, achieving an average AUROC of 0.917. The model used data such as air temperature, relative humidity, dew point, and CO_2 concentration, which are also relevant for nematode dynamics under climate change.

Additionally, Yuan et al. (2024) used LSTM for classifying nematode behavioral patterns based on time-series data, achieving 73.49% accuracy in a five-classification task. While focused on detection, this highlights the potential for time-series analysis in nematode research.

17.6.3 Climate Change Impacts on Plant Nematodes

Climate change affects nematodes both directly and indirectly. Direct effects include changes in developmental rates and survival strategies due to elevated temperatures and CO_2 levels. For example, Somasekhar and Prasad (2012) noted that nematodes show a neutral or positive response to CO_2 enrichment, with some species potentially building up rapidly. Indirect effects involve altered host plant physiology, which can influence nematode–host interactions.

Specific studies, such as one by Colagiero and Ciancio (2011), discussed expected effects such as altered reproductive cycles and geographical dispersion due to global warming, particularly in Mediterranean environments. These changes underscore the need for forecasting tools to predict nematode pressures under future climate scenarios.

17.6.4 Integrating ML and Climate Change for Nematode Forecasting

While no single study directly combines ML time-series forecasting with plant nematodes under climate change, the methodologies from pest forecasting can be adapted. For instance, Pun et al. (2023) mentioned AgroDSS, which uses time-series clustering and predictive modeling for pest population prediction, potentially integrating climate variability, suggesting the feasibility of incorporating climate change into decision support systems (DSSs).

A potential approach involves collecting historical time-series data on nematode populations and climate variables, training an ML model (e.g., LSTM or CNN), and then using climate change projections as input to forecast future nematode dynamics. This could help identify vulnerable regions or crops, aiding in proactive management strategies.

17.6.5 Challenges and Future Directions

Developing such a forecasting system faces challenges, including the need for large, annotated datasets on nematode populations and climate variables, which can be difficult to obtain. Agarwal et al. (2023) highlighted the potential of digitizing nematode diagnostic lab data for national management, which could support time-series forecasting. Additionally, integrating climate models with ML requires interdisciplinary collaboration between nematologists, climatologists, and data scientists.

Future research should focus on creating comprehensive datasets, developing hybrid models that combine statistical and AI approaches, and validating forecasts under various climate scenarios. The growing complexity of pest-related challenges due to climate change underscores the urgency for such advancements (Kontopoulou et al. 2023).

17.6.6 Detailed Examples and Metrics

Table 17.3 summarizes key studies relevant to ML and nematode forecasting.

This table highlights the overlap between ML capabilities and climate impact studies, suggesting a pathway for integration.

In conclusion, the development of ML time-series forecasting for plant nematodes under climate change is an emerging field with significant potential. While specific studies are limited, the combination of existing research on climate impacts on nematodes and ML for pest forecasting provides a foundation. By leveraging

Table 17.3 Summary of key studies relevant to ML and nematode forecasting

Model/algorithm	Task type	Input type	Accuracy/ metric	Relevance to climate change	References
ALIC (LSTM+ICNN)	Forecasting	Meteorological, pest data	Not specified	Direct, uses climate data	Yuan et al. (2024)
Deep learning	Risk prediction	Environmental data	AUROC 0.917	Indirect, environmental factors	Lee and Yuan (2023)
Experimental	Impact assessment	Temperature	Increased virulence	Direct, temperature effects	Khanal and Land (2023)

historical data and climate projections, such systems can help predict future nematode pressures, supporting sustainable agricultural practices in the face of climate change.

17.7 Climate Scenarios and Data Inputs

Predictive models rely on standardized greenhouse gas concentration trajectories—RCPs—to drive downscaled climate projections. The four widely used RCPs (2.6, 4.5, 6.0, and 8.5 W/m^2) span low- to high-emission futures and are adopted across impact studies (Wikipedia 2025). Key climate predictors include the following:

- Temperature metrics such as monthly averages and degree-day accumulation;
- Moisture variables such as soil moisture and relative humidity;
- Extreme events such as heat waves and drought frequency.

High-resolution climate layers (e.g., 1 km^2 grid) from CMIP6 ensembles allow models to capture local variability critical for soil-borne nematode life cycles.

17.7.1 Model Calibration, Validation, and Uncertainty

Robust forecasts require rigorous calibration against independent occurrence or monitoring datasets, often using k-fold cross-validation or area-under-curve metrics in SDMs (El-Gabbas and Dormann 2018). Ensembles of multiple GCMs and RCPs help quantify projection uncertainty, with consensus mapping highlighting areas of high and low confidence. Sensitivity analyses on physiological parameters (e.g., lower and upper temperature thresholds) further delineate which life-history traits most influence model outputs (MedCrave 2014).

17.8 Benefits

17.8.1 Improved Disease Management

Early warnings about potential outbreaks can allow for timely interventions and reduce crop losses.

17.8.2 Development of Climate-Smart Agriculture

Models can guide the development of climate-resistant crop varieties and sustainable management practices.

17.8.3 Understanding the Impact of Climate Change

Models can help quantify the potential impacts of climate change on nematode diseases and their consequences for global food security.

17.9 Detailed Research Findings

Recent studies provide specific examples of prediction models and their findings:

17.9.1 Virus-Vector Nematodes

Neilson and Boag (Nielson and Boag 1996) predicted that a 1 °C temperature increase could expand the range of virus-transmitting nematodes in Great Britain, emphasizing the need for monitoring in northern regions.

17.9.2 Potato Cyst Nematode

Carter et al. (1996) used simulated models to predict an expanded distribution and increased generations per year for *Globodera rostochiensis* in Finland by 2050, highlighting the potential for accelerated reproduction under warming.

17.9.3 SCN in Québec

St-Marseille et al. (2019) projected that under RCP 4.5, SCN could complete up to four generations per year by 2050, and under RCP 8.5, up to five generations, while soybean cultivation could expand, necessitating resistant varieties.

17.9.4 General Impacts on PPNs

Research suggests that climate change can increase PPN virulence and distribution, with higher temperatures potentially leading to greater crop damage. For example, a study on *Rotylenchulus reniformis* and *Meloidogyne floridensis* showed that while reproduction may decrease at higher temperatures, virulence increases, leading to greater crop losses (Khanal and Land 2023).

17.9.5 Integration With Management Strategies

Forecast maps of future suitability can be incorporated into DSSs to guide:

- Surveillance prioritization for targeting high-risk regions for soil sampling;
- Temporal interventions for timing of nematicide applications based on predicted generation peaks;
- Host-plant deployment for selecting resistant cultivars in areas projected to become more favorable to nematodes.

17.10 Challenges and Limitations

While promising, models face challenges. Data on nematode distributions can be limited, and climate projections have uncertainties. Nematodes might also adapt in unexpected ways, affecting model accuracy. Despite this, they offer valuable insights for managing risks.

Despite their potential, prediction models face several challenges:

17.10.1 Data Availability

Comprehensive data on current nematode distributions and their responses to environmental factors are often limited, especially for less-studied species or regions. This can affect model accuracy and reliability.

17.10.2 Model Complexity

Integrating multiple factors, such as climate, host plants, soil conditions, and nematode interactions, into a single model can be computationally intensive and challenging. This complexity may introduce errors or require significant resources.

17.10.3 Uncertainty in Climate Projections

Climate models have inherent uncertainties, which propagate into nematode population predictions. For example, differences between RCP 4.5 and 8.5 scenarios can lead to varying outcomes, complicating decision-making.

17.10.4 Nematode Adaptation

Nematodes may adapt to changing conditions in ways that are not yet fully understood, potentially altering model outcomes. For instance, studies suggest that some nematodes may develop resistance to environmental stressors, affecting long-term predictions.

17.11 Implications for Management Strategies

The insights from prediction models are crucial for developing effective management strategies, particularly in agriculture. By forecasting changes in nematode populations and disease risk, these models enable proactive measures to protect crops and ensure food security. Examples include the following.

17.11.1 Breeding-Resistant Varieties

Developing nematode-resistant crop lines, such as soybeans with resistance sources other than PI88788, to counter increased pressure under future climates.

17.11.2 Adjusting Agricultural Practices

Implementing cultural practices such as crop rotation, cover cropping, and biofumigation to reduce nematode populations. For instance, agro-ecological practices have been shown to enhance nematode diversity and reduce PPN abundance.

17.11.3 Timing and Scheduling

Adjusting planting and harvesting dates to avoid peak nematode activity, especially under scenarios with earlier planting due to warmer temperatures.

These strategies are informed by the detailed predictions from models, ensuring that management efforts are targeted and effective.

17.12 Comparative Analysis of Model Approaches

Table 17.4 highlights the variety of modeling approaches, from ecological niche modeling to bioclimatic simulations, and their specific contributions to understanding nematode responses under climate change.

17.13 Future Directions

In the future, the following aspects need more emphasis:

- Limited long-term monitoring of nematode populations hinders time-series modeling (data gaps).
- Models calibrated in one region may perform poorly elsewhere without local parameterization (model transferability).
- Downscaling GCM outputs can introduce errors in soil moisture projections, affecting process-based predictions (climate model bias).
- Land-use change and crop management adaptations under different SSPs add complexity beyond climatic drivers (socio-economic factors).

Future research should focus on integrating multi-model ensembles, improving mechanistic understanding of nematode–soil–host interactions, and leveraging emerging AI approaches for real-time outbreak prediction.

17.14 Conclusion

Prediction models for nematode pathogens under climate change are essential tools for forecasting changes in nematode populations and disease risk, enabling the development of effective management strategies. By integrating nematode biology, climate projections, and ecological modeling, these models provide valuable

Table 17.4 Specific contributions of modeling approaches to understanding nematode responses under climate change

Focus	Methodology	Key finding	References
Virus-vector nematodes	Ecological niche modeling	1 °C temperature increase expands the range by 160–200 km	Neilson and Boag (1996)
Potato cyst nematode	Simulated distribution models	Increased generations per year by 2050 in Finland	Carter et al. (1996)
SCN	Bioclimatic models, RCP 4.5 and 8.5	Up to five generations by 2050, the need for resistant varieties	St-Marseille et al. (2019)
R. reniformis and *M. floridensis*	Direct and indirect heat exposure assays	Higher temperatures increase virulence, crop losses	Khanal and Land (2023)

insights for agriculture and ecosystem management. Despite challenges such as data limitations and uncertainties, they offer a proactive approach to mitigating the impacts of climate change on nematode-related diseases, ensuring food security and sustainable land use.

By combining diverse modeling frameworks with high-resolution climate projections and empirical validation, researchers and practitioners can better anticipate nematode-driven disease risks and implement timely, location-specific management strategies.

17.15 Suggested Readings

- Clavero-Camacho, I., Archidona-Yuste, A., Cantalapiedra-Navarrete, C. et al. 2024. Prevalence and ecological factors affecting the distribution of plant-parasitic nematodes in Prunus groves in Spain. *Journal of Integrative Agriculture* 23(2): 566–589.
- Chen, L., Lu, W., Lamont, B.B. et al. 2024. Modeling the distribution of pine wilt disease in China using the ensemble models MaxEnt and CLIMEX. *Ecology and Evolution* 14: e70277.
- El-Gabbas, A. and Dormann, C.F. 2018. Improved species-occurrence predictions in data-poor regions: using large-scale data and bias correction with down-weighted Poisson regression and Maxent. *Ecography*. https://doi.org/10.1111/ecog.03149.
- Holajjer, P., Harish, G., Khan, Z. et al. 2024. Predicting potential climatic suitability for *Ditylenchus dipsaci* in India using MaxEnt modelling. *Indian Journal of Nematology* 54: 242–248.
- Jones, L.M., Koehler, A.K., Trnka, M. et al. 2017. Climate change is predicted to alter the current pest status of *Globodera pallida* and *G. rostochiensis* in the United Kingdom. *Glob. Chang. Biol.* 23: 4497–4507.
- Nielson, R. and Boag, B. 1996. Predicted impact of climate change on virus-vector nematodes in Great Britain. *European J. Plant Pathol.* 102: 193199.
- Xiao, Y, Guo, Q., Xie, N. et al. 2024. Predicting the global potential distribution of *Bursaphelenchus xylophilus* using an ecological niche model: expansion trend and the main driving factors. *BMC Ecology and Evolution* 24: 48.
- Zhang, Z., Zhou, J., Molinos, J.G. et al. 2024. Incorporating physiological knowledge into correlative SDMs minimizes bias introduced by the choice of calibration area. *Mar Life Sci Technol* 6(2): 349–362.

17.16 Review Questions

- Which climatic factors most strongly influence the distribution of plant parasitic nematodes?
- How can a correlative SDM prediction model be used to forecast nematode shifts under climate change?

- How can mechanistic niche models (e.g., CLIMEX) be used to forecast nematode shifts under climate change?
- How can ML time-series forecasting be used to forecast nematode shifts under climate change?
- What are the major challenges in validating and applying nematode prediction models across regions?
- How can prediction modeling support climate-resilient nematode management strategies?
- Explain the role of ML ensemble models in predicting shifts in nematode populations, such as northward migrations, under future climate scenarios.

References

Agarwal S, Curran ZC, Yu G et al (2023) Plant parasitic nematode identification in complex samples with deep learning. J Nematol 55(1):20230045

Ankrom KE, Franco AL, Fonte SJ et al (2020) Ecto-and endoparasitic nematodes respond differently across sites to changes in precipitation. Oecologia 193:761–771

Ankrom KE, Franco AL, Fonte SJ et al (2022) Ecological maturity and stability of nematode communities in response to precipitation manipulations in grasslands. Appl Soil Ecol 170:104263. https://doi.org/10.1016/j.apsoil.2021.104263

Carter TR, Saarikko RA, Niemi KJ (1996) Assessing the risks and uncertainties of regional crop potential under a changing climate in Finland. Agric Food Sci Finl 5:329–350

Chen L, Lu W, Lamont BB et al (2024) Modeling the distribution of pine wilt disease in China using the ensemble models MaxEnt and CLIMEX. Ecol Evol 14:e70277

Colagiero M, Ciancio A (2011) Climate changes and nematodes: expected effects and perspectives for plant protection. Archieves 94:113–118

Del Lesto I, Magliano A, Casini R et al (2024) Ecological niche modelling of *Culicoides imicola* and future range shifts under climate change scenarios in Italy. Med Vet Entomol 38(4):416–428

Dutta TK, Phani V (2023) The pervasive impact of global climate change on plant-nematode interaction continuum. Front Plant Sci 14:20–23

El-Gabbas A, Dormann CF (2018) Improved species-occurrence predictions in data-poor regions: using large-scale data and bias correction with down-weighted Poisson regression and Maxent. Ecography. https://doi.org/10.1111/ecog.03149

Fleming TR, McGowan NE, Maule AG, Fleming CC (2016) Prevalence and diversity of plant parasitic nematodes in Northern Ireland grassland and cereals, and the influence of soils and rainfall. Plant Pathol 65:1539–1550

Ghini R, Hamada E, Pedro Junior MJ et al (2008) Risk analysis of climate change on coffee nematodes and leaf miner in Brazil. Pesq Agrop Brasileira 43:187–194

Holajjer P, Harish G, Khan Z et al (2024) Predicting potential climatic suitability for *Ditylenchus dipsaci* in India using MaxEnt modelling. Indian J Nematol 54:242–248

Ikoyi I, Grange G, Finn JA, Brennan FB (2023) Plant diversity enhanced nematode-based soil quality indices and changed soil nematode community structure in intensively-managed agricultural grasslands. Eur J Soil Biol 118:103542

Jones LM, Koehler AK, Trnka M et al (2017) Climate change is predicted to alter the current pest status of *Globodera pallida* and *G. rostochiensis* in the United Kingdom. Glob Chang Biol 23:4497–4507

Khanal C, Land J (2023) Study on two nematode species suggests climate change will inflict greater crop damage. Sci Rep 13:14185

Kim MS, Hanna JW, Stewart JE et al (2021) Predicting present and future suitable climate spaces (potential distributions) for an Armillaria root disease pathogen (*Armillaria solidipes*) and its host, Douglas-fir (*Pseudotsuga menziesii*), under changing climates. Front For Glob Change 4:740994

Kontopoulou VI, Panagopoulos AD, Kakkos I, Matsopoulos GK (2023) A review of ARIMAvs. Machine learning approaches for time series forecasting in data driven networks. Future Internet 15:255

Lee S, Yun EM (2023) A deep learning model for predicting risks of crop pests and diseases from sequential environmental data. Plant Methods 19:145

Li W, Kang J, Wang Y (2023) Spatiotemporal changes and driving forces of ecological security in the Chengdu-Chongqing urban agglomeration, China: quantification using health-services-risk framework. J Clean Prod 389:136135. https://doi.org/10.1016/j.jclepro.2023.136135

MedCrave (2014) Advances in Plants & Agriculture Research. medcraveonline.com. eISSN: 2373–6402

Mugniery D, Bossis M (1988) *Heterodera carotae* Jones, 1950. 1. Gamme d'hôtes, vitesse de développement, cycle. Revue de Nématologie 11(3):307–313

Nielson R, Boag B (1996) Predicted impact of climate change on virus-vector nematodes in Great Britain. Eur J Plant Pathol 102(193):199

Pires D, Orlando V, Collett RL et al (2023) Linking nematode communities and soil health under climate change. Sustainability 15(15):11747. https://doi.org/10.3390/su151511747

Pridannikov M, Zinovieva S, Khudyakova EA et al (2022) Range dynamics of potato cyst nematode *Globodera rostochiensis* (Wollenweber, 1923) (Nematoda, Heteroderidae) under conditions of global climate change in Russia. Russian J Biol Invasions 13:510–529. https://doi.org/10.1134/S2075111722040099

Pun TB, Neupane A, Koech R (2023) A deep learning-based decision support tool for plant-parasitic nematode management. J Imaging 9(11):240. https://doi.org/10.3390/jimaging9110240

Rebetez M, Dobbertin M (2004) Climate change may already threaten scots pine stands in the Swiss Alps. Theor Appl Climatol 79:1–9

Rose H, Caminade C, Bolajoko MB et al (2016) Climate-driven changes to the spatio-temporal distribution of the parasitic nematode, *Haemonchus contortus*, in sheep in Europe. Glob Chang Biol 22(3):1271–1285

Siebert J, Ciobanu M, Schädler M, Eisenhauer N (2020) Climate change and land use induce functional shifts in soil nematode communities. Oecologia 192(1):281–294

Singh S, Darren JK, Gavin A, Mike H (2014) Surveillance strategies for early detection of nematode incursions (Abstr.). J Nematol 46(2):236

Somasekhar N, Prasad JS (2012) Plant–nematode interactions: consequences of climate change. In: Venkateswarlu B, Shanker AK, Shanker C, Maheswari M (eds) Crop stress and its management: perspectives and strategies. Springer, Dordrecht, pp 547–564

St-Marseille AFG, Bourgeois G, Brodeur J, Mimee B (2019) Simulating the impacts of climate change on soybean cyst nematode and the distribution of soybean. Agric For Meteorol 264:178–187

Wikipedia (2025) Representative Concentration Pathway (en.wikipedia.org)

Xiao Y, Guo Q, Xie N et al (2024) Predicting the global potential distribution of *Bursaphelenchus xylophilus* using an ecological niche model: expansion trend and the main driving factors. BMC Ecol Evol 24:48

Yeates GW, Tate KR, Newton PCD (1997) Response of the fauna of a grassland soil to doubling of atmospheric carbon dioxide concentration. Biol Fertil Soils 25:307–315

Yuan Z, Nasamu Musa N, Dybal K et al (2024) Quantifying nematodes through images: datasets, models, and baselines of deep learning. arXiv:2404.19748v1 [cs.CV] 30 Apr 2024

Zhang Z, Zhou J, Molinos JG et al (2024) Incorporating physiological knowledge into correlative species distribution models minimizes bias introduced by the choice of calibration area. Mar Life Sci Technol 6(2):349–362

Nematode Management Strategies to Climate Variability

18

Abstract

Climate change may worsen plant-parasitic nematode (PPN) damage by altering their reproduction and distribution, requiring adaptive management. The integrated strategies, such as cover crops and resistant varieties, can help manage PPNs under changing conditions. The adaptive management, including monitoring and biological controls, is crucial for sustainability. Sustainable practices, such as organic amendments, enhance soil health. Controversy exists around chemical nematicides due to environmental concerns, pushing for alternatives.

Keywords

Adaptation · Mitigation

18.1 Learning Objectives

- Recognize limitations of conventional nematode management strategies under climate stress conditions.
- Assess risks of increased nematode outbreaks and resistance development due to climate change.
- Explore climate-resilient cultural, biological, and chemical methods for nematode management.
- Examine the role of host plant resistance and transgenic/genome-editing approaches in a changing climate.
- Formulate integrated nematode management (INM) strategies considering future climate scenarios.
- Promote ecological and environmentally safe approaches to mitigate nematode-induced yield losses.
- Identify research gaps in nematode ecology and management under climate change.
- Propose innovative approaches for long-term nematode management in climate-smart agriculture.

R. P. Parvatha, *Plant Parasitic Nematodes*,
https://doi.org/10.1007/978-981-95-9306-4_18

18.2 Introduction

Plant-parasitic nematodes (PPNs) pose a significant threat to global agriculture, causing substantial economic losses estimated at US$173 billion annually, damaging a wide variety of crops. Climate change, characterized by rising temperatures, altered precipitation patterns, and increased CO_2 levels, exacerbates this challenge by influencing nematode biology, distribution, and interactions with host plants. Warmer soils and altered precipitation patterns are likely to enhance nematode survival, reproduction, and virulence, necessitating adaptive management strategies. This chapter synthesizes recent research to provide a detailed overview of management strategies, highlighting the need for integrated and adaptive approaches to ensure sustainable crop production under changing environmental conditions.

Climate change is profoundly altering the ecology of PPNs, with rising temperatures and shifting precipitation patterns driving changes in their abundance, community composition, and geographic distribution. Moderate warming has been shown to increase PPN densities in many agro-ecosystems, although extreme heat may eventually suppress them; overall, erratic rainfall tends to favor plant parasites over other nematode trophic groups in the longer term. Additionally, elevated atmospheric CO_2 levels can indirectly boost PPN populations by increasing root biomass and altering plant physiology, thereby providing more resources for nematode reproduction (Dutta and Phani 2023). These dynamics underscore the urgency of adapting nematode management strategies to a rapidly changing climate.

Below is a concise overview of strategies to manage PPNs in the context of climate change, based on current scientific understanding.

18.3 Impact of Climate Change

18.3.1 Temperature Effects

Research indicates that warmer soils can accelerate nematode life cycles, increasing reproduction rates and potentially expanding their geographic range to higher latitudes. For instance, species like *Meloidogyne* (root-knot nematodes) thrive in warmer conditions, with studies showing that soil temperatures between 32 °C and 34 °C can enhance survival for *Rotylenchulus reniformis* by 10% at 34 °C, while reducing survival for *Meloidogyne floridensis* by 12% at the same temperature. However, disease severity, such as root gall index for *M. floridensis*, can increase by 46% at 34 °C, highlighting complex temperature-dependent outcomes. These changes underscore the need for management strategies that account for warmer soils.

18.3.2 Moisture and Precipitation Changes

Altered rainfall patterns influence soil moisture, affecting nematode movement and survival. Drought stress may weaken plants, making them more susceptible to PPN damage, while excessive moisture can facilitate nematode dispersal.

18.3.3 CO_2 Levels

Elevated CO_2 can enhance plant growth but may also increase root biomass, providing more resources for PPNs, potentially exacerbating infestations. This effect is noted in studies discussing the impact on nematode abundance and distribution. Climate change enables PPNs to invade new regions, including temperate areas previously unsuitable due to cooler climates.

18.3.4 Distribution Shifts

Climate change enables PPNs to invade new regions, including temperate areas previously unsuitable due to cooler climates, necessitating region-specific management strategies.

18.4 Management Strategies

Under climate change—characterized by rising temperatures, shifts in precipitation patterns, and elevated atmospheric CO_2—traditional cultural controls for PPNs must be adapted to remain effective. Managing PPNs under climate change requires a multifaceted approach that integrates cultural, biological, chemical, and technological methods. To address these challenges, integrated pest management (IPM) approaches tailored to climate change are critical. Key strategies include the following:

- Cultural practices
- Novel less toxic chemicals
- Biological control
- Resistant and tolerant varieties
- Precision agriculture
- Integrated management strategies

18.5 Cultural Practices

Cultural practices, as part of INM, offer sustainable alternatives to chemical controls, especially as restrictions on nematicides like methyl bromide increase. These practices must be adapted to climate change to maintain efficacy, considering factors such as temperature shifts, water availability, and soil health.

Cultural tactics such as cover cropping, crop rotation, biofumigation, flooding, and soil solarization are fundamental to PPN suppression but must be adapted to minimize greenhouse gas (GHG) emissions and maintain efficacy under changing conditions. Incorporating antagonistic cover crops (e.g., oil radish and sorghum–Sudan grass hybrids) can suppress nematodes while enhancing soil resilience and carbon sequestration (Dutta and Phani 2023). On the other hand, traditional flooding may exacerbate methane release, and fallowing can spike N_2O emissions, so alternative rotation schedules, reduced tillage, and solarization—using clear films during the hottest months—offer more sustainable options (Dutta and Phani 2023).

18.5.1 Soil Solarization and Deep Summer Plowing

Soil solarization, using solar heat to create a "biological vacuum," is effective for minimizing initial nematode populations (Phani et al. 2021). Its efficacy may vary with rising temperatures, requiring adjustments in timing based on local climate projections. This practice is particularly relevant in protected agriculture, where controlled environments can enhance its impact.

Higher ambient temperatures can make solarization—covering moist soil with transparent plastic—more effective at raising soil temperatures to lethal levels for nematodes. Similarly, deep summer plowing exposes nematodes to scorching sun and desiccation; both tactics can be synchronized with projected heat waves to maximize mortality. However, there is a need to balance these with erosion risks under intense rainfall events.

18.5.2 Crop Rotation With Non-host and Resistant Varieties

Crop rotation is a well-established practice to disrupt nematode life cycles, with specific rotations like rice or taro in Taiwan decreasing nematode populations and increasing strawberry yields (Chen and Tsay 2006). Planting corn can reduce northern root-knot nematode (*M. hapla*), but other *Meloidogyne* spp. may increase with persistent cultivation, highlighting the need for tailored rotations. Climate change necessitates customizing rotations to local agro-climatic regimes, as shifting seasons affect both crop growth and nematode dynamics.

Rotating crops with non-host or resistant varieties disrupts nematode life cycles. Biofumigation, using crops like mustard, releases nematicidal compounds when incorporated into the soil, improving carbon sequestration under changing climates (Somasekhar and Prasad 2012). Rotate crops with non-host or resistant varieties to disrupt nematode life cycles. For example, rotating cereals with legumes can reduce *Meloidogyne* populations.

Rotating cash crops with non-hosts or varieties bred for nematode resistance remains foundational. Under warming trends, select rotations that also tolerate heat and drought. For example, incorporating cereals (e.g., oat against *Heterodera avenae*) or legumes such as cowpea and *Crotalaria* spp. (trap crops for *Meloidogyne* spp.) can suppress nematode buildup while fitting altered growing seasons (Barker 1997). Recent work shows that resistant soybean rotations can reduce *Rotylenchulus reniformis* densities enough to protect subsequent cotton even under warmer soil conditions (Afzal and Mukhtar 2024).

18.5.3 Cover and Trap Cropping With Climate-Smart Species

Cover cropping with PPN-antagonistic species, such as grain sorghum and leguminous crops such as *Mucuna pruriens* and *Crotalaria spectabilis*, can reduce nematode incidence while replenishing soil nutrients (Barker 1997). These crops benefit environmental quality under different cropping systems (Grant et al. 2002; Asmus et al. 2008; Khanal and Harshman 2022). Under climate change, cover crops enhance carbon sequestration, crucial for agro-ecosystem resilience, but selection must consider adaptability to higher temperatures and altered precipitation patterns.

Utilizing PPN-antagonistic cover crops such as grain sorghum, millets, oil radish, Mulato grass, and sorghum-Sudan grass can suppress nematode populations while benefiting environmental quality. These crops release nematicidal compounds and enhance soil biodiversity (Dutta and Phani 2023).

Use cover crops like marigold (*Tagete*s spp.) or Sudan grass, which produce nematicidal compounds, to suppress PPN populations.

18.5.4 Adjusting Planting Dates and Irrigation Management

Climate-driven shifts in nematode life-cycle timing require planting early or late to avoid peak juvenile activity. For instance, planting potatoes and sugar beets in cooler windows can escape cyst nematode infection when soils warm too quickly (Harini 2021). Drip irrigation or regulated deficit irrigation helps maintain soil moisture below nematode-optimal levels without stressing the crop.

18.5.5 Organic Amendments and Vermicompost

Adding farmyard manure (FYM), neem oilcakes, or chitinous residues builds soil organic matter (SOM), buffering against moisture extremes and fostering antagonistic microbes. Under elevated CO_2, faster residue decomposition can boost microbial activity—strengthening biological control of PPNs (Barker 1997). Tailor amendment timing to avoid nutrient leaching during heavy rains and to maximize microbial colonization prior to nematode emergence.

Incorporate organic matter (e.g., compost and biochar) to improve soil health and promote antagonistic microorganisms that suppress nematodes. Organic soil amendments, particularly vermicompost, have emerged as dual-purpose tools for nutrient enrichment and PPN suppression. Vermicompost enhances soil structure, enriches beneficial microfauna, and releases metabolites that inhibit root-knot nematode hatching and development across diverse horticultural crops such as tomato, strawberry, and carrot (Dutta and Phani 2023). Similarly, chitin-fortified black soldier fly frass not only boosts plant growth but also significantly reduces cyst nematode reproduction, offering an integrated organic amendment approach compatible with smallholder systems (Dutta and Phani 2023).

Applying FYM improves soil health and promotes antagonistic microorganisms, offering a safer option with lower GHG emissions compared to other organic amendments (Mori 2018). It improves soil structure and supports beneficial microorganisms, enhancing soil health under climate change while minimizing environmental impact.

18.5.6 Tillage System Adaptations

Conservation tillage can maintain soil structure and microbial diversity, reducing PPN favorability under variable climate conditions. While deep tillage can reduce nematode populations by bringing them to desiccating surfaces, no-till or reduced-till systems conserve soil structure and moisture under erratic precipitation. Under lighter rainfall, limited disturbance preserves SOM; under heavy rains, minimal tillage reduces erosion and runoff that can spread nematodes (Afzal and Mukhtar 2024; Barker 1997).

18.5.7 Avoidance of Monoculture

Continuous monoculture aids PPN proliferation, especially in protected agriculture, due to stable microclimates (Phani et al. 2021). Diversifying crops breaks this cycle, reducing nematode build-up. Under climate change, this practice becomes critical as pest pressure may increase, necessitating adaptive crop diversification strategies.

18.5.8 Sanitation and Use of Clean Planting Material

Ensuring nematode-free planting materials (seed tubers or transplants) is vital, particularly in protected agriculture, to prevent contamination (Phani et al. 2021). Rigorous inspection and removal of symptomatic material, coupled with hot-water or biological treatments, prevent introduction into fields. Climate change may alter pest distributions, increasing the risk of new nematode introductions, making biosecurity measures like clean planting material essential.

18.5.9 Optimizing Microclimate

In protected structures, managing humidity and temperature is responsible for reducing conditions favorable for nematodes (Phani et al. 2021). With rising temperatures, this practice becomes challenging but necessary, requiring adaptive management of ventilation, shading, and irrigation to maintain less hospitable environments for PPNs.

18.5.10 Flooding and Fallowing

Flooding, as seen in rice cultivation, suppresses nematode lifestyles, with rotations involving rice or taro in Taiwan showing reduced nematode populations (Chen and Tsay 2006). However, flooding amplifies GHG emissions by disrupting carbon and nutrient cycling, and fallowing increases N_2O emissions without nitrogenous fertilizer (Sánchez-Rodríguez et al. 2019; Mosier et al. 1991). These practices require cautious use under climate change, balancing nematode control with environmental impacts.

18.5.11 Crop Scheduling

Adjusting crop schedules is necessary due to altered summer and winter durations under climate change, affecting PPN population dynamics (Dutta and Phani 2023). Adaptive management based on local climate data ensures that planting and harvesting avoid peak nematode activity and maintain crop health.

Adjusting planting dates to avoid peak nematode activity, especially in regions with extended growing seasons due to warmer climates, is crucial for minimizing damage.

18.5.12 Key Considerations Under Climate Change

The effectiveness of these cultural practices depends on adaptability to local climate projections. Rising temperatures may increase nematode reproduction rates, necessitating adjustments in practices such as soil solarization and cover cropping. Water availability changes may affect flooding and fallowing, while shifting seasons require flexible crop scheduling. INM, combining cultural, biological, physical, botanical, and chemical measures, is recommended for climate-smart nematode management, aiming for reduced crop loss, enhanced ecosystem services, minimal GHG emissions, and agricultural system resilience (Dutta and Phani 2023).

In Table 18.1, we detail the cultural practices identified, supported by recent studies, and discuss their relevance under climate change scenarios.

18.5.13 Integration Into Climate-Smart IPM

All cultural measures should feed into a broader IPM framework. Real-time monitoring of nematode densities, weather-driven risk models, and decision support

Table 18.1 Summary of key practices, their descriptions, and climate-related considerations

Cultural practice	Description	Climate change relevance
Cover cropping	Use PPN-antagonistic cover crops (e.g., grain sorghum, millets, oil radish, Mulato grass, sorghum-Sudan-grass, *Mucuna pruriens*, *Crotalaria spectabilis*) to suppress nematodes and improve soil health	Enhances carbon sequestration, critical under climate change. Select cover crops adaptable to higher temperatures and altered rainfall
Crop rotation	Rotate crops (e.g., rice, taro, and corn) to disrupt nematode life cycles, reducing population density	Must be customized to agro-climatic regimes, considering shifts in growing seasons and nematode behavior under warmer conditions
Soil solarization	Cover soil with plastic to trap solar heat, killing nematodes and other pathogens	Efficacy may vary with temperature changes; timing adjustments needed based on local climate projections
Avoidance of monoculture	Diversify crops to prevent nematode build-up, breaking the continuous host availability	Crucial as climate change may increase pest pressure; diversification helps adapt to new pest dynamics
Use of clean planting material	Ensure seeds and seedlings are nematode-free to prevent infestations	Essential as climate change alters pest distributions, requiring strict biosecurity measures
Optimizing microclimate	Manage humidity and temperature in protected agriculture to reduce nematode-friendly conditions	Challenging with rising temperatures; adaptive management of ventilation and irrigation needed
Flooding and fallowing	Flooding suppresses nematodes by disrupting life cycles; fallowing allows soil rest to reduce populations	Flooding increases GHG emissions (e.g., disrupting carbon cycling); fallowing may increase N_2O emissions without nitrogenous fertilizer, requiring cautious use
Use of FYM	Apply FYM as a safer organic amendment with lower GHG production compared to others	Supports soil health under climate change, minimizing environmental impact while managing nematodes
Crop scheduling	Adjust planting and harvesting times to avoid peak nematode activity, considering seasonal shifts	Necessary due to altered summer and winter durations, requiring adaptive management based on local climate data

tools allow dynamic adjustment of cultural tactics to on-ground conditions, ensuring that practices remain effective as climates continue to change.

By selecting climate-resilient crop species, timing cultural operations with shifting nematode phenology, and enhancing soil health to buffer extremes, farmers can maintain—and even improve—the efficacy of cultural management against PPNs under future climate scenarios.

18.6 New Generation Nematicides

Traditional chemical nematicides, such as halogenated hydrocarbons, organophosphates, and carbamates, have proven effective but pose substantial risks to human health, non-target organisms, and the environment. Chemical nematicides remain a critical short-term tool, but temperature- and moisture-driven changes in soil chemistry can alter their persistence and efficacy. Consequently, there is an urgent need to develop novel, less toxic chemical controls that remain effective under shifting climatic conditions. Current-generation compounds (e.g., fluensulfone, fluopyram, and fluazaindolizine) provide targeted action, yet label restrictions and environmental concerns limit broad-scale use under future climates (Dutta and Phani 2023). Integrating low-risk biopesticides (e.g., neem-based formulations) with reduced-rate fumigants—tailored by real-time soil monitoring—can optimize chemical interventions while mitigating GHG emissions and off-target impacts (Yadav and Patil 2021).

Use nematicides judiciously, as climate change may reduce their efficacy due to faster degradation in warmer soils. Eco-friendly alternatives like biopesticides need to be explored. Nanotechnology-based solutions, such as nanoparticle-delivered nematicides, which may offer targeted control with reduced environmental impact, should be investigated.

Chemical controls are necessary but must be used judiciously:

- Nematicides such as fluensulfone, fluopyram, and fluazaindolizine are effective but should be used sparingly due to environmental concerns and global pesticide reduction goals (Gatto et al. 2016; Jacquet et al. 2022). Higher temperatures may reduce their persistence, requiring careful application timing.
- Heat treatments such as hot water (32–34 °C for 7 h) and anaerobic soil disinfestation are explored as safer options, with studies showing varied effects on nematode survival and reproduction.

Recent advances have yielded a new generation of fluorinated non-fumigant nematicides characterized by high target specificity and low non-target toxicity. The primary compounds in commercial use or advanced development are as follows.

18.6.1 Fluensulfone (Nimitz®)

A fluoroalkenyl thioether registered in the USA in 2014, fluensulfone irreversibly affects nematode motility without the hypercontraction typical of organophosphates and carbamates (Kearn et al. 2014). It exhibits low acute toxicity to mammals and minimal impact on beneficial soil fauna (Kawanobe et al. 2019).

This nematicide depletes lipid storage in nematodes, disrupting their energy reserves. It is effective against root-knot and lesion nematodes, with a toxicity profile that is safer for the environment compared to older chemicals.

18.6.2 Fluopyram (Velum®)

Originally developed as a fungicide, fluopyram also demonstrates nematostatic action against root-knot nematodes (e.g., *Meloidogyne incognita* and *Rotylenchulus reniformis*), likely through disruption of mitochondrial respiration.

A succinate dehydrogenase inhibitor is effective against a range of PPNs and is used in seed treatments to reduce application rates. It is less toxic to non-target organisms, making it suitable for IPM.

18.6.3 Fluazaindolizine (Salibro™ in Development)

A novel sulfonamide nematicide with broad activity against root-knot species; in comparative field trials on carrots, fluazaindolizine at 2.24 kg ha^{-1} provided the most consistent marketable yields, outperforming fluensulfone and fluopyram under high disease pressure (Becker et al. 2019).

Registered by the EPA on September 12, 2023, this nematicide is effective against root-knot nematodes with an LC50 of 177 mg/L. It has a low LD50 of 1187 mg/kg in rats, indicating reduced toxicity, though it is carcinogenic at high doses, affecting the urinary tract, hematopoietic tissue, and liver/gall bladder. Application rates range from 0.25 to 2.0 kg/ha or 5–500 mg/L, with efficacy shown in reducing galls at 1.12 kg/ha, though ovicidal activity is limited below 400 mg/L.

18.6.4 Advantages

These fluorinated chemistries share several advantageous traits over legacy nematicides:

- *Selective Targeting*: They act on specific nematode biochemical pathways, reducing collateral effects on non-target organisms.
- *Reduced Toxicity*: Acute toxicity to mammals and beneficial invertebrates is substantially lower; restricted-entry intervals are ≤12 h versus ≥48 h for many organophosphates/carbamates (Becker et al. 2019).
- *Environmental Safety*: Field studies confirm that fluensulfone maintains soil free-living nematode diversity and abundance at levels comparable to untreated controls (Kawanobe et al. 2019).

18.6.5 Mode of Action

Fluensulfone's mechanism diverges markedly from cholinesterase inhibitors and macrocyclic lactones. In *Caenorhabditis elegans*, fluensulfone induces pleiotropic effects—impairing locomotion, feeding, development, and egg-laying—without causing the hypercontraction seen with organophosphates, and remains effective against carbamate- and avermectin-resistant mutants (Kearn et al. 2014). Fluopyram, by inhibiting succinate dehydrogenase, disrupts nematode energy metabolism, while fluazaindolizine's precise biochemical target is under investigation but appears to involve fatty acid biosynthesis pathways.

18.6.6 Field Efficacy and Non-target Impacts

- *Crop Yield Improvements*: In sweet potato field trials, a pre-plant incorporated drench of fluensulfone (1.96 kg ha^{-1}) more than doubled marketable yields compared to untreated and metam-sodium treatments, with over 80% reduction in root infestation (Ploeg et al. 2019).
- *Non-Target Fauna Preservation*: Across diverse soil types and host crops (tomato, sweet potato), fluensulfone applications maintained free-living nematode community indices (Shannon–Wiener, Simpson's D, richness, evenness) on par with untreated soils, underscoring its limited ecological disruption (Kawanobe et al. 2019).

18.6.7 Challenges Under Climate Change Scenarios

Climate change variables can influence both PPN dynamics and nematicide performance:

- *Temperature Effects*: Elevated soil temperatures can accelerate PPN life cycles but may also alter nematicide degradation rates and soil sorption characteristics. Historical data show that methyl bromide efficacy increased at 30 °C versus 15 °C; similar studies are needed for novel nematicides.
  Temperature increases may accelerate nematode life cycles, potentially requiring higher application rates or more frequent treatments, though novel nematicides like tioxazafen, with prolonged root activity, may mitigate this.
- *Soil Moisture Variability*: Changes in precipitation patterns affect soil moisture, impacting nematicide diffusion and nematode behavior; entomopathogenic nematode (EPN) efficacy is known to vary with moisture regimes (Frankenstein et al. 2024).
  Precipitation changes can affect soil persistence, with studies suggesting that organophosphates and carbamates require moderate temperature and moisture for toxicity, and climate-induced fluctuations may reduce their effectiveness (Delcour et al. 2014).
- *Elevated CO_2*: Increased atmospheric CO_2 indirectly modifies root exudation, potentially altering nematode attraction and penetration, and may influence nematicide uptake by plants.
  However, the design of these new nematicides, often aided by *in silico* approaches and nanoencapsulation, aims to improve stability and efficacy under varying conditions. For example, nano-encapsulation reduces costs and environmental

impact, enhancing delivery under changing climates. Technological advances such as RNA-Seq and CRISPR (clustered regularly interspaced short palindromic repeat)-PN2 (CRISPR-Cas9) are also being explored to enhance biocontrol and gene editing for PPN resistance, complementing chemical controls.

18.6.8 Future Research Thrusts

To ensure sustained efficacy, research must address:

- *Fate and Transport*: How do higher temperatures and variable moisture alter sorption–desorption, volatilization, and microbial degradation of fluorinated nematicides?
- *Dose Optimization*: Should application rates or timing be adjusted seasonally to match PPN peak activity under warmer or wetter conditions?
- *Resistance Management*: Monitoring for PPN adaptation to these novel modes of action, especially under the accelerated life cycles driven by warming.

18.6.9 Integrated Nematode Management

Given the complexity of climate change impacts, relying solely on chemical control is insufficient. Judicious use of novel nematicides, integrated with other methods to reduce environmental impact. INM is recommended, combining:

- *Cultural practices*: Crop rotation, cover cropping, and biofumigation to disrupt nematode life cycles.
- *Biological controls*: Use of bionematicides and beneficial microorganisms, such as *Beauveria bassiana* and *Metarhizium anisopliae*, which have shown efficacy against *M. incognita* in tomatoes and cucumbers (Karabörklü et al. 2022).
- *Host resistance*: Breeding or selecting nematode-resistant crop varieties, evaluated through host suitability studies (e.g., summer cover crops for *M. enterolobii*) (Khanal and Harshman 2022).
- This holistic approach ensures sustainable nematode management, reducing crop loss, enhancing ecosystem services, and minimizing GHG emissions, aligning with climate-smart agriculture goals.
- *Integrated Modeling*: Combine PPN population models under climate scenarios with nematicide fate and transport simulations to optimize application strategies.
- *Precision Delivery*: Utilize sensor networks and variable-rate applicators to tailor nematicide doses based on real-time soil moisture and temperature profiles.
- *Synergistic Approaches*: Explore formulations combining low-toxicity chemicals with biocontrol agents or botanical extracts to broaden the spectrum of action and delay resistance.

18.6.10 Conclusion

Fluorinated nematicides such as fluensulfone, fluopyram, and fluazaindolizine represent a promising shift toward safer, target-selective chemical control of PPNs. Their low non-target toxicity and demonstrated field efficacy make them well suited for integration into sustainable nematode management programs. However, to maintain their effectiveness under the unpredictable influences of climate change, focused research on their environmental fate, dose optimization, and integration with precision agriculture is critical. Such proactive measures will safeguard crop productivity and ecological health in a warming world.

The management of PPNs under climate change scenarios relies on the integration of novel, less toxic nematicides with broader INM strategies. While these chemicals offer promising solutions, ongoing research is needed to optimize their efficacy under varying climatic conditions, particularly through technological advances such as nano-encapsulation and gene editing. The evidence suggests a balanced approach, combining chemical, biological, and cultural methods, to ensure sustainable agriculture in the face of climate change.

18.7 Biological Control

Biological control of PPNs is an environmentally sustainable approach to managing these pests, especially under climate change scenarios that exacerbate their impact on crops. PPNs, such as *Rotylenchulus reniformis* and *Meloidogyne* species, cause significant agricultural losses (estimated at $358 billion annually) by feeding on plant roots, and climate change—through rising soil temperatures, altered precipitation, and elevated CO_2 levels—can intensify their reproduction, virulence, and geographic spread.

Biological control—using living organisms or their metabolites to suppress PPN populations—has gained prominence as a sustainable alternative. Harnessing soil microbiota provides a powerful, eco-friendly layer of nematode management. Indigenous nematophagous fungi (e.g., *Arthrobotrys* spp.) and bacteria (e.g., *Bacillus thuringiensis* and *Pseudomonas fluorescens*) deploy diverse mechanisms—trapping structures, lytic enzymes, antibiotics, and induction of plant systemic resistance—to reduce PPN viability and reproduction. Conservation agriculture practices that support microbial diversity can amplify these natural biocontrol services, making biological agents more reliable under variable climate conditions.

Promotion of natural enemies such as fungi (*Purpureocillium lilacinum and Pochonia chlamydosporia*), bacteria (*Pasteuria penetrans*), and predatory nematodes helps to control PPNs. Enhancing soil biodiversity through organic farming supports nematode-suppressive ecosystems, which are more resilient to climate variability.

Biological control leverages natural enemies to manage PPNs, with considerations for climate impacts. Below is an overview of biological control strategies for

PPN management, tailored to address challenges posed by climate change, based on current research.

18.7.1 Nematophagous Fungi

Trichoderma spp., *Pochonia chlamydosporia*, and *Purpureocillium lilacinum* are widely studied for their ability to parasitize PPN eggs, juveniles, or females, reducing population densities. These fungi also produce lytic enzymes, toxins, or induce plant resistance (e.g., systemic acquired resistance or induced systemic resistance). Mycorrhizal fungi (e.g., arbuscular mycorrhizal fungi) enhance plant nutrient and water uptake, counteracting PPN damage by improving root health and modifying rhizosphere interactions. However, they may also promote PPN populations by disarming plant defenses in some cases. Endophytic fungi colonize plant tissues and reduce nematode invasion or reproduction by competing for resources or inducing plant defense mechanisms.

Nemaphagous fungi encompass a diverse group, including trapping fungi (e.g., *Arthrobotrys* spp.), egg-parasitic fungi (e.g., *Purpureocillium lilacinum*), and endoparasitic fungi (e.g., *Hirsutella rhossiliensis*). These organisms employ mechanisms such as adhesive nets, specialized spores, and toxin secretion to capture, penetrate, and kill nematodes. Under elevated temperatures and CO_2, studies report variable impacts: while some *Trichoderma* strains maintain antagonism under heat stress, others show diminished spore germination and volatile organic compounds (VOC) production, affecting volatile-mediated nematode suppression (Ayaz et al. 2024). Similarly, moisture deficits can reduce the formation of trapping structures by *Arthrobotrys* spp., highlighting the need for robust isolates adapted to drier soils.

The use of biocontrol agents is increasing, with studies showing potential for *Beauveria bassiana* and *Metarhizium anisopliae* in controlling *Meloidogyne incognita* in tomato and cucumber.

18.7.2 Nematophagous Bacteria

Pasteuria penetrans, Bacillus subtilis, and *Pseudomonas fluorescens* suppress PPNs through parasitism, toxin production, or competition for nutrients. *Pasteuria* adheres to nematode cuticles, inhibiting movement and reproduction, while rhizobacteria like *Pseudomonas* induce systemic resistance in plants. Bacteria can also produce nematicidal metabolites, offering a targeted approach with minimal environmental impact.

Rhizobacteria and endophytes such as *Bacillus* spp., *Pseudomonas fluorescens*, and *Pasteuria penetrans* are among the most studied bacterial BCAs. They employ multiple modes of action: direct parasitism (e.g., *P. penetrans* attaching to J2 nematodes), antibiosis via secondary metabolites (e.g., hydrogen cyanide and lytic enzymes), competition for niche and nutrients, and induction of systemic resistance in host plants. Multi-omics approaches have identified novel nematicidal metabolites—such as dimethyl disulfide and 2-undecanone—from *Bacillus atrophaeus*, which retain activity at varied temperature ranges and soil moisture levels (Ayaz et al. 2024). Yet, the interplay between elevated CO_2-driven root exudation changes and bacterial colonization patterns warrants further investigation to ensure consistent performance under future climate scenarios.

18.7.3 Predatory Nematodes

Species from orders such as Mononchida, Dorylaimida, Diplogasterida, and Aphelenchida prey on PPNs, reducing their populations (Figs. 18.1 and 18.2). Mononchids, for example, lack prey specificity, while diplogasterids show higher specificity, making them effective in specific contexts. Their efficacy depends on traits such as predation efficiency, reproductive potential, and adaptation to environmental conditions, which are critical under changing climates.

18.7.4 Entomopathogenic Nematodes

EPNs such as *Steinernema* spp. and *Heterorhabditis* spp. have been reported to suppress root-knot nematodes (*Meloidogyne* spp.) on tomato and boxwood; *Tylenchorhynchus* spp., *Meloidogyne* sp., *Belonolaimus longicaudatus,* and *Criconemella* sp. on turf grass; and the burrowing nematodes (*Radopholus similis*) on banana. Hence, EPN can be effectively used as a biological control agent for the management of PPNs.

Nematodes like *Phasmarhabditis* spp. have been proposed to control non-nematode pests (e.g., invasive gastropods), which indirectly reduces PPN transmission by limiting vector movement.

EPN like *Steinernema* can interfere with the infection process of a PPN (*Meloidogyne* spp.). The EPNs are seen entering the insect host cadaver, where they

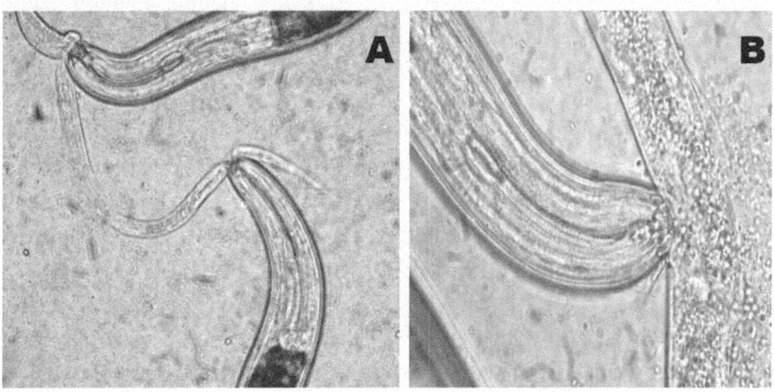

Fig. 18.1 Predatory nematodes feeding activities. (**a**) Two individuals of *Mononchoides gaugleri* (diplogasterid) feeding on the same prey. (**b**) *M. gaugleri* feeding on a prey

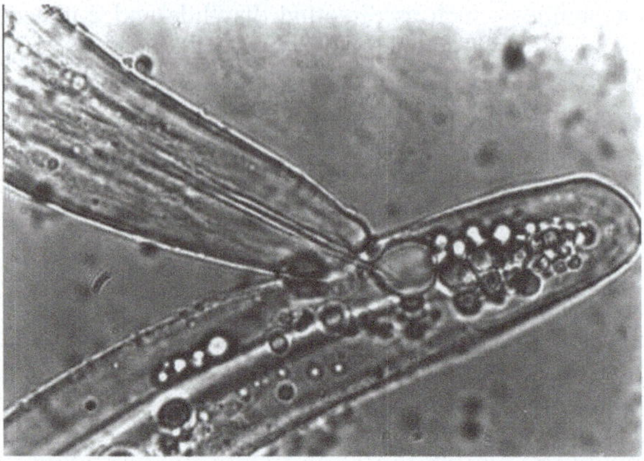

Fig. 18.2 *Seinura* (Aphelenchina-Tylenchida, Secernentea) feeding on nematode

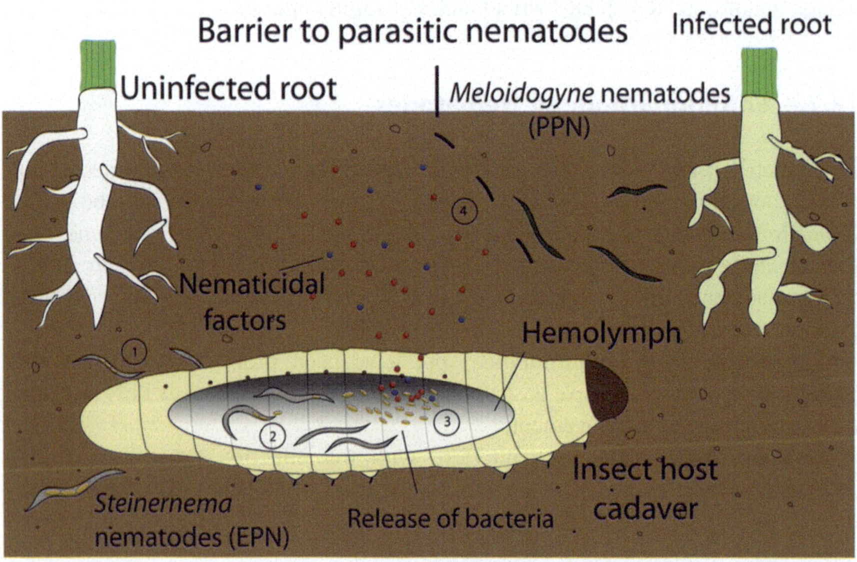

Fig. 18.3 Mode of action of EPN on root-knot nematodes

then release their mutualistic *Xenorhabdus* bacteria into the hemolymph. A number of bacterial factors are then produced, some of which are selectively nematicidal, resulting in the repulsion of, or interference with, the plant pathogenic species. The plant root to the right that is not protected by nematicidal factors is shown to be burdened with nematode-induced root gall formation (Fig. 18.3).

18.7.5 Influences of Climate Change on Biological Control

- *Temperature*: Higher soil temperatures (e.g., 32–34 °C) increase PPN virulence, as seen with *R. reniformis*, which shows greater aggressiveness at elevated temperatures, while *M. floridensis* reproduction decreases above 33 °C. BCAs such as fungi and bacteria may also face altered activity; for instance, warmer soils could enhance fungal reproduction but disrupt their antagonistic effects if temperatures exceed their optima.

- *Precipitation and Moisture*: Altered rainfall patterns affect nematode movement and survival, particularly for species with soil-dwelling infective stages. Reduced water availability in mesic habitats may decrease predation pressure, allowing PPNs to proliferate. BCAs like predatory nematodes thrive in moist conditions, so drought could limit their efficacy.
- *Elevated CO_2*: Increased atmospheric CO_2 promotes plant growth but alters root exudates, potentially attracting more PPNs or modifying BCA interactions with plants. This necessitates BCAs that can adapt to changing rhizosphere dynamics.
- *Geographic Spread*: Warming climates enable PPNs like *Globodera* spp. to invade new regions, requiring BCAs with broad adaptability to diverse climates.

18.7.6 Viral and Nematode–Nematode Interactions

Emerging research highlights the potential of bacteriophages targeting bacterial endosymbionts of nematodes, as well as predatory nematodes that feed on PPNs. While still in early stages, these approaches offer complementary modes of action that could buffer against climate-induced declines in fungal or bacterial BCA efficacy. However, the temperature sensitivity of viral stability and predator–prey dynamics under fluctuating soil conditions remains largely uncharacterized.

18.7.7 Integrating Multi-Omics for Climate-Resilient BCAs

Advances in metagenomics, transcriptomics, proteomics, and metabolomics facilitate the discovery and optimization of BCAs with enhanced stress tolerance and nematicidal potency. Functional metagenomics enables the mining of soil microbiomes for novel biosynthetic gene clusters encoding heat-stable enzymes or VOCs with broad-spectrum nematicidal activity (Ayaz et al. 2024). Transcriptomic profiling under simulated climate stress can identify candidate genes for engineering or selecting BCA strains that maintain biocontrol functions at high temperatures or low moisture. Proteomic and metabolomic analyses further elucidate the modes of action, enabling formulation improvements to protect active compounds in variable field conditions.

18.7.8 Climate-Smart Implementation and Field Validation

The successful deployment of BCAs under climate change requires rigorous field validation across diverse agro-ecological zones and under projected climate scenarios. Integrating BCAs into climate-smart agricultural practices—such as precision irrigation, organic amendments to improve soil moisture retention, and cover cropping to moderate soil temperature—can enhance BCA persistence and efficacy. Decision support systems incorporating climatic and soil data can guide optimal timing and choice of BCAs for region-specific conditions.

18.7.9 Effective Biological Control Under Climate Change

- *INM*: Combining BCAs with cultural practices (e.g., crop rotation, cover cropping, and biofumigation) and resistant plant varieties suppresses PPNs below economic thresholds. For example, biofumigation with organic amendments like FYM enhances BCA activity while improving soil health with lower GHG emissions. Minimize tillage and chemical inputs to preserve soil microbial communities and natural enemies of PPNs.
- *Climate-Smart Approaches*: Develop BCAs with robust nematicidal activity under diverse climatic conditions. Genetic engineering of bacteria or fungi to overexpress nematicidal genes (e.g., *Pasteuria* spp.) could enhance their resilience to temperature and moisture fluctuations. Use precision monitoring and modeling to predict PPN population dynamics under climate scenarios, enabling timely BCA application. Remote sensing and disease forecasting models can identify PPN hotspots for targeted interventions.
- *Enhancing BCA Efficacy*: Select BCAs with traits suited to climate stressors, such as high reproductive rates or tolerance to temperature extremes. For instance, diplogasterid predatory nematodes with short life cycles are promising for rapid response. Applying BCAs in combination (e.g., *Bacillus subtilis* + *Purpureocillium lilacinum*) leverages synergistic effects, as demonstrated in banana crops against *Radopholus similis*. Use organic amendments to provide supplementary food sources for BCAs, boosting their populations while maintaining soil food web stability.
- *Sustainable Practices*: Promote soil health through organic matter additions, which enhance BCA activity and suppress PPNs by fostering antagonistic microbial communities. Avoid practices like excessive tillage or broad-spectrum biocides that disrupt beneficial soil biota.

18.7.10 Future Directions

Climate change may alter the effectiveness of these agents, necessitating research into climate-resilient biocontrol options. Key challenges include regulatory hurdles for BCA registration, variable performance in heterogeneous field conditions, and limited understanding of microbe–plant–nematode–climate interactions. Future research should focus on the following:

- *Knowledge Gaps*: Limited field studies on BCA performance under realistic climate scenarios hinder predictive models. More research is needed on how temperature and moisture interactions affect BCA-nematode dynamics.
- *Regional Variability*: PPN responses to climate change vary by species and region, requiring tailored BCA strategies. For example, *R. reniformis* adapts better to higher temperatures than *M. floridensis*, necessitating species-specific approaches.
- *Scalability*: Inundative BCA applications are often inconsistent or uneconomical due to the need for high concentrations and long-term protection. Conservation biological control, which enhances natural enemies through environmental modifications, offers a more sustainable alternative but requires site-specific validation.
- *Molecular Insights*: Advances in metagenomics and genomics (e.g., sequencing *Pasteuria* genomes) can identify novel BCAs and their nematicidal mechanisms, improving targeted applications.
- *Screening and Selection*: BCA consortia with complementary stress tolerances.
- *Formulation Technologies*: For example, encapsulation, carrier development to protect BCAs, and their metabolites under abiotic stress.
- *Long-term Monitoring*: BCA persistence and non-target effects in field settings.
- *Economic Assessments*: To evaluate the cost benefit under changing climatic conditions and to incentivize adoption.

18.7.11 Conclusion

Biological control offers a promising, environmentally benign strategy for managing PPNs against the backdrop of climate change. Leveraging multi-omics tools to identify and optimize BCAs, coupled with climate-smart agronomic practices and supportive regulatory frameworks, will be critical to developing resilient nematode management systems. Continued interdisciplinary research and stakeholder engagement are essential to translate laboratory insights into field-level solutions, safeguarding crop productivity and sustainability under a warming world.

Biological control of PPNs under climate change scenarios requires a shift to climate-smart, integrated management strategies that leverage fungi, bacteria, and predatory nematodes while minimizing environmental impact. By combining BCAs with sustainable practices like crop rotation, organic amendments, and precision monitoring, farmers can mitigate PPN damage and enhance crop resilience. Ongoing research into BCA adaptability and molecular mechanisms will be crucial to address the evolving challenges posed by climate change.

18.8 Resistant and Tolerant Varieties

Developing nematode-resistant crop varieties remains a cost-effective, long-term strategy. Traditional resistance genes, however, risk breakdown under heightened nematode pressure and may disrupt beneficial microbiomes. Innovative breeding programs that harness plant–microbe interactions—via organic breeding or CRISPR/Cas9—aim to create cultivars that both resist infection and promote suppressive soil microbiomes (Dutta and Phani 2023). This dual-focus breeding is especially vital in arid and semi-arid zones, where combined abiotic stresses intensify nematode impacts.

Development and deployment of crop varieties resistant to key PPNs, such as *Meloidogyne* or *Pratylenchus,* is essential. Breeding programs should focus on resilience to both nematodes and climate stressors (e.g., drought and heat). Use tolerant varieties that maintain yield despite nematode presence, especially in regions facing increased PPN pressure due to warming.

Climate change, characterized by rising global temperatures, altered precipitation, and extreme weather events, is expected to exacerbate nematode-related challenges. The average global temperature has increased by approximately 1.09 °C since the late nineteenth century, with predictions of a 1.5–2 °C rise by 2040 and potentially 2.4–2.6 °C by the end of the century. These changes can influence nematode biology, including survival, reproduction, and virulence, potentially reducing the effectiveness of resistant varieties.

18.8.1 Impact of Climate Change on Host Resistance

Research suggests that climate change, particularly global warming, may affect R-gene-mediated resistance in plants against PPNs. A critical finding is that resistance governed by the *Mi* gene in tomatoes is lost above 28 °C (El-Sappah et al. 2019; Williamson and Kumar 2006). Heat-stable *Mi* genes have been identified, but their efficacy under high heat stress remains unevaluated, highlighting a research gap. Additionally, warmer environments may alter general plant immune responses, potentially leading to the emergence of new PPN biotypes with altered parasitic potential (Venkatesh and Kang 2019).

Studies on specific nematodes, such as *Rotylenchulus reniformis* and *Meloidogyne floridensis*, indicate that higher soil temperatures can increase virulence, even if reproduction rates decrease slightly. For example, a study published by Khanal and Land (2023) found that *R. reniformis* survival increased by 10% at 34 °C compared to 32 °C, while *M. floridensis* showed increased disease severity (root gall index) by 46% at 34 °C. This suggests that while resistant varieties may still offer some protection, their effectiveness could diminish under future climate scenarios, necessitating adaptive strategies.

18.8.2 Specific Resistant Varieties and Their Application

Several resistant varieties have been identified for various crops, particularly for root-knot nematodes, and are recommended for use in integrated management systems. Below is a list of vegetable varieties resistant to root-knot nematodes, based on resources from the Alabama Cooperative Extension System (Table 18.2).

Table 18.2 Vegetable varieties resistant to root-knot nematodes

Crop	Resistant varieties
Sweet potato	Bayou Belle, Bonita, Covington, Hernandez, Jewel, Murasaki
Southern pea	California Blackeye #5, Magnolia Blackeye, Mississippi purple, Mississippi shipper, Mississippi silver, zipper cream
Lima beans	Nemagreen
Cayenne-type pepper	Carolina Cayenne, Charleston hot
Tomato	Beefmaster VFN, Burpee's Super Beefmaster VFN, Better Boy hybrid VFN, Amelia VR, Celebrity, Crista, Mountain Merit, Mountain Rouge, Red Bounty, Red Mountain
Tomato (Roma)	Granadero, Mariano

Additionally, crops such as asparagus, onion, and strawberry are noted to be resistant to most root-knot nematode populations in Alabama (Pams Blog 2022).

18.8.3 Breeding Strategies for Climate-Resilient Resistance

Developing resistant crop varieties is a key strategy, though climate impacts must be considered. Deploying crops resistant to key PPNs like *Meloidogyne* or *Pratylenchus* is vital. However, climate change may reduce resistance efficacy. Breeding programs should focus on resilience to both nematodes and climate stressors such as drought and heat.

Novel methods such as RNA interference or CRISPR-based technologies are suggested for enhanced sustainability (Fuller et al. 2008).

- *Gene Pyramiding*: Stacking multiple R-genes with complementary spectra can delay virulence adaptation by nematodes. For instance, pyramiding *Mi-1* with novel *Mi* homologs may extend durability even when one locus is compromised by heat stress (Dutta and Phani 2023).
- *MAS for Heat-Stable Alleles*: Kompetitive Allele-Specific PCR (KASP) markers linked to heat-stable resistance loci—such as the newly identified *RRKN1* in tomato line MT12—enable rapid introgression of these alleles into elite backgrounds without extensive phenotyping under variable field conditions (Han et al. 2024).
- *Genome-Wide Association Studies (GWAS) in Soybean*: GWASs have pinpointed quantitative trait nucleotides underlying soybean cyst nematode (SCN) resistance across diverse environments. By incorporating climate-interaction terms, genomic selection models can predict cultivar performance under future temperature regimes, accelerating the breeding cycle (Han et al. 2024).
- *CRISPR/Cas9-Mediated Gene Editing*: Targeted mutation of susceptibility (S) genes offers an alternative route to resistance. Huang et al. (2023) used CRISPR/Cas9 to knock out *OsHPP04* in rice, enhancing resistance to *Meloidogyne*

graminicola without yield penalties. Such edits can be designed for loci whose expression remains stable under heat or drought, creating inherently climate-proof resistance (Han et al. 2024).
- *Transgenic R-Gene Stacking*: Co-expression of multiple R-genes under stress-inducible promoters may ensure sustained defense activation only under nematode attack, potentially reducing fitness costs under abiotic stress (Han et al. 2024).

18.8.4 Case Studies

- Tomato (*Solanum lycopersicum*): *RRKN1* mapped on chromosome 6 confers stable resistance to RKN at soil temperatures up to 30 °C. Introgression into commercial cultivars via KASP-assisted backcrossing has shown field performance with <5% yield penalty under heat stress (Han et al. 2024).
- Rice (*Oryza sativa*): CRISPR-edited *OsHPP04* knockouts display up to 70% reduction in gall formation by *M. graminicola* under both optimal (25 °C) and elevated (32 °C) conditions, with no adverse agronomic traits noted over two growing seasons (Han et al. 2024).
- Soybean (*Glycine max*): Peking-type resistance at *rhg1/Rhg4* remains effective across 25–30 °C but shows partial breakdown above 32 °C. GWAS-guided breeding is ongoing to combine novel minor-effect loci that buffer R-gene expression under heat (Han et al. 2024).

18.8.5 Climate Adaptation and INM

Given the potential reduction in effectiveness of resistant varieties under climate change, INM is essential. INM combines multiple strategies to ensure sustainability, including:

- *Crop Rotation*: Rotating with non-host crops, such as cereals or legumes not susceptible to nematodes, can reduce population levels. For example, the University of Florida EDIS (2023) recommends cover crops such as cowpea and sorghum-Sudan grass, which are not hosts for root-knot nematodes (Gill et al. 2023).
- *Cover Crops*: Nematode-resistant cover crops, such as marigolds and sunhemp, can suppress nematode populations by starving them, as they cannot feed on these plants. A study by López-Pérez et al. (2010) found that incorporating broccoli as a winter cover crop reduced root galling on summer tomatoes by 36% and increased yields by 19%.
- *Cultural Practices*: Practices such as fallowing, soil solarization (using plastic tarping to raise soil temperatures), and applying organic amendments can help

manage nematodes, especially in home gardens, as noted by the Alabama Cooperative Extension System.

- *Biocontrol and Safer Approaches*: Increasing reliance on biocontrol agents and biofumigation (using plants like mustard for natural fumigation) can reduce dependence on chemical nematicides, which are often environmentally harmful and less effective under changing conditions.

These strategies are particularly important as climate change may lead to the geographical expansion of PPNs, with modeling studies predicting increased disease risks in new regions due to warmer climates (Somasekhar and Prasad 2012).

18.8.6 Challenges and Research Gaps

Several challenges remain in managing nematodes under climate change:

- *Emergence of Virulent Biotypes*: Climate-mediated shifts in nematode phenology can select for alleles that overcome existing R-genes. For example, global warming has facilitated the northward migration of *M. incognita*, with field isolates at 28–30 °C showing enhanced virulence on traditional *Mi-1* cultivars.
- *Trade-Offs With Abiotic Stress Tolerance*: R-gene expression often incurs metabolic costs; under drought or heat stress, plants may downregulate defense pathways, reducing nematode resistance when it is most needed (Dutta and Phani 2023).
- *Integration With Agronomic Practices*: Reliance on single resistance sources can lead to rapid resistance breakdown. Integrating resistant varieties with crop rotation, cover cropping, and biological controls creates a multi-tactic system that is more robust to climate variability (Westerdahl 2021).
- *Nematode Adaptation*: Nematodes may adapt to warmer soils, potentially increasing their virulence and reproduction, as seen with *R. reniformis* showing greater adaptability at higher temperatures. This could lead to the breakdown of traditional management tactics, including resistance.
- *Farmer Adoption*: The adoption of resistant varieties, especially climate-resilient ones, can be hindered by factors such as seed availability, access to extension services, and socio-economic status. A scoping review by Acevedo et al. (2020) found that 60% of papers highlighted seed access as a major issue for small-scale producers in low- and middle-income countries, emphasizing the need for local seed systems and outreach.
- *Research Needs*: There is a need for more studies evaluating the performance of resistant varieties under simulated climate change scenarios, particularly focusing on temperature thresholds for resistance breakdown. The efficacy of heat-stable *Mi* genes under high heat stress, for example, remains unevaluated (Dutta and Phani 2023).

18.8.7 Future Directions and Recommendations

To ensure effective nematode management under climate change, the following steps are recommended:

- *Breeding Programs*: Plant breeding programs should prioritize developing varieties with heat-stable resistance genes and traits that enhance overall climate resilience. Modern molecular technologies and field phenotyping techniques can support this (Kopeć 2024).
- *Climate-Smart Packages*: Implementing climate-smart agricultural packages that integrate resistant varieties with other INM practices can enhance food security. This includes precise monitoring and modeling of nematode population dynamics in relation to climatic fluctuations (Dutta and Phani 2023).
- *Farmer Education and Support*: Enhancing extension services and farmer education, particularly in vulnerable regions, can improve adoption rates. Local seed systems and social networks should be leveraged to ensure access to climate-resilient, nematode-resistant varieties.

18.8.8 Multi-dimensional Breeding Pipelines

- *Phenotyping Under Simulated Climate Regimes*: High-throughput platforms that mimic combined heat, drought, and nematode stress will enable screening of breeding populations for truly climate-resilient resistance traits.
- *Leveraging Microbiomes*: Breeding that preserves or enhances root microbiomes associated with nematode suppression (e.g., *Bacillus* spp.) could synergize with genetic resistance (Elhady et al. 2024).
- *Predictive Modeling*: Projects like NEMatode Ecological Modelling for Emerging Risks, Global distribution, and Environmental change (NEM-EMERGE) use spatial sampling and climate projections to forecast nematode emergence hotspots. Coupling these models with regional cultivar deployment recommendations can optimize the placement of heat-stable resistance genes.
- *Gene Editing Innovation*: Beyond knocking out S-genes, CRISPR prime editing may allow precise tuning of R-gene promoters for enhanced expression under heat stress, creating "smart" resistance alleles.
- *International Collaboration*: Climate change impacts on nematode resistance are global. Sharing germplasm, molecular markers, and field data through networks such as consultative group on international agricultural research (CGIAR) and national programs will accelerate the development of universally adapted resistant varieties.

18.8.9 Conclusion

Host resistance through resistant varieties is a cornerstone of nematode management, but climate change poses significant challenges by potentially reducing their effectiveness, especially under rising temperatures. By integrating resistant varieties with crop rotation, cover crops, and other cultural practices, and by investing in breeding programs for climate-resilient varieties, it is possible to mitigate nematode-related crop losses. Ongoing research and farmer support are crucial to address adaptation gaps and ensure sustainable agriculture in the face of environmental changes.

Under climate change, host plant resistance to PPNs must evolve from single-locus, temperature-sensitive approaches to integrated, multi-layered strategies combining pyramided R-genes, heat-stable alleles, genome editing, and beneficial microbiomes. By aligning breeding priorities with projected climate scenarios and incorporating advanced phenotyping and modeling tools, the next generation of resistant varieties can sustain crop health and productivity in a warming world.

18.9 Precision Agriculture

Employment of remote sensing, soil sensors, and geographic information system (GIS) is essential to monitor nematode hotspots and optimize resource use (e.g., water and fertilizers) under changing climate conditions. Using predictive models to forecast PPN outbreaks based on temperature, moisture, and crop data enables timely interventions.

Accurate, near-real-time monitoring of PPN populations and environmental conditions underpins adaptive management. Hyperspectral imaging holds promise for early detection of nematode infestation amid drought or nutrient stress, enabling precision interventions before economic damage occurs (Dutta and Phani 2023). Meanwhile, robotic soil diagnostics automate the extraction and quantification of cyst nematodes, streamlining data for decision-support systems that forecast outbreaks under future climate scenarios (Legner et al. 2021).

18.10 Integrated Management Strategies

This approach combines multiple tactics to keep PPNs below economic thresholds. It is particularly relevant under climate change, as it allows for flexible, ecosystem-based responses to environmental shifts. For example, cover cropping with Mulato grass can benefit environmental quality while suppressing nematodes (Grant et al. 2002).

A combination of methods is recommended, including crop rotation to break nematode life cycles, cover cropping with nematode-antagonistic plants like grain sorghum, and biological controls using natural enemies such as bacteria (*Pasteuria penetrans*) and fungi (*Pochonia chlamydosporia*). These approaches help suppress nematode populations while being environmentally friendly.

INM remains the cornerstone of sustainable PPN control, combining cultural, physical, botanical, biological, and chemical measures to keep nematode populations below economic thresholds (Dutta and Phani 2023). However, climate-driven shifts in nematode phenology and agro-ecosystem processes necessitate tuning these strategies to new environmental realities. For instance, practices developed under historical climate regimes—such as fixed crop rotation schedules or standard fumigation timings—may no longer align with nematode life cycles or soil microclimate, calling for dynamic, locally calibrated INM plans (Yadav and Patil 2021).

The summary of management strategies and climate impacts is presented in Table 18.3.

This table summarizes key strategies, their descriptions, the climate impacts they address, and relevant examples, providing a structured overview for practitioners.

18.11 Specific Research Insights

Recent studies provide detailed insights into PPN management under climate change:

18.11.1 Temperature Effects on Nematode Biology

A study on *R. reniformis* and *M. floridensis* found that direct heat exposure at 32–34 °C for 7 h increased *R. reniformis* survival by 10% at 34 °C but decreased reproduction by 45–49% (from 89,080 to 45,416 eggs/root system). For *M. floridensis*, survival decreased by 12% at 34 °C, but disease severity (root gall index) increased by 46.

18.11.2 Root Exudates and Plant Defense

Research on *Globodera* spp. suggests that root exudates can modulate nematode behavior, offering potential for sustainable management strategies based on transcriptional changes in pre-parasitic juveniles.

18.11.3 Raffinose Family Oligosaccharides

In cucumber plants infected with *M. incognita*, raffinose family oligosaccharides initially protect roots but are later used by nematodes as a carbon source. This insight can guide breeding programs for resistant cultivars by manipulating sugar withdrawal processes.

18.12 Challenges and Considerations

Several challenges remain, including the lack of host-plant resistance in high-yielding cultivars, variability in nematode reproduction and virulence among geographic isolates, and the limited season-long protection offered by nematicides (Bell et al. 2014; Khanal et al. 2018b, 2019, 2021). Additionally, while agro-ecological practices show promise, their impacts in real agro-ecosystems under

Table 18.3 Summary of management strategies and climate impacts

Strategy	Description	Climate impact addressed	Example species/crops
Reduced tillage	Minimizes soil disturbance to preserve food webs	Supports longer, complex food chains for better ecosystem regulation under stress	All crops
Cover crops	Use PPN-antagonistic crops to suppress populations	Enhances soil health, suppresses nematodes, and benefits environmental quality under changing climates	*Meloidogyne*, grain sorghum
Crop rotation and scheduling	Breaks nematode life cycles by alternating host and non-host crops. Rotate with non-host varieties to disrupt life cycles	Distribution shifts. Adjusts to altered seasonal durations, averting damage during peak nematode activity	Cereals, legumes
Crop diversity	Implements intercropping, polycropping, or rotations	Dilutes herbivory, leverages soil feedback, and increases plant productivity under climate stress	Cereals, legumes
Biofumigation	Incorporates plants releasing biocidal compounds into soil	Enhances carbon sequestration, supporting climate resilience	Brassicaceous crops
Organic amendments	Adds FYM to improve soil biodiversity	Reduces GHG emissions, supports beneficial nematodes, and enhances soil stability	All crops
Biological control	Uses natural enemies like *Pasteuria penetrans* and *Pochonia chlamydosporia*	Efficacy may vary with temperature and UV radiation, requiring climate-specific selection	*Rotylenchulus*, tomatoes
Host resistance	Develop resistant varieties, considering climate impacts. Utilizes R-genes in crops; e.g., *Mi* gene effective below 28 °C	Temperature impacts efficacy, necessitating biotechnological innovations for stable resistance	*Pratylenchus*, cucumbers
Chemical nematicides	Use sparingly, explore heat treatments as alternatives. Uses alternatives like fluensulfone, fluopyram; limited due to regulations	Persistence may be affected by climate fluctuations, pushing for reduced reliance	*Meloidogyne floridensis*, various
Adaptive management	Monitoring, scenario planning, and precision agriculture	All climate factors	All crops
INM	Combines cultural, physical, botanical, biological, and chemical measures	Adapts to environmental fluctuations, ensuring resilience; emphasizes ecosystem-based approaches	All crops

climate change require further research (Pörtner et al. 2022). The need for coordinated efforts among researchers and policymakers is also highlighted to develop future nematode management programs for warmer soils.

Several challenges complicate PPN management under climate change.

18.12.1 Regional Variability

Management must be tailored to local conditions, as climate impacts vary by region. For example, tropical areas may face intensified PPN pressure, while temperate regions may see new invasions.

18.12.2 Sustainability

Over-reliance on chemical controls can degrade soil health and exacerbate climate-related issues. Integrated approaches are essential.

18.12.3 Knowledge Gaps

Limited data on PPN responses to specific climate scenarios (e.g., extreme weather events) necessitates further research and monitoring.

18.13 Future Directions

To enhance PPN management under climate change, future efforts should focus on the following.

18.13.1 Research Needs

Invest in studies on PPN-climate interactions, including long-term field trials and modeling to predict distribution shifts.

18.13.2 Farmer Education

Train farmers on climate-adapted IPM practices through extension services, focusing on sustainable and cost-effective methods.

18.13.3 Policy Support

Encourage policies that promote climate-resilient agriculture, such as subsidies for organic amendments or resistant crop varieties.

By combining these strategies, farmers can mitigate the impact of PPNs under climate change, ensuring sustainable crop production.

18.14 Recommendations

Adapting PPN management to climate change requires a holistic, systems-based approach that integrates biological, cultural, chemical, and genetic tactics within flexible, data-driven frameworks. Key recommendations include the following.

18.14.1 Dynamic INM Planning

Regularly recalibrate rotation schedules and amendment timings using local climate forecasts.

18.14.2 Boost Soil Health

Prioritize cover crops and organic amendments to sustain microbial antagonists.

18.14.3 Selective Chemical Use

Employ reduced-risk nematicides in combination with real-time soil monitoring.

18.14.4 Next-Generation Cultivars

Invest in microbiome-aware breeding for durable resistance.

18.14.5 Leverage Technology

Deploy hyperspectral sensors and automated diagnostics for early-warning and tailored responses.

By synchronizing these climate-smart strategies, growers can sustainably suppress PPN populations, safeguard crop yields, and reduce environmental footprints in an era of unprecedented climatic uncertainty.

18.15 Conclusion

Managing PPNs under climate change requires a holistic, adaptive approach that integrates sustainable practices, leverages biological and cultural controls, and reduces reliance on chemicals. As climate conditions evolve, continuous research and innovation will be essential to develop resilient agricultural systems that mitigate nematode impacts while supporting food security and environmental sustainability.

Managing PPNs under climate change requires a proactive and integrated approach that adapts to shifting environmental conditions. By combining cultural practices, biological control, host resistance, and adaptive management strategies, farmers can mitigate PPN damage while ensuring sustainable crop production. Ongoing research and collaboration are essential to develop climate-resilient management practices tailored to specific regions and nematode species, addressing the complex interplay between climate, pests, and agriculture.

18.16 Suggested Readings

- Abd-Elgawad MMM (2025) Integrated nematode management strategies: optimization of combined nematicidal and multi-functional inputs. Plants 14(7):1004. https://doi.org/10.3390/plants14071004.
- Afzal A, Mukhtar T (2024) Revolutionizing nematode management to achieve global food security goals—an overview. Heliyon 10(3):e25325. https://doi.org/10.1016/j.heliyon.2024.e25325.
- Dutta TK, Phani V (2023) The pervasive impact of global climate change on plant-nematode interaction continuum. Front Plant Sci 14:20–23.
- Pires D, Orlando V, Collett RL et al. (2023) Linking nematode communities and soil health under climate change. Sustainability 15(15):11747. https://doi.org/10.3390/su151511747.
- Pörtner HO, Roberts DC, Adams H et al (2022). Climate change 2022: impacts, adaptation and vulnerability. Technical summary. Cambridge University Press, Cambridge, UK/New York, NY.
- Sikora RA, Helder J, Molendijk LPG et al (2023) Integrated nematode management in a world in transition: constraints, policy, processes, and technologies for the future. Ann Rev Phytopathol 61:209–230.
- Somasekhar N, Prasad JS (2012) Plant–nematode interactions: consequences of climate change. In: Venkateswarlu B, Shanker AK, Shanker C, Maheswari M (eds) Crop stress and its management: perspectives and strategies. Springer, Dordrecht, pp 547–564.
- Westerdahl BB (2021) Scenarios for sustainable management of plant parasitic nematodes. Ind Phytopathol 74:469–475.

18.17 Review Questions

- How does climate change (elevated CO_2, temperature rise, and altered rainfall patterns) influence the population dynamics and pathogenicity of PPNs?
- Why might traditional nematode management strategies (chemical nematicides, crop rotation, and resistant varieties) become less effective under changing climatic conditions?
- Discuss the role of climate-smart agriculture in nematode management. Which practices could be adapted to mitigate nematode problems under future climates?
- How can soil health management (e.g., organic amendments, biofumigation, and cover crops) contribute to sustainable nematode suppression under climate change scenarios?
- What potential does biological control (nematophagous fungi, bacteria, and predatory nematodes) hold in managing nematodes when climatic conditions shift in favor of pests?
- Evaluate the challenges and opportunities of developing nematode-resistant or climate-resilient crop varieties.
- How can predictive modeling and decision-support systems assist in planning nematode management strategies under uncertain climate futures?
- Compare the environmental and economic sustainability of chemical versus non-chemical nematode management approaches in the context of climate change adaptation.
- What INM strategies would you recommend for smallholder farmers versus commercial growers under climate stress conditions?

References

Acevedo M, Pixley K, Zinyengere N et al (2020) A scoping review of adoption of climate-resilient crops by small-scale producers in low- and middle-income countries. Nat Plants 6:1231–1241

Afzal A, Mukhtar T (2024) Revolutionizing nematode management to achieve global food security goals—an overview. Heliyon 10(3):e25325. https://doi.org/10.1016/j.heliyon.2024.e25325

Asmus GL, Inomoto MM, Cargnin RA (2008) Cover crops for reniform nematode suppression in cotton: greenhouse and field evaluations. Trop Plant Pathol 33:85–89

Ayaz M, Zhao W, Chi YK et al (2024) Biocontrol of plant parasitic nematodes by bacteria and fungi: a multi-omics approach for the exploration of novel nematicides in sustainable agriculture. Front Microbiol 15:1433716. https://doi.org/10.3389/fmicb.2024.1433716

Barker KR (1997) Opportunities for integrated management of plant-parasitic nematodes in the near east. In: Maqbool MA, Kerry B (eds) Plant nematode problems and their control in the near east region (FAO plant production and protection paper—144), Proceedings of the expert consultation on plant nematode problems and their control in the near east region Karachi, Pakistan, 22–26 November 1992. FAO, Rome

Becker JG, Ploeg A, Nuñez JJ (2019) Multi-year field evaluation of fluorinated nematicides against *Meloidogyne incognita* in carrots. Plant Dis 103(9):2392–2396

Bell AA, Robinson AF, Quintana J et al (2014) Registration of LONREN-1 and LONREN-2 germplasm lines of upland cotton resistant to reniform nematode. J Plant Reg 8:187–190

Chen P, Tsay T (2006) Effect of crop rotation on *Meloidogyne* spp. and *Pratylenchus* spp. populations in strawberry fields in Taiwan. J Nematol 38:339–344

Delcour I, Spanoghe P, Uyttendaele M (2014) Impact of climate change on pesticide use. Food Res Int 68. https://doi.org/10.1016/j.foodres.2014.09.030

Dutta TK, Phani V (2023) The pervasive impact of global climate change on plant-nematode interaction continuum. Front Plant Sci 14:20–23

Elhady A, Alghanmi L, Abd-Elgawad MMM et al (2024) Plant parasitic nematode research in the arid desert landscape: a systematic review of challenges and bridging interventions. Front Plant Sci 15:1432311. https://doi.org/10.3389/fpls.2024.1432311

El-Sappah AH, Islam MM, El-Awady HH et al (2019) Tomato natural resistance genes in controlling the root-knot nematode. Genes 10:925

Frankenstein D, Luu MS, Luna-Ayala J et al (2024) Soil moisture conditions alter behavior of entomopathogenic nematodes. J Sci Food Agric 104:4383–4390

Fuller VL, Lilley CJ, Urwin PE (2008) Nematode resistance. New Phytol 180:27–44

Gatto MP, Cabella R, Gherardi M (2016) Climate change: the potential impact on occupational exposure to pesticides. Ann Ist Super Sanità 52:374–385

Gill HK, Grabau ZJ, McSorley R (2023) Cover crops for managing root-knot nematodes. IFAS Extension, University of Florida, Gainesville, FL

Grant CA, Peterson GA, Campbell CA (2002) Nutrient considerations for diversified cropping systems in the northern great plains. Agron J 94:186–198

Han S, Schliemann W, Liu S (2024) Editorial: resistance of plants to parasitic nematodes and its application in breeding. Front Plant Sci 15:1439535. https://doi.org/10.3389/fpls.2024.1439535

Harini P (2021) Cultural method in nematode control. slideshare.net

Huang Q, Lin B, Cao Y (2023) CRISPR/Cas9-mediated mutagenesis of the susceptibility gene OsHPP04 in rice confers enhanced resistance to rice root-knot nematode. Front Plant Sci 14:1134653. https://doi.org/10.3389/fpls.2023.1134653

Jacquet F, Jeuffroy MH, Jouan J et al (2022) Pesticide-free agriculture as a new paradigm for research. Agron Sustain Dev 42:8. https://doi.org/10.1007/s13593-021-00742-8

Karabörklü S, Volkan V, Dura O (2022) The potential of *Beauveria bassiana* and *Metarhizium anisopliae* in controlling the root-knot nematode *Meloidogyne incognita* in tomato and cucumber. J Asia Pac Entomol 25(1):101846. https://doi.org/10.1016/j.aspen.2021.101846

Kawanobe M, Toyota K, Fujita T, Hatta D (2019) Evaluation of nematicidal activity of fluensulfone against non-target free-living nematodes under field conditions. Agronomy 9(12):853. https://doi.org/10.3390/agronomy9120853

Kearn J, Ludlow E, Dillon J et al (2014) Fluensulfone is a nematicide with a mode of action distinct from anticholinesterases and macrocyclic lactones. Pestic Biochem Physiol 109:44–57

Khanal C, Harshman D (2022) Evaluation of summer cover crops for host suitability of *Meloidogyne enterolobii*. Crop Prot 151:105821. https://doi.org/10.1016/j.cropro.2021.105821

Khanal C, Land J (2023) Study on two nematode species suggests climate change will inflict greater crop damage. Sci Rep 13:14185

Khanal C, McGawley EC, Overstreet C, Stetina SR (2018a) The elusive search for reniform nematode resistance in cotton. Phytopathology 108:532–541

Khanal C, Mcgawley EC, Overstreet C et al (2018b) Reproduction and pathogenicity of endemic populations of *Rotylenchulus reniformis* on cotton. Nematropica 48:68–81

Khanal C, Kularathna MT, Ray JD et al (2019) Single nucleotide polymorphism analysis using KASP assay reveals genetic variability in *Rotylenchulus reniformis*. Plant Dis 103:1835–1842

Khanal C, Galbieri R, Timper P (2021) Rotations with *Crotalaria* spp. do not suppress populations of *Meloidogyne incognita* in cotton. Nematology 23:929–937

Kopeć P (2024) Climate change-the rise of climate-resilient crops. Plants 13(4):490. https://doi.org/10.3390/plants13040490

Legner CM, Tylka GL, Pandey S (2021) Robotic agricultural instrument for automated extraction of nematode cysts and eggs from soil to improve integrated pest management. Sci Rep 11:3212. https://doi.org/10.1038/s41598-021-82261-w

López-Pérez JA, Roubtsova T, de Cara García M, Ploeg A (2010) The potential of five winter-grown crops to reduce root-knot nematode damage and increase yield of tomato. J Nematol 42(2):120–127

Mori A (2018) Application of farmyard manure rather than manure slurry mitigates the net greenhouse gas emissions from herbage production system in Nasu, Japan. Atmosphere 9:261. https://doi.org/10.3390/atmos9070261

Mosier A, Schimel D, Valentine D et al (1991) Methane and nitrous oxide fluxes in native, fertilized and cultivated grasslands. Nature 350:330–332

Pams Blog (2022) Nematode-resistant food crops and cover crops. *Farmerpam*, September 14, 2022

Phani V, Khan MR, Dutta TK (2021) Plant-parasitic nematodes as a potential threat to protected agriculture: current status and management options. Crop Prot 144:105573. https://doi.org/10.1016/j.cropro.2021.105573

Ploeg A, Stoddard S, Becker JO (2019) Control of *Meloidogyne incognita* in sweet potato with fluensulfone. J Nematol 51:1–8

Pörtner, H.O., Roberts, D.C., Adams, H. *et al.* 2022. Climate change 2022: impacts, adaptation and vulnerability. Technical Summary; Cambridge University Press: Cambridge, UK/New York, NY

Sánchez-Rodríguez AR, Nie C, Hill PW et al (2019) Extreme flood events at higher temperatures exacerbate the loss of soil functionality and trace gas emissions in grassland. Soil Biol Biochem 130:227–236

Somasekhar N, Prasad JS (2012) Plant–nematode interactions: consequences of climate change. In: Venkateswarlu B, Shanker AK, Shanker C, Maheswari M (eds) Crop stress and its management: perspectives and strategies. Springer, Dordrecht, pp 547–564

Venkatesh J, Kang BC (2019) Current views on temperature-modulated *R* gene-mediated plant defense responses and trade-offs between plant growth and immunity. Curr Opin Plant Biol 50:9–17

Westerdahl BB (2021) Scenarios for sustainable management of plant parasitic nematodes. Indian Phytopathol 74:469–475

Williamson VM, Kumar A (2006) Nematode resistance in plants: the battle underground. Trends Genet 22:396–403

Yadav S, Patil JA (2021) The emerging nematode problems in horticultural crops and their management. In: Belle C, Kaspary TE (eds) Nematodes—recent advances, management and new perspectives. Intechopen.com. https://doi.org/10.5772/intechopen.99292

Part VII

Artificial Intelligence

Artificial Intelligence for Nematode Counting and Diagnosis

19

Abstract

Accurate counting, diagnosis, and identification of plant-parasitic nematodes (PPNs) are critical for effective management of these cryptic pests. Traditional counting methods and diagnosis based on morphological and molecular assays, while reliable, are labor-intensive and time-consuming and require specialized expertise. Recent advances in artificial intelligence (AI)—particularly machine learning and deep learning (DL)—offer automated, high-throughput alternatives for nematode counting, detection, classification, and quantification. This chapter synthesizes current AI-driven approaches in PPN counting, diagnosis, and identification, covering convolutional neural networks, object detection architectures (You Only Look Once variants), image-based methods, hyperspectral and remote sensing integration, and decision-support tools. Key algorithms, performance metrics, implementation challenges, and future directions for AI in nematology have also been discussed.

Keywords

AI-driven approaches · Image-based methods · Hyperspectral methods · Remote sensing integration · Decision-support tools

19.1 Learning Objectives

- Explain the need for artificial intelligence (AI) applications in nematode detection, counting, and diagnosis.
- Discuss the limitations of conventional microscopic and molecular methods.
- Identify major AI approaches used (e.g., computer vision, machine learning (ML), deep learning (DL), and image recognition).
- Describe how AI models automate nematode counting in soil, root, and cyst samples.

R. P. Parvatha, *Plant Parasitic Nematodes*,
https://doi.org/10.1007/978-981-95-9306-4_19

- Compare the accuracy, speed, and cost-effectiveness of AI versus manual counting.
- Learn how AI differentiates between plant-parasitic nematode (PPN) genera/species using morphological and imaging data.
- Analyze the potential of AI in precision nematode management and sustainable agriculture.
- Identify gaps, limitations, and future research needs in AI-based nematode diagnostics.

19.2 Introduction

PPNs, microscopic worms that feed on plant roots and other parts, pose significant challenges to agriculture by causing crop damage and yield losses. The integration of AI, particularly DL, has emerged as a transformative approach for both counting and diagnosing these pests, enhancing efficiency and precision in nematological studies and pest management. Accurate quantification of PPNs, especially root-knot nematodes (RKNs) (*Meloidogyne* spp.) and cyst nematodes (*Heterodera* spp.), is critical for assessing infestation levels and guiding management decisions. Manual microscopic counting, the traditional standard, can require hours per sample and suffers from inter-operator variability. Automation promises to accelerate counting, improve consistency, and integrate seamlessly with decision-support systems for timely interventions (Pun et al. 2023).

19.3 Nematode Counting

Efficient counting and identification of the nematodes are crucial for crop disease management, agricultural productivity, and scientific research. Traditionally, nematode counting has been a labor-intensive and time-consuming process, often requiring specialized expertise. However, the advent of AI, particularly DL, has revolutionized this field by automating and enhancing the precision of nematode counting. This survey note explores the multifaceted role of AI in this domain, drawing from recent research, surveys, and practical applications.

19.3.1 Traditional Counting Methods

Conventional approaches involve flotation or centrifugation to extract nematodes, followed by manual identification under a microscope. Though accurate in expert hands, these methods limit throughput to a few samples daily and are unsuitable for large-scale surveys (Pun et al. 2023). Image analysis with simple thresholding and morphological operations has been attempted, but these techniques falter in complex backgrounds and when nematode morphologies overlap (Saikai et al. 2024).

19.3.2 Emergence of DL in Nematode Counting

- *Early CNN Architectures:* Akintayo et al. (2016) introduced a convolutional selective autoencoder (CSAE) to detect soybean-cyst nematode eggs in cluttered images, demonstrating improved accuracy over threshold-based methods. Kalwa et al. (2019) adapted the U-Net architecture for *Heterodera glycines* egg detection in microfluidic flow-chip images, achieving reliable segmentation despite staining requirements. These foundational studies established DL's potential for precise nematode egg identification.
- *You Only Look Once (YOLO)-Based Object Detection Models:* The YOLO family of one-stage detectors has become the de facto standard for rapid, accurate object detection in nematology:
  - *YOLOv5*: Pun et al. (2023) integrated YOLOv5-640 into NemDST, reporting a precision of 0.992, a recall of 0.959, an F1-score of 0.975, and a mean average precision (mAP) of 0.979 for RKN egg detection, with inference times under 4 ms per frame.
  - *YOLOv7 and v6*: Comparative analyses revealed YOLOv7-640 achieved $R^2 = 0.964$ (mean absolute percentage error (MAPE) = 3.58%), slightly outperforming YOLOv6-480 ($R^2 = 0.957$), in counting RKN eggs on complex datasets (Pun et al. 2023).
  - *YOLOv8x*: Saikai et al. (2024) developed Spatial Environment for Graphical NEmatode Modeling and Analysis (SEGNEMA), a graphical user interface (GUI) embedding YOLOv8x for simultaneous detection of *Meloidogyne* eggs and juveniles. The model achieved >93% classification accuracy and a correlation coefficient >0.70 on unseen images, underscoring its robustness across imaging platforms (Fig. 19.1).
- *Segmentation and Hybrid Approaches:* Beyond detection, segmentation networks such as Mask R-CNN and enhanced U-Nets have been applied to delineate nematode boundaries and count overlapping individuals. Chen et al. (2020) extended the U-Net for worm-shaped object counting, improving pipeline repro-

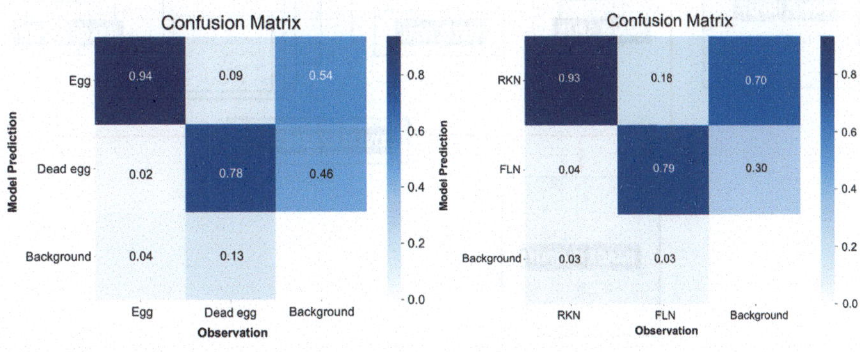

Fig. 19.1 Confusion matrix for the detection of nematode eggs (*Left*) and juveniles (*Right*) using the YOLOv8x deep-learning model

ducibility. However, segmentation often demands pixel-wise annotations, raising dataset creation costs.

- *Decision-Support Tools and User Interfaces:* Translational deployment of AI models requires accessible interfaces. NemDST presents a web-based portal allowing growers to upload images, receive real-time population estimates, and obtain management recommendations (Pun et al. 2023). Similarly, SEGNEMA offers cross-platform GUIs with model update capabilities. Open-source availability (e.g., GitHub repositories) promotes community-driven enhancements and genus expansion.
- *Performance Metrics and Benchmarking:* Standard evaluation metrics include precision, recall, F1-score, mAP, correlation coefficient (R^2), and MAPE. Across studies, YOLO variants consistently exceed 90% accuracy, with throughput enabling hundreds of images per minute. Benchmark datasets remain limited; publicly shared annotated image sets would standardize comparisons.

19.3.3 Benefits of AI for Nematode Counting

- *Automation and Efficiency:* One of the primary roles of AI in PPN counting is to automate a process that is inherently slow and manual. Research from Wageningen University, published by Saikai et al. (2024), developed a GUI named SEGNEMA, which leverages the YOLOv8x convolutional neural network (CNN) architecture for simultaneous detection and classification of nematode eggs and juveniles. This tool addresses the challenge of counting *Meloidogyne* spp., a common PPN, by processing images from advanced imaging systems such as the Leica Stellaris 5 Confocal LSM and LEITZ DM IRB (Fig. 19.2).

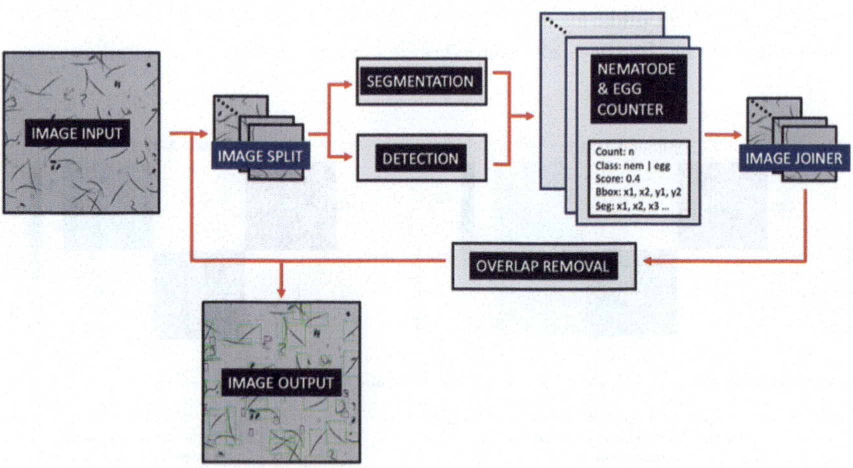

Fig. 19.2 The operational workflow of SEGNEMA

The study reported that SEGNEMA achieves processing times of approximately 4–5 s per image, a stark contrast to manual counting, which can take 2.5 h for 10 stitched images (Saikai et al. 2024). This efficiency is critical for large-scale agricultural and research applications, where rapid analysis can inform timely interventions. The tool is open-source and available at GitHub, requiring Pytorch for neural network processing, thus fostering collaboration and further development.

- *Accuracy and Precision:* AI's ability to achieve high accuracy in nematode counting is another significant contribution. The SEGNEMA study found that the models correctly identified eggs as eggs and *Meloidogyne* juveniles as *Meloidogyne* juveniles in 94% and 93% of instances, respectively, with correlation coefficients greater than 0.70 on unseen images (0.81 and 0.98 for juveniles and 0.72 and 0.96 for eggs on Stellaris and LEITZ, respectively) (Saikai et al. 2024). This high precision is vital for distinguishing nematodes from free-living organisms and debris, especially in complex soil samples.

A broader survey from arXiv, published by Yuan et al. (2024), validated seven state-of-the-art object detection models on multiple datasets, including the AgriNema dataset specifically designed for agricultural nematode detection. The survey established baselines for performance, with YOLOv6 achieving the highest mAP at 0.5 Intersection over Union (IOU) of 96.53% on AgriNema, demonstrating the potential of AI to handle diverse nematode types.

The performance metrics across different models and datasets are detailed in Table 19.1.

This table underscores the variability in model performance across datasets, with YOLOv5 and YOLOv6 standing out for their high accuracy on agricultural nematode datasets.

- *Accessibility and Democratization:* AI also plays a crucial role in making nematode counting accessible to non-experts, thereby democratizing the process. A practical example is Veridi Technologies, a company highlighted by Marston (2025). Veridi has developed an AI-powered automated microscopy system that extracts nematode solutions from soil samples, scans them, and provides results within three clicks. This system is reported to be 10 times cheaper than tradi-

Table 19.1 Performance metrics across different models and datasets

Models	AgriNema (%)	*C. elegans* (%)	BBBC010 (%)	Microorganism (%)
Faster RCNN	81.27	19.8	79.97	86.29
YOLOv5	95.6	93.1	85.6	90.4
YOLOX	53.75	91.26	36.54	75.8
YOLOv6	96.53	84.75	90.31	89.27
YOLOv7	53.9	35.7	79.7	22.2
YOLOv8	94.5	84.9	89	88.1
DETR	86.1	56.1	72.3	16.1

tional methods and 10–20% more accurate, supported by a €2.5 million ($2.7 million) grant from the EU's EIC Accelerator for R and D, market launch, and pre-market activities (Marston 2025).

This accessibility is particularly beneficial for small-scale farmers and laboratories, including government labs, as it integrates with existing agronomic models to recommend crop protection products, soil additives, and crop choices based on nematode presence (Marston 2025). The system is user-friendly in nature, reduces the dependency on specialized nematologists, and makes soil biodiversity analysis more widespread.

19.3.4 Research Advancement

AI also advances scientific research by providing precise data on nematode populations, which is essential for studying their behavior, impact, and management strategies. The SEGNEMA study plans to expand its capabilities to include other economically significant genera such as *Globodera* spp., *Heterodera* spp., and *Pratylenchus* spp., requiring global collaboration for diverse image datasets (Saikai et al. 2024). The arXiv survey identifies key challenges, such as the lack of large-scale data (e.g., soil extract samples), the need for lightweight models for field use, and optimization for tiny object detection (Pun et al. 2023). It also suggests potential in zero-shot and few-shot learning to address data scarcity, indicating a promising future for AI in this field.

19.3.5 Datasets and Supporting Infrastructure

The availability of datasets is critical for AI model training. The AgriNema dataset, constructed for agricultural nematode detection, includes 525 images of *Meloidogyne hapla*, *Globodera pallida*, *Pratylenchus*, *Ditylenchus*, and cyst nematodes (Pun et al. 2023). Other public datasets mentioned include I-Nema (9215 images, 40 classifications), *C. elegans* (1908 images for instance segmentation), BBBC010 (100 images, 2 classifications), and Microorganism (498 images for instance segmentation) (Pun et al. 2023). These datasets enable the validation and benchmarking of AI models, ensuring robustness across different scenarios.

19.3.6 Challenges and Limitations

- *Domain Shift:* Models trained on specific imaging setups often underperform on new hardware or sample preparations. Domain adaptation techniques—feature-based, adversarial methods, and discrepancy minimization—are crucial for generalizability (Pun et al. 2023).

- *Annotation Burden:* High-quality bounding boxes and segmentations require an expert's time. Semi-supervised learning and synthetic data augmentation (e.g., mosaic techniques) can mitigate annotation demands, as shown in mosaic-augmented YOLO studies (Zeng 2024).
- *Hardware Constraints:* While GPU acceleration is ideal, lightweight models optimized for CPU inference ensure broader adoption in resource-limited settings.

19.3.7 Future Directions and Conclusions

To realize universal AI-driven nematode counting, collaborative efforts should focus on the following:

- *Shared Datasets:* Establishing centralized repositories of multi-genus, multi-platform annotated images
- *Modular Frameworks:* Developing plug-and-play architectures enabling rapid integration of new genera via transfer learning
- *Explainability:* Incorporating attention maps and confidence heat maps to aid expert validation
- *Integration with Internet of Things:* Coupling image-based counts with soil sensors and remote monitoring for predictive nematode population modeling

In conclusion, AI, particularly DL, is playing a transformative role in PPN counting by automating the process, achieving high accuracy, enhancing accessibility, and providing practical benefits for agriculture. While challenges such as data scarcity and model optimization remain, ongoing research and practical applications, such as those by Veridi and SEGNEMA, demonstrate the field's potential. This integration of AI not only improves efficiency but also supports global efforts to mitigate crop losses and advance scientific understanding.

AI-powered automation, led by CNN-based object detection, has dramatically advanced the efficiency and accuracy of PPN counting. YOLO variants, in conjunction with user-friendly interfaces, deliver real-time population estimates crucial for precision agriculture. Overcoming domain adaptation and annotation challenges through community-driven datasets and hybrid learning paradigms will pave the way for universally accessible nematode-counting platforms, ultimately enhancing sustainable crop protection.

19.4 Nematode Diagnosis and Identification

Accurate diagnosis and identification are critical for effective management. Traditional diagnosis relies on morphological examination of extracted nematodes or eggs under a microscope—an approach that is labor-intensive and time-consuming, requires taxonomic expertise, and often takes several hours per sample. Molecular methods such as polymerase chain reaction (PCR) and quantitative PCR (qPCR) enhance specificity and sensitivity but involve costly reagents and laboratory infrastructure (Boagle et al. 2020). The emergence of AI-powered assays promises rapid, objective, and scalable solutions for in-field and laboratory diagnostics. AI has emerged as a transformative tool in PPN diagnosis and identification, offering rapid, accurate, and scalable solutions. The role of AI in this domain, highlighting key applications, techniques, challenges, and future prospects based on recent advancements, is discussed.

19.5 Limitations of Traditional Diagnostic Methods

19.5.1 Morphological Identification

- *Pros:* Direct visualization of diagnostic features (stylet length and perineal patterns).
- *Cons:* Requires experienced taxonomists; differentiation of closely related species is challenging; throughput is limited to tens of samples per day (Boagle et al. 2020).

19.5.2 Molecular Diagnostics

- *PCR/qPCR*: High sensitivity; can quantify nematode densities (Fig. 19.3).
- *Next-generation sequencing and metabarcoding*: Offers community profiling but incurs high sequencing costs and bioinformatics demands (Pun et al. 2024).
- *Isothermal amplification Loop-mediated isothermal amplification (LAMP) and recombinase polymerase amplification (RPA)*: Field-deployable, rapid (<30 min), versatile when coupled with lateral flow detection; however, primer design complexity and occasional non-specific amplification remain hurdles (Fig. 19.3).

19.6 AI Techniques and Methodologies

19.6.1 Machine Learning

- Supervised ML models (e.g., support vector machine and decision trees) are trained on labeled datasets of nematode images or genetic data for classification tasks.
- Unsupervised ML (e.g., clustering) identifies patterns in unlabeled data, aiding in the discovery of new PPN species or strains.

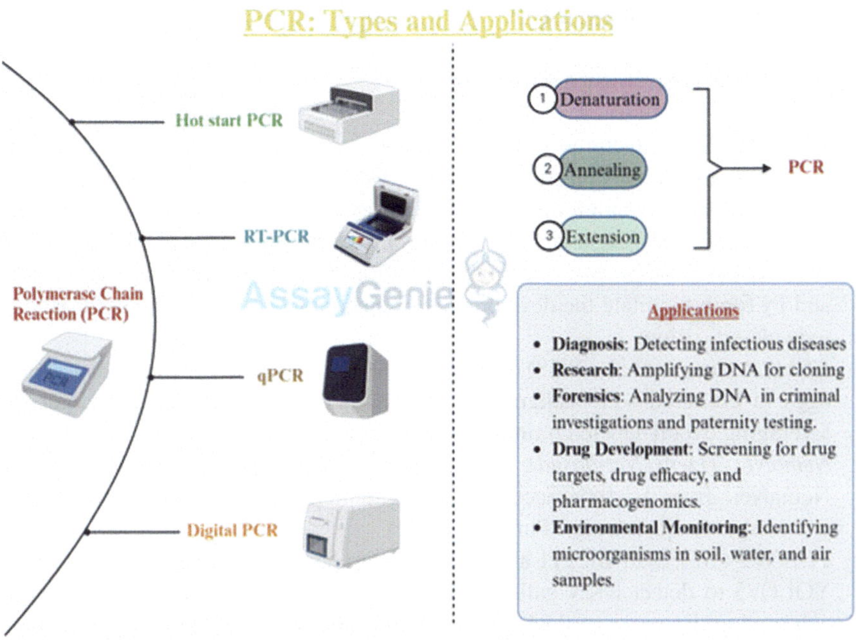

Fig. 19.3 Demonstration of the principles and working processes involved in traditional PCR methods and isothermal amplification techniques

19.6.2 Deep Learning

- CNNs, such as VGG16, ResNet, and YOLO, dominate image-based PPN identification due to their ability to extract complex features automatically.
- Transfer learning leverages pre-trained DL models to overcome limited nematode-specific datasets, improving model performance.

19.6.3 Natural Language Processing

- Natural language processing (NLP) tools extract relevant information from taxonomic literature and databases, aiding in species identification and updating PPN classifications.
- Example: NLP pipelines have been used to curate data from sources like NEMBASE for AI model training.

19.6.4 Data Fusion

- AI integrates multimodal data (e.g., images, spectral data, and genomics) to enhance diagnostic accuracy. Ensemble models combine outputs from multiple algorithms for robust predictions.

19.7 AI-Driven Methods

19.7.1 AI-Driven Image Analysis

- *CNNs:* CNN architectures have revolutionized object detection and classification tasks in computer vision. In nematology, CNNs automate the identification of eggs, juveniles, and adult forms from microscopic images. Studies have reported CNN models achieving over 90% accuracy in identifying genera like *Meloidogyne* (RKNs) and *Pratylenchus* from high-resolution images. Tools such as TensorFlow and PyTorch facilitate the development of custom CNN architectures for nematode classification.
- *CSAE:* Akintayo et al. (2018) developed CSAE to detect soybean cyst nematode eggs in cluttered soil backgrounds, achieving >95% precision and recall by leveraging expert-labeled training data to suppress noise.
- *NemaNet (DenseNet-based):* A customized DenseNet-121 model, termed NemaNet, gives c > 98% accuracy on augmented datasets (rotation, mirroring, and contrast adjustments) (Pun et al. 2024).
- *YOLOv5-CMS:* San-Blas et al. (2020) integrated a dual attention module into YOLOv5 to detect RKN galls on cucumber roots. The YOLOv5-CMS model achieved an F1-score of 0.91 and mAP of 94.8%, demonstrating real-time performance suitable for high-throughput screening.

Model performance and benchmarking are presented in Table 19.2.

19.7.2 Hyperspectral and Remote Sensing Integration

AI combined with hyperspectral imaging enables non-invasive estimation of nematode infestation by detecting stress-induced spectral signatures in host plants (Fig. 19.4).

Sensors integrated into robotic platforms can analyze soil texture, pH levels, and microbial activity without manual intervention (Fig. 19.5).

- *Early RKN Detection in Cotton:* Hyperspectral data (350–1000 nm) classified RKN-infested versus healthy cotton with 95% accuracy using principal component analysis and linear discriminant analysis (Pun et al. 2024).

Table 19.2 Model performance and benchmarking

Model	Task	Precision	Recall	F1-Score	mAP	References
CSAE	SCN egg detection	0.96	0.94	0.95	–	Boagle et al. (2020)
NemaNet	Multi-genera classification	–	–	–	–	Pun et al. (2024)
YOLOv5-CMS	RKN gall detection	0.94	0.89	0.91	0.948	San-Blas et al. (2020)

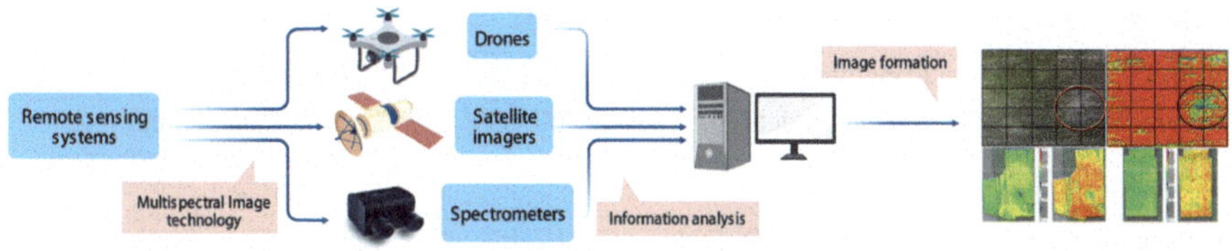

Fig. 19.4 Demonstration of the principles and working processes involved in remote sensing techniques

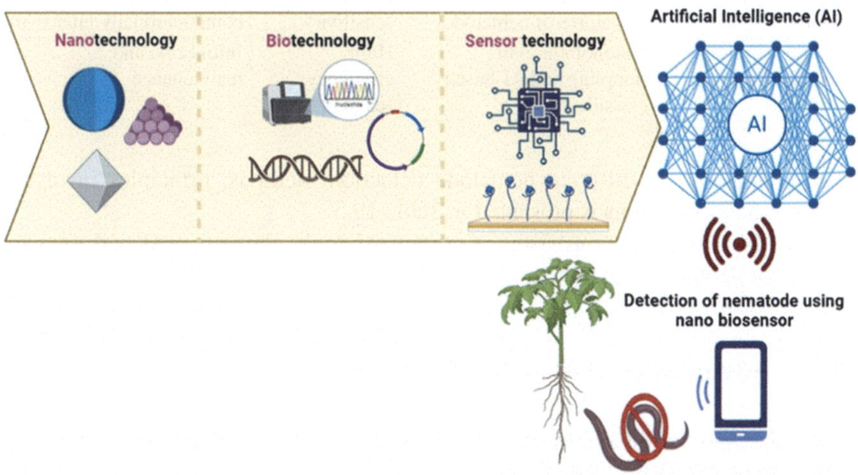

Fig. 19.5 Nano-biosensors: A powerful technology for early detection of plant parasitic nematodes

- *Abiotic Versus Biotic Stress Differentiation:* EfficientNet models distinguished RKN infestation from water stress in tomatoes, achieving >92% classification accuracy when coupled with brightness augmentation and transfer learning (Pun et al. 2024).
- *Field-Scale Monitoring:* CRISPR/LAMP assays paired with drone-mounted spectral sensors (Fig. 19.5) offer proof-of-concept for landscape-level surveillance, though integration of AI for real-time mapping remains in early development (Shao et al. 2023).
- *AI-Based Decision-Support Tools:* The translation of AI models into user-friendly applications is pivotal for end-users. A prototype web application, NemDST, uses YOLOv5-640 to detect RKN eggs and juveniles from uploaded images with mAP = 0.979 and inference time < 4 ms per image. It provides population estimates, tracks dynamics over time, and recommends management actions based on threshold guidelines (San-Blas et al. 2020).

Table 19.3 Comparison of nematode detection methods

Method	Principle	Accuracy	Limitations
Manual microscopy	Morphological identification using a microscope	60–80%	Time-consuming, labor-intensive, requires expertise
Molecular techniques (PCR and qPCR)	Detects specific DNA markers of nematodes	High	Expensive, requires specialized equipment
AI-based image recognition	Uses ML to classify nematodes	Up to 95% accuracy	Requires high-quality training data
Hyperspectral imaging	Analyses the spectral signatures of nematodes	High sensitivity	High costs, computationally intensive
Robotic sampling and analysis	Autonomous soil sampling and AI-based classification	High efficiency, scalable	Initial cost and maintenance

A comparison of different nematode detection methods, principles involved, accuracy, and limitations is presented in Table 19.3.

Table 19.4 gives a comparison of current AI tools, datasets, and performance metrics for nematode counting and diagnosis.

19.7.3 Advantages of AI in PPN Diagnosis

- *Speed and Scalability*: AI processes thousands of samples in seconds, compared to days or weeks for manual methods.
- *Accuracy*: DL models often outperform human experts in identifying subtle morphological or genetic differences.
- *Accessibility*: AI-powered mobile apps and cloud-based platforms democratize PPN diagnosis for farmers and non-experts.
- *Cost-Effectiveness*: Automation reduces labor and equipment costs, making diagnostics affordable for resource-limited regions.
- *Early Detection*: AI predicts infestations before visible symptoms, enabling timely interventions.

19.7.4 Challenges and Limitations

- *Data Scarcity and Quality*
  - Publicly available, annotated image datasets of PPNs are limited.
  - Variation in soil background, lighting, and sample preparation impacts model generalization (Pun et al. 2024).
- *Computational Resources*
  - Deep models require GPUs for training; edge deployment on smartphones or low-power devices remains challenging.
  - Real-time inference in the field may be constrained by connectivity and power availability.

Table 19.4 Comparison of current AI tools, datasets, and performance metrics for nematode counting and diagnosis

Task	Target/scope	AI tool/model	Dataset (size; availability)	Accuracy	References
Counting (eggs)	RKN eggs (*Meloidogyne* spp.) in microscope images	SEGNEMA GUI + YOLOv8x (detection)	600 egg images; 8655 eggs +5379 dead eggs annotated; lab images; not yet public repo for data	0.86	Saikai et al. (2024)
Counting (juveniles J2)	RKN versus free-living nematodes (FLN) segmentation	SEGNEMA GUI + YOLOv8x (segm)	1289 juvenile images; 4742 RKN + 1153 FLN segmented; lab images	0.87	Saikai et al. (2024)
Counting (juveniles) and detection	RKN J2 and FLN in microscope images; overlapping objects	YOLOv4/5/7 with mosaic augmentation; best YOLOv5–608/ YOLOv7–608	Custom datasets (dataset-II/III) incl. Overlap simulation; not public	0.994 (RKN) 0.991 (FLN)	Pun et al. (2023)
Detection + counting (eggs & J2)/ decision support	General PPN (RKN eggs and juveniles)	NemDST (web app) + YOLOv5-640	Pretrained weights; dataset details limited in paper	0.992	Pun et al. (2023)
Species identification (diagnosis)	Pale cyst versus yellow cyst nematodes (*Globodera pallida* versus *G. rostochiensis*) from cyst images	CNN (Thevenoux et al. 2021)	Image set of cyst morphometrics; research dataset (not openly linked)	0.880	Thevenoux et al. (2021)
Species/ genus classification (diagnosis)	11 genera of PPNs (Indonesia soil)	EfficientNetV2B, ResNet101v2, EfficientNetV2M, CoAtNet-0 (best reported)	957 samples; Indonesian soil images	0.900	Shabrina et al. (2023)
Species identification (diagnosis)	5 soybean-relevant PPN species	NemaNet (custom CNN)	NemaDataset: 3063 microscope images; 5 species; public (paper claims)	0.988	Abade et al. (2021)

- *Standardization and Validation*
  - Lack of universally accepted performance benchmarks hinders cross-study comparisons.
  - Field validation across different cropping systems and geographies is sparse.

19.7.5 Future Prospects

- *Transfer Learning and Few-Shot Learning:* Adapting pre-trained vision models (e.g., EfficientNet and ResNet) for nematode tasks with minimal new annotations can accelerate deployment in resource-limited settings.
- *Edge AI and Mobile Apps:* Integration of optimized, lightweight models (e.g., MobileNet) into smartphone apps can empower farmers with on-device diagnostics.
- *Multimodal AI:* Fusion of image, spectral, and molecular data within a unified AI framework may enhance specificity and early detection capabilities.
- *Crowdsourcing and Open Datasets:* Community-driven initiatives to share labeled nematode images and metadata will bolster model robustness and democratize access.

In conclusion, AI technologies—particularly DL-based image analysis and hyperspectral integration—are revolutionizing PPN diagnosis and identification. While current models demonstrate high accuracy in controlled settings, translation to field-ready tools requires addressing data scarcity, hardware constraints, and validation across diverse agro-ecosystems. Future research should prioritize open-access datasets, edge AI deployment, and user-centric decision-support platforms to realize the full potential of AI for nematode management and global food security.

19.8 Case Studies

19.8.1 Root-Knot Nematodes

PPNs, especially sedentary endoparasitic species like RKNs, pose a threat to major crops and vegetables, leading to substantial yield losses and economic consequences (Pun et al. 2023). AI enables early detection, allowing growers to take proactive measures to protect plants. For instance, Veridi's system scans for nematodes posing danger to crops, facilitating timely interventions. Additionally, a study by Pun et al. (2023) developed a DL-based decision-support tool using the YOLOv5 model, achieving a precision of 0.992, a recall of 0.959, and an mAP of 0.979 for detecting RKN eggs, with an inference time of 3.9 ms. This speed and accuracy are crucial for real-time field applications.

19.9 Challenges and Future Directions

Despite these advancements, challenges remain, such as handling overlapped or irregularly shaped nematodes in counting and ensuring model robustness across diverse imaging systems. The studies emphasize the need for global collaboration, particularly in nematology, to enhance model accuracy and expand to other economically significant genera. Funding, such as internal grants from Wageningen Research, supports these efforts, with supplementary materials available for further exploration.

Future research aims to leverage transfer learning for broader species coverage and improve diagnostic precision, potentially revolutionizing PPN management. The open-source nature of tools like SEGNEMA and the availability of algorithms on platforms such as GitHub and Zenodo facilitate this collaborative approach, ensuring accessibility for the global research community.

19.10 Conclusion

AI, particularly DL, is actively transforming the counting and diagnosis of PPNs, offering high accuracy and efficiency compared to traditional methods. Tools such as SEGNEMA, EfficientNetV2M, and custom computer vision algorithm (CCVA)/CNN exemplify these advancements, with ongoing research poised to expand their applicability. This integration not only enhances agricultural nematode management but also supports sustainable farming practices by reducing manual labor and improving decision-making.

19.11 Suggested Readings

- Abade AS, Porto LF, Ferreira PA, Vidal FDB (2021) NemaNet: a convolutional neural network model for identification of nematodes soybean crop in Brazil. Res Gate. https://doi.org/10.48550/arXiv.2103.03717.
- Loveth RG (2025) AI-powered robotic systems for high-throughput nematode detection and soil health monitoring. Res Gate, March 2025. https://www.researchgate.net/publication/389988206.
- Masasi J, Ng'ombe JN, Masasi B (2024) Artificial intelligence in agriculture: current trends and innovations. Big Data Agric 6(2):96–99.
- Saikai KK, Bresilla T, Kool J et al (2024) Counting nematodes made easy: leveraging AI-powered automation for enhanced efficiency and precision. Front Plant Sci 15: https://doi.org/10.3389/fpls.2024.1349209.

19.12 Review Questions

- What are the major challenges in traditional nematode counting and diagnosis methods, and how can AI help overcome them?
- Explain the role of image analysis in AI-based nematode identification.

- What are the commonly used AI-driven methods for nematode detection and counting?
- Compare the performance of AI-based diagnostic tools with conventional microscopic methods in terms of speed, accuracy, and cost.
- Discuss the role of AI in *automated cyst nematode counting* from soil or root samples.
- How can AI help to improve trust and adoption of AI-driven nematode diagnostics among researchers and practitioners?

References

Abade AS, Porto LF, Ferreira PA, Vidal FDB (2021) NemaNet: a convolutional neural network model for identification of nematodes soybean crop in Brazil. Res Gate. https://doi.org/10.48550/arXiv.2103.03717.2

Akintayo A, Lee A, Chawla V et al (2016) An end-to-end convolutional selective autoencoder approach to soybean cyst nematode eggs detection. https://arxiv.org/abs/1603.07834 [cs.CV]

Akintayo A, Tylka GL, Singh AK et al (2018) A deep learning framework to discern and count microscopic nematode eggs. Sci Rep 8(1): 9145. https://doi.org/10.1038/s41598-018-27272-w

Boagle M, Baniya A, DiGennaro P (2020) Nematode identification techniques and recent advances. Plants 9(10). https://doi.org/10.3390/plants9101260

Chen L, Strauch M, Daub M et al (2020) A CNN framework based on line annotations for detecting nematodes in microscopic images. In: 2020 IEEE 17th international symposium on biomedical imaging (ISBI), USA. IEEE, Iowa City, pp 508–512. https://doi.org/10.1109/ISBI45749.2020.9098465

Kalwa U, Legner C, Wlezien E et al (2019) New methods of removing debris and high-throughput counting of cyst nematode eggs extracted from field soil. PLoS One 14(10):e0223386

Marston J (2025) Machine learning to tackle Europe's $55bn soil biodiversity problem. AgFunderNews, April 9, 2025

Pun TB, Neupane A, Koech R (2023) A deep learning-based decision support tool for plant-parasitic nematode management. J Imaging 9(11). https://doi.org/10.3390/jimaging9110240

Pun TB, Thapa Magar R, Koech R et al (2024) Emerging trends and technologies used for the identification, detection, and characterisation of plant-parasitic nematode infestation in crops. Plants 13(21). https://doi.org/10.3390/plants13213041

Saikai KK, Bresilla T, Kool J et al (2024) Counting nematodes made easy: leveraging AI-powered automation for enhanced efficiency and precision. Front Plant Sci 15. https://doi.org/10.3389/fpls.2024.1349209

San-Blas F et al (2020) Infrared spectra analysis and AI-based assay for early detection of root-knot nematodes. Sensors 20:1234

Shabrina N, Lika R, Indarti S (2023) Deep learning models for automatic identification of plant-parasitic nematode. Artif Intell Agric 7. https://doi.org/10.1016/j.aiia.2022.12.002

Shao H, Zhang P, Peng D et al (2023) Current advances in the identification of plant nematode diseases. Plant Sci 14:1106784. https://doi.org/10.3389/fpls.2023.1106784

Thevenoux R, Van Linh LE, Villessèche H et al (2021) Image based species identification of *Globodera* quarantine nematodes using computer vision and deep learning. Comput Electron Agric 186. https://doi.org/10.1016/j.compag.2021.106058

Yuan Z, Musa N, Dybal K et al (2024) Quantifying nematodes through images: datasets, models, and baselines of deep learning. arXiv:2404.19748 [cs.CV]. https://doi.org/10.48550/arXiv.2404.19748

Zeng W (2024) Image data augmentation techniques based on deep learning: a survey. Math Biosci Eng 21(6):6190–6224

Artificial Intelligence for Nematode Management

20

Abstract

Plant-parasitic nematodes inflict significant crop losses worldwide, estimated at over US $173 billion annually. Traditional diagnostic and management approaches are laborious and often lack precision. Artificial intelligence (AI) and machine learning (ML) can improve sustainable plant nematode diagnosis and management by automating identification and reducing chemical use. AI helps to predict outbreaks early, supporting targeted interventions and minimizing environmental impact. AI enhances precision agriculture, integration with emerging technologies, information technology, and smart farming, making farming practices more efficient and eco-friendly. There is ongoing research, and some methods are still being refined, so results may vary by context and crop type. This chapter synthesizes key developments in AI applications—ranging from deep learning-based image analysis to hyperspectral sensing and ML-driven decision support—highlighting achievements, remaining challenges, and future research directions.

Keywords

Artificial intelligence (AI) · Machine learning (ML) · Image analysis · Hyperspectral sensing · ML-driven decision support

20.1 Learning Objectives

- Understand the concept and scope of artificial intelligence (AI) in agriculture, with specific emphasis on plant health and nematode management.
- Identify AI-driven diagnostic methods (such as image recognition, sensor-based systems, and data analytics) for accurate and rapid identification of nematode species and symptoms.

R. P. Parvatha, *Plant Parasitic Nematodes*,
https://doi.org/10.1007/978-981-95-9306-4_20

- Examine the integration of AI with complementary technologies (remote sensing, drones, molecular diagnostics, and decision support systems (DSSs)) in nematode monitoring and management.
- Evaluate case studies and real-world applications of AI in nematode management, including early warning systems, risk mapping, and site-specific management practices.
- Assess the advantages and limitations of AI-based approaches compared to traditional nematode management methods.
- Discuss the potential role of AI in climate-smart nematode management, addressing challenges of climate change, sustainability, and resource-use efficiency.
- Explore future research needs and opportunities in applying AI for developing integrated nematode management (INM) strategies.

20.2 Introduction

Plant nematodes, such as *Meloidogyne* spp. (root-knot nematodes) and *Heterodera* spp. (cyst nematodes), can lead to billions in annual crop losses. Traditional management often relies on chemical nematicides, which can harm ecosystems and beneficial organisms. Sustainable management seeks to minimize environmental impact through precision, reduced chemical use, and integrated pest management (IPM) strategies. AI, with its capabilities in machine learning (ML), deep learning (DL), and computer vision (CV), offers a promising solution by enhancing efficiency and accuracy in nematode control.

AI and ML are revolutionizing sustainable plant nematode management by addressing the challenges posed by plant-parasitic nematodes (PPNs), which are microscopic worms causing significant crop losses globally. By leveraging AI and ML, farmers can manage these pests more sustainably, with less harm to the environment and fewer chemicals. This chapter provides a comprehensive analysis of AI's role, drawing from recent research and applications, focusing on identification, predicting their spread, precision agriculture, sustainability, supporting eco-friendly farming practices, to enhance efficiency, accuracy, and proactive nematode management. AI techniques—especially DL and ML—have been increasingly adopted to automate detection, quantify populations, and provide data-driven management recommendations, thereby enhancing precision agriculture and IPM.

AI, when integrated with nano-biosensors, can detect nematode-specific biochemical markers like volatile organic compounds in real time. This early detection is crucial for timely interventions to prevent crop damage. AI algorithms analyze data from weather, soil, and crop health to predict nematode outbreaks weeks in advance with high precision (81–95%). This enables farmers to implement preemptive management strategies, reducing potential losses. AI supports precision agriculture by providing data-driven insights for targeted control measures. It can recommend optimal pesticide use, automated irrigation, or smart pest control systems, promoting sustainable farming practices.

20.3 Nematode Diagnosis: Prerequisite for Management

20.3.1 Automated Identification and Counting

AI uses advanced image recognition, like DL models, to quickly and accurately identify and count nematodes from microscopic images. For example, tools like spatial environment for graphical NEmatode modeling and analysis (SEGNEMA), a user-friendly interface, can detect nematode eggs and juveniles with up to 94% accuracy, cutting down the time from hours to seconds compared to manual methods. This speed and precision help farmers make timely decisions, reducing the need for broad chemical applications and supporting sustainability.

AI facilitates rapid diagnosis by identifying specific nematode species from soil samples, a critical step for effective management. Researchers at the University of Florida, for example, are developing AI algorithms to classify nematodes, aiming to reduce diagnosis turnaround time and costs. This is particularly important given the annual economic impact of parasitic nematodes, estimated at $173 billion globally. The UF/IFAS Nematode Assay Lab processes around 7000 samples yearly, generating hundreds of thousands of images, which AI can analyze to distinguish harmful species, potentially saving growers from costly management methods or crop losses.

One of the primary roles of AI is automating the identification and counting of nematodes, a traditionally labor-intensive and time-consuming task. Saikai et al. (2024) highlighted the development of a DL algorithm using convolutional neural networks (CNNs), specifically YOLOv8x, for detecting and classifying nematode eggs and juveniles of *Meloidogyne* spp. This model achieved 94% accuracy for egg detection and 93% for juvenile detection, with a processing time of 4–5 s per image, compared to 2.5 h for manual counting of 10 stitched images. The practical application, SEGNEMA, is a user-friendly graphical user interface freely available at GitHub, enabling simultaneous detection and classification. Plans to expand to other genera such as *Globodera* spp., *Heterodera* spp., and *Pratylenchus* spp. suggest scalability. This automation reduces labor costs, improves consistency, and supports sustainable management by enabling accurate population density assessments for pot, micro-plot, and field trials, crucial for effective nematode management strategies.

Another study, published by Shabrina et al. (2023), evaluated four DL models (ResNet101v2, CoAtNet-0, EfficientNetV2B0, and EfficientNetV2M) for nematode classification, with EfficientNetV2M achieving a test accuracy of 97.94% and a mean class accuracy of 98.66%. Data augmentation techniques such as image flip, blurring, and brightness adjustment were used, enhancing model robustness. This high accuracy supports precise identification, reducing the risk of misdiagnosis and unnecessary chemical use.

AI also integrates with DNA analysis for enhanced identification accuracy, as seen in efforts to identify guava root-knot nematode using molecular techniques alongside AI. Real-time monitoring is another key application, with Internet-of-Things (IoT) sensors collecting data on soil conditions, temperature, moisture, and humidity. These data are analyzed by AI algorithms to track nematode populations, providing farmers with alerts via mobile apps or farm management software, ensuring timely interventions.

20.3.2 Early Detection and Prediction

The early warning system allows farmers to take preventive actions, like adjusting crop rotation or using biological controls, before damage occurs. By minimizing reactive chemical use, AI helps protect the environment and promotes sustainable farming.

AI's ability to predict nematode outbreaks is another critical contribution to sustainability. By integrating data from soil samples, weather conditions, and crop health, AI models can forecast spatial-temporal outbreak risks weeks in advance with 81–95% precision (González-Rodríguez et al. 2024). For instance, studies like Garrett et al. (2022) used random forest and support vector machines (SVMs) for climate-pathogen interplay, achieving over 90% accuracy for infection risk predictions. This predictive capability allows farmers to implement preventive measures, such as crop rotation or biological controls, reducing the need for reactive chemical treatments. Early detection, as seen in Zen et al. (2022), achieving 100% accuracy for early blight and late blight in tomatoes using CNN and recurrent neural network (RNN) with TensorFlow.js, minimizes environmental impact by limiting pesticide use, aligning with sustainable agriculture goals. These AI tools can also be used for nematode detection with slight modifications.

20.3.3 High-Throughput Detection With Robotic Systems

Recent advancements include AI-powered robotic systems for high-throughput nematode detection and soil health monitoring (Loveth 2025). These systems integrate AI-driven CV, DL, and spectral imaging for precise identification of nematode species, differentiating between beneficial and harmful varieties. CNNs achieve 94.7% accuracy in nematode classification, outperforming human experts (92.0% accuracy) and other models (SVM: 88.2% and random forest: 85.6%), as shown in Table 20.1. Hyperspectral imaging analyzes spectral signatures, achieving 92% accuracy in distinguishing nematode species, as noted by Sun et al. (2022).

Table 20.1 AI model performance metrics for nematode classification

Metric	CNN model (%)	SVM model (%)	Random forest (%)	Human expert (%)
Accuracy	94.7	88.2	85.6	92.0
Precision	95.1	87.5	84.3	93.5
Recall	94.2	86.8	83.7	92.8
F1-score	94.6	87.1	84.0	93.1

Robotic systems, equipped with GPS, LiDAR, vision sensors, robotic arms, and AI processing units like NVIDIA Jetson AGX Xavier, automate soil sampling, imaging, and analysis. They cover 10 ha in 4 h, a significant improvement over manual methods (2–3 days), and process ~200 samples/h compared to 50–70 samples/day manually. This efficiency reduces nematode detection time by 85% and operational costs by ~70% over one growing season, with AI systems costing ~$1500/10 ha versus traditional methods at ~$5000/10 ha. Hyperspectral imaging cameras and automated microscopes enhance non-invasive soil quality assessment, correlating nematode infestations with soil pH (<5.5 for higher PPNs and 6.5–7.5 for beneficial nematodes), moisture (>30% favors proliferation), and nutrient levels (low nitrogen increases infestations) (Loveth 2025).

20.4 Nematode Management

AI appears to support sustainable management by optimizing data-driven strategies for nematode control. It informs decision-making for targeted interventions, such as automated irrigation systems or smart pest control, reducing reliance on broad-spectrum pesticides. This aligns with precision agriculture, where AI integrates with smart farming systems to monitor real-time data and recommend tailored crop protection strategies, boosting resilience and minimizing environmental impact.

A study on DL-based decision support tools highlights the use of YOLOv5 models for detecting root-knot nematode eggs with 99.2% precision and 95.9% recall, processing images in milliseconds, faster than traditional methods. This efficiency supports IPM, reducing economic losses and promoting biodiversity conservation. Additionally, AI's role in minimizing chemical usage contributes to lower carbon footprints, aligning with eco-friendly agricultural practices.

20.4.1 AI-Driven Management Recommendations

AI-based platforms can analyze the collected data and recommend tailored management methods. For example, depending on the severity of the infestation and soil health, AI systems might recommend:

- *Biological Control Methods:* Introducing biopesticides (such as beneficial fungi or bacteria) that specifically target nematodes.
- *Crop Rotation Strategies:* AI can suggest crop rotation schedules based on local conditions and historical data, minimizing nematode populations in the soil.

20.4.2 Automated Soil Treatment

In advanced setups, autonomous robots or drones can be equipped with AI algorithms to apply nematicides (biological or chemical) precisely to the areas of the field where nematodes have been detected. This precision farming approach minimizes the use of chemicals and reduces costs for farmers while ensuring maximum effectiveness of the treatment.

20.4.3 Smart Irrigation Management

Nematodes thrive in specific soil moisture conditions. By integrating AI with smart irrigation systems, it is possible to create an unfavorable environment for nematode development. For instance, AI can adjust watering schedules to avoid over-saturation or drought, conditions that may exacerbate nematode infestation. These smart systems not only help control nematodes but also conserve water.

20.4.4 Predictive Modeling and Management

AI uses historical data—such as nematode population trends, weather patterns, soil conditions, and crop types—to predict future infestations. These predictive models enable farmers and researchers to implement preventive measures, like crop rotation or targeted treatments, before outbreaks occur. Furthermore, by integrating AI with Geographic Information Systems (GISs), it is possible to map nematode distribution across large areas, pinpointing hotspots and understanding their spread for more effective regional management.

Predictive modeling is a growing area where AI shows significant promise. ML models such as random forests and SVMs analyze historical data on nematode outbreaks, crop health, and environmental factors to forecast future infestations. According to a detailed report from indiaai.gov.in, these models can predict outbreaks by identifying patterns in soil, crop, and environmental data, enabling preemptive actions. For instance, ML can estimate the economic impact of infestations by correlating root damage severity with yield loss, aiding cost–benefit analyses for control measures.

Specific applications include the collaboration between Agmatix and BASF, announced on March 12, 2025, to develop an AI tool for detecting and predicting soybean cyst nematode (SCN) infestations (Sagester and Myers 2025). This initiative uses Agmatix's Axiom AI engine to process vast agronomic datasets, offering real-time, scalable insights to mitigate yield losses caused by SCN, one of the most damaging soybean pests.

Management strategies are also enhanced by AI, which recommends tailored control methods based on data analysis. For example, AI can suggest biological controls like biopesticides (beneficial fungi or bacteria) or crop rotation, informed by local and historical data. Precision agriculture is further supported by autonomous robots and drones equipped with AI, which apply nematicides precisely to infested areas, minimizing chemical use. Smart irrigation systems, integrated with AI, adjust watering to create unfavorable conditions for nematodes, conserving water while protecting crops.

20.4.5 Decision Support and Management Applications

- *Variable-Rate Nematicide Application:* By overlaying risk maps onto farm fields via GIS, AI advises where—and at what dosage—to apply nematicides, reducing chemical use by up to 40% while maintaining control.
- *Precision Irrigation and Cover Crop Recommendations:* Models optimize soil moisture regimes or identify cover-crop species that suppress nematode reproduction, guided by AI-forecasted risk.
- *Threshold-based Recommendations:* AI-powered models analyze high-resolution images captured via minirhizotron systems—transparent tubes inserted into the soil that enable non-destructive, real-time imaging of root systems. These models quantify root traits such as length, density, branching, discoloration, necrosis, or deformation indicative of nematode infestations. The growers are notified when root-health images exceed pest thresholds, triggering targeted soil sampling.
- *IPM Optimization:* Reinforcement-learning agents simulate multi-year rotations, biocontrol releases, and chemical interventions to recommend robust, cost-effective IPM schedules.

20.4.6 INM Strategies

- *Integration With Precision Agriculture:* AI optimizes nematode control by determining the exact timing, location, and dosage of treatments, such as nematicides or biological control agents. This precision reduces costs and minimizes environmental impact while improving effectiveness. AI-powered robots can also automate tasks such as soil sampling, nematode extraction, and targeted application of control measures, decreasing reliance on manual labor and enhancing accuracy.

AI integrates with precision agriculture to map nematode hotspots in fields, enabling targeted treatments instead of widespread chemical use. This approach optimizes resource use, such as water and fertilizers, and reduces environmental impact. For instance, AI can recommend specific areas for intervention, helping farmers save costs and protect ecosystems.

AI's integration with precision agriculture enhances sustainable nematode management by optimizing resource use and reducing environmental footprints. Piddubna (2024) discusses how AI explores soil health, monitors weather, and recommends targeted applications of fertilizers and pesticides. For nematode management, AI can map infestation hotspots, as seen in Virginia Tech's work at the Tidewater Agricultural Research and Extension Center, where AI algorithms analyze over 40,000 microscopic images to identify soybean nematodes (Esterhuizen 2022). This precision allows for targeted interventions, reducing non-selective pesticide use, which can harm beneficial nematodes vital for carbon cycling. The development of executable tools for growers, including GIS infestation heat maps, further supports data-driven decision-making, enhancing farm resilience and sustainability.

AI's integration into precision agriculture helps in boosting precision crop protection toward agriculture 5.0 via ML and emerging technologies. A contextual review involves optimizing data-driven, tailored crop protection strategies. This includes managing pests like nematodes by analyzing large datasets to uncover multivariate relationships and predict disease risks. The evolution from agriculture 1.0 (manual control) to agriculture 5.0 (AI-driven, data-centric) underscores AI's role in enhancing resilience against pests, including nematodes, by enabling informed decision-making and resource allocation.

- *Integration With Biological Control:* Recent research, highlighted by Jiue-in Yang (2025), explores integrating AI co-pilot systems with nematophagous microbes from natural suppressive soils. This approach aims to create innovative strategies for managing PPNs, emphasizing sustainable farming practices in response to climate change and emerging pests. Yang's work at UC Riverside focuses on microbe-plant nematode interactions, biological control mechanisms, and nematode-suppressive soils, suggesting a promising synergy between AI and biological methods.
- *Breeding for Resistance:* AI also plays a role in plant breeding for disease resistance (Ma et al. 2025). It leverages DL, large language models, and multi-modal models to enhance nematode disease detection and omics prediction, accelerating genomic and phenomic selection. This can include predicting resistance-associated traits against nematodes, harmonizing multi-omics data to forecast plant disease-resistant phenotypes, and supporting high-throughput analysis to improve crop resilience.
- *Integration With Emerging Technologies:* AI is increasingly integrated with remote and proximal sensing technologies for large-scale detection. Remote sensing, using satellites and drones, and proximal sensing, with soil sensors, provide data on plant health and infestation foci, which AI analyzes to optimize sampling and management strategies. This is particularly useful for identifying nematode hotspots in fields, as discussed in emerging technologies for INM.
- *Integration With IoT and Smart Farming:* When paired with IoT devices, AI enables continuous monitoring of soil and plant conditions, detecting changes in moisture, temperature, or nutrient levels that correlate with nematode activity. AI-driven DSSs then provide farmers with real-time recommendations—such as when to irrigate, fertilize, or apply treatments—promoting sustainable and efficient farming practices.

20.4.7 Accessibility and Democratization

Accessibility is a key factor in ensuring widespread adoption of AI for sustainable nematode management. Tools like SEGNEMA, freely available at GitHub, and the executable tools developed by Virginia Tech for growers, democratize access to advanced technologies (Esterhuizen 2022). This is particularly important for small-scale farmers in regions with limited expertise. By making AI accessible, it ensures that sustainable practices can be implemented globally, enhancing agricultural resilience and reducing economic disparities.

20.4.8 Benefits of AI in Nematode Management

- *Increased Accuracy:* AI minimizes human error in identifying and counting nematodes, leading to more reliable diagnoses.
- *Time and Cost Efficiency:* Automation and predictive tools streamline tasks, saving both time and resources.
- *Scalability:* AI can process vast amounts of data from multiple sources, making it ideal for large-scale agricultural or regional programs.
- *Sustainability:* By reducing the overuse of chemicals and supporting precision agriculture, AI lowers the environmental footprint of nematode control.

AI is transforming the predictive modeling and management of PPNs, offering tools for early detection, automated counting, outbreak prediction, and sustainable management. These advancements enhance agricultural efficiency, reduce crop losses, and promote eco-friendly practices, positioning AI as a cornerstone of modern precision agriculture. By fusing high-throughput sensing, advanced machine-learning algorithms, and decision-support tools, AI is transforming both the predictive modeling of PPN risk and the precision with which interventions are applied—paving the way toward more sustainable, data-driven nematode management.

20.5 Sustainability and Environmental Benefits

AI is transforming plant nematode management by enabling sustainable practices that minimize environmental impact while maintaining crop health and yield. By integrating data-driven insights from sensors, imaging, and environmental monitoring, AI systems facilitate targeted interventions that reduce chemical inputs, conserve resources, and enhance ecosystem resilience. AI plays a significant role in advancing sustainability and environmental benefits in plant nematode management by enabling precise, efficient, and eco-friendly approaches.

By reducing chemical dependency and supporting IPM, AI contributes to sustainable agriculture. Tools like SEGNEMA are freely available, making advanced nematode management accessible to small-scale farmers. This democratization helps ensure that sustainable practices can be adopted widely, enhancing global food security.

AI's role in sustainability is evident in its potential to minimize chemical inputs and promote ecological plant health management. By providing accurate and timely information, AI supports IPM strategies that combine biological, cultural, and chemical controls. For example, reducing herbicide usage through robotic weeders (González-Rodríguez et al. 2024) can be extended to nematode management, minimizing environmental harm. For instance, studies by Garrett et al. (2022) used random forest and SVMs for climate-pathogen interplay, achieving over 90% accuracy for infection risk predictions. This predictive capability allows farmers to implement preventive measures, such as crop rotation or biological controls, reducing the need for reactive chemical treatments. Early detection, as seen in Zen et al. (2022), achieving 100% accuracy for early blight and late blight in tomatoes using CNN and RNN with TensorFlow.js, minimizes environmental impact by limiting pesticide use, aligning with sustainable agriculture goals. The Virginia Tech study highlights that AI improves environmental sustainability by reducing non-selective pesticide use, protecting beneficial organisms, and supporting global food security amidst climate change pressures. This aligns with the review's emphasis on AI as a strategic innovation pathway for next-generation phytopathology.

20.5.1 Early Detection and Monitoring

AI-powered imaging and sensor platforms can continuously scan soil and root systems to detect nematode presence at incipient stages, generating real-time alerts for farmers through mobile or cloud-based dashboards. Early detection not only prevents widespread infestations but also allows for site-specific treatment, avoiding blanket pesticide applications that can harm non-target organisms. Studies have demonstrated the effectiveness of ML algorithms in identifying nematode eggs and juvenile stages from microscopic images with high accuracy, enabling prompt, precise responses (Gupta and Kumar Pal 2025).

20.5.2 Precision Application of Treatments

AI optimizes the application of nematicides or biocontrol agents by analyzing soil health, nematode populations, and environmental conditions. This reduces chemical overuse, preserving soil ecosystems and reducing runoff into water bodies. Tools like AI-driven DSSs recommend site-specific treatments, minimizing environmental impact.

Autonomous drones and ground robots equipped with AI-guided nozzles can apply biocontrol agents or reduced-dose nematicides directly to infested zones, cutting overall chemical usage by up to 50%. This precision approach ensures that only hotspots receive treatment, preserving beneficial soil microbiota and reducing off-target contamination. For instance, AI-driven platforms can analyze spatial infestation maps to optimize spray patterns, significantly lowering environmental pesticide load while maintaining nematode suppression efficacy (Bluewhite 2024).

20.5.3 Smart Irrigation and Soil Health Management

Nematode development is closely linked to soil moisture regimes. AI-integrated irrigation systems adjust watering schedules based on real-time soil moisture, weather forecasts, and crop stage, creating conditions less favorable for nematode proliferation. Such systems not only suppress nematode activity but also conserve water—studies report up to 30% reduction in irrigation volumes with AI control—benefiting both plant health and regional water sustainability (BIA 2024). Additionally, AI recommendations for balanced fertilizer and organic amendment placement help maintain soil structure and microbial diversity, further inhibiting nematode survival.

20.5.4 Biological Control and Alternatives

AI models predict the efficacy of biocontrol agents (e.g., beneficial microbes or fungi) by analyzing environmental and soil data, promoting sustainable alternatives to chemical nematicides. ML identifies resistant crop varieties or optimizes crop rotation schedules to suppress nematode populations naturally.

20.5.5 Predictive Modeling and Decision Support

ML models trained on historical infestation, climate, and management data can forecast nematode outbreaks weeks or months in advance. These predictions inform crop rotation schedules, cover cropping choices, and nematicide timing, reducing reliance on calendar-based sprays and preventing over-application. DSSs aggregate multi-source data—soil tests, remote sensing, and field reports—to generate tailored management plans, driving more efficient resource allocation and lowering environmental risks (Gupta and Kumar Pal 2025).

20.5.6 Environmental and Climate Benefits

By minimizing chemical runoff, preserving beneficial organisms, and reducing greenhouse gas emissions through optimized field operations, AI-enhanced nematode management contributes to broader environmental goals. Targeted interventions reduce non-point source pollution in waterways, while data-driven conservation tillage and cover cropping strategies foster carbon sequestration in soils. Collectively, these practices support biodiversity, improve farm resilience to climate extremes, and align nematode control with sustainable agriculture frameworks (Synthetix 2024).

- *Reduced Chemical Use:* AI's precision targeting lowers the reliance on toxic nematicides, protecting pollinators, soil microbiota, and water quality.
- *Soil Health Preservation:* AI-driven practices like crop rotation or organic amendments enhance soil biodiversity and long-term fertility.
- *Carbon Footprint Reduction:* Optimized farming practices, guided by AI, reduce energy-intensive interventions and promote sustainable land use.

20.5.7 Economic and Environmental Impact

AI's role in effective nematode management includes significant economic and environmental benefits. Hartley (2024) reported that AI-powered systems reduce operational costs by ~70% over one growing season, with costs dropping from $5000 to $1500 per 10 ha. This cost saving, combined with efficiency gains, allows farmers to allocate resources more effectively, particularly benefiting small-scale farmers. Environmentally, AI enables precision agriculture by mapping nematode hotspots, reducing chemical use by up to 50%, protecting beneficial organisms, and minimizing environmental harm. Tools like SEGNEMA, being open-source and freely available, democratize access to advanced techniques, enhancing global agricultural resilience and food security.

By leveraging AI, farmers can manage plant nematodes sustainably, balancing productivity with environmental conservation.

20.6 Enhanced Research and Development

AI is transforming research and development in plant nematode management, offering tools that make processes faster, more accurate, and environmentally friendly. This technology helps in counting and identifying nematodes, detecting them early, and integrating with modern farming systems for better management.

In research, AI processes large genomic datasets to identify genetic markers linked to nematode resistance in plants or virulence in nematodes, speeding up the development of resistant crops and new biocontrol methods. AI can also analyze video footage to study nematode behavior—such as feeding or movement patterns—providing valuable insights into their biology that inform better management strategies.

20.6.1 Sustainable Management Practices

AI-driven technologies promote sustainable practices by minimizing environmental impact. The nano-biosensor study notes that AI integration reduces chemical usage, conserves biodiversity, prevents pollution, and lowers carbon footprints, contributing to eco-friendly agriculture. This aligns with global efforts to enhance agricultural sustainability, protecting resources and ecosystem resilience. The automation of nematode management reduces the need for broad-spectrum pesticides, focusing on precision-based solutions that target specific infestations, thus supporting long-term environmental health.

20.6.2 Smart Farming and Future Potential

AI's role extends to smart farming systems, integrating with IoT for remote monitoring and data sharing, facilitating global nematode management networks. The future outlook includes multifunctional, autonomous AI-driven systems that predict outbreaks and recommend optimal management strategies, enhancing precision agriculture. The SEGNEMA study plans to expand to other genera such as *Globodera* spp., *Heterodera* spp., and *Pratylenchus* spp. through transfer learning, requiring global collaboration among nematology teams. This ongoing development promises broader applications, improving research and practical outcomes in nematode management.

20.6.3 Supporting Evidence and Examples

- The SEGNEMA tool exemplifies AI's impact, processing images rapidly with high accuracy (Saikai et al. 2024). Its open-source availability at GitHub ensures accessibility for researchers worldwide.
- The nano-biosensor study, published by Mani et al. (2025), underscores AI's transformative potential, with detailed algorithms and environmental benefits, supporting sustainable practices.

20.6.4 Challenges and Considerations

While AI offers significant advancements, challenges remain, such as the need for diverse image datasets for broader genera and collaboration among global research teams. The SEGNEMA study notes difficulties with overlapped objects and free-living nematodes detection (79% accuracy), mitigated by expanding datasets. These challenges highlight the need for continued research and development to refine AI models for universal application.

20.6.5 Conclusion

AI enhances plant nematode management research and development by automating counting and identification, enabling early detection via nano-biosensors, promoting sustainable practices, and integrating into smart farming systems. Key tools such as SEGNEMA and advanced AI algorithms, supported by studies from 2024 and 2025, demonstrate significant improvements in efficiency, accuracy, and environmental sustainability, with ongoing efforts to expand applications globally.

20.7 AI in Farmer Education and Extension Services

AI-powered platforms enhance knowledge dissemination and farmer training.

20.7.1 Chatbots and Virtual Assistants

AI chatbots, like those integrated with mobile apps, provide farmers with real-time advice on nematode management, including identification guides and control recommendations.

20.7.2 Automated Diagnostics

Mobile apps using AI image recognition allow farmers to upload images of affected plants or soil for instant nematode diagnosis and management advice.

20.8 Comparative Analysis: AI Versus Traditional Methods

To illustrate AI's impact, Table 20.2 compares traditional and AI-based nematode management.

This table underscores AI's superiority in efficiency, accuracy, and sustainability, supporting its role in modern agriculture.

20.9 Case Studies and Real-World Applications

20.9.1 Virginia Tech Research

Esterhuizen (2022) used AI to analyze over 40,000 microscopic images to identify soybean nematodes, enabling the creation of GIS infestation heat maps. This

Table 20.2 Comparison of traditional and AI-based nematode management technologies

Aspect	Traditional methods	AI-based methods
Identification time	2.5 h for manual counting of 10 images	4–5 s per image using DL models
Accuracy	Variable, depends on expertise (often <90%)	Up to 94–97% for detection and classification
Chemical use	Often requires broad-spectrum pesticides	Targeted applications, reducing chemical use by up to 50%
Environmental impact	High, due to non-selective pesticide use	Low, protects beneficial organisms and ecosystems
Accessibility	Limited to regions with nematode specialists	Widely accessible via free tools like SEGNEMA

precision helps farmers target nematode hotspots, reducing chemical use and improving crop yields.

20.9.2 AI-Powered Robots

Saikai et al. (2024) highlighted the use of AI-powered robotic systems for high-throughput nematode detection, demonstrating their ability to cover large areas efficiently while providing real-time soil health data.

20.10 Challenges and Limitations

As AI technology advances, its role in nematode management will grow even more significant. Sophisticated models, such as DL, could further improve detection and classification accuracy. AI might also integrate with biotechnology, such as clustered regularly interspaced short palindromic repeats/CRISPR-associated protein 9 (CRISPR/Cas9), to develop nematode-resistant crops or beneficial nematodes for biological control. Additionally, AI-powered platforms could enable global collaboration by sharing data and strategies, enhancing nematode management worldwide.

While AI shows promise, challenges remain, including dataset limitations and the need for global collaboration to enhance model accuracy with diverse imaging systems. González-Rodríguez et al. (2024) recommend open-access platforms and transfer learning to improve generalization, addressing technical barriers. Ethical concerns, such as data privacy and farmer autonomy, necessitate inclusive stakeholder participation to ensure equitable benefits. Ongoing research, as seen in Virginia Tech's work with host differential bioassays and molecular tools, suggests a dynamic field with potential for further advancements in sustainable nematode management (Esterhuizen 2022).

While AI shows promise, challenges remain, including high initial costs, data quality issues, model generalization, and sensor calibration drift. Shoaib et al. (2023) recommend open-access platforms and transfer learning to improve generalization, addressing technical barriers. Ethical concerns, such as data privacy and farmer autonomy, necessitate inclusive stakeholder participation to ensure equitable benefits. Future trends include Edge AI with 5G connectivity, quantum AI for soil data processing, and cloud-based predictive analytics for proactive pest management, as outlined by Jain (2025).

Despite these advancements, challenges remain. Developing robust AI models requires large, diverse datasets, which can be difficult to obtain due to the microscopic nature of nematodes and variability in environmental conditions. Current models are often specialized for specific species, such as *Meloidogyne* spp., with future research aiming to expand to other genera such as *Globodera* spp., *Heterodera* spp., and *Pratylenchus* spp. through transfer learning. Ethical considerations, such as data privacy and model biases, also need to be addressed, alongside the need for global collaboration to enhance datasets and model performance.

20.11 Future Prospects

Advancements in AI offer promising avenues for nematode management:

20.11.1 Multi-Modal AI Integration

Fusion of imaging (RGB, hyperspectral), spectroscopic, and omics data (e.g., soil metagenomics) via advanced DL architectures (e.g., transformers) can yield holistic nematode surveillance platforms.

20.11.2 Edge Computing and Mobile Apps

Deployment of lightweight DL models on smartphones or low-cost edge devices will democratize access, enabling on-farm diagnostics without internet dependence.

20.11.3 Community Data Repositories

Establishing global standardized nematode image and spectral libraries will accelerate model development and benchmarking.

20.11.4 Automated Robotics

Integration of AI with robotic platforms for in-field sample collection and automated egg extraction will further streamline diagnostics.

20.11.5 Integration With IoT and Big Data

Combining AI with IoT and big data analytics will enable real-time, field-specific management strategies.

20.11.6 DL Innovations

Emerging DL techniques, such as generative adversarial networks, could enhance nematode detection by generating synthetic training data to overcome dataset limitations.

20.11.7 Climate-Resilient Strategies

AI can model the impact of climate change on nematode populations, aiding in the development of adaptive management practices.

20.12 Conclusion

AI is revolutionizing nematode diagnosis and management by automating labor-intensive tasks, enabling rapid and accurate identification, providing real-time monitoring, and offering predictive insights. These advancements help farmers make informed decisions, reduce crop losses, and minimize environmental impact. The integration of AI with traditional pest management methods is seen as essential for comprehensive control, with ongoing research promising further innovations in precision agriculture.

AI plays a pivotal role in sustainable plant nematode management by automating identification, enabling early detection and prediction, integrating with precision agriculture, and promoting environmental stewardship. Its accessibility and ongoing research promise to enhance global agricultural resilience, making it a cornerstone for sustainable farming practices.

AI-driven plant nematode management represents a transformative approach to agriculture, combining efficiency, accuracy, and sustainability. Recommendations include using AI-powered counting tools like SEGNEMA, implementing AI-integrated nano-biosensors for early detection, deploying AI-powered robotic systems for high-throughput detection, and integrating these with precision agriculture systems. These strategies, grounded in cutting-edge research, offer farmers high-tech, eco-friendly solutions to protect crops and enhance ecosystem resilience. By combining high-resolution data collection, robust AI modeling, and seamless integration into decision-support workflows, you can shift from calendar-based nematicide use to a truly data-driven, eco-friendly nematode management paradigm.

By integrating AI with IoT, remote sensing, and genomic tools, agriculture can move toward sustainable and efficient nematode management. However, addressing challenges such as data availability and accessibility is crucial to ensure widespread adoption, particularly in resource-limited settings. Continued research and collaboration between AI experts, agronomists, and farmers will unlock the full potential of AI in combating PPNs.

20.13 Suggested Readings

- Esterhuizen M (2022) Virginia Tech researchers use artificial intelligence to combat soybean nematodes. Virginia Tech News, 24th October 1922.
- Garrett KA, Bebber DP, Etherton BA et al (2022) Climate change effects on pathogen emergence: artificial intelligence to translate big data for mitigation. Annu Rev Phytopathol 60:357–378.

- Jiue-in Yang (2025) Artificial intelligence and nematodes. UC ANR/Bug Squad, January 24, 2025.
- Ma J, Cheng Z, Cao Y (2025) Artificial intelligence-assisted breeding for plant disease resistance. Int J Mol Sci 26. https://doi.org/10.3390/ijms26115324.
- Mani J, Nagachandrabose S, Somasundaram P, Deenan S (2025) Artificial intelligence integrated nano biosensor technology: a breakthrough in early detection and sustainable management of phytonematodes. Physiol Mol Plant Pathol 139. https://doi.org/10.1016/j.pmpp.2025.102756.

20.14 Review Questions

- Define AI and explain its relevance in nematode disease management.
- Discuss the major AI-based techniques used in nematode management.
- Explain how expert systems can be integrated into nematode management decision-making.
- What role does predictive modeling play in forecasting nematode outbreaks under changing agro-climatic conditions?
- How can AI-driven DSSs improve INM?
- In what ways can AI help optimize nematicide application and reduce environmental hazards?
- What are the limitations of current AI applications in plant nematology, and how might they be overcome?
- Predict the future scope of AI in nematode research—what emerging technologies.

References

BIA (2024) AI in agriculture: *enhancing productivity and sustainability*. Boston Institute of Analytics. bluewhite.aiBoston Institute of Analytics

Bluewhite (2024) Sustainable farming practices with AI: reducing environmental impact. INDIAaibluewhiteai

Esterhuizen M (2022) Virginia Tech researchers use artificial intelligence to combat soybean nematodes. Virginia Tech News, 24th October 1922

Garrett KA, Bebber DP, Etherton BA et al (2022) Climate change effects on pathogen emergence: artificial intelligence to translate big data for mitigation. Annu Rev Phytopathol 60:357–378

González-Rodríguez VE, Izquierdo-Bueno I, Cantoral JM et al (2024) Artificial intelligence: a promising tool for application in phytopathology. Horticulturae 10. https://doi.org/10.3390/horticulturae10030197

Gupta G, Kumar Pal S (2025) Applications of AI in precision agriculture. Discov Agric 3:61. https://doi.org/10.1007/s44279-025-00220-9

Hartley M (2024) AI and automation in reducing IT operational costs. Int J Adv Innov Res 9(1):548–555

Jain R (2025) Artificial intelligence in precision agriculture: advanced systems for crop management and farm optimization. Int J Sci Res Comput Sci Eng Inf Technol 11(1):1849–1857

Loveth RG (2025) AI-powered robotic systems for high-throughput nematode detection and soil health monitoring. Res Gate. https://www.researchgate.net/publication/ 389988206

Ma J, Cheng Z, Cao Y (2025) Artificial intelligence-assisted breeding for plant disease resistance. Int J Mol Sci 26. https://doi.org/10.3390/ijms26115324

Mani J, Nagachandrabose S, Somasundaram P, Deenan S (2025) Artificial intelligence integrated nano biosensor technology: a breakthrough in early detection and sustainable management of phytonematodes. Physiol Mol Plant Pathol 139. https://doi.org/10.1016/j.pmpp.2025.102756

Piddubna A (2024) AI in agriculture – the future of farming. Intellias. https://Intellias.com/artificial-intelligence-in-agriculture

Sagester P, Myers S (2025) AI meets agriculture: Agmatix and BASF partner to develop soybean cyst nematode detection tool. Trade News, March 12, 2025

Saikai KK, Bresilla T, Kool J et al (2024) Counting nematodes made easy: leveraging AI-powered automation for enhanced efficiency and precision. Front Plant Sci 15:1349209

Shabrina NH, Lika RA, Indarti S (2023) Deep learning models for automatic identification of plant-parasitic nematode. Artif Intell Agric 7:1–12

Shoaib M, Shah B, EI-Sappagh S et al (2023) An advanced deep learning models-based plant disease detection: a review of recent research. Front Plant Sci 14. https://doi.org/10.3389/fpls.2023.1158933

Sun Z, Ibrayim M, Hamdulla A (2022) Detection of pine wilt nematode from drone images using UAV. Sensors 22. https://doi.org/10.3390/s22134704

Synthetix L (2024) Harvesting tomorrow: AI's role in sustainable agriculture. AI Slackers, October 22, 2024

Yang J-I (2025) Artificial intelligence and nematodes. UC ANR/Bug Squad, January 24, 2025

Zen BP, Kresna I, Fransisca DC (2022) Applications for detecting plant diseases based on artificial intelligence. Sinkron: Jurnal dan Penelitian Teknik Informatika 7(4):2539–2546

Part VIII

Outlook

A Vision for Future

21

Abstract

As agriculture faces increasing pressures from rising global populations, climate variability, and the need for sustainable practices, the challenge posed by plant-parasitic nematodes (PPNs) demands innovative and integrative solutions. The future plant nematode management will focus on sustainable, integrated strategies. The new chemical nematicides and bionematicides will play a key role, alongside nanotechnology. The evidence leans toward increased use of molecular tools like clustered regularly interspaced short palindromic repeats (CRISPR) and host plant resistance for long-term solutions. There is an ongoing debate about balancing chemical use with environmental impact, but collaboration aims to address this. Looking ahead, we anticipate that advances in technology, data integration, and collaborative frameworks will transform both diagnostics and management of PPNs.

Keywords

Bionematicides • Predictive modeling • Integrated nematode management • Nanotechnology • Precision agriculture

21.1 Learning Objectives

- Examine the role of cutting-edge tools (e.g., artificial intelligence (AI), machine learning, CRISPR-based diagnostics, portable sequencing devices, and biosensors) in nematode detection and monitoring.
- Assess the potential of advanced management strategies, including RNA interference (RNAi), transgenic crops, biocontrol, and precision agriculture.
- Discuss future strategies that align with ecological sustainability, reduced chemical reliance, and climate-smart agriculture.
- Formulate approaches for strengthening integrated nematode management (INM) with emphasis on farmer-friendly, cost-effective solutions.

R. P. Parvatha, *Plant Parasitic Nematodes*,
https://doi.org/10.1007/978-981-95-9306-4_21

- Appreciate the importance of international data sharing, harmonized diagnostic protocols, and collaborative surveillance systems.
- Explore the role of policies, capacity building, and extension networks in shaping the future of nematode diagnosis and management.
- Anticipate future research needs and gaps in plant nematology.
- Encourage innovative thinking to design next-generation diagnostic platforms and management practices for nematode control.

21.2 Introduction

Plant-parasitic nematodes (PPNs) pose persistent, significant, and often underestimated threats to global agriculture, causing substantial yield losses (billions in crop losses annually) across a wide range of crops. Their hidden, soil-borne nature makes early detection and effective control challenging, while the over-reliance on chemical nematicides raises environmental and regulatory concerns. Rapid advances in biotechnology, digital agriculture, and ecological intensification now offer unprecedented opportunities to revolutionize nematode diagnosis and management. By integrating molecular diagnostics, AI-powered detection systems, precision-targeted control methods, and climate-resilient agro-ecological practices, the future promises a shift from reactive, chemical-heavy approaches to proactive, sustainable, and farmer-friendly solutions. INM combines cultural, biological, chemical, and technological approaches to sustainably control these pests while minimizing environmental impact. Looking ahead, the future of nematode diagnosis and management envisions a shift toward precision, sustainability, and resilience, driven by advancements in technology, genetics, and data-driven decision-making. This chapter emphasizes holistic agro-ecological systems that adapt to emerging challenges such as climate change, soil degradation, and regulatory pressures.

21.3 Diagnosis: Molecular and Technological Advances

The future of diagnosing PPNs is heavily reliant on molecular and technological advancements, aiming to enhance precision and early detection. Research suggests that understanding the molecular basis of plant–nematode interactions is crucial for developing specific control methods. Omics approaches, including genomics, proteomics, and metabolomics, combined with *in silico* analyses, are expected to play a pivotal role. These methods analyze genes and proteins involved in pathogenesis activities such as invasion, feeding, and colonization, as highlighted in recent literature (Desaeger et al. 2020).

A specific technique, loop-mediated isothermal amplification (LAMP), is already in use for diagnosis, offering rapid and accurate detection. A review by Ahuja and Somvanshi (2021) details its application, noting its effectiveness in identifying nematode infestations. Beyond molecular methods, remote sensing and proximal sensing technologies are emerging as game-changers. Hyperspectral imaging, which captures detailed spectral data, and unmanned aerial vehicles (UAVs) for spectral detection are being utilized to identify nematode-induced plant stress. For instance, studies have demonstrated their application in detecting stress in sugar beet and soybean (Hillnhütter et al. 2011; Nutter et al. 2002).

Machine learning further enhances these technologies, enabling precision detection by analyzing large datasets from remote sensing. Examples include detecting pine wood nematode disease using UAV images (Qin et al. 2021). These advancements promise non-invasive, large-scale monitoring, potentially revolutionizing early diagnosis in agricultural settings.

21.4 Management: Sustainable and Integrated Strategies

Management strategies for PPNs are shifting toward sustainability, integrating new technologies with traditional practices to reduce environmental impact and enhance efficacy. The development of new nematicides is a key focus, with compounds such as fluopyram, fluensulfone, and fluazaindolizine being less toxic and having different mechanisms of action compared to older non-fumigant nematicides such as carbamates and organophosphates (Desaeger et al. 2020). These innovations are supported by market growth projections, with the global nematicides market valued at US$1.78 billion in 2022 and expected to reach USD 4.28 billion by 2032 at a compound annual growth rate of 9.1%, according to Spherical Insights LLP (Desaeger et al. 2020).

INM is another cornerstone of future strategies, combining traditional methods such as crop rotation, genetic improvement, and biological control with modern technologies. Cover crops, for instance, can be modified to include resistant varieties or those with biofumigation potential, reducing nematode populations (Desaeger and Rao 2001). Genetic solutions, such as genome editing, target nematode genes involved in parasitism, offering a promising avenue for control through methods like RNAi (Banerjee et al. 2017). The microbiome is also being leveraged for suppression, with studies showing its potential to enhance soil health and reduce nematode damage (Topalovic et al. 2020).

Precision agriculture concepts, supported by decision support systems, will enable site-specific management, optimizing the application of nematicides and other inputs. This approach is detailed in recent reviews, emphasizing its role in holistic management across agricultural scales (Molendijk and Sikora 2022). Biological control agents, such as nematophagous bacteria (*Burkholderia cepacia*, *Paenobacillus macerans*, and *Bacillus firmus*), are also gaining traction, with commercial products such as Deny, Bio Yield™, and BioNem showing efficacy in reducing nematode populations and increasing crop yields (Etesami et al. 2023). These agents act synergistically, promoting plant growth and facilitating rhizosphere colonization by microbial antagonists (Etesami et al. 2023).

The innovative technologies and some of the solutions that will influence INM and fill the knowledge gap in the future are presented in Table 21.1. The table includes the suggestions and projections made by many of the researchers when

Table 21.1 Current and anticipated progress in technologies that will have major impacts on INM and soil and plant health

Technologies	2000–2020	2020–2050
Sanitation	Hot water treatment of plant material Certified seed and planting material but limited equipment cleaning	New sanitation techniques (e.g., controlled atmosphere temperature treatment, steam, and accelerated electrons) Seed certification more widespread Quarantine expanding
Rotation and cover crops	Standard but often narrow, decreasing in importance as an INM tool Often limited info on the specific nematode host status	Clever rotations, intercropping, mixed and relay cropping increasingly designed for nematode management More nematode-targeted crop and cover crop cultivars available
Biological soil disinfestation	Anaerobic soil disinfestation, solarization, biofumigation not widely used, and results often inconsistent	Integrated programs including a biological soil component more common More effective agents and methods (e.g., microbial volatiles)
Soil-free production	Substrate production for vegetables (mostly in the European Union and Canada) Ultraviolet radiation treatment of irrigation water	Expanded to include more crops and regions Ozone generators for decontamination of irrigation water in closed systems
Cultivar selection	Resistant cultivars and rootstocks used where available, relatively few crops with nematode resistance. Tolerant cultivars often preferred over resistant	New sources of resistance, increasing use of resistant and tolerant cultivars, and rootstocks Expanding the success and acceptance of modern genetic editing (e.g., CRISPR/Cas9)
Seed-based control	Combined fungicide/nematicides seed coating for row crops introduced	Increasing availability of chemical and biological seed treatment options, combined with nematode resistance/ tolerance
Chemical control	Old chemistry still widely used but being replaced First new reduced-risk nematicides becoming available	Older nematicides and fumigants removed from the market New nematicides becoming available (including basipetal treatments) and replacing old nematicides, especially in high-value crops
Biological control	Several products with limited or scattered levels of success, and relatively few field data Effective seed-based treatments still lacking Potential use of biochemical treatments still untested	More competitive agents and strains, including plant- and microbial-derived products Seed and seedling-based treatments becoming standard for many crops Increased use of biologicals for targeted post-plant treatment, and partnered with chemicals and bio-chemicals

(continued)

Table 21.1 (continued)

Technologies	2000–2020	2020–2050
Soil antagonistic potential	Presence acknowledged but impact unsatisfactory and not used on a practical scale Organic agriculture still leading promoter Soil microbiome research expanding	Soil microbiome advances leading to targeted methods to stimulate soil antagonistic potential Metagenomics and holistic approaches available for monitoring Holobiont concept used to help monitor nematode populations and improve soil antagonistic potential
Soil sampling	Field scouting, manual soil and root sampling	Remote sensing, drone-driven mapping, use of robotics to aid sampling Analysis of samples using molecular and machine-learning tools, including holistic soil analysis approaches
Precision agriculture	Technology developed; field application still limited	Site-specific treatment using sensors and robotics adopted in the field and horticultural crops
Modeling population dynamics	Lack of data on population dynamics limits use Initiation of remote sensing for the detection of nematode distribution tested	Mathematical modeling available to calculate population development over rotations and *pi* estimation Remote sensing coupled with practical models used in precision INM Satellite imaging allows detection of nematode hotspots and nematode "densities"
Other supportive tactics	Limited number of decision-support tools (DSTs)/decision-support systems (DSSs) with a low level of implementation	DST with IT and AI expanding, especially in horticultural crops High-resolution data platforms providing the farmer/adviser with detailed information and INM recommendations, facilitating integrated crop management

anticipating how INM will need to change in the coming decades. There is a need to adjust our approaches to offset the drivers of change: human population growth, food insecurity, soil degradation, climate change and volatility, and the anticipated increase in nematode impact on soil and crop health.

21.5 Future Plant Nematode Management Strategies

The management of PPNs is at a pivotal juncture, driven by the need to address significant economic losses—estimated at $358.24 billion annually for the top 40 crops—and ensure global food security amidst environmental challenges. The economic burden underscores the urgency for effective management strategies.

21.5.1 Emerging Chemical and Biological Nematicides

The future of chemical control is shifting toward safer alternatives due to regulatory restrictions on traditional nematicides such as aldicarb and methyl bromide. Recent developments include the following:

- *New-Generation Nematicides*: Fluensulfone (Nimitz, Adama), registered by the EPA in 2014, and Fluopyram (Velum®, Bayer), registered in 2013, target specific nematode processes such as cellular energy inhibition and succinate dehydrogenase, reducing environmental impact.
- Biological control agents such as *Bacillus firmus* (Nortica®), *Myrothecium verrucaria* (DiTera® DF, WP), *Pasteuria usgae* (Econem™), and *Purpureocillium lilacinum* (MeloCon®, BioAct WP) are gaining market demand. These are safe for beneficial organisms and align with organic farming practices. Nemguard®, based on garlic extract, is being tested for expansion on crops like grapes, offering a natural soil application method.

21.5.2 Integrated Multi-omics and Real-Time Diagnostics

- *Holistic Pest Profiling*: Future diagnostic platforms will combine genomics, transcriptomics, proteomics, metabolomics, and epigenomics to deliver a systems-level portrait of nematode–host interactions. Portable field-deployable sequencers (e.g., handheld nanopore devices) coupled with rapid sample-prep chemistries will allow on-site identification of species, virulence genes, and resistance-breaking alleles within hours.
- *Biosensor Networks*: Advances in microfluidics and nanotechnology will yield low-cost biosensors capable of detecting nematode effectors or host phytohormone signatures in soil or root washings. Integration into Internet-of-Things networks will enable continuous monitoring, early warnings, and dynamic risk mapping at farm and landscape scales.

21.5.3 AI and Predictive Modeling

- *Machine Learning Diagnostics*: Image-based AI models trained on vast libraries of root and soil imagery will rapidly classify nematode species and life stages with high accuracy, even under field conditions. Deep-learning algorithms will discern subtle symptom patterns before visible damage manifests.
- *DSSs*: Next-generation DSS will leverage weather forecasts, soil moisture sensors, crop phenology models, and nematode population dynamics to predict outbreak windows. These tools will recommend optimized timing and combinations of biological, cultural, and chemical controls tailored to local conditions.

21.5.4 Precision Agriculture and Digital Soil Health

- *Variable-Rate Application*: Coupling Global Positioning System-guided machinery with site-specific nematode risk maps will allow variable-rate delivery of biocontrol agents, amendments (e.g., biofumigants and organic mulches), and nematicides—minimizing inputs while maximizing efficacy.
- *Remote Sensing*: UAVs and multispectral satellites will detect early stress signatures associated with nematode infestation (e.g., canopy temperature anomalies and chlorophyll fluorescence), enabling targeted ground-truth sampling and intervention.

21.5.5 Nanotechnology

Nanotechnology is revolutionizing PPN control by enhancing nematicide effectiveness and reducing production costs. Nanoencapsulation addresses manufacturing defects, leading to cleaner production processes. Recent research explores plant-based nanoparticles that disrupt nematode cuticles and induce oxidative stress, providing novel control mechanisms.

21.5.6 Biotechnological Innovations for Sustainable Management

Molecular tools are at the forefront of future management strategies:

- *Gene-Edited Crops*: CRISPR/Cas and other genome-editing platforms will be used to engineer broad-spectrum resistance by modifying key host susceptibility genes or introducing synthetic immune receptors. Stacking multiple defense traits will reduce the risk of resistance breakdown.
- *Designer Microbiomes*: Synthetic consortia of antagonistic bacteria, fungi, and nematophagous organisms—formulated for compatibility and stability—will be developed as "living" soil amendments. Metabolic-engineering approaches will enhance the production of anti-nematode compounds directly *in situ*.
- *RNA-Based Biopesticides*: Sprayable double-stranded RNAs targeting essential nematode genes offer species-specific control with minimal off-target impacts. Advances in nanoparticle carriers and formulation chemistry will improve stability and uptake in field conditions.
- *RNAi and CRISPR-Cas9*: These techniques target nematode genes involved in parasitism, with CRISPR-PN2 software facilitating targeted genome editing. Recent breakthroughs include identifying vulnerabilities in cyst nematodes' communication with plant roots, particularly for soybean cyst nematodes (SCNs), which could reduce damage.

- *Metagenomic Tools and AI*: AI is being integrated for PPN identification, with high-throughput sequencing (HTS) and meta-DNA barcoding, analyzing soil food webs. These tools enhance diagnostic precision, moving toward field-based detection systems.

21.5.7 Host Plant Resistance and Genetic Engineering

Breeding for nematode resistance is a cornerstone of long-term management:

- Pangenome analysis of wild relatives and grafting onto resistant rootstocks, such as tomato–melon–pepper–watermelon on resistant varieties, is being explored. For example, bell pepper cultivars like "Charleston Belle" have shown effectiveness in reducing nematode levels in double-cropping systems.
- Epigenetic studies via transcriptome analyses are providing insights into plant–nematode interactions, aiding in developing more resilient crops. Recent research identified new genes that could strengthen soybean crops against SCN resistance, addressing the pest's ability to skirt genetic resistance.

21.5.8 Integrated Pest Management and Agricultural Practices

Integrated pest management (IPM) is evolving to integrate multiple strategies for holistic management:

- Cultural practices such as crop rotation, soil solarization, cover crops, and reduced tillage are being combined with nematicide applications. Site-specific application reduces dosage, minimizing environmental impact.
- DSTs like Nematool, LoRa Soil sensors, and DSS platforms like Akkerweb/FarmMaps enable real-time monitoring and decision-making, enhancing management efficiency.

21.5.9 Predictive Modeling and Management

Predictive modeling is a growing area where AI shows significant promise. Machine learning models such as random forests and support vector machines analyze historical data on nematode outbreaks, crop health, and environmental factors to forecast future infestations. According to a detailed report from indiaai.gov.in, these models can predict outbreaks by identifying patterns in soil, crop, and environmental data, enabling pre-emptive actions. For instance, ML can estimate the economic impact of infestations by correlating root damage severity with yield loss, aiding cost–benefit analyses for control measures.

Specific applications include the collaboration between Agmatix and BASF, announced on March 12, 2025, to develop an AI tool for detecting and predicting

SCN infestations. This initiative uses Agmatix's Axiom AI engine to process vast agronomic datasets, offering real-time, scalable insights to mitigate yield losses caused by SCN, one of the most damaging soybean pests.

Management strategies are also enhanced by AI, which recommends tailored control methods based on data analysis. For example, AI can suggest biological controls like biopesticides (beneficial fungi or bacteria) or crop rotation, informed by local and historical data. Precision agriculture is further supported by autonomous robots and drones equipped with AI, which apply nematicides precisely to infested areas, minimizing chemical use. Smart irrigation systems, integrated with AI, adjust watering to create unfavorable conditions for nematodes, conserving water while protecting crops.

- *Precision Agriculture*: AI's integration into precision agriculture, as discussed in Boosting precision crop protection toward agriculture 5.0 via machine learning and emerging technologies: A contextual review involves optimizing data-driven, tailored crop protection strategies. This includes managing pests like nematodes by analyzing large datasets to uncover multivariate relationships and predict disease risks. The evolution from agriculture 1.0 (manual control) to agriculture 5.0 (AI-driven, data-centric) underscores AI's role in enhancing resilience against pests, including nematodes, by enabling informed decision-making and resource allocation.
- *Breeding for Resistance*: AI also plays a role in plant breeding for disease resistance, as outlined in Artificial Intelligence-Assisted Breeding for Plant Disease Resistance. It leverages deep learning, large language models, and multi-modal models to enhance disease detection and omics prediction, accelerating genomic and phenomic selection. This can include predicting resistance-associated traits against nematodes, harmonizing multi-omics data to forecast plant disease-resistant phenotypes, and supporting high-throughput analysis to improve crop resilience.

21.5.10 Comparative Analysis: Current Versus Future Approaches

To illustrate the shift, consider Table 21.2 comparing current and future approaches to diagnosis and management.

This table underscores the transition toward more sustainable and technology-driven solutions, addressing both efficacy and environmental concerns.

21.6 Integration With Emerging Technologies

AI is increasingly integrated with remote and proximal sensing technologies for large-scale detection. Remote sensing, using satellites and drones, and proximal sensing, with soil sensors, provide data on plant health and infestation foci, which

Table 21.2 Comparison of current and future approaches for diagnosis and management

Aspect	Current approaches	Future approaches
Diagnosis	Traditional soil sampling, microscopy, and polymerase chain reaction-based methods	Molecular techniques (omics, LAMP), remote sensing (hyperspectral imaging and UAVs), and machine learning for precision detection
Management	Chemical nematicides (e.g., carbamates and organophosphates), crop rotation, and resistant cultivars	New, less toxic nematicides, INM (genome editing and microbiome leverage), precision agriculture, and biological control agents
Challenges	Environmental impact of chemicals and limited scalability of traditional methods	Adapting to climate variability, ensuring global technology access, and regulatory compliance

AI analyzes to optimize sampling and management strategies. This is particularly useful for identifying nematode hotspots in fields, as discussed in emerging technologies for INM.

21.7 Global Collaboration and Capacity Building

Addressing the North–South divide is critical for equitable management solutions.

Expanding laboratories and research centers in developing countries, along with cooperative programs for farmer training, is essential. Partnerships for bioinformatics, nanoencapsulation, and RNA-Seq are fostering innovation and knowledge sharing.

21.7.1 Participatory Research Networks

Engaging farmers, extension agents, and researchers through digital platforms will accelerate co-development of locally adapted strategies and feedback loops, ensuring both relevance and adoption.

21.7.2 Policy Frameworks

Regulatory harmonization for biocontrol agents, RNA-based products, and genome-edited crops will be essential to streamline approvals while maintaining safety. Subsidies and incentives for IPM adoption can catalyze transitions away from reliance on broad-spectrum nematicides.

21.7.3 Data Sharing and Open Science

Global repositories of nematode genomic data, diagnostic assay protocols, field trial results, and sensor datasets will fuel innovation and avoid duplication. Open-access platforms will democratize knowledge and tools, particularly benefiting smallholders and resource-limited regions.

21.7.4 Climate Change and Adaptive Strategies

Climate change, particularly rising soil temperatures, is expected to exacerbate nematode damage. Research is focusing on the effects on nematode survival, reproduction, and virulence, with adaptive strategies being developed to mitigate impacts.

21.8 Challenges and Controversies

Challenges remain in spite of several advancements in AI. Developing robust AI models requires large, diverse datasets, which can be difficult to obtain due to the microscopic nature of nematodes and variability in environmental conditions. Current models are often specialized for specific species, such as *Meloidogyne* spp., with future research aiming to expand to other genera such as *Globodera* spp., *Heterodera* spp., and *Pratylenchus* spp. through transfer learning. Ethical considerations, such as data privacy and model biases, also need to be addressed, alongside the need for global collaboration to enhance datasets and model performance.

Despite these advancements, several challenges must be addressed to realize this vision. Adapting INM strategies to climate variability is critical, as changing environmental conditions can intensify nematode damage (Sikora et al. 2022). For instance, increased temperatures and altered precipitation patterns may affect nematode life cycles and host susceptibility, necessitating flexible management approaches. Additionally, inequalities in technology access between the Global North and South pose significant barriers. The review by Annual Reviews highlights the need to improve INM across all scales, addressing disparities that influence access to advanced tools and knowledge (Sikora et al. 2023).

Policy and regulatory frameworks also play a role, with evolving legislation on pesticide use in regions like the European Union influencing control strategies (Zhou et al. 2025). The integration of these technologies requires collaboration among researchers, policymakers, and farmers to ensure equitable benefits and sustainable outcomes.

There is an ongoing debate about balancing chemical nematicide use with environmental sustainability, given the phase-out of older chemicals such as aldicarb and methyl bromide. While new nematicides offer safer alternatives, their long-term ecological effects are still under study. The integration of GM crops for resistance also faces regulatory and public acceptance challenges, particularly in regions with strict biotech policies.

21.9 Future Strategies

21.9.1 Nematode Diagnosis With Speed, Precision, and Accessibility

The future of nematode diagnosis is set to be defined by speed, precision, and accessibility. Breakthroughs in molecular biology will enable the development of portable "lab-on-a-chip" devices that can identify nematode species in real time using CRISPR-based biosensors, microfluidics, and isothermal amplification techniques such as RPA and LAMP. AI will play a central role, with deep learning models capable of analyzing images captured through smartphone microscopes or drone-mounted cameras to detect nematode symptoms at an early stage. Advances in remote sensing and GIS will allow nematode hotspots to be mapped at field and regional scales, providing predictive insight into where and when outbreaks are likely to occur. These tools will be supported by cloud-connected diagnostic networks where farmers and extension workers can upload data and receive immediate, AI-generated recommendations.

21.9.2 Targeted, Eco-friendly, and Integrated Management Strategies

Management strategies will evolve to become more targeted, eco-friendly, and integrated. Genome editing through CRISPR-Cas systems will accelerate the development of crop varieties with durable, multi-nematode resistance. Host-induced gene silencing and RNAi sprays will provide non-transgenic, species-specific suppression of nematode virulence genes. Biological control will advance beyond traditional agents to include engineered microbiomes and endophytic organisms that enhance plant defenses. Chemical control will shift toward precision nematicide delivery systems, including nanoformulations and smart-release capsules that reduce environmental impact while maintaining efficacy. Agro-ecological practices such as intercropping, organic amendments, and trap cropping will be fine-tuned and integrated with modern technologies to ensure long-term nematode suppression while improving soil health.

21.9.3 Data-Driven Decision-Making

Data-driven decision-making will underpin this transformation. By integrating soil health indicators, climate data, and nematode population dynamics into unified platforms, predictive analytics will allow for proactive interventions rather than reactive treatments. Mobile dashboards and farmer-friendly applications will deliver site-specific recommendations in real time, enabling even smallholder farmers to make informed decisions. Importantly, these systems will be adaptable to climate change scenarios, helping agricultural systems remain resilient in the face of shifting pest dynamics, extreme weather events, and evolving nematode species distributions.

21.9.4 Capacity Building and Global Collaboration

Capacity building and global collaboration will be crucial to ensuring that these innovations reach the farmers who need them most. Digital extension platforms, multilingual training materials, and immersive virtual reality modules will support knowledge transfer across diverse farming communities. International partnerships will facilitate open sharing of genomic data, field trial results, and diagnostic resources, accelerating innovation and reducing duplication of effort. Public–private collaborations will help scale up technologies from the lab to the field, ensuring affordability and accessibility.

Table 21.3 summarizes future strategies and their description along with examples.

Table 21.4 summarizes the key trends and their implications for plant nematode management.

This comprehensive approach ensures that future plant nematode management is sustainable, precise, and collaborative, addressing both current challenges and emerging threats like climate change.

Table 21.3 Summary of future nematode management strategies

Strategy	Description	Examples
Safe chemical nematicides	New, safer chemicals targeting specific nematode processes	Fluensulfone (Nimitz, Adama), Fluopyram (velum®, Bayer)
Bionematicides	Biological agents for natural control, safe for beneficial organisms	*Bacillus firmus*, Nemguard® (garlic extract)
Nanotechnology	Nanoencapsulated nematicides and plant-based nanoparticles for enhanced control	Nanoencapsulation, plant-based NPs
Molecular tools	RNAi, CRISPR for targeted gene editing and disruption of nematode processes	CRISPR-PN2, RNA-Seq
Host plant resistance	Breeding and genetic engineering for resistant crop varieties	Resistant bell peppers, new SCN resistance genes
IPM	Combining cultural practices, nematicides, and real-time monitoring	Crop rotation, Nematool, LoRa soil sensors
Biotechnological diagnostics	AI, HTS, and field-based diagnostics for rapid identification	Metagenomic tools, AI for PPN identification
Global collaboration	Expanding research and training in developing countries, fostering partnerships	Cooperative programs, bioinformatics partnerships
Climate adaptation	Strategies to mitigate increased nematode damage due to climate change	Research on soil temperature effects, adaptive practices

Table 21.4 Key trends and their implications for plant nematode management

Trend	Description	Implications
Genetic innovations	MAS, gene editing (CRISPR-Cas9), RNAi for resistant crops	Faster breeding, ethical debates, varied regulations
Biological control	Microbial biopesticides, nematode-based biocontrol	Reduced chemical use, challenges in production, and shelf life
INM	Cultural practices, biological controls, selective chemicals	Sustainable, long-term management, soil health focus
Climate resilience	Research on nematode responses to climate change, resilient crop development	Adaptation to environmental changes, enhanced crop survival
Advanced sensing	Remote and proximal sensing for early detection	Precision interventions, reduced environmental impact
Sustainable practices	Organic methods, waste management for nematode control	Environmental benefits, circular economy integration
Research and development	Investment in nematology, new nematicides, nanoencapsulation	Innovation, practical solutions, global food security

21.10 Conclusion

AI is revolutionizing nematode diagnosis and management by automating labor-intensive tasks, enabling rapid and accurate identification, providing real-time monitoring, and offering predictive insights. These advancements help farmers make informed decisions, reduce crop losses, and minimize environmental impact. The integration of AI with traditional pest management methods is seen as essential for comprehensive control, with ongoing research promising further innovations in precision agriculture.

The vision for future diagnosis and management of PPNs is one of integration and innovation, leveraging molecular insights, remote sensing, and sustainable practices to mitigate their impact on agriculture. While challenges such as climate variability and global inequalities remain, the projected advancements offer a pathway to enhanced food security and environmental stewardship.

The convergence of high-resolution diagnostics, data-driven decision support, and sustainable control modalities promises a paradigm shift in PPN management. By harnessing the power of multi-omics, AI, precision agriculture, and biotechnological breakthroughs—in concert with robust policy and participatory extension models—we can anticipate a future where nematode threats are detected early, managed precisely, and kept at economically and environmentally acceptable levels. This integrative, forward-looking framework will be pivotal in safeguarding global food security and ecosystem health in the decades to come.

In the coming decades, the successful management of PPNs will depend on our ability to merge scientific innovation with practical, field-ready solutions. The integration of advanced diagnostics, genome-based resistance breeding, biological control agents, and data-driven decision tools will enable farmers to detect threats early, respond with precision, and maintain healthy soils for future generations. By fostering global collaboration, investing in farmer capacity building, and prioritizing environmentally responsible practices, we can transform nematode management from a hidden challenge into a model of sustainable crop protection. This vision not only safeguards agricultural productivity but also supports food security, climate resilience, and ecological integrity worldwide.

21.11 Suggested Readings

- Bellé, C. and Edu Kaspary, T. (eds). 2022. Nematodes – Recent advances, management and new perspectives. *IntechOpen*. doi:10.5772/intechopen.94719.
- Desaeger, J., Sikora, R.A. and Molendijk, L.P.G. 2021. Outlook: A vision of the future of integrated nematode management. In: Desaeger, J., Sikora, R.A. and Molendijk, L.P.G. (eds.) *Integrated Nematode Management: State-Of-The-Art and Visions for the Future*, pp. 475–483. CABI Publishing.
- Parvatha Reddy, P. 2024. *Advances in Diagnosis and Management of Phytonematodes*. Elite Publishing House, New Delhi.
- Sikora, R.A., Padgham, J. and Desaeger, J. 2022. The unpredictability of adapting integrated nematode management to climate variability. In R.A. Sikora, J. Desaeger and L.P.G. Molendijk (eds.) *Integrated Nematode Management: State-of-the-Art and Visions for the Future*, pp. 463–471. CABI Publishing.

21.12 Review Questions

- What role can AI and machine learning play in the early detection and management of plant nematode infestations?
- Discuss the potential of nanotechnology in developing innovative diagnostic tools and nematicide delivery systems.
- How can portable, field-friendly diagnostic kits contribute to precision nematode management in agriculture?
- In what ways could climate change influence the evolution of nematode diagnostic and management strategies?
- How can integrated data platforms (combining soil health, nematode population dynamics, and crop performance) support predictive nematode management?
- How could remote sensing and imaging technologies (e.g., drones and hyperspectral imaging) be used in large-scale nematode monitoring?
- How might "next-generation nematicides" differ from conventional ones in terms of specificity, environmental safety, and regulatory acceptance?
- What role will international collaborations and open-access databases play in strengthening nematode diagnostics and management research?
- How can farmer-centric digital advisory platforms enhance the adoption of innovative nematode management practices?

- Predict the potential of microbiome engineering (beneficial microbes and endophytes) as a sustainable strategy against PPNs.
- What are the key barriers (economic, technical, and policy-related) that need to be addressed for the widespread adoption of futuristic nematode diagnostic and management tools?

References

Ahuja A, Somvanshi VS (2021) Diagnosis of plant-parasitic nematodes using loop-mediated isothermal amplification (LAMP): a review. Crop Prot 147:105459. https://doi.org/10.1016/j.cropro.2020.105459

Banerjee S, Banerjee A, Gill S et al (2017) RNA interference: a novel source of resistance to combat plant parasitic nematodes. Front Plant Sci 8. https://doi.org/10.3389/fpls.2017.00834

Desaeger J, Rao MR (2001) The potential of mixed cover crops of *Sesbania*, *Tephrosia* and *crotalaria* for minimizing nematode problems on subsequent crops. Field Crop Res 70:111–125

Desaeger J, Wram C, Zasada I (2020) New reduced-risk agricultural nematicides – rationale and review. J Nematol 52. https://doi.org/10.21307/jofnem-2020-091

Etesami H, Jeong BR, Glick BR (2023) Biocontrol of plant diseases by *Bacillus* spp. Physiol Mol Plant Pathol 126. https://doi.org/10.1016/j.pmpp.2023.102048

Hillnhütter C, Mahlein AK, Sikora RA, Oerke EC (2011) Remote sensing to detect plant stress induced by *Heterodera schachtii* and *Rhizoctonia solani* in sugar beet fields. Field Crop Res 122:170–177

Molendijk PG, Sikora RA (2022) Decision support systems in integrated nematode management: the need for a holistic approach. In: Sikora RA, Desaeger J, Molendijk LPG (eds) Integrated nematode management: state-of-the-art and visions for the future. CABI Publishing, Wallingford, pp 428–438

Nutter JFW, Tylka GL, Guan J et al (2002) Use of remote sensing to detect soybean cyst nematode-induced plant stress. J Nematol 34:3222

Qin J, Wang B, Wu Y et al (2021) Identifying pine wood nematode disease using UAV images and deep learning algorithms. Remote Sens 13:2162

Sikora RA, Padgham J, Desaeger J (2022) The unpredictability of adapting integrated nematode management to climate variability. In: Sikora RA, Desaeger J, Molendijk LPG (eds) Integrated nematode management: state-of-the-art and visions for the future. CABI Publishing, Wallingford, pp 463–472

Sikora RA, Helder J, Molendijk LPG et al (2023) Integrated nematode management in a world in transition: constraints, policy, processes, and technologies for the future. Annu Rev Phytopathol 61:209–230

Topalovic O, Hussain M, Heuer H (2020) Plants and associated soil microbiota cooperatively suppress plant-parasitic nematodes. Front Microbiol 11:313

Zhou W, Li M, Achal V (2025) A comprehensive review on environmental and human health impacts of chemical pesticide usage. Emerging Contaminants 11(1). https://doi.org/10.1016/j.emcon.2024.100410

The manufacturer's authorised representative in the EU is Springer Nature Customer Service Centre GmbH, Europaplatz 3, 69115 Heidelberg, Germany. If you have any concerns regarding our products, please contact ProductSafety@springernature.com

Printed and bound by CPI Group (UK) Ltd, Croydon, CR0 4YY
21/09/2026
02220012-0018